Botany: A Comprehensive Study of Plant Biology

Botany: A Comprehensive Study of Plant Biology

Editor: Jerry Miller

R CALLISTO REFERENCE

www.callistoreference.com

Callisto Reference,
118-35 Queens Blvd., Suite 400,
Forest Hills, NY 11375, USA

Visit us on the World Wide Web at:
www.callistoreference.com

ISBN: 978-1-64116-104-6 (Hardback)

Cataloging-in-Publication Data

Botany : a comprehensive study of plant biology / edited by Jerry Miller.
  p. cm.
Includes bibliographical references and index.
ISBN 978-1-64116-104-6
1. Botany. 2. Plants. 3. Biology. I. Miller, Jerry.
QK45.2 .B68 2019
580--dc23

Table of Contents

Preface

This book has been a concerted effort by a group of academicians, researchers and scientists, who have contributed their research works for the realization of the book. This book has materialized in the wake of emerging advancements and innovations in this field. Therefore, the need of the hour was to compile all the required researches and disseminate the knowledge to a broad spectrum of people comprising of students, researchers and specialists of the field.

Botany or plant biology is a broad multidisciplinary science that studies the structure, growth, reproduction, taxonomy and epidemiology of plants. The principles of botany have varied applications in agriculture, horticulture, forestry, plant breeding, energy production, environmental management and biodiversity maintenance. Research in botany expands across diverse areas of molecular genetics, epigenetics, plant evolution, phylogenetics, plant behavior and conservation among others. This book discusses the modern theories, concepts and applications of plant biology in extensive details. Some of the diverse topics covered herein address the varied branches that fall under the domain of botany. In this book, using case studies and examples, constant effort has been made to make the understanding of the difficult aspects of botany as easy and informative as possible for the readers.

At the end of the preface, I would like to thank the authors for their brilliant chapters and the publisher for guiding us all-through the making of the book till its final stage. Also, I would like to thank my family for providing the support and encouragement throughout my academic career and research projects.

Editor

1

Molecular cloning, structure, phylogeny and expression analysis of the invertase gene family in sugarcane

Liming Wang[1,2], Yuexia Zheng[2], Shihui Ding[1], Qing Zhang[1,2], Youqiang Chen[2] and Jisen Zhang[1,3,4*] (ID)

Abstract

Background: Invertases (INVs) are key enzymes regulating sucrose metabolism and are here revealed to be involved in responses to environmental stress in plants. To date, individual members of the invertase gene family and their expression patterns are unknown in sugarcane due to its complex genome despite their significance in sucrose metabolism.

Results: In this study, based on comparative genomics, eleven cDNA and twelve DNA sequences belonging to 14 non-redundant members of the invertase gene family were successfully cloned from sugarcane. A comprehensive analysis of the invertase gene family was carried out, including gene structures, phylogenetic relationships, functional domains, conserved motifs of proteins. The results revealed that the 14 invertase members from sugarcane could be clustered into three subfamilies, including 6 neutral/alkaline invertases (ShN/AINVs), and 8 acid invertases (ShAINVs). Faster divergence occurred in acid INVs than in neutral/alkaline INVs after the split of sugarcane and sorghum. At least a one-time gene duplication event was observed to have occurred in the four groups of acid INVs, whereas ShN/AINV1 and ShN/AINV2 in the β8 lineage were revealed to be the most recently duplicated genes among their paralogous genes in the β group of N/AINVs. Furthermore, comprehensive expression analysis of these genes was performed in sugarcane seedlings subjected to five abiotic stresses (drought, low temperature, glucose, fructose, and sucrose) using Quantitative Real-time PCR. The results suggested a functional divergence of INVs and their potential role in response to the five different treatments. Enzymatic activity in sugarcane seedlings was detected under five abiotic stresses treatments, and showed that the activities of all INVs were significantly inhibited in response to five different abiotic stresses, and that the neutral/alkaline INVs played a more prominent role in abiotic stresses than the acid INVs.

Conclusions: In this study, we determined the INV gene family members of sugarcane by PCR cloning using sorghum as a reference, providing the first study of the INV gene family in sugarcane. Combining existing INV gene data from 7 plants with a comparative approach including a series of comprehensive analyses to isolate and identify INV gene family members proved to be highly successful. Moreover, the expression levels of INV genes and the variation of enzymatic activities associated with drought, low temperature, glucose, fructose, and sucrose are reported in sugarcane for the first time. The results offered useful foundation and framework for future research for understanding the physiological roles of INVs for sucrose accumulation in sugarcane.

Keywords: Sugarcane, Invertase, Gene expression pattern, Abiotic stress, Quantitative RT-PCR

* Correspondence: zjisen@126.com
[1]Center for Genomics and Biotechnology, Fujian Provincial Key Laboratory of Haixia Applied Plant Systems Biology, Haixia Institute of Science and Technology (HIST), Fujian Agriculture and Forestry University, Fuzhou 350002, China
[3]College of Life Sciences, Fujian Normal University, Fuzhou 350117, China
Full list of author information is available at the end of the article

Background

Sugarcane is one of the most economically valuable crops worldwide and accounts for up to 80% of the global sucrose production. It serves as an important model crop to study sucrose accumulation due to its remarkable ability to accumulate large amounts of sucrose in its stems that can reach close to 700 mM or in excess of 50% of the dry weight (DW) [33]. With the increasing demand due to biofuel production and challenges with biomass production [7], more and more attention has been devoted to increase sucrose yield in sugarcane. Therefore, the mechanism of sucrose accumulation in sugarcane is now considered to be one of the top priorities in sugarcane research. However, the modern sugarcane cultivar has one of the most complex genomes being both aneupoid and autopolypoid with an extreme ploidy level ranging from octoploid (x = 8) to dodecaploid (x = 12). To date, the studies of the genes in sucrose metabolism have been especially limited.

In plants, invertases (EC 3.2.1.26, INV) catalyze the irreversible hydrolysis of sucrose into glucose and fructose, and are thus considered to be a pivotal enzyme in the regulation of sucrose metabolism [4]. In addition, INVs have also been demonstrated to contribute to numerous aspects of plant growth and development [49, 51], organ formation, sugar transport, stress response [56], carbon partitioning [41, 51], phloem unloading and source/sink regulation [37], and adjusting the composition and levels of sugar in sink tissue [37]. Plant INVs are encoded by large genes families, which can be divided into an acid INV sub-family and a neutral/alkaline INV sub-family according to their optimal pH for activity [46, 55]. The acid INV subfamily includes cell wall invertases (CWINVs) as cell wall-bound forms and vacuole invertases (VINVs) as soluble forms [4]; cell wall invertase originated from respiratory eukaryotes, whereas vacuole invertases derive from aerobic bacteria [47]. Vacuole invertases as soluble acid invertases (SAINVs) generally branch off from cell wall-bound acid invertases in evolution [23]. Cell wall INVs may play versatile regulatory roles in reproductive development, phloem unloading, carbon partitioning [41, 51] and sink development [10, 46]. Vacuolar INVs regulate osmotic pressure, sugar signals, sucrose accumulation, and sucrose concentration especially during the expansion phases of sink organs [25]. The gene structure of acid invertases (AINVs) is highly conserved and contains six to eight exons. In almost all acid invertase genes, the second exon codes for only three amino acids, DPN, belonging to the conserved NDPNG motif of the catalytic domain, and is the smallest functional motif known in plant biology [1]. The molecular weights of mature acid invertases, which are N-glycosylated at multiple sites, range from 55 to 70 kD [52]. Acid invertases contain an N-terminal

domain structure, a mature polypeptide and a C-terminal region [46]. The N-terminal domain structure comprises a signal peptide and a propeptide that is in total about 100 amino acid residues long [47]. Neutral/alkaline invertases (N/AINVs), which only exist in plants and in photosynthetic bacteria, are believed to have originated from cyanobacteria [4]. In contrast to acid INVs, neutral/alkaline INVs localize to multiple subcellular compartments including mitochondria, plastids [34], and the nucleus [41]. Neutral/alkaline INVs differ from acid INVs as they do not contain an N-terminal signal peptide, are not glycosylated [36], and are therefore less stable [39]. The role of neutral/alkaline INVs is less clear than that of acid INVs [42]. So far, extensive characterization of invertases including cDNAs, protein purification and/or genes have been reported from several plants including agave, rice, wheat, tomato, carrot, maize, *Populus trichocarpa* and Arabidopsis [4, 9–11, 16, 23, 27, 30, 46, 47, 53].

In sugarcane, the enzyme invertases are well documented key regulators of the accumulation of sucrose in the stems [13, 17, 18, 58]; N/AINVs are found in the cytoplasm or in metabolic compartments of cells, have low activity in meristematic cells, and are involved in sugar accumulation in storage tissue [17]. In 18 month-old sugarcane cultivars, N/AINVs have been suggested to be more important in sucrose hydrolysis than SAINVs in the early storage of sugar in the stems after harvest [44]. Recently, a reduction of N/AINVs activity in transgenic sugarcane plants was revealed to cause a decrease in respiration and sucrose cycling, and an increase in the sucrose to hexose ratio, demonstrating the essentiality of N/AINVs in directing carbon towards respiratory processes in the sugarcane culm [40]. SAINVs were found primarily in the vacuoles of storage parenchyma cells [13, 18], and presented high activity in tissues that are rapidly growing [17]. Sucrose accumulation in the sugarcane stalk has been suggested to be regulated by the difference between the activities of SAINVs and sucrose phosphate synthase [58]. A positive correlation between the changes in transcript levels and enzyme activity in sugarcane cultivars, and between SAINV activity and content of hexose sugars was observed, whereas a negative correlation was found between SAINV activity and sucrose content in mature and immature internodes [44]. In transgenic sugarcane, the intracellular and extracellular sugar composition was highly sensitive to the changes in INV activity, and acid INV activity was negatively correlated to sucrose accumulation [14]; in another study of transgenic sugarcane, 70% reduction in the level of acid INV activity did not alter sucrose load or purity [5].

Although these long-term studies implicated INVs as the principal enzymes regulating sugarcane growth and sucrose accumulation, elucidating the molecular mechanism for invertase function lags behind the guidelines of sucrose genetic modification in sugarcane. A comprehensive understanding of the molecular mechanism and evolution of the gene family in a plant species is the first key step to understand the physiological roles and metabolic mechanism regulated by invertases. The available genome of *Sorghum bicolor*, the closest diploid relative of sugarcane in the Andropogonae tribe, provides an excellent model for sugarcane genomic studies [35]. In this study, the invertase gene members were predicted, based on comparative genomics approaches, further verified by PCR cloning and sequencing, and gene expression levels were investigated by real time PCR. The aims of this study were to: (1) identify the members of the invertase gene family in sugarcane; (2) characterize the expression patterns of the invertase gene family under different abiotic stresses and (3) analyze the genetic diversity and the function to differentiate invertase gene families.

Methods

Plant material
Sugarcane cultivar FN41 was maintained in the campus of Fujian Agriculture and Forestry University (Fuzhou, China). According to the method of Moore [32], fresh leaf tissue (the third mature leaf, counted from the leaf rolls) was harvested from 7 to 9-month-old field-grown sugarcane plants for DNA and RNA isolation, which were further used for genomic and cDNA cloning of INV genes

Sugarcane seedlings from callus culture at four leaf stage were used for PEG treatment, cold treatment and sugar treatment experiments. To avoid the background effect of stalk storage nutrition, the seedling plants were recovered using sugar-free MS culture solution for 24 h prior to experimental treatment. PEG stress treatment: seedlings were incubated in sugar-free culture solution containing 10% (*W/V*) PEG6000 for a photoperiod of 16-h light at 28 °C/ 8-h dark at 24 °C. Low temperature treatment: seedlings were incubated in sugar-free culture solution and grown in 16 h light at 15 °C/8 h dark at 10 °C. Sugar treatment: three groups of seedlings were incubated in MS culture solution for 4 h dark at 24 °C with 3% (*W/W*) sucrose, 3% (*W/W*) glucose, 3% (*W/W*) fructose respectively. Control treatment: seedlings were incubated in sugar-free culture solution for a photoperiod of 16 h light at 28 °C/8 h dark at 24 °C. Fresh leaf tissue from each of the five treatment experiments was immediately snap-frozen in liquid nitrogen and stored at −80 °C prior to RNA isolation and enzymes extraction. RNA was used for RT-qPCR analysis of INVs' gene expression.

BLAST searches of the INV gene families in seven plant species
Genomic sequences of seventeen known INV genes (Additional file 1: Table S1) from *Arabidopsis* (http://

www.arabidopsis.org/) and nineteen from rice [23] (Additional file 1: Table S1) were used as queries to search the full set of INV genes in the genomes of grape, papaya, *Brachypodium distachyon*, maize and sorghum. BLAST matches achieved similarity scores of >50.0 and probability scores of <10^{-4} were collected as candidate sequences. These candidate sequences of INV genes were further verified by their annotated database (http://www.phytozome.net/) through BLAST and BLASTX. Furthermore, candidate INV proteins were confirmed by searching for conserved domains of invertase using CD-Search tool (http://www.ncbi.nlm.nih.gov/Structure/cdd/docs/cdd_search.html).

Cloning of sugarcane invertase genes
To obtain the ORF and DNA sequence of the sugarcane INV genes, primer pairs were designed based on the pile up of the INV gene sequences from sorghum and sugarcane EST resource from Genbank (http://www.ncbi.nlm.nih.gov/genbank/). A forward primer and reverse primer were designed to be at opposite ends of the ORF (Additional file 2: Table S2, available as Supplementary Material to this paper).

Total RNA was extracted using TRIzol reagent (Invitrogen Co., Carlsbad, CA, USA) from fresh leaf samples of mature sugarcane from the field, then treated with RNase-free DNaseI (Ambion, AM1906) prior to being used for reverse transcription. Integrity of the RNA sample was analyzed by agarose gel electrophoresis. Genomic DNA was isolated from fresh leaf samples in mature sugarcane according to the TIANcombi DNA Lyse&Det PCR Kit manufacturer's instructions (TIANGEN, China).

For cDNA cloning, the first-strand cDNA was synthesized from 1 g of total RNA according to the instructions of the Revert Aid™ First Strand cDNA Synthesis Kit (Thermo Fisher Scientific.). cDNA fragments covering the whole open reading frame of INV genes were amplified by PCR using gene specific primers. Similar PCR protocols were used for both RT-PCR and PCR for genomic DNA. PCR was performed in a 10 μL reaction volume containing 1 μL template of total DNA or first-strand cDNA, 5 μL 2 × GC LA Taq Buffer, 0.2 μL of each PCR primer, 0.2 μL LA Taq, 0.8 μL dNTP (2.5 mmol/L) and 2.6 μL ddH$_2$O. The amplified DNA fragments and cDNA fragments were cloned into pMD19-T Vector Kit (TaKaRa) and subsequently sequenced by BGI Tech Solutions Co., Ltd. (BGI-Tech).

Sequence analysis of invertase family members
Gene annotation: The cloning sequences of potential INVs were BLAST/BLASTx to Genbank database. The determined genomic sequences were annotated with the FGENESH program (http://www.softberry.ru/berry.phtml)

with references of cDNA sequences and EST of *Saccharum*. The annotated genes were further manually examined.

Sequencing analysis: The sequences of genomic and cDNA were BLAST/BLASTX to Genbank to confirm that these sequences were INVs. The determined cDNA sequences were translated into protein sequences using the online tool (http://web.expasy.org/translate/). Furthermore, their theoretical isoelectric point (pI) and molecular weight (Mw) were analyzed (http://web.expasy.org/compute_pi/). The putative conserved domains were detected using the CD-Search tool (http://www.ncbi.nlm.nih.gov/Structure/cdd/docs/cdd_search.html), and subcellular localizations were predicted by the subcellular location Prediction Servers (Plant-mPLoc, http://www.csbio.sjtu.edu.cn/bioinf/plant-multi/; SignalP 4.1 Server, http://www.cbs.dtu.dk/services/SignalP/; MitoProt, http://ihg.gsf.de/ihg/mitoprot.html; ChloroP 1.1 Server, http://www.cbs.dtu.dk/services/ChloroP/). The motifs of INV proteins were analyzed by using MEME (http://alternate.meme-suite.org/tools/meme, MEME Suite 4.10.1) with the parameters of maximum motif number with 15, minimum motif width with 6, maximum motif width with 50, and distribution of motif occurrences with Zero or one per sequence. The gene schematic structures were drawn by using the Gene Structure Display Server (http://gsds.cbi.pku.edu.cn/index.php) [19].

Phylogenetic analysis
The amino acid sequences of INV genes from eight plant species (Additional file 1: Table S1) were used to construct an unrooted phylogenetic tree by MEGA5.1 [50]. Neighbor-joining topologies were generated as the consensus of 1000 bootstrap alignment replicates by running MEGA 5.2 with ClustalW alignment.

RT-qPCR analysis of gene expressions
1 μg of total RNA from each sample of different treatments was reverse transcribed using the RevertAid™ First Strand cDNA Synthesis Kit (Fermentas). Based on the annotated sugarcane *INVs* genomic sequences, RT-qPCR primers (Additional file 3: Table S3) were designed using software program Beacon Designer 7 to amplify sequences spanning at least one intron, and primers' specificity was tested via regular RT-PCR for experimental quality control. Real-time PCR was performed in three technical replicates from three biological replicates. To determine the amplification efficiency for each primer set, the calibration curve for each gene was obtained by performing real-time PCR with four dilutions of cDNA $(4^0, 4^{-1}, 4^{-2}, 4^{-3}, 4^{-4})$. RT-qPCR was performed in ABI Prism®7300HT Fast Real-Time PCR machine (Applied Biosystems, USA). PCR reactions contained 10 μL of 2X

SYBR Green Master Mix (Takara), 2 μL of template cDNA (10X dilution), 0.4 μL of primers mixed (20 mM of each) and 7.6 μL ddH$_2$O. The PCR cycle was: 3 min at 95 °C followed by 40 cycles of 95 °C for 15 s and 60 °C for 45 s and the specificity of the individual PCR amplification was checked using a heat dissociation protocol from 65 to 95 °C following the final cycle of the PCR. 25SrRNA (E1:5′-CCTATTGGTGGGTGAACAATCC-3′; E2:5′-GCAGCCAAGCGTTCATAGC-3′) were used as reference gene, which was verified to exhibit stable levels of expression in a broad range of sugarcane tissues [15, 20, 31]. All the genes from each sample were compared with the expression level of 25SrRNA from leaves of sugarcane and the relative expression level of each INV in different treatments was calculated based on normalized relative quantities.

Invertase activity
Extraction of enzymes
The leaf tissue from each sample of the different treatments was ground in liquid nitrogen and the subsequent procedure for sample extraction was conducted at 4 °C or lower.

Extraction of soluble INVs
0.5 g tissues were homogenized with 5 ml 50 mmol/L HEPES extraction buffer, containing of 12 mmol/L MgCl$_2$, 1 mmol/L EDTA, 1 mmol/L EGTA,10 mmol/L DTT, 2 mmol/L benzamidine,0.05% Triton-X 100,0.05% BSA,2% PVPP [58]. Homogenates were filtered through a microfiltration membrane and were centrifuged at 9366 g for 10 min. The supernatant was desalted and desugared immediately using Sephadex G-25 (Pharmacia PD-10) and kept on ice until the assay was performed.

Extraction of cell wall INVs
0.5 g sugarcane leaf tissue was homogenized with 10 ml 50 mmol/L HEPES buffer excluding 2% PVPP) [58] and kept on ice for 10 min. Homogenates were centrifuged at 25151 g rpm for 15 min. The sediments (containing the cell wall fraction) were homogenized with 1.7 ml 50 mmol/L HEPES buffer(the same as above, and were subsequently centrifuged at 25151 g for 15 min at. The supernatants (enriched for the CWI fractions) were desalted twice using Sephadex G-25 (Pharmacia PD-10) and kept on ice until use.

INV activity assay
Similarly to Tang et al. [51], 0.4 mL of desalted extracts were homogenized at 37 °C with the reaction mixture (2.4 mL) (1.2 mL 0.1 mol/L phosphate/citric and 0.8 mL 0.1 mol/L sucrose, pH 4.6 for acid INV and pH 7.5 for neutral/alkaline INV) and incubated for 60 min. For the cell wall INV activity assay, 1 mL of desalted extracts

were homogenized with 6 mL reaction mixture (3 mL 0.1 mol/L phosphate/citric buffer (pH 4.6), 2 mL 0.1 mol/L of desalted extracts sucrose) and incubated at 37 °C for 120 min. The boiled desalted extracts and concentration gradient of glucose (Sigma-Aldrich) were used as background control and standard, respectively. The reaction was stopped by adding 2.4 mL DNS followed by boiling for 5 min. The liberated reducing sugars were quantified by measuring the absorbance at 540 nm. Micrograms of product formed per gram of total protein per minute(μg.Glc.g-1.Pr.min-1)were used as the enzymatic activity units.

Results

Identification of INV genes in the genomes of six plant species

To obtain the reference sequence of the INV gene in sugarcane for comparative genomics analysis, 19 and 17 well-annotated *INV*s from *Oryza sativa* [23] and *Arabidopsis thaliana* (https://www.arabidopsis.org) respectively were used to search these family members from *Vitis vinifera*, *Carica papaya*, *Brachypodium distachyon*, *Zea mays* and *Sorghum bicolor* (Additional file 1: Table S1). 17 *INV*s were found from *Vitis vinifera*, 8 from *Carica papaya*, 19 from *Brachypodium distachyon*, 21 from *Zea mays*, and 19 from *Sorghum bicolor* (Additional file 1: Table S1). The conserved domains and the chromosomal location of these INV genes from seven plant species (including rice and *Arabidopsis*) were analyzed and are listed in Additional file 1: Table S1.

Being the closest diploid relative of sugarcane, sorghum INVs (referred from here on as *SbINV*s) (Additional file 1: Table S1) are described specifically here for further references. In our study, genome-wide identification of the INV gene family in sorghum revealed that there are 19 INVs in the sorghum genome. Of the 19 SbINVs, seven are neutral/alkaline INVs (*SbN/AINV*s), twelve are acid INVs containing ten cell wall INVs (*SbCWINV*s), and two vacuolar INVs (*SbVINV*s). The seven SbN/AINV proteins contain a conserved domain of Glyco_hydro_100, while acid INV proteins contain a conserved domain of both Glyco_hydro_32 N and Glyco_hydro_32 C. There are two sets of SbCWINV, with one set containing SbCWINV2/3/5/6, and the other set containing SbCWINV8/9/10, which are located in unassembled supercontig_67 and Chromosome 6, respectively. Both of the two sets of genes were observed to have originated from tandem duplications (Additional file 1: Table S1).

Cloning and sequence analysis of INV gene family in sugarcane

Using the 19 *SbINV*s combined with sugarcane ESTs as reference for primer design, 11 cDNAs of the homologous INVs in sugarcane were cloned by RT-PCR. The 11 cDNAs are referred to as ShN/AINV1, 2–2, 3–2, 4–2, 5, 6–2, ShCWINV6, 7–3, 8–2, 10 and ShVINV1 according to both primers reference from SbINVs and sequences similar to SbINVs (Table 1). 3 sequences (ShCWINV6, ShCWINV8–2, ShVINV1) only harbour partial open reading frames (ORF), the remaining 8 were predicted to contain full ORFs (Table 2). Furthermore, to examine the gene structure of INVs in sugarcane, genomic PCRs were performed to clone sugarcane INVs. Twelve DNA fragments corresponding to 9 ShINVs were obtained, 8 of these sequences (in addition to ShCWINV3, ShCWINV8–1, ShCWINV9–1 and ShCWINV9–2) were determined to contain full coding regions (Table 2). Among the 9 ShINVs, ShCWINV7 and ShCWINV9 had 2 and 3 gene alleles, respectively. Overall, the 23 sequences including 11 cDNA and 12 genomics fragments corresponded to 14 INV genes including 6 N/AINVs, 7 CWINVs and 1 VINV. These DNA and cDNA sequences were submitted to Genbank: *ShN/AINV1* (KC145794), *ShN/AINV2–1* (KC145808), *ShN/AINV2–2* (KC145795), *ShN/AINV3–1* (KC145809), *ShN/AINV3–2* (KC145796), *ShN/AINV4–1* (KC145810), *ShN/AINV4–2* (KC145797), *ShN/AINV5* (KC145799), *ShN/AINV6–1* (KC145807), *ShN/AINV6–2* (KC145798), *ShCWINV1* (KC145815), *ShCWINV3* (KC145801), *ShCWINV6* (KC145800), *ShCWINV7–1* (KC145811), *ShCWINV7–2* (KC145812), *ShCWINV7–3* (KC145802), *ShCWINV8–1* (KC145814), *ShCWINV8–2* (KC145816), *ShCWINV9–1* (KC145803), *ShCWINV9–2* (KC145804), *ShCWINV9–3* (KC145813), *ShCWINV10* (KC145805), *ShVINV1* (KC145806) (Table 1).

These INV sequences were translated into amino acid sequences for computational analysis of protein characteristics (Table 2). Of these 23 ShINV sequences, 16 ShINV sequences containing complete ORFs (open reading frames) were predicted to have molecular weights ranging from 62.60 to 70.23 kDa (Table 2). Comparative analysis of the protein sequences of orthologous sorghum genes showed, neutral/alkaline INVs shared higher identities (ranging from 97 to 99%) than the acid INVs (ranging from 87 to 94%), indicating that faster divergence occurred in neutral/alkaline INVs than in acid INVs after the split of sugarcane and sorghum (Table 2). The molecular mass of the homologous INV proteins in sugarcane and sorghum were similar excluding those sequences without full ORFs (Table 2). Based on predictions of the subcellular localization by Plant-mPLoc, INVs were divided into three types: cell-wall, vacuolar, and cytoplasmic. For neutral/alkaline INVs, except ShN/AINV2–1 and ShN/AINV4–1, the other neutral/alkaline INVs were predicted to localize to the chloroplast. However, the localization probability predicted by ChloroP/MitoProt suggested that five neutral/alkaline INV

Table 1 The information on PCR products of the invertase genes in sugarcane

Sorghum		Sugarcane DNA clone		Sugarcane cDNA clone		Protein coverage and similarity (%)
Gene name	Gene ID	Gene name	Gene ID	Gene name	Gene ID	
SbN/AINV1	Sobic.004G172700	N/A	N/A	ShN/AINV1	KC145794	N/A
SbN/AINV2	Sobic.004G255600	ShN/AINV2–1	KC145808	ShN/AINV2–2	KC145795	99%/98%
SbN/AINV3	Sobic.005G058800	ShN/AINV3–1	KC145809	ShN/AINV3–2	KC145796	100%/99%
SbN/AINV4	Sobic.004G024500	ShN/AINV4–1	KC145810	ShN/AINV4–2	KC145797	100%/98%
SbN/AINV5	Sobic.004G163800	N/A	N/A	ShN/AINV5	KC145799	N/A
SbN/AINV6	Sobic.003G153800	ShN/AINV6–1	KC145807	ShN/AINV6–2	KC145798	100%/99%
SbN/AINV7	Sobic.001G391600	N/A	N/A	N/A	N/A	N/A
SbCWINV1	Sobic.001G099700	ShCWINV1	KC145815	N/A	N/A	N/A
SbCWINV2	Sobic.K040900	N/A	N/A	N/A	N/A	N/A
SbCWINV3	Sobic.K041100	ShCWINV3	KC145801	N/A	N/A	N/A
SbCWINV4	Sobic.004G166700	N/A	N/A	N/A	N/A	N/A
SbCWINV5	Sobic.K041000	N/A	N/A	N/A	N/A	N/A
SbCWINV6	Sobic.K041200	N/A	N/A	ShCWINV6	KC145800	N/A
SbCWINV7	Sobic.003G440900	ShCWINV7–1	KC145811	ShCWINV7–3	KC145802	99%/92%
		ShCWINV7–2	KC145812			99%/93%
SbCWINV8	Sobic.006G255500	ShCWINV8–1	KC145814	ShCWINV8–2	KC145816	85%/98%
SbCWINV9	Sobic.006G255400	ShCWINV9–1	KC145803	N/A	N/A	N/A
		ShCWINV9–2	KC145804	N/A	N/A	N/A
		ShCWINV9–3	KC145813	N/A	N/A	N/A
SbCWINV10	Sobic.006G255600	N/A	N/A	ShCWINV10	KC145805	N/A
SbVINV1	Sobic.004G004800	N/A	N/A	ShVINV1	KC145806	N/A
SbVINV2	Sobic.006G160700	N/A	N/A	N/A	N/A	N/A

proteins (SbN/AINV5, ShN/AINV5, SbN/AINV6, ShN/AINV6–1, ShN/AINV6–2) most likely localize to the mitochondria, since the probability of mitochondrial targeting was higher than chloroplast targeting (Table 2). In addition, using the online tool SignalP, four sets of orthologous genes of sugarcane and sorghum (CWINV1s, CWINV7s CWINV9s and CWINV10s) were predicted to possess the hydrophobic N-terminal signal peptide required for secretory proteins (Table 2).

Because *Saccharum* hybrids are highly allopolyploid with genetic backgrounds from *S.officinarum and S.spontaneum*, the sequences for gene alleles could derive from either of these two *Saccharum* species. Of the 23 sequences for 14 INV genes, 7 *ShINVs* (*ShN/AINV2*, *ShN/AINV3*, *ShN/AINV4*, *ShN/AINV6*, *ShCWINV7*, *ShCWINV8* and *ShCWINV9*) had 2–3 gene alleles. Of these 7 *ShINVs*, the alleles of *ShN/AINV2*, *ShN/AINV3*, *ShN/AINV4* and *ShN/AINV6* shared protein sequence similarities ranging from 98% to 99% (Table 3). The alleles of *ShCWINV7* shared protein sequence identities ranging from 92 to 93%, and *ShCWINV9–1*, *ShCWINV9–2* and *ShCWINV9–3* shared protein sequences ranging from 91 to 97%, while the protein

sequence identity of the alleles *ShCWINV8–1* and *ShCWINV8–2* was 98%. These results indicated that the alleles of *ShN/AINV2*, *ShN/AINV3*, *ShN/AINV4*, *ShN/ AINV6*, and *ShCWINV8* probably originated from *S. officinarum*, which contributed approximately 80% of genetic background, while the gene alleles of *ShCWINV7* and *ShCWINV9* could be derived from the two *Saccharum* species since the alleles within each genes presented sequence variation (Table 4).

Gene structural and phylogenetic analysis of the INV genes in sugarcane and sorghum

Phylogenetic analysis of the INV gene family from sugarcane and sorghum showed that INVs can be divided into two branches (acid INV branch and neutral/alkaline INV branch), and each of these branches can be further subdivided into α and β subgroups (Fig. 4). Acid INV branch contained *CWINV* and *VINV*, the group of *CWINV* could be further divided into two subgroups (α and β). In subgroup α, three *CWINVs* (*ShCWINV1*, *ShCWINV3* and *ShCWINV7*) from sugarcane had a very conserved gene structure with their orthologous genes in sorghum, and among them ShCWINV7 had two gene

Table 2 Comparison of the characterisation of the invertases between sugarcane and sorghum

Sorghum				Subcellular localization		Sugarcane				Subcellular localization		Coverage/ identity
Gene name	Protein size(aa)	MW (kDa)	pI	SignalP/Chlor-oP/ MitoProt	Plant-mPLoc	Gene name	Protein size(aa)	MW (kDa)	pI	SignalP/Chlor-oP/ MitoProt	Plant-mPLoc	
SbN/AINV1	559	63.16	6.3	N/44.6[c]/22.3[m]	Chloroplast	ShN/AINV1	559	63.13	6.1	N/44.6[c]/23.8[m]	Chloroplast	100%/98%
SbN/AINV2	572	64.17	6.5	N/44.8[c]/3.1[m]	Chloroplast	ShN/AINV2-1	567	63.58	6.0	N/44.3[c]/2.3[m]	Chloroplast Mitochondrion	99%/97%
						ShN/AINV2-2	575	64.58	6.5	N/45.6[c]/4.1[m]	Chloroplast Cytoplasm	100%/97%
SbN/AINV3	558	63.65	6.2	N/43.4[c]/6.2[m]	Chloroplast	ShN/AINV3-1	557	63.58	6.2	N/43.5[c]/22.1[m]	Chloroplast	100%/99%
						ShN/AINV3-2	557	63.58	6.2	N/43.5[c]/2.1[m]	Chloroplast	100%/99%
SbN/AINV4	627	70.51	7.9	N/43.0[c]/0.3[m]	Chloroplast	ShN/AINV4-1	563	64.06	6.7	N/43.1[c]/0.1[m]	Chloroplast Nucleus	100%/98%
						ShN/AINV4-2	563	63.90	6.7	N/43.1c/0.3 m	Chloroplast	100%/97%
SbN/AINV5	603	67.93	6.3	N/54.9[c]/89.7[m]	Chloroplast	ShN/AINV5	607	68.28	6.3	N/55.2[c]/97.9[m]	Chloroplast	100%/97%
SbN/AINV6	627	70.06	5.4	N/55.3[c]/98.6[m]	Chloroplast	ShN/AINV6-1	623	69.67	5.4	N/53.1[c]/99.6[m]	Chloroplast	99%/98%
						ShN/AINV6-2	629	70.23	5.4	N/54.5[c]/99.2[m]	Chloroplast	100%/98%
SbCWINV1	579	63.25	5.9	Y/46.2[c]/56.1[m]	Cell wall	ShCWINV1	572	62.79	6.2	Y/45.7[c]/37.2[m]	Cell wall	98%/94%
SbCWINV3	594	65.67	9.3	N/43.8[c]/3.8[m]	Cell wall	ShCWINV3[*]	453	50.21	9.4	N/44.1[c]/12.2[m]	Cell wall	100%/92%
SbCWINV6	599	66.90	9.5	N/44.5[c]/2.5[m]	Cell wall	ShCWINV6[*]	349	39.84	9.5	N/44.0[c]/53.5[m]	Cell wall	66%/90%
SbCWINV7	643	73.06	6.7	Y/50.9[c]/87.6[m]	Cell wall	ShCWINV7-1	588	66.40	6.4	Y/50.4[c]/72.9[m]	Cell wall	99%/90%
						ShCWINV7-2	598	67.84	6.2	Y/50.7[c]/69.2[m]	Cell wall	99%/90%
						ShCWINV7-3	593	67.13	6.3	Y/51.1[c]/85.3[m]	Cell wall	99%/90%
SbCWINV8	556	61.44	5.4	N/44.6[c]/49.4[m]	Cell wall	ShCWINV8-1[*]	501	55.60	5.4	N/44.1[c]/41.5[m]	Cell wall	88%/92%
						ShCWINV8-2[*]	448	49.46	5.0	N/46.7[c]/9.4[m]	Cell wall	76%/93%
SbCWINV9	590	64.62	5.3	Y/46.3[c]/9.7[m]	Cell wall	ShCWINV9-1[*]	211	23.08	5.3	N/44.3[c]/8.9[m]	Cell wall	40%/94%
						ShCWINV9-2[*]	503	54.76	4.9	N/44.6[c]/5.3[m]	Cell wall	89%/89%
						ShCWINV9-3	574	62.60	5.1	Y/48.1[c]/9.1[m]	Cell wall	100%/91%
SbCWINV10	625	68.30	5.2	Y/51.5[c]/25.3[m]	Cell wall	ShCWINV10	592	64.39	6.0	Y/50.1[c]/0.5[m]	Cell wall	99%/87%
SbVINV1	638	69.32	5.7	N/46.3[c]/8.3[m]	Vacuole	ShVINV1[*]	487	52.94	5.0	N/46.3[c]/5.0[m]	Vacuole	100%/94%

Note: [m] Probability (%) of targeting to mitochondrion, [c] Probability (%) of targeting to chloroplast; N–Non-secretory protein, Y– Secretory protein, [*] represented truncated gene

Table 3 Amino acid sequence pairwise comparisons (% similarity) between neutral/ alkaline INV members in sugarcane

	ShN/AlNV1	ShN/AlNV2–1	ShN/AlNV2–2	ShN/AlNV3–1	ShN/AlNV3–2	ShN/AlNV4–1	ShN/AlNV4–2	ShN/AlNV5	ShN/AlNV6–1	ShN/AlNV6–2
ShN/AlNV1										
ShN/AlNV2–1	77									
ShN/AlNV2–2	79	98								
ShN/AlNV3–1	75	69	70							
ShN/AlNV3–2	75	69	70	99						
ShN/AlNV4–1	65	67	67	69	69					
ShN/AlNV4–2	65	66	66	67	67	98				
ShN/AlNV5	52	57	57	61	61	54	54			
ShN/AlNV6–1	60	60	61	62	62	57	57	74		
ShN/AlNV6–2	60	60	60	62	62	57	56	73	99	

alleles whereas ShCWINV7–1 has 10 amino acids fewer than ShCWINV7–2 and one exon more than ShCWINV7–2 (Table 2, Fig. 4). In subgroup β, ShCWINV8 shared a similar gene structure with its orthologous genes in sorghum, although its homologous gene SbCWINV8 contains one more 9 bp exon and one more last exon, while the three gene alleles of ShCWINV9 were observed to have exon splits in the third corresponding exon of SbCWIN9 in sorghum as shown by the gene structure of ShCWINV9–3 with full CDS.

In neutral/alkaline INV branch, the genes from subgroups α and β contained 6 and 4 exons, respectively (Fig. 4). In the α subgroup, genes were observed to present variation in intron size, whereas *ShN/AlNV6–1* had the same gene structure as its orthologous gene *SbN/AlNV6* in sorghum. In the β subgroup, genes showed conserved exon size, consequently, the three *ShN/AlNVs* (*ShN/AlNV2–1*, *ShN/AlNV3–1* and *ShN/ AlNV4–1*) shared same intron-exon structures with their sorghum orthologous genes. *N/AlNVs* were more conserved than *AlNVs* according to the comparative analysis of sorghum and sugarcane, which is consistent with the above comparative analysis for orthologous gene pairs between sorghum and sugarcane based on sequence similarity.

Motif distribution in sugarcane and sorghum invertases
To compare the INV functional domain between sugarcane and sorghum, we employed the MEME web server combined with DNAMAN to identify the motifs of INVs from sugarcane and sorghum. There were 15 conserved motifs identified in the INVs (Additional file 4: Table S4), of these acid INV motifs, motif 12, motif 7 and motif 14 contained the catalytic residues NDPN, RDP and WECP/VD respectively [29]. In general, excluding the sugarcane INVs lacking full CDS, the ShINVs harbored motif sequences similar to the orthologous INVs in sorghum except for a slight variation

in one pair of orthologous genes (ShCWINV10/ SbCWINV10). In ShCWINV10/SbCWINV10, motif 10 was absent in ShCWINV10 but present in SbCWINV10 (Fig. 5, Additional file 5: Figure S2 and Additional file 6: Figure S3). DPN in the NDPN motif is encoded by the mini-exon and is susceptible to alternative splicing under cold stress in potato with transfructosylating capabilities [6, 43], but it was absent from ShCWINV3*, ShCWINV6*, SbCWINV7/ShCWINV7–1/ShCWINV7–2, ShCWINV8– 1*/ShCWINV8–2*, ShCWINV9–1*/ShCWINV9–2*/ ShCWINV9–3 and SbVINV1/ShVINV1* (Additional file 5: Figure S2).

Neutral/alkaline INVs from sugarcane and sorghum were conserved for the putative functional motifs of their orthologous genes. Similarly, among the paralogous genes, the neutral/alkaline INVs from both sugarcane and sorghum were generally conserved for the motif distributions except for three motifs (motifs 13, 14 and 15) in the N-terminus. 12 motifs, including motif 3 and motif 6 which contained catalytic residues (two Asps) [21, 23], were observed to be consistent in their sizes and distributions (Fig. 5b, Additional file 6: Figure S3). Of the three variant motifs (motifs 13, 14 and 15) at the N-terminus, motif 15 was distributed in the N-terminus of SbN/AlNV1/ShN/AlNV1, SbN/AlNV2/ShN/AlNV2– 1,-2 and SbN/AlNV3/ShN/AlNV3–1,-2; motif 13 was specifically distributed in the N-terminus of SbN/ AlNV4/ShN/AlNV4–1,-2; motif 14 was specifically distributed in the N-terminus of SbN/AlNV6/ShN/ AlNV6–1,-2 (Fig. 5b). Based on these differences from motif comparison, N/AlNVs could be classed into four subfunctional divergences. These motif variations may lead to different biological characteristics and functions.

Phylogenetic analysis of the invertase gene family and other plant invertase homologs
To understand the evolutionary relationship among the INV genes in sugarcane, unrooted phylogenetic trees were constructed for acid INVs and neutral/alkaline

Table 4 Amino acid sequences pairwise comparisons (% similarity) between acid INV members in sugarcane

	ShCWINV1	ShCWINV3*	ShCWINV6*	ShCWINV7-1	ShCWINV7-2	ShCWINV7-3	ShCWINV8-1*	ShCWINV8-2*	ShCWINV9-1*	ShCWINV9-2*	ShCWINV9-3	ShCWINV10	ShVINV1*
ShCWINV1													
ShCWINV3*	59												
ShCWINV6*	61	80											
ShCWINV7-1	51	49	53										
ShCWINV7-2	51	49	53	93									
ShCWINV7-3	51	49	53	92	93								
ShCWINV8-1*	48	46	47	51	52	50							
ShCWINV8-2*	47	47	47	50	51	50	98						
ShCWINV9-1*	63	49	57	56	56	56	72	71					
ShCWINV9-2*	49	46	48	48	48	48	61	62	97				
ShCWINV9-3	49	49	49	50	50	50	64	64	93	91			
ShCWINV10	50	49	51	52	51	51	66	65	70	64	64		
ShVINV1*	42	40	47	38	37	37	40	41	50	40	41	40	

INVs using protein sequences from eight plant species (Additional file 1: Table S1, Fig. 2 and Fig. 3). In addition, 8 phylogenetic trees were constructed using INVs from *Arabidopsis thaliana*, *Vitis vinifera*, *Carica papaya*, *Oryza sativa*, *Brachypodium distachyon*, *Zea mays*, *Sorghum bicolor* and *Saccharum*, respectively (Additional file 7: Figure S1). Comparison of the phylogenetic trees from these plants demonstrated that the phylogenetic relationships of INVs in these plants are consistent and conserved (Additional file 7: Figure S1). All INVs from these plants fall into the acid and neutral/ alkaline INV classes. Acid INVs can be divided into vacuolar INVs and two cell-wall subgroups (α and β groups). Alkaline/neutral INVs can also be further subdivided into α and β subgroups (Additional file 7: Figure S1). The phylogenetic tree of acid INV genes from eight plant species formed four evident branches that were designated as group I to IV (Fig. 2). In the phylogenetic tree, vacuolar INVs (VINVs) were distributed in the distinct branch I, which consists of genes from both dicotyledons and monocotyledons, whereas cell-wall INVs (CWINVs) could be classed into three branches (referred to as II, III and IV respectively). Of the three branches for cell-wall INVs, branch II and IV were specifically comprised of genes from monocotyledons and dicotyledons respectively, while, branch III contained genes from both dicot and monocot. These results indicated that the genes in branch III were more ancient than the genes in the other two branches. Furthermore, monocotyledonous genes specific to branch IV contained two subgroups that were more distinct than the dicotyledonous genes specific to branch II, suggesting that the gene duplications that occurred in monocotyledons predated those in dicotyledons. Alternatively, this result may indicate that monocot plants have a higher gene evolutionary rate than dicotyledons. In sugarcane, ShVINV1 was distributed in group I, ShCWINV1, ShCWINV3 and ShCWINV6 were distributed in group III, and ShCWINV7–1/2/3, ShCWINV8–1/2, ShCWINV9–2/3 and ShCWINV10 were in group IV.

The phylogenetic tree of neutral/alkaline INV genes from eight plant species could be separated into two distinct groups, referred to as α group and β group (Fig. 3). In the α group, the genes were subdivided into two subgroups, α1 and α2, which contained N/AINVs from dicot and monocot. In the α2 subgroup, the genes could be separated into two branches, whereas the genes from dicotyledons were distributed in one of these branches, suggesting the gene duplications were more ancient in monocot than in dicot in this subgroup. The β group contained 8 subgroups (β1-β8), of these 8 subgroups, β1, β4 and β8 were only comprised of monocotyledon N/ AINVs, the remaining 5 subgroups only contained

dicotyldeon N/AINVs. It is interesting that each of the monocotyledon specific subgroup (β1, β4 and β8) contained genes from all of the examined monocotyldeon species, while the dicotyledon genes were more divergent and distributed in different subgroups. These results suggested that N/AINVs in the β group were more recent in monocotyledons than in dicotyledons. In sugarcane, ShN/AINV5 and ShN/AINV6–1/2 were distributed in α1 and α2, respectively. ShN/AINV4–1/2, ShN/ AINV3–1/2 and ShN/AINV1 were distributed in β1, β2, and β3, respectively.

Expression of INVs under PEG stress, cold stress and sugar treatments

To test how INVs respond to drought and low temperature stress, and to illustrate whether the expression of INVs is regulated by the hydrolysis products (glucose and fructose) or by its substrate (sucrose), we examined the transcription levels of the 13 INVs in 1-month-old sugarcane seedlings under PEG stress, cold stress and sugar (glucose, fructose and sucrose) treatments. This could also provide insight into the potential function divergence of the INV gene family members. Of the examined genes (Fig. 6), *ShCWINV3*, *ShCWINV7*, *ShCWINV9* and *ShN/AINV4* were found to be up-regulated (at least as two fold as controls) under PEG stress, cold stress and sugar treatments, *ShCWINV6* and *ShN/AINV6* were down-regulated under PEG stress, cold stress and sugar treatments. As for *ShVINV1*, its expression decreased 0.5 fold under cold treatment and increased nearly 2 folds under fructose treatment compared to the control, while it showed no change under other treatments (Fig. 6). *ShN/AINV3* showed no significant difference in expression in any of the treatments except for cold treatment. The other 5 INVs were observed to be remarkably dissimilar in terms of the relative expression levels under five treatments (Fig. 6). In PEG treatment, compared to the control, the expression of *ShCWINV3*, *ShCWINV7*, *ShCWINV8*, *ShCWINV9*, *ShN/AINV3*, *ShN/AINV4* and *ShN/AINV5* increased, with the most significant being *ShCWINV3* with an increase in excess of 3 fold. The expression of *ShCWINV6*, *ShCWINV10*, *ShN/AINV1*, *ShN/AINV2* and *ShN/AINV6* decreased, and with *ShN/AINV6* decreasing most significantly at 70% compared to the control (Fig. 6). In response to the cold treatment, acid INVs *ShCWINV3*, *ShCWINV7*, *ShCWINV8* and *ShCWINV9* were induced, while *ShCWINV6*, *ShCWINV10* and *ShVINV1* were repressed. In particularly, *ShCWINV3* had the largest up-regulation (greater than 5.5 fold) after cold treatment. All neutral/alkaline INVs were up-regulated (at least as two fold as the controls) in response to the cold treatment except *ShN/AINV6* that showed a slight degradation (Fig. 6). In treatments with three different sugars, the expressions of *ShCWINV3*, *ShCWINV7*, *ShCWINV9* and *ShN/AINV5* was up-regulated, in particular the expression of *ShCWINV7* and *ShCWINV9* increased about 4-fold, 6.5-fold and 4-fold expression under glucose, fructose and sucrose treatments respectively. In addition, *ShCWINV6* and *ShN/AINV6* were down-regulated under three sugar treatments, whereas no obvious change in expression was observed for *ShN/AINV1* and *ShN/AINV3* (Fig. 6).

Overall, transcripts of neutral/alkaline INVs were more abundant than transcripts of acid INVs. Among neutral/alkaline INVs, *ShN/AINV5* was the most abundant form and the expression level of *ShN/AINV2* under five treatments was the lowest. Among the acid INVs across the five treatments and control, *ShCWINV8* had significantly higher expression levels than the other genes, whereas *ShCWINV6* had the lowest expression levels. In addition, *ShCWINV1* displayed no expression in any of the five treatments and in the control (Fig. 6).

Variation of enzymatic activity under drought, low temperature and sugar treatments in sugarcane

The activity of cell-wall INVs, soluble acid INVs and neutral/alkaline INVs was assayed in sugarcane seedlings under drought, low temperature and sugar treatments. Changes in the activity of all these INVs exhibited the same trend in response to the five different abiotic stresses (Fig. 7), in which the activity of all INVs decreased (Fig. 7). More specifically, compared to the control, cell-wall INV activity showed a gradual decrease (about 48%–66%) after five treatments. Soluble acid INV activity was reduced by 87% under PEG and fructose treatment, and up to 92% following sucrose treatment. In addition, neutral/alkaline INV activity revealed the smallest decrease among the three INV classes, with only a 53% decrease occurring upon sucrose treatment compared to the control. However, its activity decreased less than 50% decrease under other treatments (Fig. 7). Three kinds of invertase activity in different experimental conditions were significantly suppressed but comparatively speaking the suppression of soluble acid INV activity was more apparent. The neutral/alkaline INVs displayed high catalytic capacity in the control, with activity being 4.16 times higher than that of cell-wall INVs and 5.75 times higher than that of soluble acid INVs. However, the neutral/alkaline INVs with the highest catalytic capacity under five different treatments was more obvious, for example it showed 8.8 times higher activity than cell-wall INVs, and 19 times higher than soluble acid INV under PEG treatment. Under cold treatment, the hydrolytic activity of neutral/alkaline INVs was 6–7 times higher than cell-wall INVs and soluble acid INVs. Moreover, neutral/alkaline INVs had 8, 9.9 folds higher activity than cell-wall INV and soluble

INVs under glucose treatment respectively and neutral/alkaline INVs had 27.6, 18.2 times higher activity than soluble acid INVs under sucrose and fructose treatment respectively (Fig. 7).

Discussion

INVs play a fundamental role in sucrose accumulation in plants, and have been well documented since half century ago in sugarcane [13, 17, 18]. Lack of a whole genome reference sequence for sugarcane made it difficult to determine the gene family members by PCR cloning using sorghum as reference, which is the closest relative of sugarcane with a reference sequence. In this study, eleven cDNAs and twelve DNAs corresponding to 14 sugarcane INVs were cloned based on the sorghum gene models (Additional file 1: Table S1, Table 1), thus providing the first study of the INV gene family in sugarcane. In sugarcane, PCR amplification of genomic and transcriptome based on the sorghum genome cannot capture highly divergent genes, making it impossible to determine the absolute number of INVs in sugarcane. Whole genome sequencing of sugarcane would be necessary for a comprehensive identification of a gene family in sugarcane (Fig. 1).

Saccharum hybrids are highly polyploid with a genetic background from *S.officinarum* and *S. spontaneum*. A typical gene in *Saccharum* can have up to 12 different alleles, each of which may be either from *S. officinarum* or *S. spontaneum* [57]. In this study, 5 of the 14 ShINVs had 2–3 gene alleles with variants for deduced amino acid sequences of each gene. Based on the sequence similarities among the gene alleles, it is difficult to determine the origin of the gene haplotype in *Saccharum* hybrids due to the close relationship between *S. officinarum* and *S. spontaneum*. Random PCR cloning from *Saccharum* hybrids for a gene functional study was not appropriate for the potential gene functional divergence between *S. officinarum* and *S. spontaneum*. Some gene alleles may have specific functions. For example, brown rust resistance gene (*Bru1*) in sugarcane was suggested to be single dose, which is the only resistant allele in Saccharum hybrid R570 [2, 12]. Future gene functional studies, should address the issue of identifying the gene alleles from *S. officinarum* contributing to the sugar characterization of *Saccharum* hybrids using the homologous genes of *S. officinarum* as a reference.

In this study, based on the phylogenetic analysis for acid INVs from eight plant species, acid INVs could be divided into four groups (I, II, III and IV, Fig. 2). All vacuolar INVs from both dicotyledons and monocotyledons were grouped in group I, which was distinctly separated from the cell-wall INV (Fig. 2), suggesting the origin of the vacuolar INVs from cell-wall INVs occurred before the last common ancestor (LCA) of dicotyledons

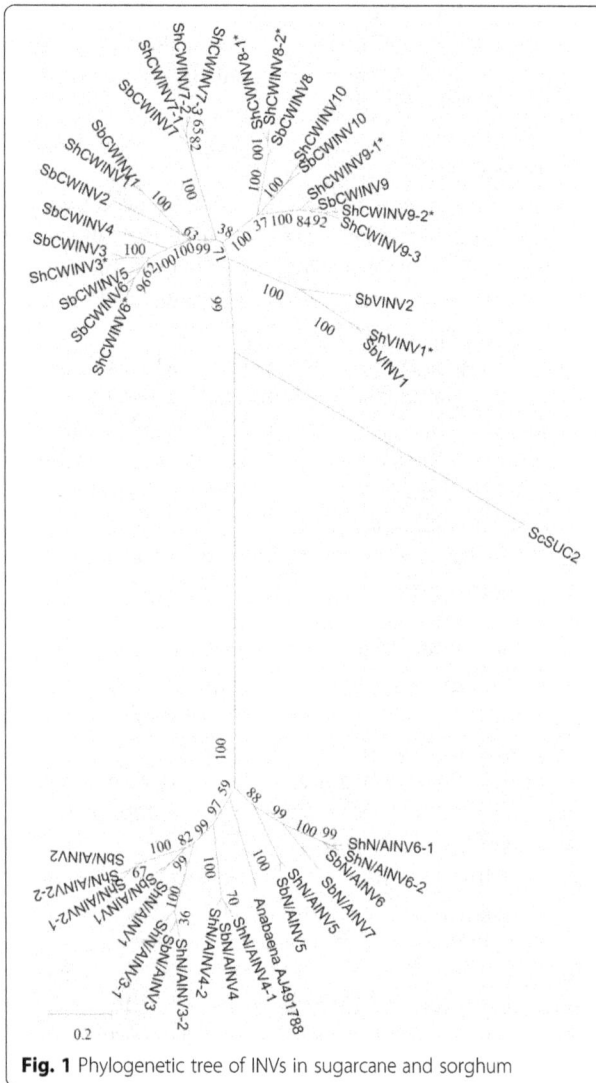

Fig. 1 Phylogenetic tree of INVs in sugarcane and sorghum

than the LCAs of group II and group IV because group III was generated before monocot/dicot divergence. The comprehensive analysis also revealed the evolutionary history of acid INVs, which was sorted by age in duplicated descending order, group I (containing ShVINV1), group III (containing ShCWINV6, ShCWINV3, ShCWINV1), group IV (ShCWINV7, ShCWINV8, ShCWINV9). At least one gene duplication event was observed to have occurred in the four groups for acid INVs, suggesting the genes have potential functional redundancy. This speculation was consistent with the evidence from the comparison between the deduced amino acids of sorghum and sugarcane (Table 2).

Analysis of the evolution of neutral/alkaline INVs for eight plants also revealed a similarly complex evolutionary relationship. In the phylogenetic tree of N/AINVs from eight plant species (Fig. 3), N/AINVs were distributed in two distinct groups (α and β groups). In the α group, the LCA for dicots and monocots contained two neutral/alkaline INV genes (boxes labeled α1 and α2). In the α1 lineage, a gene from grape (VvN/AINV6) was grouped together with monocot plant genes, while, in the α2 lineage, the genes from dicot and monocot were separated into two clear branches, and genes from monocot could also be subdivided into two branches. Therefore, the sugarcane genes (ShN/AINV6, and ShN/AINV7 (absent in sugarcane)) in the α2 lineage were suggested to be recent duplications after the split of dicot/monocot plants and were older than the sugarcane genes in the α1 lineage (ShN/AINV5). In the β group, LCAs of N/AINVs possessed eight genes (boxes labeled β1–β8), which occurred after the divergence of dicot and monocot plant species. Of the eight LCAs, β2 and β3 were the forerunners of grape N/AINVs, β5, β6 and β7 were the forerunners of dicot specific N/AINVs, β1, β4 and β8 were monocot specific forerunners of N/AINVs. It is interesting to note that all orthologous N/AINVs from monocot plants were grouped together, whereas, the dicotyledonous genes were grouped into β2 and β3 lineages which only contained grape genes, and β5 lineages which only contained one *Arabidopsis* gene. These results indicated that N/AINVs from the β group in dicot plants were more divergent than those in monocots. ShN/AINV1 and ShN/AINV2 in the β8 lineage were the most recently duplicated genes among their paralogous genes in the β group.

The orthologous gene pairs of N/AINVs between sorghum and sugarcane shared higher identities (97%–99%) than those of soluble acid INVs (87%–94%), demonstrating acid INVs had undergone a faster divergence than N/AINVs after the split of sorghum and sugarcane (Table 2). Sequence comparison of the paralogous gene in sugarcane also revealed that N/AINVs presented lower divergence than acid soluble INVs (Table 3). These

and monocotyledons. This result broadened the previous deduction that the origin of vacuolar INVs from cell-wall INVs predated the LCA of rice and Arabidopsis [23]. The separation of the dicotyledon vacuolar INVs from the monocotyledon vacuolar INVs suggests that the LCA may contain a single vacuolar INV gene by two different pathways (gained or lost intron) to produce two different genes in the lineages to dicotyledons and monocotyledons, or the LCA may have possessed two vacuolar INV genes and the two precursors in LCA respectively underwent duplication events in each of the lineages to dicots and monocots. Both monocotyledons and dicotyledons had genes in group III, but monocots had higher numbers of the orthologous genes than dicots in group III, suggesting both a duplication exclusive to monocots and no potential gene functional redundancy in dicots. Furthermore, phylogenetic analysis suggested that the LCA of group III were more ancient

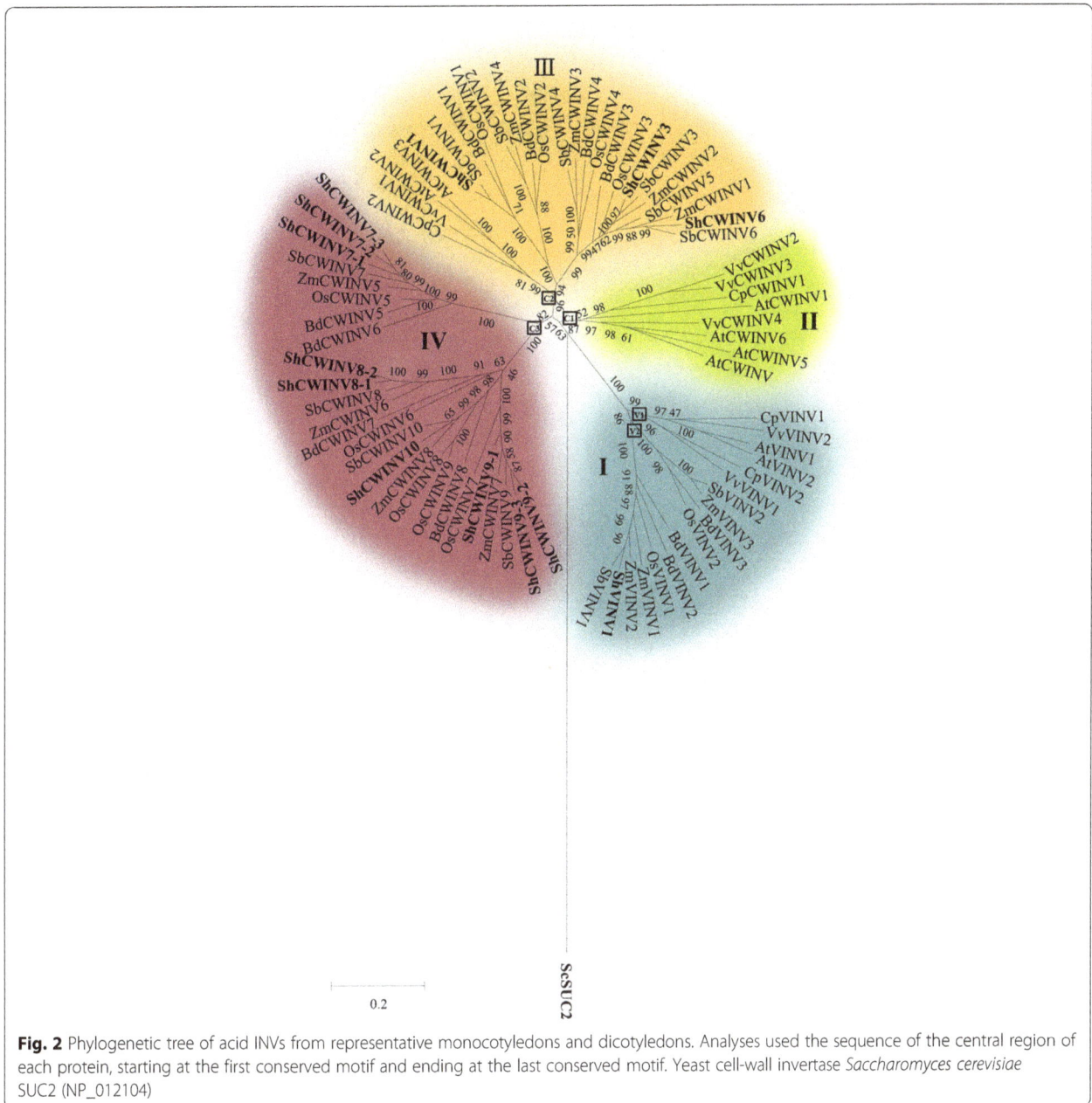

Fig. 2 Phylogenetic tree of acid INVs from representative monocotyledons and dicotyledons. Analyses used the sequence of the central region of each protein, starting at the first conserved motif and ending at the last conserved motif. Yeast cell-wall invertase *Saccharomyces cerevisiae* SUC2 (NP_012104)

results indicated that N/AINVs had undergone stronger functional constraint than acid soluble INVs. Sugarcane and sorghum had undergone 1 to 2 rounds gene duplications for LCAs of acid INVs after the splits of dicot/monocot, but only two sets of paralogous N/AINVs (SbN/AINV6/SbN/AINV6, ShN/AINV7 (absent in sugarcane) /SbN/AINV7) were recent duplications. Therefore, sorghum and sugarcane presented higher sequence variation for acid INVs than N/AINVs, which may be caused by functional redundancy of these acid INVs ancestors.

To further understand the gene evolution of INVs, we analyzed the pattern of the exon–intron structure in sorghum and sugarcane showing that ShCWINV3, ShCWINV8–1, ShCWINV9–1 and ShCWINV9–2 underwent exons-loss (Fig. 4). Thus this evidence further confirmed that they were truncated genes, which was consistent with the analysis of amino acid sequences discussed above (Fig. 4, Additional file 5: Figure S2). Except these truncated genes, the exon–intron organizations of other genes were divergent among the CWINV gene families. The main motifs were kept in all the CWINV genes (Fig. 5), which suggested that the gene structure variation was caused by exon splitting or intron length variation but not pseudo-exonization as this would have resulted in motif deletion. The gene organization is

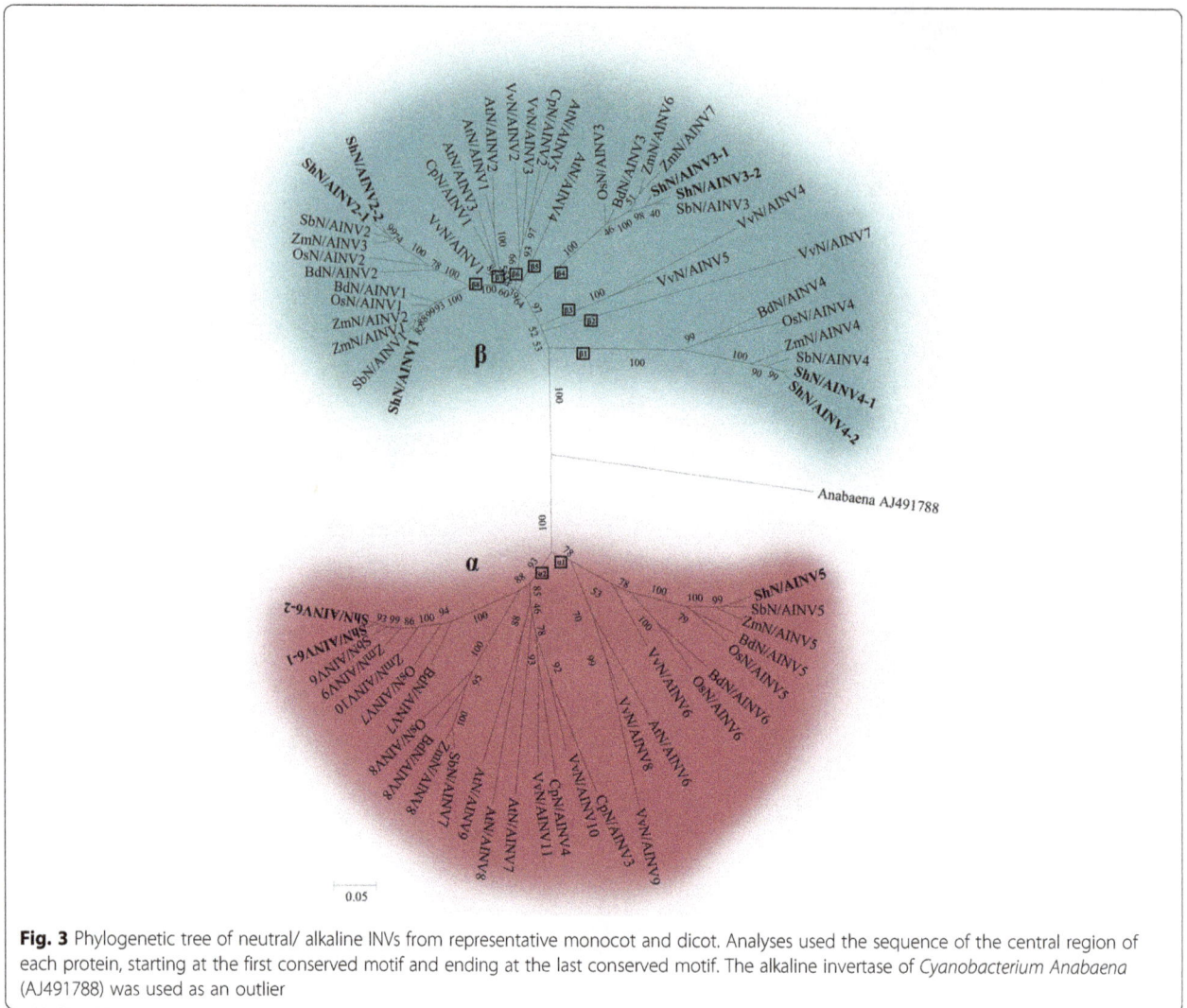

Fig. 3 Phylogenetic tree of neutral/ alkaline INVs from representative monocot and dicot. Analyses used the sequence of the central region of each protein, starting at the first conserved motif and ending at the last conserved motif. The alkaline invertase of *Cyanobacterium Anabaena* (AJ491788) was used as an outlier

highly conserved within the N/AINV gene family, and the evolution of neutral/alkaline INVs for sorghum and sugarcane was particularly clear in the sense that the locations of the exon–intron junctions in α group were distinct from the β group. All α group members had 6 exons and all β group members had 4 exons, with the exon–intron junctions being fully conserved. Therefore, the different intron-exon junctions and the different number of exons proved that the α and β groups of neutral/alkaline INV genes in sugarcane and sorghum derived from different ancestral genes with 4 and 6 exons, which is consistent with the findings reported in Fig. 3. Based on motif comparison, exonization may have occurred for the first exon of N/AINV gene because their first exons encode variant motifs. Overall, the N/AINVs have a more conserved gene structure than acid INVs, which also supported the above conclusion that acid INVs had undergone a faster divergence than N/AINVs after the split of sorghum and sugarcane.

The ability of sugarcane to accumulate sugar is impaired under drought stress and low temperature stress. We examined the expression patterns and activities of sucrose cleavage enzymes involved in sucrose metabolism under drought and low temperature stresses. These short-term physiological changes lead to sugar concentration changes, which may be significant enough to efficiently regulate gene expression. The expression of nearly all INV genes was affected by PEG and low-temperature treatment, except ShVINV1, whose expression was not altered in response to PEG. The observed up-regulation of expression of INV genes in sugarcane leaves in response to drought stress or cold treatment is likely due to more INVs being required to cleave sucrose into hexose sugars and subsequently provide cells with more energy to sustain increased respiration activity in addition to liberating more carbon and energy to synthesis different compounds [46], and enhance resistance to environmental stresses [53]. It is also possible that raised

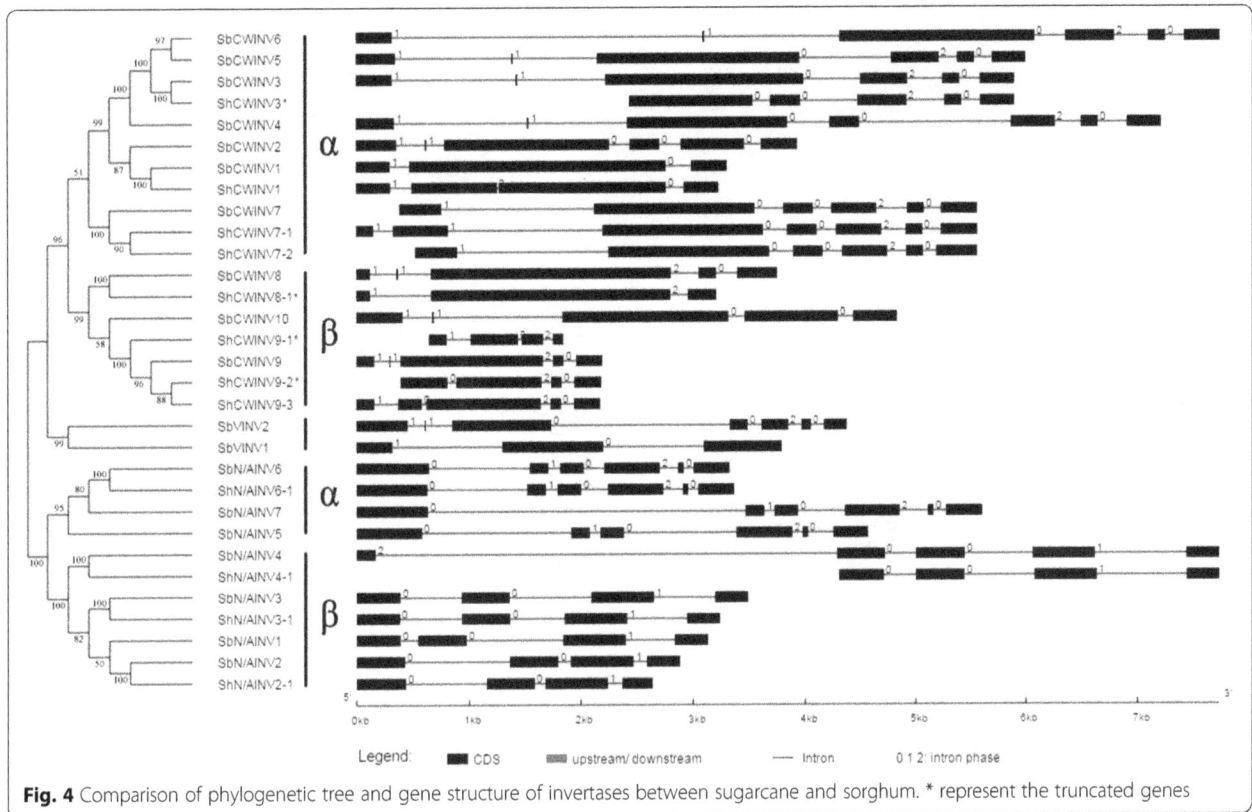

Fig. 4 Comparison of phylogenetic tree and gene structure of invertases between sugarcane and sorghum. * represent the truncated genes

levels of INVs are required to cleave more Suc into Glc and Fru to greatly increase in the osmotic pressure of cells to cope with stresses, which indirectly provides evidence for INVs producing an osmoregulation substrate as in *Arabidopsis* [24]. Thus, it is possible that the expression of some INVs were up-regulated to cope with the need to cleave more sucrose under drought and cold stress, whereas the expression of other INVs could be down-regulated to maintain sucrose homeostasis. As shown in Fig. 6, four of the examined genes (ShCWINV3, ShCWINV7, ShCWINV8 and ShCWINV9) were up-regulated (especially ShCWINV3), and two genes (ShCWINV6 and ShCWINV10) were down-regulated under PEG and cold stress. Cell-wall INVs were thought to play a role in establishing metabolic sinks through irreversible cleavage of sucrose to glucose and fructose to metabolize sugars for both downstream metabolic functions and sugar modulation signaling [10, 38]. Whilst up-regulation of cell-wall INVs can easily be explained as a need to increase the ability for cleaving Suc into Glc and Fru, and enhance the osmotic pressure of cells, the down-regulated expression of ShCWINV6 might be due to a requirement for its distribution in specific subcellular compartments and interconnected with sugar modulation signal for blocking the downstream metabolism to adapt to the drought and cold stresses. The expression of ShN/AINV6 decreased under drought

and cold stress, more specifically it decreased 3 fold under drought stress (Fig. 6). This result could be explained by the finding that the neutral/alkaline INVs may function as maintenance enzymes involved in sucrose degradation and maintenance of sucrose concentration [55]. In previous studies, soluble acid INVs were revealed to localize to the vacuole to control sucrose storage and sugar composition [48]. Under cold stress, all vacuolar INVs were observed to be up-regulated under cold stress in Populus [8] and the total INV mRNA levels were substantially upregulated in tulip (*Tulipa gesneriana* L. cv. Apeldoorn) bulbs [3]. However, in this study, ShVINV1 showed no changes in expression levels under drought stress, but was almost halved under cold stress. Thus, the down regulation of ShVINV1 and ShN/AINV6 under cold stress suggested that the difference of the molecular regulation mechanism between cold stress and PEG stress (Fig. 6). Sugar participates in numerous cellular processes. In addition to being a source of energy and form structural components during plant growth and development, it acts in signal transduction pathways to modulate the gene expression of sugar metabolism [46]. Glucose, sucrose as well as fructose have long been known as important signal molecules in the regulation of sugar accumulation [37]. To respond to changes in availability of sugars, cells can adjust the amount of invertases involved in sugar metabolism. Sugarcane grown on

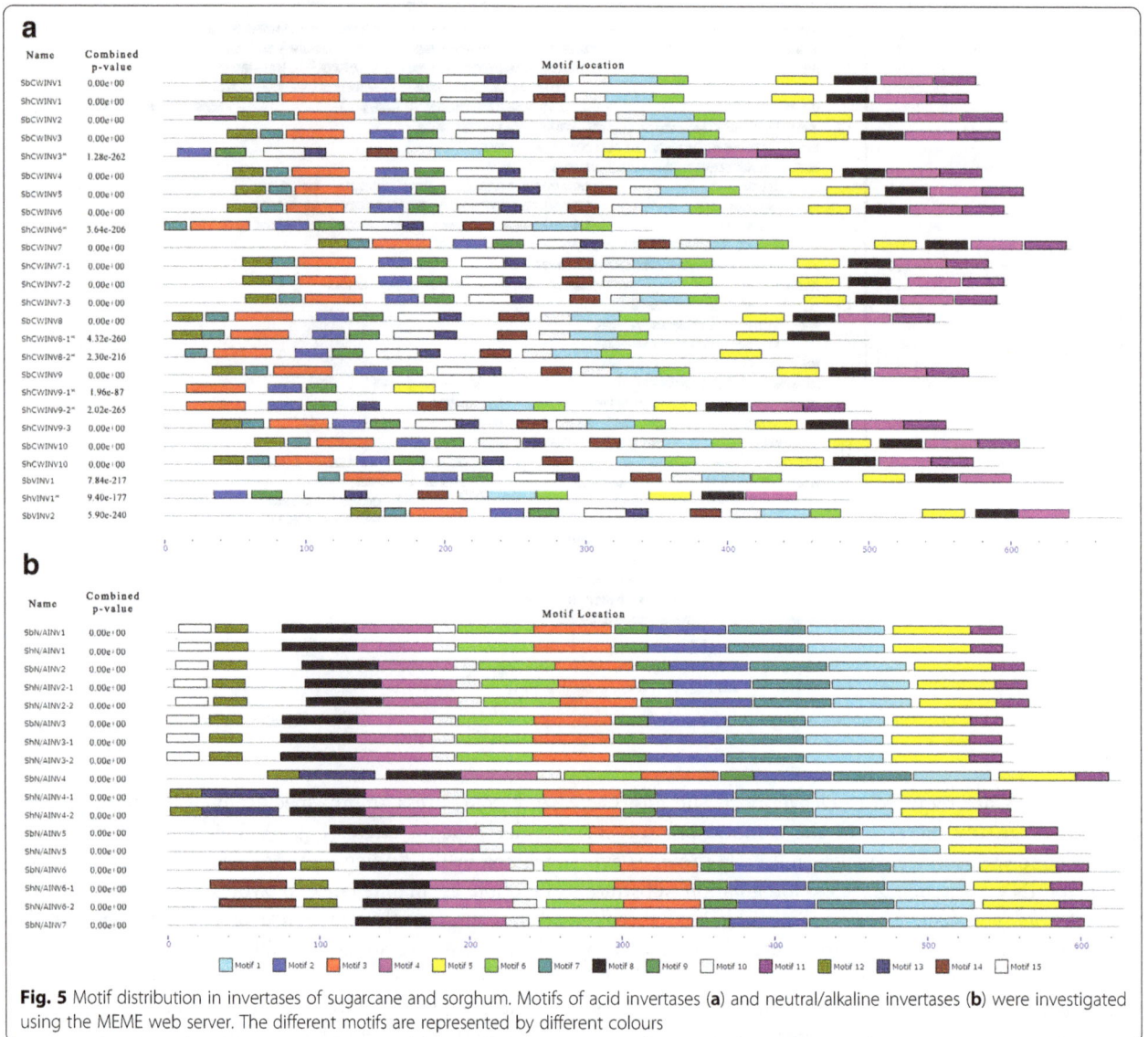

Fig. 5 Motif distribution in invertases of sugarcane and sorghum. Motifs of acid invertases (**a**) and neutral/alkaline invertases (**b**) were investigated using the MEME web server. The different motifs are represented by different colours

glucose, or fructose or sucrose as the sole carbon source display altered patterns of INV gene expression (Fig. 6). In response to sugar treatments, the INV genes in sugarcane presented different expression patterns consistent with the acid INV genes from maize in response to sucrose, glucose as well as other metabolizable sugars [56]. Previous studies on sucrose induction of INV expression have not addressed whether sucrose itself or its components (glucose and fructose) were the actual inducer [26]. However, recent studies on the regulation of INV by the nature of sugar signal molecules have revealed that sugar-specific pathways may be differentiated. For example, glucose, but not sucrose, induced the expression of cell wall INVs in Arabidopsis roots [55] and *C. rubrum* [38]. Moreover, the regulation of acid INV activity was repressed by hexose sugars, in particular by fructose [54]. However, in this study, both sucrose as a substrate of the

INVs and also glucose induced the expression of 5 genes (ShN/AINV4, ShN/AINV5, ShCWINV3, ShCWINV7 and ShCWINV9). Exogenous sugars and sucrose repressed the expression of 2 genes (ShN/AINV6 and ShCWINV6) (Fig. 6).

These results suggested that sugars could directly regulate the expression of the majority of INV genes. In addition, nearly equally high levels of all invertase mRNAs, except ShCWINV10, were detected when the carbon source was either glucose or sucrose, which was lower than their expression with fructose, except for *ShCWINV3*, *ShCWINV8* and *ShN/AINV3* (Fig. 6). This finding indicated that the INV gene members have different sugar-specific response mechanisms; with the response of some INV genes to fructose being more sensitive than to sucrose or glucose. The general picture emerging from previous studies was that sugar signals

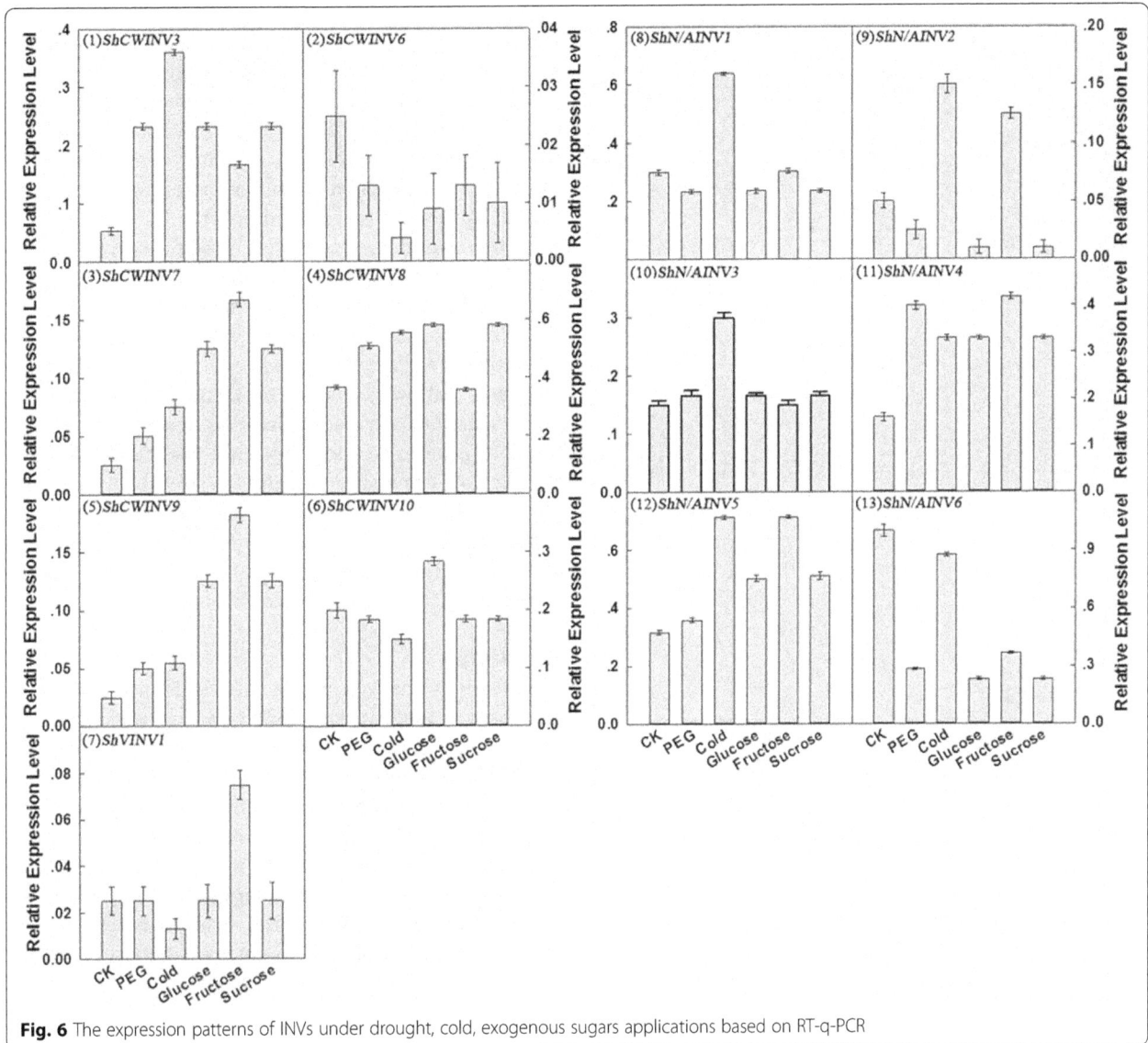

Fig. 6 The expression patterns of INVs under drought, cold, exogenous sugars applications based on RT-q-PCR

decrease the transcription of genes for sugar synthesis and sucrose metabolism, and increase the transcription of genes involved in sugar storage and utilization [22, 28, 45, 55]. In agreement with this conclusion as well as our experimental results, we propose that in the presence of sufficient carbon sources regulating gene expression, ShN/AINV6 and ShCWINV6 may play a pivotal role in sugar synthesis and sucrose metabolism, and ShN/AINV4, ShN/AINV5, ShCWINV3, ShCWINV7 and ShCWINV9 may participate in sugar storage and utilization. Investigating gene expression and activities of INVs in sugar storage will be necessary to for further study these aspects. In general, the contributions of different INVs to the overall enzymatic activity of the INV family were different. In this study, activity of all INVs was significantly inhibited under five different abiotic stresses (Fig. 7). Figure 7 showed that neutral/alkaline INV had a much higher activity than soluble acid INV and cell-wall INV activity compared to the controls and other treatments. ShN/AINV6 with highest transcription levels in controls decreased sharply under the five abiotic stresses, leading to a possible decrease of neutral/alkaline INV activities (Fig. 6), suggesting that, it may provide the greatest contribution to the enzymatic activity of the whole neutral/alkaline INV family. Also in cell-wall INVs, ShCWINV8 with highest transcription levels in control may provide the largest contribution to the activity of the whole cell-wall INV family (Fig. 6). The expression of ShCWINV1 in control and under all stresses was not detected, and it cannot be excluded that its spatiotemporal expression pattern is very specific and was not captured in our experiments. We cloned only one member of the soluble

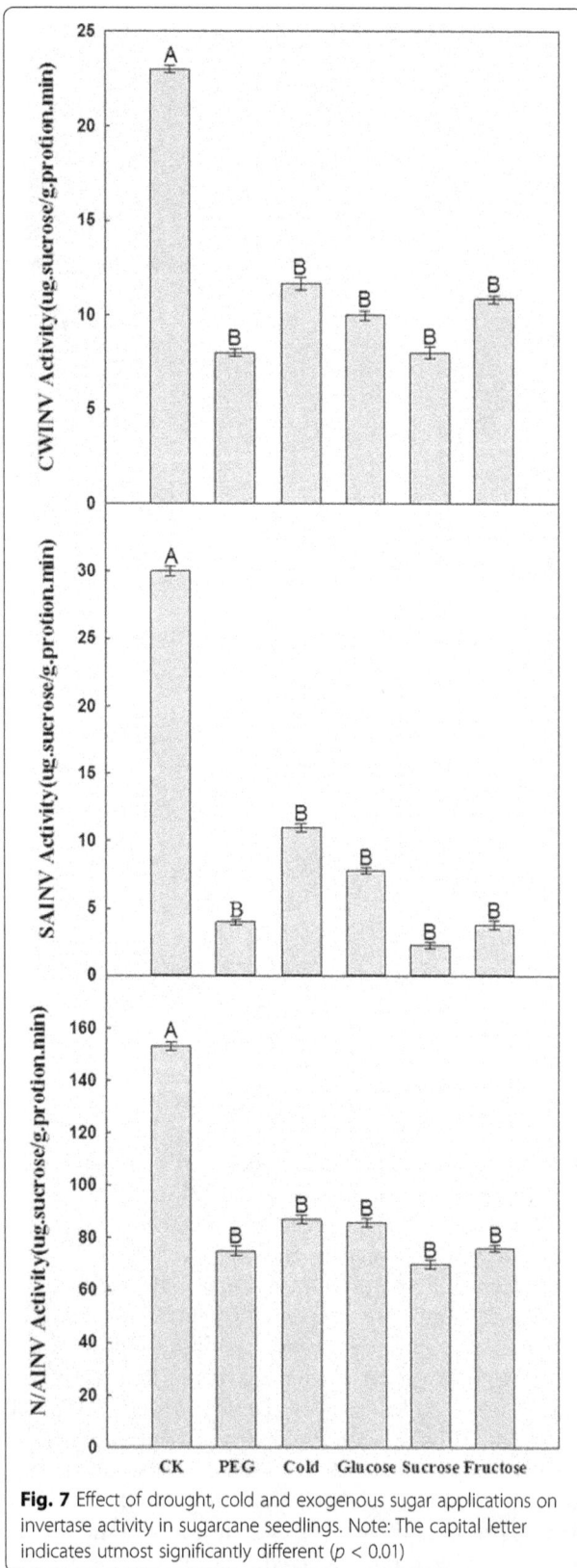

Fig. 7 Effect of drought, cold and exogenous sugar applications on invertase activity in sugarcane seedlings. Note: The capital letter indicates utmost significantly different ($p < 0.01$)

acid INV genes, ShVINV1, and its expression levels were the same in all treatments except under cold and fructose stress (Fig. 6). Irrespective of the different experimental condition, its enzymatic activity was suppressed (Fig. 7). Thus it can be speculated that the decline in ShVINV1 enzymatic activity may be due to the expression of other genes, which are induced by abiotic stress and inhibit INV activity, or may be caused by structural modification(s) to the ShVINV1 of sugarcane. Together, both the gene expression patterns and enzyme activity changes under the biotic stress could help to further understand the interactive regulatory network of INV genes and sugar signaling pathways under drought, low temperature and exogenous sugars stresses. Noteworthy, cold may regulate gene expression, but cold itself also decreases the actual enzymatic activity in the plant. In our study, the activity of SAINVs under cold stress was higher than the other stresses in vitro. It is possible that the increase in enzyme occurs to maintain actual enzymatic activity under cold stress.

Conclusion

This study is the first report for the 14 non-redundant members of the invertase gene family in sugarcane. We speculated that there were 6 neutral/alkaline invertases (ShN/AINVs) and 8 acid invertases (ShAINVs). We provided a comprehensive analysis of the gene allelic haplotypes, phylogenetic relationships, gene structure, functional domains, conserved motifs of proteins, gene expression patterns, and the variation of enzymatic activity under five abiotic stresses treatments for the INV gene family in sugarcane. Sequence comparison of the paralogous genes in sugarcane presented higher sequence variation for acid INVs than N/AINVs indicating that N/AINVs had undergone stronger functional constraint than acid soluble INVs. Furthermore, the N/AINVs have a more conserved gene structure than CWINVs, which also supports the idea that acid INVs have undergone a faster divergence than N/AINVs after the split of sorghum and sugarcane. Despite the high polyploidy level, the examined INV genes exhibited conserved gene structures and high similarity of amino acid sequences among the allelic haplotypes. Transcripts of neutral/alkaline INVs were more abundant than transcripts of acid INVs under drought stress, cold stress and sugar treatments. The expression of ShCWINV3, ShCWINV7, ShCWINV8 and ShCWINV9 were upregulated to cope with the need to cleave more sucrose under drought and cold stress, whereas the expression of ShCWINV6 and ShCWINV10 could be downregulated to maintain sucrose homeostasis. According to our experimental results, we also propose that sugars could directly regulate the expression of the majority of

INV genes. In the presence of sufficient carbon sources regulating gene expression, ShCWINV9, ShCWINV3, ShCWINV7, ShN/AINV4 and ShN/AINV5 may critically participate in sugar storage and utilization, and ShN/AINV6 and ShCWINV6 may play a pivotal role in sugar synthesis and sucrose metabolism. In addition, all INVs' activities were inhibited significantly under five different abiotic stresses. Based on the accrued data, we speculated that the contributions of neutral/alkaline INV to the overall enzymatic activity of the INV family were more than the sum of those of soluble acid INV and cell-wall INV activity. Further confirmatory experiments such as gene editing through the CRISPR-Cas9 system would be necessary to confirm this hypothesis. This study represents the first investigation of the INV gene family in sugarcane, providing the foundation to understand the physiological roles for each INV gene and unravel the molecular mechanism of sugar accumulation in sugarcane.

Additional files

Additional file 1: Table S1. Basic information on invertase genes in 8 plant species.

Additional file 2: Table S2. PCR primer sequences and annealing temperatures of invertase genes in sugarcane.

Additional file 3: Table S3. Real time PCR primers of 13 invertase genes in sugarcane.

Additional file 4: Table S4. Motif distribution in the invertase family.

Additional file 5: Figure S2. Alignment of the conserved regions from known acid invertases of sorghum and sugarcane.

Additional file 6: Figure S3. Alignment of the conserved regions from neutral/alkaline invertases.

Additional file 7: Figure S1. Phylogenetic tree of INVs in monocotyledons and dicotyledons.

Abbreviations
AINV: Acid invertase; BACs: Bacterial artificial chromosomes; CWINV: Cell-wall invertase; INV: Invertase; N/AINV: Neutral/alkaline invertase; SAINV: Soluble acid invertase; VINV: Vacuole invertases

Acknowledgements
The authors would like to thank Ping Zhou and Juan Liu (Fujian Agriculture and Forestry University, Fuzhou, China) for assistance with technical aspects of the project. We would also like to thank Hanyang Cai (Fujian Agriculture and Forestry University, Fuzhou, China) for providing the reference.

Funding
This project was supported by grants from the 863 program (2013AA102604), and NSFC (31201260) and the funding from the Fujian Agriculture and Forestry University.

Authors' contributions
LW and JZ conceived and designed the experiments. LW, YZ, SD, and QZ performed the experiments. LW, JZ performed data analysis. YC provided materials and reference. LW and JZ drafted the manuscript. LW, MR and JZ revised the manuscript. All authors read and approved the final manuscript.

Competing interests
The authors declare that they have no competing interests.

Author details
Center for Genomics and Biotechnology, Fujian Provincial Key Laboratory of Haixia Applied Plant Systems Biology, Haixia Institute of Science and Technology (HIST), Fujian Agriculture and Forestry University, Fuzhou 350002, China. ²College of Life Science, Fujian Agriculture and Forestry University, Fuzhou 35002, China. ³College of Life Sciences, Fujian Normal University, Fuzhou 350117, China. ⁴Key Laboratory of Sugarcane Biology and Genetic Breeding Ministry of Agriculture, Fujian Agriculture and Forestry University, Fuzhou 350002, China.

References
1. Alberto F, Bignon C, Sulzenbacher G, Henrissat B, Czjzek M. The three-dimensional structure of invertase (beta-fructosidase) from Thermotoga maritima reveals a bimodular arrangement and an evolutionary relationship between retaining and inverting glycosidases. J Biol Chem. 2004;279(18):18903–10.
2. Asnaghi C, D'Hont A, Glaszmann J, Rott P. *Resistance of sugarcane cultivar R570 to Puccinia Melanocephala* isolates from different geographic locations. Plant Dis. 2001;85(3):282–6.
3. Balk PA, de Boer AD. Rapid stalk elongation in tulip (*Tulipa gesneriana* L. cv. Apeldoorn) and the combined action of cold-induced invertase and the water-channel protein γTIP. Planta. 1999;209(3):346–54.
4. Bocock PN, Morse AM, Dervinis C, Davis JM. Evolution and diversity of invertase genes in *Populus trichocarpa*. Planta. 2008;227(3):565–76.
5. Botha F, Sawyer B, Birch R. Sucrose metabolism in the culm of transgenic sugarcane with reduced soluble acid invertase activity. Proceedings of the XXIV congress, Brisbane, Australia, 17-21 September 2001. International Society of Sugar Cane Technologists. 2001;2:588–91.
6. Bournay A-S, Hedley PE, Maddison A, Waugh R, GC M. Exon skipping induced by cold stress in a potato invertase gene transcript. Nucleic Acids Res. 1996;24(12):2347–51.
7. Cheavegatti-Gianotto A, de Abreu HM, Arruda P, Bespalhok Filho JC, Burnquist WL, Creste S, et al. Sugarcane (Saccharum X officinarum): a reference study for the regulation of genetically modified cultivars in Brazil. Trop Plant Biol. 2011;4(1):62–89.
8. Chen Z, Gao K, Su X, Rao P, An X. Genome-wide identification of the Invertase Gene family in Populus. PLoS One. 2015;10(9):1–21.
9. Cheng WH, Taliercio EW, Chourey PS. The Miniature1 seed locus of maize encodes a Cell Wall Invertase required for normal development of endosperm and maternal cells in the pedicel. Plant Cell. 1996;8(6):971–83.
10. Chourey PS, Jain M, Li QB, Carlson SJ. Genetic control of cell wall invertases in developing endosperm of maize. Planta. 2006;223(2):159–67.
11. Cortes-Romero C, Martinez-Hernandez A, Mellado-Mojica E, Lopez MG, Simpson J. Molecular and functional characterization of novel fructosyltransferases and invertases from *Agave tequilana*. PLoS One. 2012;7(4):e35878.
12. Daugrois JH, Grivet L, Roques D, Hoarau JY, Lombard H, Glaszmann JC, et al. A putative major gene for rust resistance linked with a RFLP markers in sugarcane cultivar 'R570'. Theor Appl Genet. 1996;92:1059–64.

13. Gayler K, Glasziou K. Physiological functions of acid and neutral invertases in growth and sugar storage in sugarcane. Physiol Plant. 1972;27(1):25–31.

14. Groenewald J, Botha F. Down-regulation of pyrophosphate: fructose 6-phosphate 1-phosphotransferase (PFP) activity in sugarcane enhances sucrose accumulation in immature internodes. Transgenic Res. 2008;17(1):85–92.

15. Guo J, Ling H, Wu Q, Xu L, Que Y. The choice of reference genes for assessing gene expression in sugarcane under salinity and drought stresses. Sci Rep. 2014;13(4):7042–52.

16. Haouazine-Takvorian N, Tymowska-Lalanne Z, Takvorian A, Tregear J, Lejeune B, Lecharny A, et al. Characterization of two members of the Arabidopsis thaliana gene family, at beta fruct3 and at beta fruct4, coding for vacuolar invertases. Gene. 1997;197(1–2):239–51.

17. Hatch MD, Glasziou KT. Sugar accumulation cycle in sugar cane. II. Relationship of Invertase activity to Sugar Content & Growth Rate in storage tissue of plants grown in controlled environments. Plant Physiol. 1963;38(3):344–8.

18. Hatch MD, Sacher JA, Glasziou KT. Sugar accumulation cycle in sugarcane. III. Physical and metabolic aspects of cycle in immature storage tissues. PLANT PHIYSIOLOGY. 1963;36:348–54.

19. Hu B, Jin J, Guo A-Y, Zhang H, Luo J, Gao G. GSDS 2.0: an upgraded gene feature visualization server. Bioinformatics. 2015;31(8):1296–7.

20. Iskandar H, Simpson R, Casu R, Bonnett G, Maclean D, Manners J. Comparison of reference genes for quantitative real-time PCR analysis of gene expression in sugarcane. Plant Mol Biol Report. 2004;22:325–37.

21. Itoh T, Akao S, Hashimoto W, Mikami B, Murata K. Crystal structure of unsaturated glucuronyl hydrolase, responsible for the degradation of glycosaminoglycan, from bacillus sp. GL1 at 1.8 a resolution. J Biol Chem. 2004;279(30):31804–12.

22. Jang JC, León P, Zhou L, Sheen J. Hexokinase as a sugar sensor in higher plants. Plant Cell. 1997;9:5–19.

23. Ji X, Van-Den-Ende W, Van-Laere A, Cheng S, Bennett J. Structure, evolution, and expression of the two invertase gene families of rice. J Mol Evol. 2005;60(5):615–34.

24. Kim JK, Bamba T, Harada K, Fukusaki E, Kobayashi A. Time-course metabolic profiling in Arabidopsis thaliana cell cultures after salt stress treatment. J Exp Bot. 2007;58(3):415–24.

25. Koch K. Sucrose metabolism: regulatory mechanisms and pivotal roles in sugar sensing and plant development. Curr Opin Plant Biol. 2004;7(3):235–46.

26. Koch KE. Carbohydrate-modulated Gene expression in plants. Annu Rev Plant Physiol Plant Mol Biol. 1996;47:509–40.

27. Koch, K. E., Y. Wu J. Xu. Sugar and metabolic regulation of genes for sucrose metabolism: potential influence of maize sucrose synthase and soluble invertase responses on carbon partitioning and sugar sensing. J Exp Bot. 1996;47 Spec No:1179–85.

28. Lalonde S, Boles E, Hellmann H, Barker L, Patrick JW, Frommer WB, et al. The dual function of sugar carriers. Transport and sugar sensing. Plant Cell. 1999;11(4):707–26.

29. Lammens W, Le Roy K, Schroeven L, Van Laere A, Rabijns A, Van den Ende W. Structural insights into glycoside hydrolase family 32 and 68 enzymes: functional implications. J Exp Bot. 2009;60(3):727–40.

30. Li Z, Palmer WM, Martin AP, Wang R, Rainsford F, Jin Y, et al. High invertase activity in tomato reproductive organs correlates with enhanced sucrose import into, and heat tolerance of, young fruit. J Exp Bot. 2012;63(3):1155–66.

31. Ling H, Wu Q, Guo J, Xu L, Que Y. Comprehensive selection of reference genes for Gene expression normalization in sugarcane by real time quantitative RT-PCR. PLoS One. 2014;9(5):10.

32. Moore PH (1987) Anatomy and morphology. In 'sugarcane improvement through breeding'. (Ed. DJ Heinz) pp. 85-142. (Elsevier: Amsterdam, The Netherlands).

33. Moore P. Temporal and spatial regulation of sucrose accumulation in the sugarcane stem. Funct Plant Biol. 1995;22(4):661–79.

34. Murayama S, Hands H. Genes for alkaline/neutral invertase in rice: alkaline/neutral invertases are located in plant mitochondria and also in plastids. Planta. 2007;225(5):1193–203.

35. Paterson AH, Bowers JE, Bruggmann R, Dubchak I, Grimwood J, Gundlach H, et al. The Sorghum bicolor genome and the diversification of grasses. Nature. 2009;457(7229):551–6.

36. Pelleschi S, Rocher JP, Prioul JL. Effect of water restriction on carbohydrate metabolism and photosynthesis in mature maize leaves. Plant Cell Environ. 1997;20:493–503.

37. Roitsch T. Source-sink regulation by sugar and stress. Curr Opin Plant Biol. 1999;2(3):198–206.

38. Roitsch T, Bittner M, Godt DE. Induction of apoplastic invertase of Chenopodium rubrum by D-glucose and a glucose analog and tissue-specific expression suggest a role in sink-source regulation. Plant Physiol. 1995;108(1):285–94.

39. Roitsch T, Gonzalez MC. Function and regulation of plant invertases: sweet sensations. Trends Plant Sci. 2004;9(12):606–13.

40. Rossouw D, Kossmann J, Botha F. Reduced neutral invertase activity in the culm tissues of transgenic sugarcane plants results in a decrease in respiration and sucrose cycling and an increase in the sucrose to hexose ratio. Funct Plant Biol. 2010;37(1):22–31.

41. Ruan Y. Sucrose metabolism: gateway to diverse carbon use and sugar signaling. Annu Rev Plant Biol. 2014;65:33–67.

42. Ruan YL, Jin Y, Yang YJ, Li GJ, Boyer JS. Sugar input, metabolism, and signaling mediated by invertase: roles in development, yield potential, and response to drought and heat. Mol Plant. 2010;3(6):942–55.

43. Schroeven L, Lammens W, Van Laere A, Van den Ende W. Transforming wheat vacuolar invertase into a high affinity sucrose:sucrose 1-fructosyltransferase. New Phytol. 2008;180(4):822–31.

44. Siswoyo TA, Oktavianawati I, Djenal UM, Sugiharto B. Changes of sucrose content and invertase activity during sugarcane stem storage. Indonesian Journal of Agricultural Science. 2007;8(2):75–81.

45. Smeekens S. Sugar regulation of gene expression in plants. Curr Opin Plant Biol. 1998;1(3):230–4.

46. Sturm A. Invertases. Primary structures, functions, and roles in plant development and sucrose partitioning. Plant Physiol. 1999;121(1):1–8.

47. Sturm A, Chrispeels MJ. cDNA cloning of carrot extracellular beta-fructosidase and its expression in response to wounding and bacterial infection. Plant Cell. 1990;2(11):1107–19.

48. Sturm A, Sebkovfi V, Lorenz K, Hardegger M, Lienhard S, Unger C. Development- and organ-specific expression of the genes for sucrose synthase and three isoenzymes of acid ,B-fructosidase in carrot. Planta. 1995;195:601–10.

49. Sturm A, Tang G. The sucrose-cleaving enzymes of plants are crucial for development, growth and carbon partitioning. Trends Plant Sci. 1999;4(10):401–7.

50. Tamura K, Peterson D, Peterson N, Stecher G, Nei M, Kumar S. MEGA5: molecular evolutionary genetics analysis using maximum likelihood, evolutionary distance, and maximum parsimony methods. Mol Biol Evol. 2011;28(10):2731–9.

51. Tang G, Lüscher M, Sturm A. Antisense repression of vacuolar and cell wall invertase in transgenic carrot alters early plant development and sucrose partitioning. The Plant Cell Online. 1999;11(2):177–89.

52. Tymowska-Lalanne Z, Kreis M. Expression of the Arabidopsis thaliana invertase gene family. Planta. 1998;207(2):259–65.

53. Vargas WA, Pontis HG, Salerno GL. Differential expression of alkaline and neutral invertases in response to environmental stresses: characterization of an alkaline isoform as a stress-response enzyme in wheat leaves. Planta. 2007;226(6):1535–45.

54. Walker RP, Winters AL, Pollock CJ. Purification and characterisation of invertases from leaves of Lolium temulentum L. New Phytol. 1997;135:259–66.

55. Winter H, Huber SC. Regulation of sucrose metabolism in higher plants: localization and regulation of activity of key enzymes. Crit Rev Biochem Mol Biol. 2000;35(4):253–89.

56. Xu J, Avigne WT, Mccarty DR, Koch KE. A similar dichotomy of sugar modulation and develpomental expression affects both path of sucrose metabolism: evidence from a maize invertase gene family Plant Cell. 1996;8:1209–1220.

57. Zhang J, Arro J, Chen Y, Ming R. Haplotype analysis of sucrose synthase gene family in three Saccharum species. BMC Genomics. 2013;14(314).

58. Zhu YJ, Komor E, Moore PH. Sucrose accumulation in the sugarcane stem is regulated by the difference between the activities of soluble acid Invertase and sucrose phosphate synthase. Plant Physiol. 1997;115(2):609–16.

microRNAs participate in gene expression regulation and phytohormone cross-talk in barley embryo during seed development and germination

Bin Bai[1], Bo Shi[1], Ning Hou[1], Yanli Cao[1], Yijun Meng[2], Hongwu Bian[1], Muyuan Zhu[1] and Ning Han[1*]

Abstract

Background: Small RNA and degradome sequencing have identified a large number of miRNA-target pairs in plant seeds. However, detailed spatial and temporal studies of miRNA-mediated regulation, which can reflect links between seed development and germination are still lacking.

Results: In this study, we extended our investigation on miRNAs-involved gene regulation by a combined analysis of seed maturation and germination in barley. Through bioinformatics analysis of small RNA sequencing data, a total of 1324 known miRNA families and 448 novel miRNA candidates were identified. Of those, 16 known miRNAs with 40 target genes, and three novel miRNAs with four target genes were confirmed based on degradome sequencing data. Conserved miRNA families such as miR156, miR168, miR166, miR167, and miR894 were highly expressed in embryos of developing and germinating seeds. A barley-specific miRNA, miR5071, which was predicted to target an *OsMLA10-like* gene, accumulated at a high level, suggesting its involvement in defence response during these two developmental stages. Based on target prediction and Kyoto Encyclopedia of Genes and Genomes analysis of putative targets, nine highly expressed miRNAs were found to be related to phytohormone signalling and hormone cross-talk. Northern blot and qRT-PCR analysis showed that these miRNAs displayed differential expression patterns during seed development and germination, indicating their different roles in hormone signalling pathways. In addition, we showed that miR393 affected seed development through targeting two genes encoding the auxin receptors TIR1/AFBs in barley, as over-expression of miR393 led to an increased length–width ratio of seeds, whereas target mimic (MIM393)-mediated inhibition of its activity decreased the 1000-grain weight of seeds. Furthermore, the expression of auxin-responsive genes, abscisic acid- and gibberellic acid-related genes was altered in miR393 misexpression lines during germination and early seedling growth.

Conclusions: Our work indicates that miRNA-target pairs participate in gene expression regulation and hormone interaction in barley embryo and provides evidence that miR393-mediated auxin response regulation affects grain development and influences gibberellic acid and abscisic acid homeostasis during germination.

Keywords: Barley (*Hordeum vulgare*), Microrna, Seed development, Germination, Embryo, Auxin response, Abscisic acid, Gibberellic acid

* Correspondence: ninghan@zju.edu.cn
[1]Key Laboratory for Cell and Gene Engineering of Zhejiang Province, Institute of Genetics and Regenerative Biology, College of Life Sciences, Zhejiang University, Zhejiang, Hangzhou 310058, China
Full list of author information is available at the end of the article

Background

Seed development and germination are two critical developmental phase transitions during a plant life cycle. As for most flowering plants, seed maturation and germination are separated by a period of quiescence, which is termed as dormancy [1]. Only after breaking dormancy, the quiescent embryo is able to germinate after imbibition. On the other hand, there is a close connection between these two processes. Genome-wide expression analysis and comprehensive metabolite studies revealed that initial protein synthesis during germination uses pre-existing mRNAs stored in mature dry seed, and metabolic preparation for germination is already initiated during the late stages of seed maturation [2, 3].

Most seeds are composed of three basic parts: embryo, endosperm and seed coat. The embryo encloses fundamental elements and patterns necessary for the new plant to develop after germination, whereas the endosperm provides nutrition for the embryo to use during germination [4]. Studies at the physiological and molecular levels have demonstrated that seed maturation and germination involve tight gene regulation and hormone control so that continuous interchange of signals between the different compartments of seeds is ensured [4–7]. Two hormones, abscisic acid (ABA) and gibberellic acid (GA), play important roles in determining the physiological state of the seed and regulating the development and germination process. The level of ABA peaks during seed maturation and dormancy, whereas GA level increases during imbibition and remains high during germination and postembryonic growth [1]. However, it seems to not be the absolute concentration but the balance of ABA and GA that determines the two events. In addition, another hormone, auxin, has also been proven to be fundamental in the first step of seed development, as well as for the determination of embryo structure and size [4, 8]. In *Arabidopsis*, it is reported that auxin plays a role in seed dormancy and germination through its crosstalk with other hormones such as ABA, indicating a coordinating network of auxin and ABA signalling in this important process [9].

An important mechanism controlling gene expression during seed development and germination is exerted by microRNAs (miRNAs) [10, 11]. miRNAs are a class of endogenous non-coding small RNAs (sRNAs) approximately 21–22 nt in length, which post-transcriptionally regulate gene expression by targeting mRNA for cleavage or translation suppression [12]. High throughput small RNA sequencing coupled with degradome analysis allows identification of conserved and novel miRNA-target pairs in seed of rice, maize, barley and *Brassica napus* [13–20]. Increasing evidence shows that miRNAs play crucial roles in diverse aspects of biological and metabolic processes in seeds including embryogenesis, pattern establishment, seed dormancy, seed germination and early seedling growth. Mutants lacking components of miRNA biogenesis and/or processing displayed severely abnormal seed development or even lethality [21, 22]. It was reported that miR172 affected seed size and yield through targeting several *APETALA2*-like transcription factors [23]; the *mir159ab* double mutant exhibited reduced seed size and altered seed shape [24]. miR159 and miR160 affected the process of germination by regulating ABA sensitivity. Transgenic plants which overexpressed a miR160-resistant form of *ARF10* were hypersensitive to ABA, indicating a point of cross-talk between ABA and auxin in imbibed seeds [25].

Barley (*Hordeum vulgare* L.) is one of the most important cereal crops worldwide, ranking fourth in terms of production and is widely used for brewing and animal feed. Seed maturation and germination are closely associated with crop yield and processing quality of barley grains. Many barley miRNAs have been identified through small RNA sequencing using samples of leaf, seedling, root, and seeds at early development stages [20, 26–30]. However, for the embryo, which is an important tissue affecting seed development and germination, a complete miRNA expression profile analysis has not yet been performed. The purpose of this study was to discover embryo-specific expression of miRNAs and their potential targets and provide vital clues for further and detailed functional studies of barley miRNAs.

Methods

Plant materials

Barley (*Hordeum vulgare* L. 'Golden Promise'), obtained from the Australian Centre for Plant Functional Genomics at the University of Adelaide, was used in this study. The plants were grown from October to May in soil under natural conditions at the Agricultural Experiment Station of Zhejiang University, Hangzhou, Zhejiang Province, China. Immature embryos at 10 days post anthesis (10 DPA) were dissected from spikes. To collect embryos at 1 DAG (days after germination) and 5 DAG, seeds were surface sterilized, rinsed, and then placed on moist filter papers in 9-cm culture dishes and kept at 4°C in the dark for 48 h. They were then grown for 1 d and 5 d in containers with 0.1 mM $CaCl_2$ solution (pH 5.8) at 24°C under a 16 h light/8 h dark photoperiod in a controlled climate chamber.

Small RNA sequencing and identification of conserved and novel miRNAs in barley

Total RNA from at least 3 g embryos for each sample, as described in Fig. 1a, was isolated using TRIzol Reagent (Invitrogen, USA) according to the manufacturer's instructions. RNA extracted from three biological replicates was pooled to constitute each sample for small

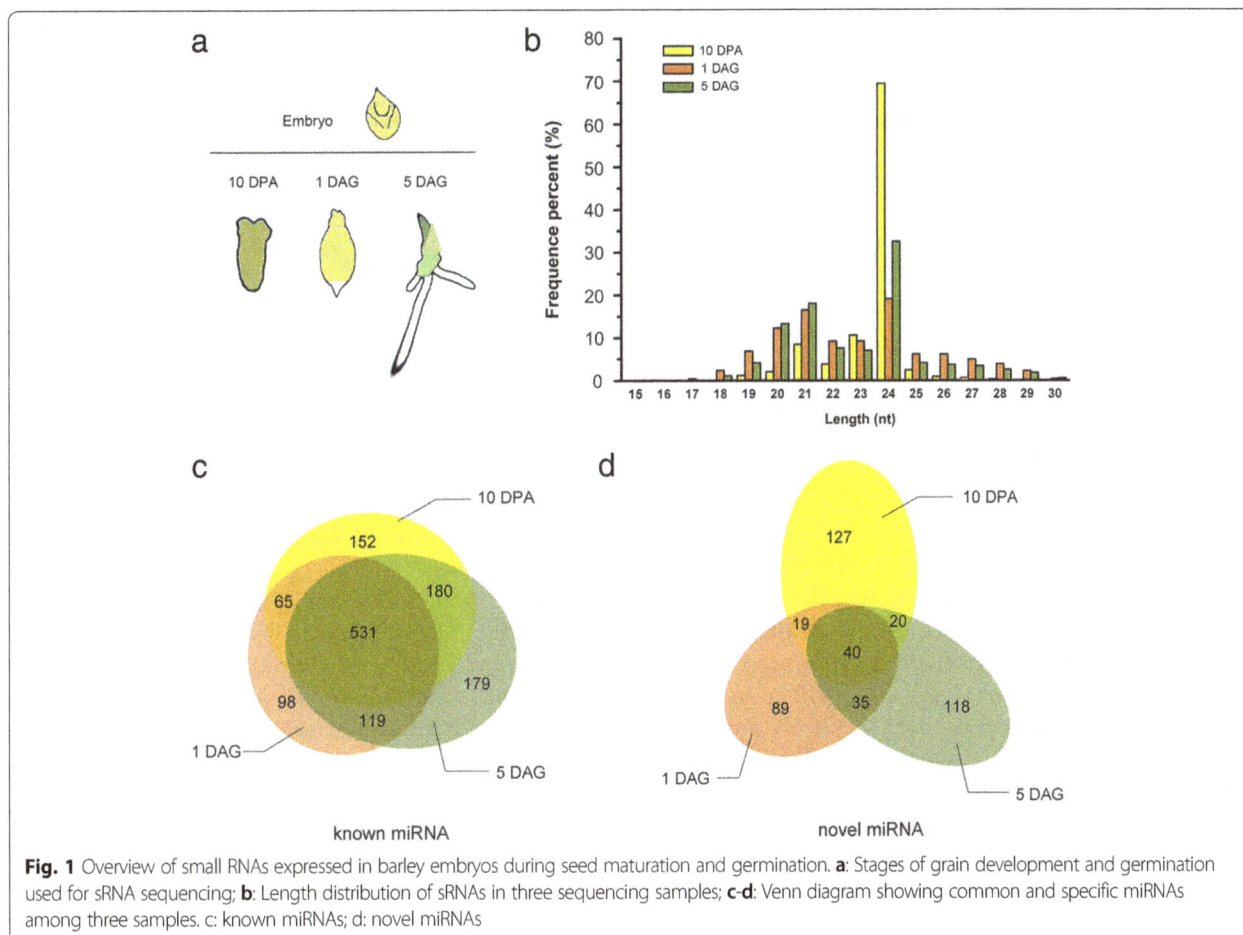

Fig. 1 Overview of small RNAs expressed in barley embryos during seed maturation and germination. **a**: Stages of grain development and germination used for sRNA sequencing; **b**: Length distribution of sRNAs in three sequencing samples; **c-d**: Venn diagram showing common and specific miRNAs among three samples. c: known miRNAs; d: novel miRNAs

RNA sequencing. The RNA quality was examined using gel-electrophoresis (28S:18S ratio > 1.5) and a Bioanalyzer (Agilent2100, RIN ≥ 8.0). The sRNA libraries were then constructed following the standard protocol used at The Beijing Genomics Institute (BGI, Shenzhen, China). The qualified total RNA samples were gel purified and fragments within the length range (18 nt to 30 nt) were selected to build a library. Then, two kinds of adapters were ligated to each end of the resulting fragments. The prepared RNA was amplified by reverse transcription PCR (RT-PCR) and RT-PCR products were then loaded on a Hiseq2000 platform for sequencing. The raw data from Hiseq sequencing went through data cleaning analysis, which included the elimination of low-quality tags and 5′ adaptor contaminants, to obtain sufficiently clean tags. After filtering, sRNAs with sizes ranging from 18 to 30 nt were aligned to the barley genome (http://plants.ensembl.org/Hordeum_vulgare/Info/Index), Genbank, and Rfam (http://rfam.xfam.org/), and the sequences matching known barley rRNAs, tRNAs, snRNAs, and snoRNAs were discarded. Unannotated small RNA tags were then aligned with miRBase 21 (http://www.mirbase.org/), and those which could be aligned to an miRNA

precursor in miRBase with no mismatch, or have at least a 16-nt overlap aligning to a known barley miRNA in miRBase, were termed as known miRNAs. If there was no miRNA information for the species in miRBase, the candidates which could be aligned to any miRNA precursor or mature miRNA from all plants in miRBase, allowing two mismatches or free gaps and with a predicted precursor, were regarded as known miRNAs. After alignment, small RNA tags corresponding to exons and introns of mRNA (to find degraded fragments of mRNA) and the unannotated tags were used for novel miRNA prediction through the Mireap software (http://sourceforge.net/projects/mireap/) developed by BGI. The secondary structures of all identified and potential pre-miRNAs in the barley genome were predicted by using RNAfold software (http://rna.tbi.univie.ac.at/cgi-bin/RNAWebSuite/RNAfold.cgi). The minimal folding energy index (MFEI) of the novel miRNAs was set to be equal or greater than 0.9.

Differentially expressed miRNAs were identified by plotting log2-ratio figures and scatter plots. The expression of miRNAs was normalized according to the expression of transcript per million (TPM). For each sample, TPM = (actual miRNA count/total count of clean reads) $\times 10^6$. The fold-change in miRNA expression was calculated as fold change = log2 (treatment/control). The differentially expressed sequence counts were analysed by the online service IDEG6 (http://telethon.bio.unipd.it/bioinfo/IDEG6_form/). We considered a fold change of at least 2 and a P value cut-off of $P < 0.00001$ as an indication of significant change and considered the given miRNA as a specific development stage-associated miRNA. The chosen potential pre-miRNA sequences obtained from the results of Blastall were subjected to the Zuker folding algorithm for in silico secondary structure generation via the web-based computational software MFOLD.

Target prediction and degradome sequencing data analysis

The prediction of putative targets for identified miRNAs was performed with Mireap and psRNATarget (http://plantgrn.noble.org/psRNATarget). The following modified parameters for the Mireap software analysis were used: (1) No more than four mismatches between sRNA and target (G-U bases count as 0.5 mismatches); (2) No more than two adjacent mismatches in the miRNA/target duplex; (3) No adjacent mismatches in positions 2–12 of the miRNA/target duplex (the 5′ end of the miRNA); (4) No mismatches in positions 10–11 of miRNA/target duplex; (5) No more than 2.5 mismatches in positions 1–12 of the of the miRNA/target duplex (the 5′ end of miRNA); and (6) The minimum free energy (MFE) of the miRNA/target duplex should be greater than or equal to 75% of the MFE of the miRNA bound to it's perfect complement. Degradome analyses were based on the barley degradome sequencing dataset (Accession number SRR513549, SRR513550 and SRR513551) downloaded from NCBI. The miRU algorithm with 'default parameters' was used in this study [31].

Gene ontology (GO) is an international standardized classification system for gene function, which supplies a set of controlled vocabulary to comprehensively describe the property of genes and gene products. There are three ontologies in GO: molecular function, cellular component and biological process. This method first maps all target gene candidates to GO terms in the database (http://www.geneontology.org/), calculating gene numbers for each term. It then uses a hypergeometric test to find significantly enriched GO terms in target gene candidates, compared to the reference gene background. Then, the GO functional analysis of the putative targets was performed by WEGO (http://wego.genomics.org.cn/cgi-bin/wego/index.pl). GO terms with a P-value <0.05 were considered to be significantly enriched.

To facilitate the understanding of the biological functions of the target gene candidates, Kyoto Encyclopedia of Genes and Genomes (KEGG) pathway analysis was used to identify significantly enriched metabolic pathways or signal transduction pathways in target genes compared with the whole reference gene background. The calculating formula was the same as that in the GO analysis.

Northern blot analysis of miRNA expression

Total RNA was extracted using TRIzol reagent (Invitrogen, USA). For RNA gel blots, 30 µg of total RNA was separated on a 17% polyacrylamide gel containing 7 M urea, blotted to HyBond-N+ membranes (Roche, Germany) and fixed by UV crosslinking. Blots were hybridized using digoxigenin end-labelled locked nucleic acid (LNA) oligonucleotide probes designed against miRNAs. The miRNA level was standardized with that of rRNA. Sequences of the oligonucleotide probes are listed in Additional file 1: Table S5.

Real-time quantitative RT-PCR analysis of miRNAs and their target levels

Total RNA from barley embryos and seedling was extracted using TRIzol reagent (Invitrogen) and 1 µg was used for first-strand cDNA synthesis using Oligo(dT) primers and a Super-Script III RT kit (TaKaRa, Dalian, China). Quantitative real-time PCR (qRT-PCR) was performed on the Mastercycler ep realplex2 system (Eppendorf, Hamburg, Germany) using a SYBR PrimeScript RT reagent Kit (Perfect Real Time, TaKaRa). Relative transcript levels were calculated with the $\Delta\Delta Ct$ method and the *ACTIN* or *UBQ* gene was used as a reference. Quantification of mature miRNA level was carried out through a poly(A)-based real-time PCR approach using Mir-X™ miRNA First Strand Synthesis Kit (Clontech Laboratories, Inc., Cat. #638315) and the miRNA level was standardized with U6. Sequences of primers are listed in Additional file 1: Table S5. The data shown were obtained from three biological replicates.

Validation of target genes through RLM-5′ RACE

RNA ligase-mediated 5′ rapid amplification of cDNA ends (RLM 5′-RACE) was performed using a SMART TM RACE cDNA Amplification kit (Promega, USA). The manufacturer's protocol was followed for 5′ end analysis. In brief, total RNA was isolated from leaf tissues at 7 DAG and ligated to a 5′ end RNA adaptor before being reverse transcribed using an oligo(dT) primer. The PCR reactions were performed using two pairs of gene specific reverse primers (Additional file 1: Table S5). The PCR products were gel purified, then cloned into the T-easy vector (Promega, USA) and sequenced.

Vector construction and barley transformation

The miR393 over-expression and MIM393 construct were generated according to Bian et al. [32] and Bai et al. [33]. The clones used for vector construction were verified by sequencing and then were electroporated into

Agrobacterium tumefaciens strain AGL1, which was used to transform Golden Promise barley plants. Barley transformation was performed using the *A. tumefaciens*-mediated co-cultivation approach described by Bai et al. [33]. The transformed plants were grown in a growth chamber (24°C, 16 h light/8 h dark) or under natural condition in 6-in. pots in a growth room (Zhejiang University).

Results

Overview of sRNAs expressed in barley embryos during seed maturation and germination

To obtain an overview of the miRNA expression profile in embryo during seed maturation and germination, we constructed sRNA libraries using samples at three different developmental stages: one time point in the grain storage phase during seed development (10 DPA), two time points during seed germination including 1 DAG (imbibition to radicle emergence), and 5 DAG (postgermination growth) (Fig. 1a). After removal of low-quality and adaptor contaminants (reads <18 nt) from the raw sequencing data, a total of 17,216,333; 17,729,069; and 20,330,126 clean reads were obtained from the libraries of 10 DPA, 1 DAG, and 5 DAG, respectively (Additional file 1: Table S1). There were two major populations of sRNAs according to their lengths (Fig. 1b). The peak of length distribution located at 24 nt accounted for 69.44%, 19.09% and 32.5% in the three libraries respectively, indicating that siRNA-associated regulation is involved in heterochromatin modification and RNA-directed DNA methylation occurring at these stages. This is consistent with a previous study in rice which showed that filling or developing grains contained the most abundant 24-nt small RNAs [15]. The 21-nt small RNAs were the second most abundant population present in the three libraries, representing 8.48%, 16.48%, and 18.05% of the total reads, respectively. After getting rid of reads which could be annotated as rRNA, repeats, exons or introns, the remaining reads were used for miRNA prediction.

Identification of known and novel miRNAs expressed in barley embryo

Through bioinformatics analysis, 1324 known miRNA families and 448 novel miRNA candidates were found to be expressed at total RPM (reads per million) up to 1 (Additional file 2: Table S2). Known miRNAs were identified through alignment to the sequences already registered in miRBase (Release 21), whereas novel miRNAs were predicted according to the characteristic hairpin structures of their precursors. As is shown in the Venn diagram, 531 conserved miRNA families and 40 novel miRNA candidates were detected in the embryo of developing and imbibed seeds at the three stages we tested. A total of 152 and 127, 98 and 89, and 179 and 118

known and novel miRNA families were found to be expressed specifically in embryos at 10 DPA, 1 DAG and 5 DAG, respectively.

The top 21 highly expressed known miRNA families by total RPM (> 2000) were grouped according to their degree of conservation across the plant kingdom (Fig. 2). Among them, four miRNAs including miR156, miR167, miR166, and miR894 are conserved in all plant species registered in miRBase; three miRNAs (miR168, miR2118, and miR164) are conserved among angiosperm; six miRNAs (miR2916, miR6441, miR2199, miR7696, miR8124, and miR6300) are detected only in dicotyledons; and five miRNAs (miR5565, miR1869, miR5813, miR5071, and miR5060) seem to be conserved families across grass families. Notably, we found that miR5565 and miR2199 were significantly and highly expressed in barley embryos during seed germination, whereas miR156, miR166, miR167, and miR168 accumulated in both the seed development and germination stages (Fig. 2).

Differential expression of miRNAs in embryo during seed development and germination

The expression levels of known miRNAs were normalized to TPM (transcript per million). Several miRNAs were found to change dramatically in embryos for each pair-wise comparison among the three development stages (Fig. 3a). Among them, significantly differentially expressed miRNAs between two samples with a fold change greater than 12 (p value <0.01) are shown in Fig. 3b. In the filling stage, miR5266, miR5252, miR5669, miR1852 and miR5814 were much more highly expressed than in the germination stage. On the contrary, miR6441 and miR1047 had elevated their expression levels at the early stage of seedling growth. miR6135 and miR401 were only expressed highly at 5 DAG, whereas miR5832 and miR7734 were expressed preferentially at 1 DAG.

Target prediction and function analysis of miRNAs

The target genes of miRNA candidates were predicted using the barley Unigene library (Esembl), Mireap, and psRNATarget (Additional file 3: Table S3). Furthermore, we searched for evidence from available degradome data from NCBI to select miRNAs that could result from cutting of the target mRNAs. A total of 40 genes targeted by 16 conserved miRNA families were identified (Table 1). Most of these targets are annotated as transcription factor-coding genes, including SQUAMOSA promoter-binding protein-like genes (*SPLs*) targeted by miR156, MYB transcription factors (*MYBs*) targeted by miR159, auxin response factors (*ARFs*) targeted by miR160, and TCP family transcription factors (*TCPs*) targeted by miR319. miR444 is predicted to target MLOC_16182, a gene encoding a RING finger protein; miR5051 targets

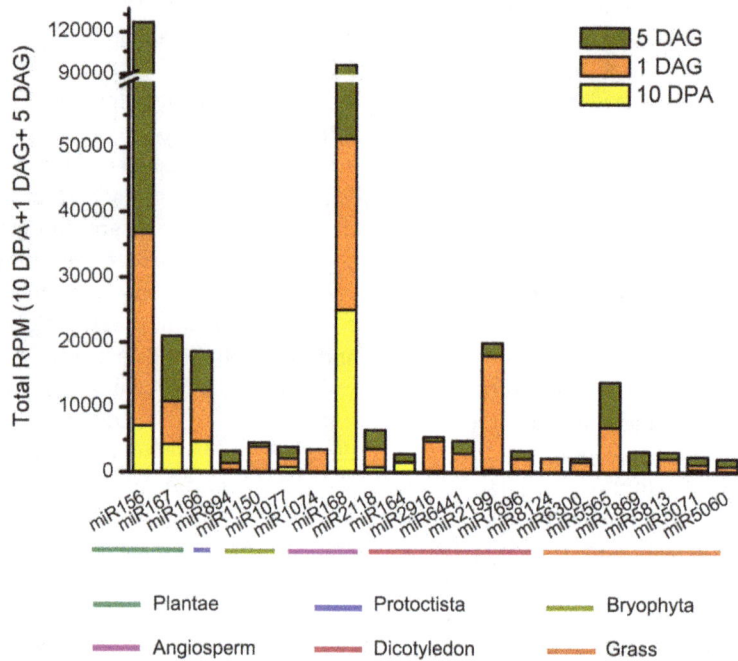

Fig. 2 Summary of highly expressed microRNAs in barley embryos. miRNA families were grouped based on their conservation level across plant kingdom according to the miRNA information in miRBase (Release 21). The total number sequences from three libraries is presented in reads per million (RPM)

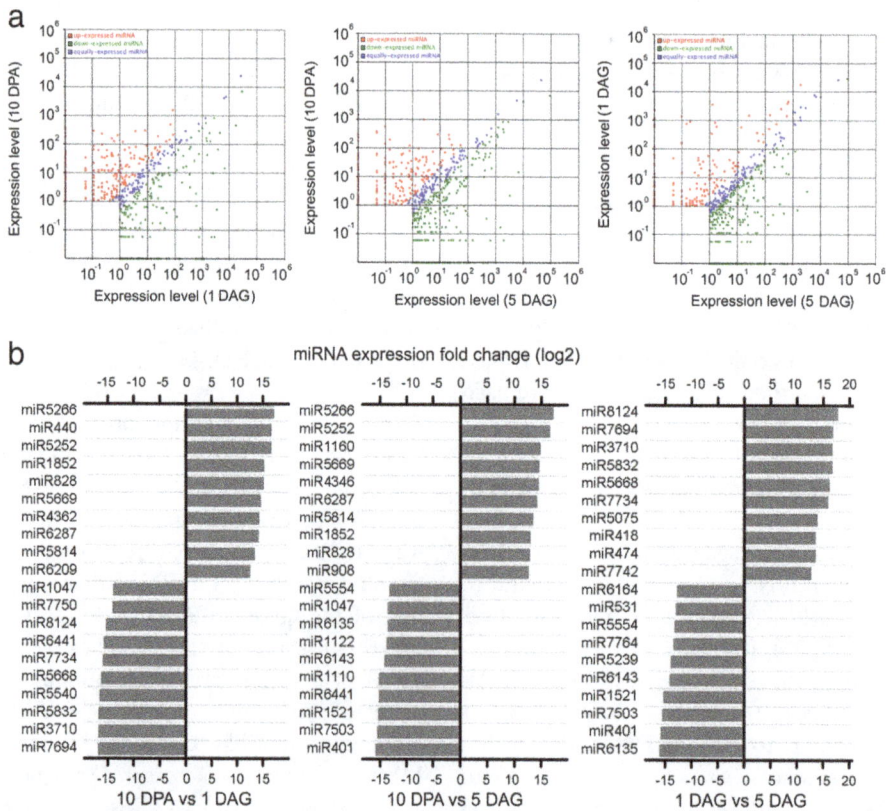

Fig. 3 Differentially expressed miRNAs between seed maturing and germination stages. **a**: Scatter plot showing differentially expressed known miRNA between two different samples. Each point in the figure represents a miRNA. The X axis and Y axis show expression level in two samples respectively; **b**: Differential expressed miRNAs between two samples

Table 1 Target prediction and degradome analysis of known miRNAs

miRNA	Target CDS_ID	Target annotation	Binding site on CDS	PARE_cut sites
miR156	AK356077	Squamosa promoter-binding protein	839–858	849
	AK374598	Squamosa promoter-binding protein	860–879	870
	MLOC_11199.2	Squamosa promoter-binding-like protein 11	1154–1173	1164
miR159	AK251726.1	MYB transcription factor	962–982	973
	AK370348	Predicted protein	1009–1028	948 and 1019
miR160	MLOC_64795.1	Auxin response factor 17	1329–1349	1340
	MLOC_69988.1	Auxin response factor 10	717–737	728
miR165	MLOC_44268.1	MATE efflux family protein	159–179	170
	AK362009	Class III homeo domain-leucine zipper	578–598	589
	AK364215	Class III homeo domain-leucine zipper	639–658	649
	MLOC_58644.1	Class III homeo domain-leucine zipper	980–1000	991
miR166	AK362009	Class III homeo domain-leucine zipper	578–598	589
	AK364215	Class III homeo domain-leucine zipper	638–658	649
	AK364215	Class III homeo domain-leucine zipper	640–658	649
	AK365312	Class III homeo domain-leucine zipper	563–583	574
	MLOC_44268.1	MATE efflux family protein	159–179	170
	MLOC_58644.1	Class III homeo domain-leucine zipper	980–1000	991
miR167	MLOC_51932.2	Auxin response factor 30	2343–2363	2354
	MLOC_58330.2	Auxin response factor 18	2528–2549	2540
	MLOC_63938.1	Auxin response factor 9	2402–2423	2414
miR171	AK371946	GRAS family transcription factor	1090–1109	1100
miR172	AK355002	AP2-like ethylene-responsive transcription factor	1214–1234	1225
miR319	AK370348	TCP family transcription factor containing protein	1009–1028	1019
	MLOC_63989.1	TCP family transcription factor containing protein	976–995	986
miR360	MLOC_69988.1	Auxin response factor 19	717–737	728
miR393	AK355927	Auxin signaling F-box 2	1560–1580	1571
	AK374984	Auxin signaling F-box 3	1974–1994	1985
miR396	MLOC_64055.4	Growth-regulating factor 2	375–395	386
	AK250947.1	Growth-regulating factor 1	480–500	491
	AK353813	Growth-regulating factor 1	387–407	398
	AK376404	Growth-regulating factor 5	417–437	428
	MLOC_67201.4	Growth-regulating factor 1	741–761	752
	MLOC_80060.1	Growth-regulating factor 1-like	507–527	518
	MLOC_12347.1	Pre-mRNA-processing protein 45	1438–1458	1449
	AK375237	Nascent polypeptide-associated complex subunit alpha-like protein	184–204	195
miR444	MLOC_16182.1	RING finger protein	1035–1055	1046
miR5051	MLOC_57965.3	Serine/threonine protein kinase	595–615	606
miR7757	MLOC_17471.2	NBS-LRR disease resistance protein, putative	194–215	206
miR9863	MLOC_24045.1	Disease resistance protein (CC-NBS-LRR class) family	1259–1280	1271

MLOC_57965 encoding a serine/threonine protein kinase; and miR7757 targets MLOC_17471 encoding a putative NBS-LRR disease resistance protein. These miRNA-target pairs are reported here in barley for the first time. The cleavage signature for miR9863 in MLOC_24045 encodes a putative disease resistance protein (CC-NBS-LRR class) family member, consistent with a previous study [33]. As for novel miRNAs, only three novel miRNA-mediated cleavage sites in four genes have been confirmed by degradome sequencing data (Table 2).

All the potential target genes were functionally annotated by ontology analysis. The predicted targets were classified into three main categories: biological processes, cellular components, and molecular functions. Of these, cell, cell part, and organelle part in cellular components, molecular binding, catalytic activity, enzyme regulator activity, molecular transducer activity, metabolic processes, binding, catalytic activity, transcription regulation in cellular components and response to stimulus in biological processes were the most observed categories (Additional file 4: Figure S1), indicating that miRNAs might play roles in diverse and important biological processes.

miRNAs involved in phytohormone signalling during seed development and germination

KEGG pathway analysis was performed for the target gene candidates. Nine miRNAs were predicted to be involved in phytohormone signalling pathways (Fig. 4a). Among them, miR156, miR159, miR390, miR164, miR396, and miR319 were predicted to regulate the ethylene pathway, miR159 regulates GA signalling; and miR172, miR396, and miR319 are likely to be associated with cytokinin signalling. Four miRNAs including miR393, miR390, miR164, and miR167, seem to be involved in the auxin pathway, whereas miR393 and miR167 regulate genes in the ABA pathway. miR319 is also predicted to be associated with the jasmonic acid pathway. Most of these miRNAs are involved in at least two phytohormone pathways, suggesting that the control of these two development stages involves cross-talk among five key phytohormones.

To determine the abundance of these miRNAs, the accumulation of nine known miRNA families was compared through sequencing data analysis combined with northern blot confirmation. As shown in Fig. 4b, miR156 and miR167 were expressed highly at the three developmental stages we tested; miR164 was expressed mainly at the seed development and postgermination stages; miR393, miR172, and miR396 were expressed preferentially in embryo at 5 DAG. The abundance of miR159 gradually increased during germination. The expression of five highly expressed

miRNAs was further validated by northern blot analysis.. Overall, the results from northern blot analysis were in accordance with the sequencing data (Fig. 4c). In embryos during seed development, miR164 accumulated from 4 DPA to 10 DPA and decreased its level at the late period of seed development. miR396 and miR156 increased their abundance when seeds entered the maturation stage. The expression of miR167 seemed to be stable throughout the whole seed development period. miR393 levels varied dramatically, and could only be detected in embryos from 8 DPA after we increased the loading amount of RNA samples. During the early seedling growth stage, miR156, miR396, miR167, and miR164 increased their abundance in root, shoot, and leaf tissues, especially at 10 DAG. miR393 and miR167 were expressed highly in shoot and leaf tissues but not so in root tissues (Fig. 4d).

To distinguish different members within known miRNA families, we performed qRT-PCR to detect the mature miRNA levels in embryo during seed imbibition. The first 24 h after imbibition is pivotal for seed germination, during which embryo cells rapidly switch from a quiescent state to a metabolically active state. All the miRNA members we tested increased their levels early in the 24-h imbibition, compared with their corresponding control (dry seed) (Fig. 5), suggesting that the expression of these miRNAs is dynamic and might participate in gene expression regulation during germination.

Verification of miRNA-induced cleavages in the target mRNAs

The targets of phytohormone-associated miRNAs mentioned above are summarized in Additional file 1: Table S4 according to our study and previous studies. Moreover, we verified the target candidates for miR396 through bioinformatics prediction coupled with degradome sequencing data-based validation (Fig. 6a). Only one gene MLOC_80060 was found to overlap from the results of the two methods. To further examine the cleavage sites in target transcripts, we conducted 5'-RACE using RNA extracted from leaves of the 1-week-old wildtype plants (WT). We detected two major cleavage products from transcripts of AK376404 after nested PCR and gel electrophoresis. Sequencing of 5'-RACE clones revealed one cleavage site in transcripts of AK376404 (Fig. 6b), which is annotated as a *GRF* transcriptional factor.

Table 2 Novel miRNA-target pairs in barley based on the degradome data sets

miRNA	Sequencing	cDNA_ID	Target annotation	miRU start-ending	PARE_cutsites
novel-mir-119	TCTTGACCTTGCAAGACCTTT	AK362090	ARF 3	1362–1382	1373
		AK369226	ARF 3	423–443	434
novel-mir-400	CGACGAGTCGGACGCGTCGAGCA	MLOC_53497.2	Predicted protein	587–609	600/733
novel-mir-205	TCACAGATAATGGTGGCCCCTG	MLOC_54213.1	Predicted protein	389–410	401

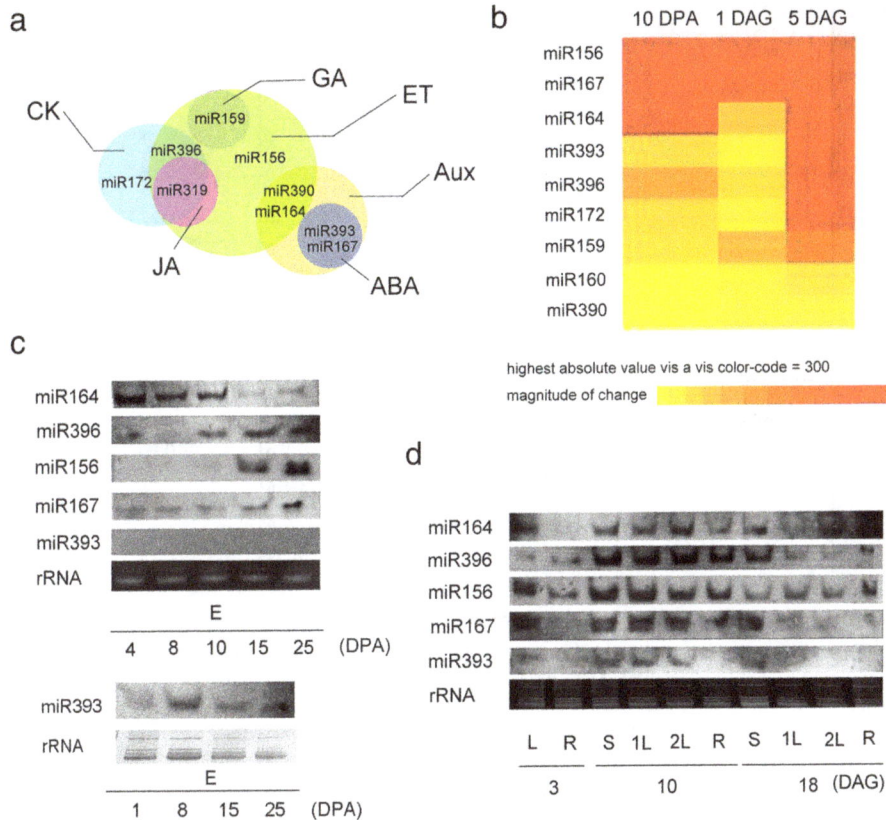

Fig. 4 Phytohormone-related miRNAs involved in seed maturation and seed germination processes. a: Overview of the miRNA-mediated regulations in phytohormone crosstalk through KEGG pathway analysis; b: sRNA sequencing data showing the expression of miRNA during seed development and germination. The expression levels are visualized by BAR HeatMapper. Numbers beneath the heat map indicate the relative expression intensities, and the higher expression intensities are indicated by more reddish colors; c: Northern blot analysis of miRNA levels in embryos during seed development. 30 μg and 50 μg (for miR393) total RNA was loaded, and rRNA was stained by EtBr as loading control. E, embryo; d: Northern blot analysis of miRNA levels during early seedling growth. Total RNA was isolated from tissues in different development stages. L, leaf; R, root; S, stem; 1 L, first leaf; 2 L, second leaf

Based on the information from target genes, the expression levels of miR160, miR159, and their targets were selected for detection using qRT-PCR (Fig. 6c, d). The transcript levels of MLOC_64795 (targeted by miR160) and MLOC_71332 (targeted by miR159) exhibited a gradually declining trend in the embryos of germinating seeds, whereas miR160 and miR159 increased their abundance when seeds started germinating.

miR393-mediated auxin signalling regulation affects seed development and germination

In our previous work, we identified two miR393 family members in barley and confirmed two target genes, *HvTIR1* (MLOC_9864) and *HvAFB* (MLOC_56088) through a modified form of 5′-RACE (rapid amplification of cDNA ends) as well as through degradome data analysis [34]. To investigate the biological function of miR393/target modules in seed development, we generated transgenic barley plant overexpressing miR393 (*35S::MIR393b*) or inhibiting miR393's activity through an artificial miRNA target mimics strategy (*35S::MIM393*). Through northern blot and qRT-PCR analysis, we showed that miR393 negatively regulates the transcript levels of its two target genes (*HvTIR1* and *HvAFB*) [34]. Moreover, we found that miR393 affects barley seed size and shape. As shown in Fig. 7a and b, compared with wildtype, the length/width ratio of the seed was greatly increased in *35S::MIR393b* plants, although the 1000-grain weight did not change significantly. In *35S::MIM393* plants, only the 1000-grain weight was significantly decreased (Fig. 7c).

Because miR393 targets two genes encoding the auxin receptor TIR1/AFBs, we then detected the expression of auxin signalling-related genes in these transgenic lines. This analysis showed that the transcripts of two *ARF* members (MLOC_77438 and MLOC_73144), one *SAUR* family member (MLOC_62887) and one *AUX/IAA* family member gene (MLOC_14320) were upregulated by more than two times in 3-d-old seedlings of the MIM393 line, and downregulated in miR393 overexpression lines (Fig. 7d), indicating that the phenotypic

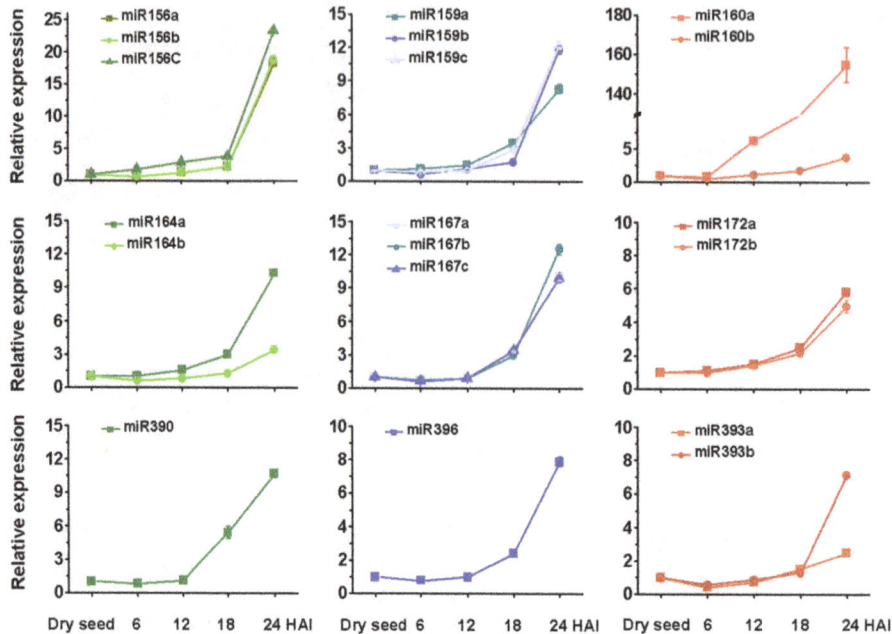

Fig. 5 Real-time qRT-PCR detection of miRNAs during seed imbibition. Total RNA was isolated from embryo at different time points in germination stage including dry seed, 6 HAI (6 h after imbibition), 12 HAI, 18 HAI, 24 HAI. The expression level in dry seed was set as 1.0. Error bars represent the SD from three independent experiments

changes in seed development might be due to the effect of miR393 on auxin response.

We then examined variation in gene expression in these transgenic lines, including for an ABA catabolic gene (*ABA8'OH-1*), ABA biosynthesis genes (*NCED1, NCED2, NCED3*) and three GA biosynthesis-associated genes (*GA20ox2, GA20ox4, GA2ox4*), all in embryo, cotyledon, and root samples during the germination stage. The analysis showed that these genes displayed differential expression in miR393 misexpression transgenic lines (Fig. 7e, f), indicating that miR393 influences ABA and GA signal pathways as well.

Discussion

So far, sRNA and degradome sequencing have been performed to identify miRNAs in seeds of crop plants including rice, maize, wheat and barley [18–20, 35]. However, detailed spatial and temporal studies of miRNA-mediated regulation are still limited. In this study, we investigated miRNA-mediated gene regulation in barley embryo during seed maturation and germination. The three samples we selected included 10 DPA, 1 DAG, and 5 DAG. The sample from 10 DPA represents transition of the seed to a storage phase characterized by dramatic transcriptional changes to mobilize energy resources and to differentiate the tissues that will constitute the mature grain. The samples from 1 DAG and 5 DAG represent the germination and post-germination stages, respectively. Via high-throughput sequencing and qRT-PCR detection, we revealed dynamic

features of the regulatory network mediated by miRNAs in embryos during barley grain development and germination.

Conserved and barley-specific miRNAs highly expressed in embryo tissues

According to bioinformatics analysis, we found that miR156 was the most highly expressed miRNA in embryos of developing and germinating seeds. miR156 is reported to be one of the most abundant miRNAs conserved among nearly all land plants including mosses [36]. In *Arabidopsis*, miR156 regulates the juvenile-to-adult phase and floral transition through targeting *SPLs* [37–39]. In rice and maize, miR156, together with miR172, acts as a regulator of inflorescence and tiller development [40–42]. Regulation of *OsSPL14* by OsmiR156 defines the ideal plant architecture in rice, whereas *OsSPL16*, another target of miR156, is involved in a regulatory module determining grain shape and can be modulated to improve rice yield and grain quality [43, 44]. In this study, we showed that miR156 exhibited a dynamic expression pattern during seed development and germination. miR156 preferentially accumulated in embryos at 15 DPA and 25 DPA (Fig. 4c) and displayed an increasing trend in the early seedling growth stage (Figs. 4d and 5). Through analysis of degradome sequencing data, three *SPL* gene family members including AK356077, AK374598 and MLOC_11199 were predicted to be the target genes of miR156 (Table 1). miR166, which is predicted to target five transcription factors belonging to the Class III homeodomain-leucine zipper (HD-ZIP III) family

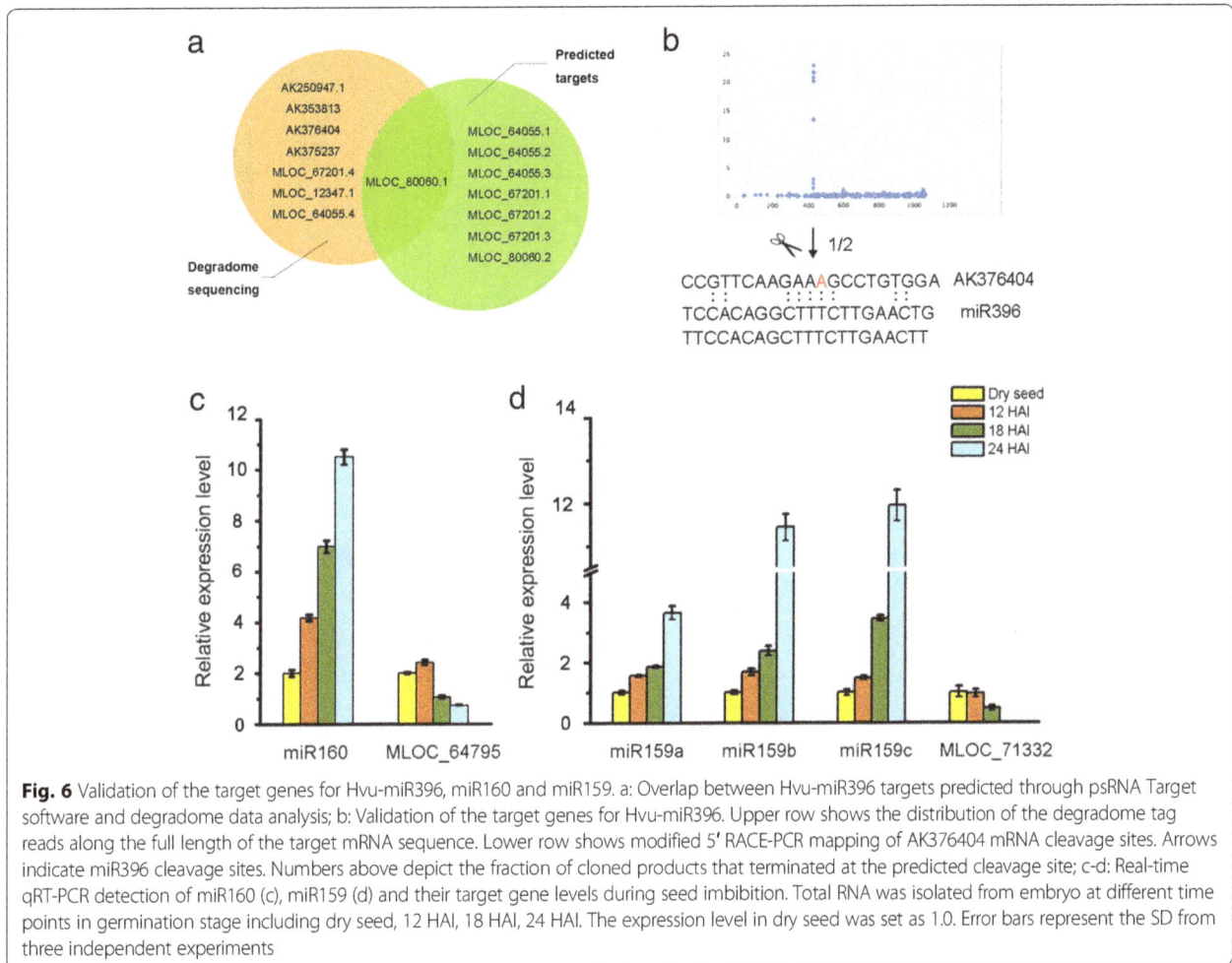

Fig. 6 Validation of the target genes for Hvu-miR396, miR160 and miR159. a: Overlap between Hvu-miR396 targets predicted through psRNA Target software and degradome data analysis; b: Validation of the target genes for Hvu-miR396. Upper row shows the distribution of the degradome tag reads along the full length of the target mRNA sequence. Lower row shows modified 5' RACE-PCR mapping of AK376404 mRNA cleavage sites. Arrows indicate miR396 cleavage sites. Numbers above depict the fraction of cloned products that terminated at the predicted cleavage site; c-d: Real-time qRT-PCR detection of miR160 (c), miR159 (d) and their target gene levels during seed imbibition. Total RNA was isolated from embryo at different time points in germination stage including dry seed, 12 HAI, 18 HAI, 24 HAI. The expression level in dry seed was set as 1.0. Error bars represent the SD from three independent experiments

(Table 1 and Additional file 1: Table S4), was expressed at a very high level in the three samples we tested. miR166/HD-ZIP III module was reported to regulate cell differentiation in root and shoot apical meristems of *Arabidopsis* [45, 46]. This indicates that miR156 and miR166-mediated repression of transcriptional factors might be conserved mechanisms to modulate cell differentiation during seed development and germination.

Another conserved miRNA family, miR894, was found to accumulate at a very low level when using samples of an entire barley seed (0–15 DPA) [20]; however, we showed that miR894 was expressed highly both in embryo of developing and germinating seeds. Moreover, the expression level was elevated more than three times at 5 DAG, when compared with that at 1 DAG (Fig. 2, Additional file 2: Table S2). miR5071, which has been identified only in barley, accumulated at 730; 1234; and 2484 RPM at 10 DPA, 1 DAG, and 5 DAG, respectively (Additional file 2: Table S2). The target of miR5071 was predicted to be an *OsMLA10-like* gene [20]. The expression of miR5071 was mostly detected in embryo, whereas *OsMLA10-like* transcript was expressed highly in endosperm dissected from the

caryopsis at 15 DPA [20]. In barley, R gene *MLA10* was reported to act as a receptor of fungal infection by recognizing avirulence proteins and confer resistance against powdery mildew fungus [47]. This suggests that miRNAs participate in defence response regulation during these two developmental stages.

Other noteworthy miRNAs, which are highly expressed in embryos, include miR5565 and miR2199. miR5565 has been identified in *Sorghum bicolor* [48] and is predicted to target a RING/U-box superfamily protein in barley [20]. As for miR2199, which has previously been reported only in common bean *Phaseolus vulgaris* and switchgrass [49], its expression and target genes in barley still need further confirmation. Overall, our results demonstrate that both conserved and barley-specific miRNAs contribute to the gene regulation in the embryo during seed development and germination.

miRNA-mediated hormone signalling regulation in embryos

ABA and GA are two major players determining seed maturation, dormancy and germination. ABA can be

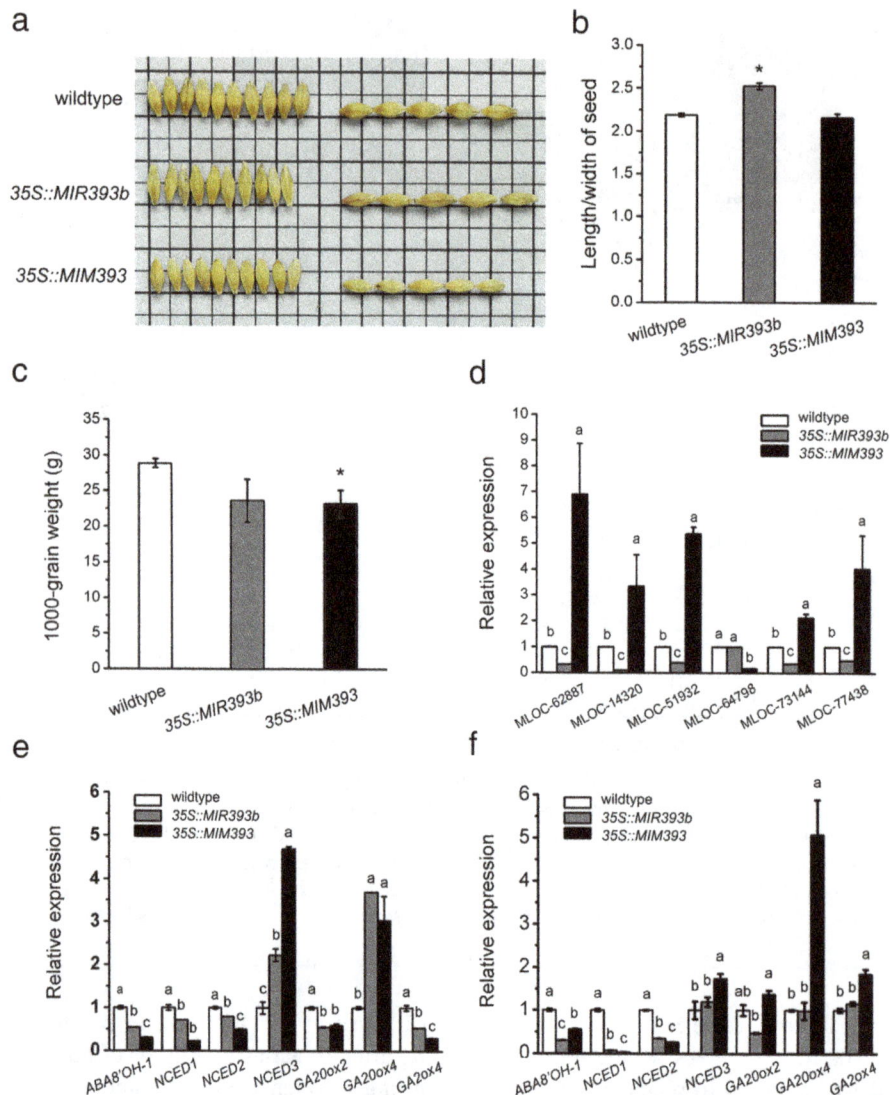

Fig. 7 miR393-mediated auxin signalling regulation affects grain phenotypes and phytohormone cross-talk during seed germination. **a**: Seed phenotype of wildtype, 35S::MIR393b and 35S::MIM393 transgenic barley; **b**: Length-width ratio of seeds in wildtype, 35S::MIR393b and 35S::MIM393 transgenic barley; **c**: 1000-grain weight of wildtype, 35S::MIR393b and 35S::MIM393 transgenic barley; Asterisks indicate significant difference compared with wildtype. *, $P < 0.05$ (Student's t-test); **d**: Relative expression levels of auxin-responsive genes in wildtype, 35S::MIR393b and 35S::MIM393 transgenic barley. Total RNA was isolated from 3 d-old seedlings. The expression level in wildtype was set as 1.0. Error bars represent the SD from three independent experiments; **e-f**: Relative expression level of ABA and GA-related genes in wildtype, 35S::MIR393b and 35S::MIM393 transgenic barley. Total RNA was isolated from embryo and cotyledon (**e**) and root (**f**) in 3 d-old seedlings. The expression level in wildtype was set as 1.0. Error bars represent the SD from three independent experiments. Different letters represent significantly different values at $P < 0.05$ (Duncan's multiple range test)

synthesized during seed maturation in the maternal tissue and the embryo, and its content decreases rapidly during imbibition [50, 51]. GA is synthesized and stored at least in the embryo and is released during imbibition to trigger the synthesis and secretion of hydrolytic enzymes in the aleurone for endosperm storage product mobilization [52]. In the present work, we showed that miRNAs are regulators of ABA and GA signalling during grain development and germination. GA-related miRNA miR159 was mainly expressed during the germination

stage (Figs. 4 and 5), which targets a gene (AK251726/ MLOC_71332) encoding a *MYB* transcription factor and another predicted protein (Table 1). qRT-PCR analysis showed that there was an inverse correlation between miR159a/b/c and their target MLOC_71332 (Fig. 6), implying that miR159-mediated target repression occurs during the first 24 h after imbibition.

miR168, which has been found to respond to salt, drought, and cold stresses or ABA treatment in previous studies, was the second most highly expressed miRNA in embryo tissue. In *Arabidopsis*, it is established that transcriptional regulation of *MIR168a* and *ARGO-NAUTE1* homeostasis plays a critical role in abscisic acid and abiotic stress responses [53]. Although we could not identify miR168-mediated target sliced products because of the different samples used in sRNA sequencing and degradome analysis, miR168 was predicted to target *AGO1* based on degradome sequencing using boron-treated barley seedling samples [54]. In addition, we found that miR168 accumulated to a high level in the three samples tested, suggesting that miR168 might be influenced by ABA in seed and probably participate in stress response during seed desiccation and germination.

Auxin, which acts as a versatile trigger in many developmental processes, also plays a critical role in seed development and germination [8]. A total of five miRNAs were found to be related to auxin signalling in our study, which included miR160 (targeting *ARF10* and *ARF17*), miR167 (targeting *ARF30, ARF18,* and *ARF9*), miR360 (targeting *ARF19*) and miR393 (targeting *AFB2* and *AFB3*) (Table 1). Two independent reports using an auxin signaling reporter, *DR5:β-glucuronidase (GUS)* showed that auxin-dependent responses increase during embryogenesis and remain present in germinating seeds [25, 55]. Accordingly, we observed differential expression of these miRNAs through northern blot and qRT-PCR analysis. During seed development, miR393 accumulated to a high level from 8 DPA, and miR167 exhibited a expression pattern opposite to that of miR164, although both the latter miRNAs seem to regulate *ARF* genes involved in auxin signalling (Fig. 4c). During the germination process, miR160, miR167, and miR393 were all upregulated within the first 24 h after imbibition (Fig. 5). We proposed that these miRNAs might be strongly induced by endogenous auxin, and affect auxin signalling in different ways.

The importance of these miRNAs in seed development and germination is supported by previous studies in other plant species. Repression of *ARF10* by miR160 is thought to be critical for seed germination and post-germination stages [25]. miR160-directed regulation of *Arabidopsis* *ARF17* is essential for proper development and modulates expression of early auxin response genes [25]. Plants expressing a miRNA-resistant version of *ARF17* altered accumulation of auxin-inducible *GH3-like* mRNAs and had dramatic developmental defects, including the embryo, floral organ development, and sterility [56]. Rice transgenic plants expressing an OsmiR160-resistant version of *OsARF18* (*mOsARF18*) exhibited pleiotropic defects in growth and development, such as dwarf stature, rolled leaves and small seeds with reduced starch accumulation [57]. *Arabidopsis* miR167 controls patterns of *ARF6* and *ARF8* expression and regulates both female and male reproduction [32]. miR393, which targets F-box genes

encoding TIR1/AFBs auxin receptors, is implicated to regulate seed development in rice. Transgenic rice plants over-expressing miR393 displayed reduced seed size, and the out glume of the spikelet had an abnormal shape and failed to close [58].

In this study, we showed that over-expression of miR393 greatly enhanced the length-width ratio of the seed, whereas over-expression of MIM393 significantly decreased the 1000-grain weight. Combined with the observation that the differential expression of auxin response genes such as *GH3, AUX/IAA* and *ARFs* in transgenic lines (Fig. 7d), this demonstrates that miR393 affects seed development through negative regulation of the auxin response, although the detailed mechanism underlying phenotypic changes in miR393 misexpression lines still need further study to elucidate.

Interplay between auxin and ABA/GA

The cross-talk of diverse hormonal signals in seed development, seed dormancy, and germination has been reported in many previous studies. For example, auxin promotes dormancy and inhibits germination through stimulation of ABA signalling by inducing *ARF*-mediated *ABI3* activation in *Arabidopsis* [9]. ABA represses embryonic axis elongation during seed germination, also potentiating auxin signalling [59]. The interactions of different hormone signals also involve miRNAs. In germinating *Arabidopsis* seeds, miR159 accumulation can be induced by ABA in an *ABI3*-dependent fashion. The targets of miR159 such as *MYB101* and *MYB33* encode transcription factors functioning as positive regulators of ABA responses, suggesting that ABA-induced accumulation of miR159 is a homeostatic mechanism to desensitize hormone signalling during seedling stress responses [41].

In this study, we found that nine miRNAs expressed in embryo regulate seed development and germination through a complex interaction of phytohormone signalling pathways based on the KEGG analysis of target genes (Fig. 4a). Moreover, we showed that miR393 regulated auxin signalling and hormone interaction. Two genes targeted by miR393 encode TIR1/AFBs auxin receptors in barley, which are components of the Skp1-Cullin1-F-box protein ubiquitin ligase complexes. The complex regulates auxin signalling via the release of ARFs from Aux/IAA (auxin/indole-3-acetic acids)-mediated heterodimerization [60, 61]. miR393 is now regarded as an important regulator of auxin signaling [62] and it affects various development processes including seed development (Fig. 7), root development and root growth response to toxic Al in barley [34]. qRT-PCR detection of auxin-related genes during germination and early seedling growth indicated that miR393 affected the expression of three putative *ARF* genes and two early auxin responsive genes in miR393 misexpression lines (Fig. 7d). This was consistent with previous data in *Arabidopsis* and barley [34, 62] and supported the notion that miR393 has a major influence on auxin homeostasis. Our data also showed that four ABA-related gene (*ABA8'OH-1* and *NCED1, 2,* and *3*) and three GA-associated genes (*GA20ox2, GA20ox4,* and *GA2ox4*) were differentially expressed in miR393 misexpression lines (Fig. 7e, f), compared with their wildtype controls. We supposed that the effect of miR393-mediated regulation on ABA and GA homeostasis might be complex, dependent on the different isoforms of related gene families and specific tissue fractions where they are expressed. Another possibility is that the variation in auxin response might influence the GA/ABA balance, rather than GA or ABA signalling alone, in the germination process. Further investigation to identify the critical factors that connect different hormone pathways will help us to better understand the molecular mechanism of hormone cross-talk.

Conclusions

Our study performed a detailed analysis of miRNA expression profiles and miRNA-target pairs in barley embryo and provided evidence for miR393-mediated regulation of auxin response and its interaction with the ABA and GA pathways during seed development and germination.

Additional files

Additional file 1: Table S1. Summary of data cleaning and length distribution of tags. **Table S4:** Highly expressed miRNAs and their target genes associated with phytohormone signaling pathways. **Table S5.** Primers and probes used in this study.

Additional file 2: Table S2. Known and novel miRNAs identified in barley embryo.

Additional file 3: Table S3. Predicted targets of known miRNAs.

Additional file 4: Figure S1. Functional distributions of predicted miRNA target genes expressed in the embryo.

Abbreviations

ABA: Abscisic acid; AFB: Auxin-signaling F-box protein; ARF: Auxin response factors; Aux/IAA: Auxin/indole-3-acetic acids; ET: Ethylene; GA: Gibberellic acid; KEGG: Kyoto Encyclopedia of Genes and Genomes analysis; microRNA: miRNA; qRT-PCR: quantitative reverse transcription PCR; RACE: Rapid-amplification of cDNA ends; SCF: Skp1-Cullin-F-box; sRNA: small RNA; TIR1: Transport inhibitor response protein

Acknowledgements

We sincerely thank Australian Centre for Plant Functional Genomics at the University of Adelaide for kindly providing wildtype Barley (*Hordeum vulgare* L. 'Golden Promise') used for small RNA sequencing and Dr. Guoping Zhang (Department of Agronomy, Zhejiang University) for providing the *Agrobacterium tumefaciens* strain AGL1 used in barley transformation.

Funding

This work was supported by the National Science Foundation of China (grant No. 31171543 and 31,571,645) and China Agriculture Research System (CARS-05).

Authors' contributions

BB and BS prepared samples for small RNA sequencing, and performed northern blot, 5' RLM-RACE and qRT-PCR detection. NHou and YC prepared the additional materials and gave a contribution for conservation analysis. YM and BB gave a contribution for bioinformatics analysis. NHan designed and coordinated the work, and wrote the manuscript with BB. HB and MZ contributed in the design and discussion of the work, and assisted in drafting the manuscript. All authors read and approved the final manuscript.

Competing interests

The authors declare that they have no competing interests.

Author details

[1]Key Laboratory for Cell and Gene Engineering of Zhejiang Province, Institute of Genetics and Regenerative Biology, College of Life Sciences, Zhejiang University, Zhejiang, Hangzhou 310058, China. [2]College of Life and Environmental Sciences, Hangzhou Normal University, Zhejiang, Hangzhou 310036, China.

References

1. Sreenivasulu N, Usadel B, Winter A, Radchuk V, Scholz U, Stein N, Weschke W, Strickert M, Close TJ, Stitt M, et al. Barley grain maturation and germination: metabolic pathway and regulatory network commonalities and differences highlighted by new MapMan/PageMan profiling tools. Plant Physiol. 2008;146(4):1738–58.
2. Nakabayashi K, Okamoto M, Koshiba T, Kamiya Y, Nambara E. Genome-wide profiling of stored mRNA in Arabidopsis thaliana seed germination: epigenetic and genetic regulation of transcription in seed. Plant J. 2005; 41(5):697–709.
3. Silva AT, Ribone PA, Chan RL, Ligterink W, Hilhorst HW. A Predictive Coexpression Network Identifies Novel Genes Controlling the Seed-to-Seedling Phase Transition in Arabidopsis thaliana. Plant Physiol. 2016;170(4):2218–31.
4. Locascio A, Roig-Villanova I, Bernardi J, Varotto S. Current perspectives on the hormonal control of seed development in Arabidopsis and maize: a focus on auxin. Front Plant Sci. 2014;5:412.
5. Gutierrez L, Van Wuytswinkel O, Castelain M, Bellini C. Combined networks regulating seed maturation. Trends Plant Sci. 2007;12(7):294–300.
6. Graeber K, Nakabayashi K, Miatton E, Leubner-Metzger G, Soppe WJ. Molecular mechanisms of seed dormancy. Plant Cell Environ. 2012;35(10):1769–86.
7. Bentsink L, Koornneef M. Seed dormancy and germination. Arabidopsis Book. 2008;6:e0119.
8. Shu K, Liu XD, Xie Q, He ZH. Two Faces of One Seed: Hormonal Regulation of Dormancy and Germination. Mol Plant. 2016;9(1):34–45.
9. Liu X, Zhang H, Zhao Y, Feng Z, Li Q, Yang HQ, Luan S, Li J, He ZH. Auxin controls seed dormancy through stimulation of abscisic acid signaling by inducing ARF-mediated ABI3 activation in Arabidopsis. Proc Natl Acad Sci U S A. 2013;110(38):15485–90.
10. Das SS, Karmakar P, Nandi AK, Sanan-Mishra N. Small RNA mediated regulation of seed germination. Front Plant Sci. 2015;6:828.
11. Curaba J, Singh MB, Bhalla PL. miRNAs in the crosstalk between phytohormone signalling pathways. J Exp Bot. 2014;65(6):1425–38.
12. Jones-Rhoades MW, Bartel DP, Bartel B. MicroRNAS and their regulatory roles in plants. Annu Rev Plant Biol. 2006;57:19–53.
13. Meng F, Liu H, Wang K, Liu L, Wang S, Zhao Y, Yin J, Li Y. Development-associated microRNAs in grains of wheat (Triticum aestivum L.). BMC Plant Biol. 2013;13:140.
14. Lu X, Chen D, Shu D, Zhang Z, Wang W, Klukas C, Chen LL, Fan Y, Chen M, Zhang C. The differential transcription network between embryo and endosperm in the early developing maize seed. Plant Physiol. 2013;162(1):440–55.
15. Peng T, Sun H, Qiao M, Zhao Y, Du Y, Zhang J, Li J, Tang G, Zhao Q. Differentially expressed microRNA cohorts in seed development may contribute to poor grain filling of inferior spikelets in rice. BMC Plant Biol. 2014;14:196.

16. Li D, Liu Z, Gao L, Wang L, Gao M, Jiao Z, Qiao H, Yang J, Chen M, Yao L, et al. Genome-Wide Identification and Characterization of microRNAs in Developing Grains of Zea mays L. PLoS One. 2016;11(4):e0153168.

17. Huang H, Long J, Zheng L, Li Y, Hu Y, Yu G, Liu H, Liu Y, Huang Z, Zhang J, et al. Identification and characterization of microRNAs in maize endosperm response to exogenous sucrose using small RNA sequencing. Genomics. 2016;108(5–6):216–23.

18. Han R, Jian C, Lv J, Yan Y, Chi Q, Li Z, Wang Q, Zhang J, Liu X, Zhao H. Identification and characterization of microRNAs in the flag leaf and developing seed of wheat (Triticum aestivum L.). BMC Genomics. 2014;15:289.

19. Lan Y, Su N, Shen Y, Zhang R, Wu F, Cheng Z, Wang J, Zhang X, Guo X, Lei C, et al. Identification of novel MiRNAs and MiRNA expression profiling during grain development in indica rice. BMC Genomics. 2012;13:264.

20. Curaba J, Spriggs A, Taylor J, Li Z, Helliwell C. miRNA regulation in the early development of barley seed. BMC Plant Biol. 2012;12:120.

21. Nodine MD, Bartel DP. MicroRNAs prevent precocious gene expression and enable pattern formation during plant embryogenesis. Genes Dev. 2010; 24(23):2678–92.

22. Schauer SE, Jacobsen SE, Meinke DW, Ray A. DICER-LIKE1: blind men and elephants in Arabidopsis development. Trends Plant Sci. 2002;7(11):487–91.

23. Tang X, Bian S, Tang M, Lu Q, Li S, Liu X, Tian G, Nguyen V, Tsang EW, Wang A, et al. MicroRNA-mediated repression of the seed maturation program during vegetative development in Arabidopsis. PLoS Genet. 2012;8(11):e1003091.

24. Allen RS, Li J, Stahle MI, Dubroue A, Gubler F, Millar AA. Genetic analysis reveals functional redundancy and the major target genes of the Arabidopsis miR159 family. Proc Natl Acad Sci U S A. 2007;104(41):16371–6.

25. Liu P-P, Montgomery TA, Fahlgren N, Kasschau KD, Nonogaki H, Carrington JC. Repression of AUXIN RESPONSE FACTOR10 by microRNA160 is critical for seed germination and post-germination stages. Plant J. 2007;52(1):133–46.

26. Hackenberg M, Gustafson P, Langridge P, Shi BJ. Differential expression of microRNAs and other small RNAs in barley between water and drought conditions. Plant Biotechnol J. 2015;13(1):2–13.

27. Ferdous J, Sanchez-Ferrero JC, Langridge P, Milne L, Chowdhury J, Brien C, Tricker PJ. Differential expression of microRNAs and potential targets under drought stress in barley. Plant Cell Environ. 2017;40(1):11–24.

28. Schreiber AW, Shi BJ, Huang CY, Langridge P, Baumann U. Discovery of barley miRNAs through deep sequencing of short reads. BMC Genomics. 2011;12:129.

29. Deng P, Wang L, Cui L, Feng K, Liu F, Du X, Tong W, Nie X, Ji W, Weining S. Global Identification of MicroRNAs and Their Targets in Barley under Salinity Stress. PLoS One. 2015;10(9):e0137990.

30. Lv S, Nie X, Wang L, Du X, Biradar SS, Jia X, Weining S. Identification and characterization of microRNAs from barley (Hordeum vulgare L.) by high-throughput sequencing. Int J Mol Sci. 2012;13(3):2973–84.

31. Shao C, Chen M, Meng Y. A reversed framework for the identification of microRNA-target pairs in plants. Brief Bioinform. 2013;14(3):293–301.

32. Bian H, Xie Y, Guo F, Han N, Ma S, Zeng Z, Wang J, Yang Y, Zhu M. Distinctive expression patterns and roles of the miRNA393/TIR1 homolog module in regulating flag leaf inclination and primary and crown root growth in rice (Oryza sativa). New Phytol. 2012;196(1):149–61.

33. Bai B, Bian H, Zeng Z, Hou N, Shi B, Wang J, Zhu M, Han N. miR393-Mediated Auxin Signaling Regulation is Involved in Root Elongation Inhibition in Response to Toxic Aluminum Stress in Barley. Plant Cell Physiol. 2017;58(3):426–39.

34. Liu J, Cheng X, Liu D, Xu W, Wise R, Shen QH. The miR9863 family regulates distinct Mla alleles in barley to attenuate NLR receptor-triggered disease resistance and cell-death signaling. PLoS Genet. 2014;10(12):e1004755.

35. Galli V, Guzman F, de Oliveira LF, Loss-Morais G, Korbes AP, Silva SD, Margis-Pinheiro MM, Margis R. Identifying microRNAs and transcript targets in Jatropha seeds. PLoS One. 2014;9(2):e83727.

36. Cuperus JT, Fahlgren N, Carrington JC. Evolution and functional diversification of MIRNA genes. Plant Cell. 2011;23(2):431–42.

37. Wang JW, Czech B, Weigel D. miR156-regulated SPL transcription factors define an endogenous flowering pathway in Arabidopsis thaliana. Cell. 2009;138(4):738–49.

38. Wang JW. Regulation of flowering time by the miR156-mediated age pathway. J Exp Bot. 2014;65(17):4723–30.

39. Wu G, Park MY, Conway SR, Wang JW, Weigel D, Poethig RS. The sequential action of miR156 and miR172 regulates developmental timing in Arabidopsis. Cell. 2009;138(4):750–9.

40. Jiao Y, Wang Y, Xue D, Wang J, Yan M, Liu G, Dong G, Zeng D, Lu Z, Zhu X, et al. Regulation of OsSPL14 by OsmiR156 defines ideal plant architecture in rice. Nat Genet. 2010;42(6):541–4.

41. Wang L, Sun S, Jin J, Fu D, Yang X, Weng X, Xu C, Li X, Xiao J, Zhang Q. Coordinated regulation of vegetative and reproductive branching in rice. Proc Natl Acad Sci U S A. 2015;112(50):15504–9.

42. Chuck G, Cigan AM, Saeteurn K, Hake S. The heterochronic maize mutant Corngrass1 results from overexpression of a tandem microRNA. Nat Genet. 2007;39(4):544–9.

43. Wang S, Li S, Liu Q, Wu K, Zhang J, Wang S, Wang Y, Chen X, Zhang Y, Gao C, et al. The OsSPL16-GW7 regulatory module determines grain shape and simultaneously improves rice yield and grain quality. Nat Genet. 2015;47(8):949–54.

44. Wang S, Wu K, Yuan Q, Liu X, Liu Z, Lin X, Zeng R, Zhu H, Dong G, Qian Q, et al. Control of grain size, shape and quality by OsSPL16 in rice. Nat Genet. 2012;44(8):950–4.

45. Carlsbecker A, Lee JY, Roberts CJ, Dettmer J, Lehesranta S, Zhou J, Lindgren O, Moreno-Risueno MA, Vaten A, Thitamadee S, et al. Cell signalling by microRNA165/6 directs gene dose-dependent root cell fate. Nature. 2010; 465(7296):316–21.

46. Zhu H, Hu F, Wang R, Zhou X, Sze SH, Liou LW, Barefoot A, Dickman M, Zhang X. Arabidopsis Argonaute10 specifically sequesters miR166/165 to regulate shoot apical meristem development. Cell. 2011;145(2):242–56.

47. Shen QH, Saijo Y, Mauch S, Biskup C, Bieri S, Keller B, Seki H, Ulker B, Somssich IE, Schulze-Lefert P. Nuclear activity of MLA immune receptors links isolate-specific and basal disease-resistance responses. Science. 2007; 315(5815):1098–103.

48. Zhang L, Zheng Y, Jagadeeswaran G, Li Y, Gowdu K, Sunkar R. Identification and temporal expression analysis of conserved and novel microRNAs in Sorghum. Genomics. 2011;98(4):460–8.

49. Pelaez P, Trejo MS, Iniguez LP, Estrada-Navarrete G, Covarrubias AA, Reyes JL, Sanchez F. Identification and characterization of microRNAs in Phaseolus vulgaris by high-throughput sequencing. BMC Genomics. 2012;13:83.

50. Jacobsen JV, Pearce DW, Poole AT, Pharis RP, Mander LN. Abscisic acid, phaseic acid and gibberellin contents associated with dormancy and germination in barley. Physiol Plant. 2002;115(3):428–41.

51. Millar AA, Jacobsen JV, Ross JJ, Helliwell CA, Poole AT, Scofield G, Reid JB, Gubler F. Seed dormancy and ABA metabolism in Arabidopsis and barley: the role of ABA 8′-hydroxylase. Plant J. 2006;45(6):942–54.

52. Bewley JD. Seed Germination and Dormancy. Plant Cell. 1997;9(7):1055–66.

53. Li W, Cui X, Meng Z, Huang X, Xie Q, Wu H, Jin H, Zhang D, Liang W. Transcriptional regulation of Arabidopsis MIR168a and argonaute1 homeostasis in abscisic acid and abiotic stress responses. Plant Physiol. 2012;158(3):1279–92.

54. Ozhuner E, Eldem V, Ipek A, Okay S, Sakcali S, Zhang B, Boke H, Unver T. Boron stress responsive microRNAs and their targets in barley. PLoS One. 2013;8(3):e59543.

55. Ni DA, Wang LJ, Ding CH, Xu ZH. Auxin distribution and transport during embryogenesis and seed germination of Arabidopsis. Cell Res. 2001;11(4):273–8.

56. Mallory AC, Bartel DP, Bartel B. MicroRNA-directed regulation of Arabidopsis AUXIN RESPONSE FACTOR17 is essential for proper development and modulates expression of early auxin response genes. Plant Cell. 2005;17(5):1360–75.

57. Huang J, Li Z, Zhao D. Deregulation of the OsmiR160 Target Gene OsARF18 Causes Growth and Developmental Defects with an Alteration of Auxin Signaling in Rice. Sci Rep. 2016;6:29938.

58. Ru P, Xu L, Ma H, Huang H. Plant fertility defects induced by the enhanced expression of microRNA167. Cell Res. 2006;16(5):457–65.

59. Belin C, Megies C, Hauserova E, Lopez-Molina L. Abscisic Acid Represses Growth of the Arabidopsis Embryonic Axis after Germination by Enhancing Auxin Signaling. Plant Cell Online. 2009;21(8):2253–68.

60. Dharmasiri N, Dharmasiri S, Weijers D, Lechner E, Yamada M, Hobbie L, Ehrismann JS, Jurgens G, Estelle M. Plant development is regulated by a family of auxin receptor F box proteins. Dev Cell. 2005;9(1):109–19.

61. Kepinski S, Leyser O. The Arabidopsis F-box protein TIR1 is an auxin receptor. Nature. 2005;435(7041):446–51.

62. Iglesias MJ, Terrile MC, Windels D, Lombardo MC, Bartoli CG, Vazquez F, Estelle M, Casalongue CA. MiR393 regulation of auxin signaling and redox-related components during acclimation to salinity in Arabidopsis. PLoS One. 2014;9(9):e107678.

3

Spatial regulation of monolignol biosynthesis and laccase genes control developmental and stress-related lignin in flax

Julien Le Roy, Anne-Sophie Blervacq, Anne Créach, Brigitte Huss, Simon Hawkins and Godfrey Neutelings* ⓘ

Abstract

Background: Bast fibres are characterized by very thick secondary cell walls containing high amounts of cellulose and low lignin contents in contrast to the heavily lignified cell walls typically found in the xylem tissues. To improve the quality of the fiber-based products in the future, a thorough understanding of the main cell wall polymer biosynthetic pathways is required. In this study we have carried out a characterization of the genes involved in lignin biosynthesis in flax along with some of their regulation mechanisms.

Results: We have first identified the members of the phenylpropanoid gene families through a combination of in silico approaches. The more specific lignin genes were further characterized by high throughput transcriptomic approaches in different organs and physiological conditions and their cell/tissue expression was localized in the stems, roots and leaves. Laccases play an important role in the polymerization of monolignols. This multigenic family was determined and a miRNA was identified to play a role in the posttranscriptional regulation by cleaving the transcripts of some specific genes shown to be expressed in lignified tissues. In situ hybridization also showed that the miRNA precursor was expressed in the young xylem cells located near the vascular cambium. The results obtained in this work also allowed us to determine that most of the genes involved in lignin biosynthesis are included in a unique co-expression cluster and that MYB transcription factors are potentially good candidates for regulating these genes.

Conclusions: Target engineering of cell walls to improve plant product quality requires good knowledge of the genes responsible for the production of the main polymers. For bast fiber plants such as flax, it is important to target the correct genes from the beginning since the difficulty to produce transgenic material does not make possible to test a large number of genes. Our work determined which of these genes could be potentially modified and showed that it was possible to target different regulatory pathways to modify lignification.

Keywords: Flax, Laccase, Lignin, microRNA, Monolignol, Phenylpropanoids, Stress

Background

Lignin is among the most abundant biological polymers on earth. It is a complex aromatic molecule synthesized during the onset of the secondary cell wall (SCW) formation in plants providing stiffness proprieties for mechanical strength, hydrophobicity for water transport and contributes to defense against pests and pathogens. Lignin is produced by a complex biosynthetic pathway (Fig. 1)

which also leads to the production of a wide range of phenylpropanoids with diverse and sometimes unknown functions such as hydroxycinnamic acids, flavonoids, coumarins, chalcones, phenylpropenes and stilbenes [1]. In brief, phenylalanine derived from the shikimate pathway is used as an initial substrate [2]. This amino acid is first deaminated by phenylalanine ammonia-lyase (PAL; EC 4.3.1.5), hydroxylated by cinammate 4-hydroxylase (C4H; EC 1.14.13.11) and then esterified with CoA by 4-coumarate:CoA ligase (4CL; EC 6.2.1.12) forming p-coumaroyl CoA. This metabolite can lead to the formation

* Correspondence: godfrey.neutelings@univ-lille1.fr
University of Lille, CNRS, UMR 8576 - UGSF - Unité de Glycobiologie Structurale et Fonctionnelle, F-59000 Lille, France

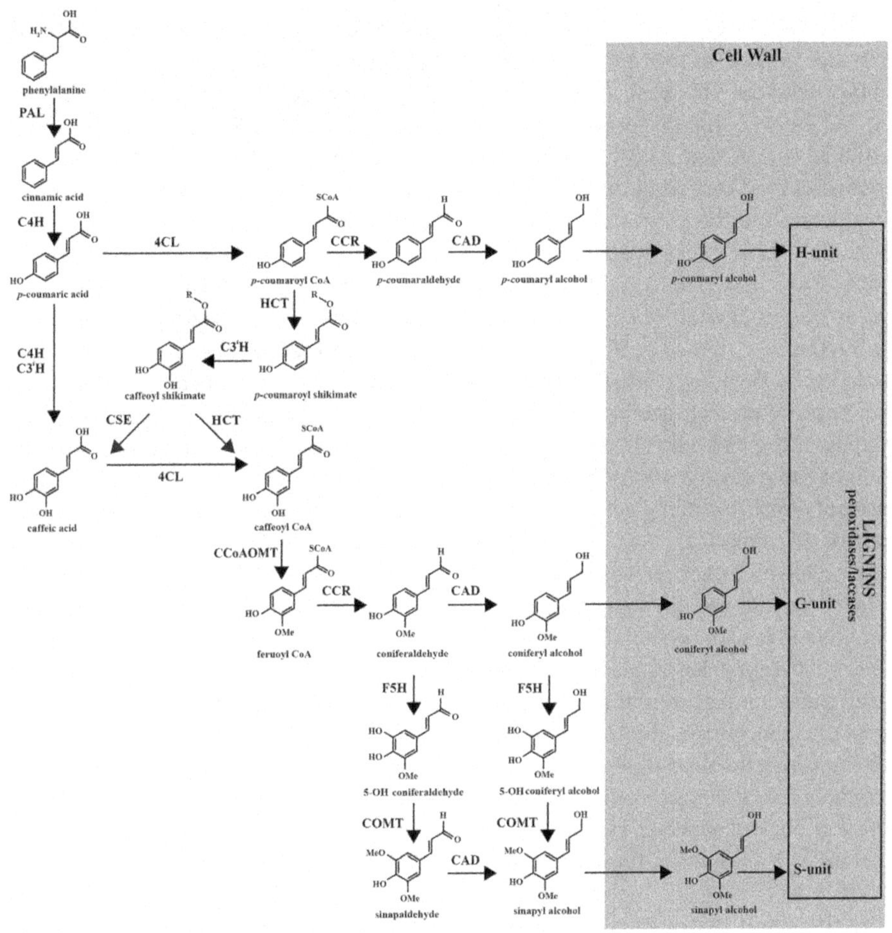

Fig. 1 The monolignol and lignin biosynthetic pathway. 4CL: 4-coumarate:CoA ligase; BGLU: beta glucosidase; C3'H: *p*-coumarate 3-hydroxylase; C4H: cinnamate 4-hydroxylase; CAD: cinnamyl alcohol dehydrogenase; CCoAOMT: caffeoyl CoA 3-O-methyltransferase; CCR: cinnamoyl CoA reductase; COMT: caffeate/5-hydroxyferulate O-methyl-transferase; CSE: caffeoyl shikimate esterase; F5H: ferulate 5-hydroxylase; HCT: hydroxycinnamoyl-CoA:shikimate hydroxycinnamoyl transferase; PAL: phenylalanine ammonia-lyase; UGT: UDP glycosyltransferase

of *p*-coumaryl alcohol by the action of cinnamoyl CoA reductase (CCR; EC 1.2.1.44) and cinnamyl alcohol dehydrogenase (CAD; EC 1.1.1.195). *p*-coumaroyl CoA can also be successively transformed by hydroxycinnamoyl-CoA:shikimate hydroxycinnamoyl transferase (HCT; EC 2.3.1.133) and *p*-coumarate 3-hydroxylase (C3'H; EC 1.14.14.1) to form caffeoyl CoA that can be methoxylated by caffeoyl CoA 3-O-methyltransferase (CCoAOMT; EC 2.1.1.104) into feruloyl CoA. This CoA ester can be reduced into coniferaldehyde by CCR and further transformed into coniferyl alcohol by CAD. Both molecules are hydroxylated by ferulate 5-hydroxylase (F5H; EC 1.14.13) and methoxy-lated by caffeate/5-hydroxyferulate O-methyl-transferase (COMT; EC 2.1.1.68) forming sinapaldehyde and sinapyl alcohols respectively. In addition to *p*-coumaryl alcohol, both alcohols are named monolignols. These can be transported to the apoplast and transformed into radicals by laccases and peroxidases [3, 4]. The oxidative coupling of these activated monolignols leads to the production of *p*-

hydroxyphenyl (H), syringyl (S) and guaiacyl (G) units respectively, present in different proportions within the lignin fractions depending on the taxonomic families, species, tissues or environment [5]. The lignin biosynthesis pathway has been comprehensively characterized in model species such as *Arabidopsis* and the identified gene models then used to search for orthologous sequences in more economically-important plants including food crops as well as woody or fiber species.

Bast fibers are present in bundles within the stem cortex between the epidermis and the xylem core in some non-woody plants. These bundles may contain only primary fibers derived from the procambium in species such as flax (*Linum usitatissimum*) or ramie (*Boehmeria nivea*) while others such as jute (*Corchorus species*), kenaf (*Hibiscus cannabinus*) or hemp (*Cannabis sativa*) contain additional secondary fibers derived from the vascular cambium. Fiber species are considered as interesting biological models because their stems contain two main populations of cells showing highly contrasted SCW compositions [6]. The xylem cells from the inner-stem tissues have a typical SCW structure with up to 30% of the cell wall weight represented by lignins whereas the bast fibers possess cellulose-rich thick SCWs with lignin contents ranging from less than 1% (ramie) to 19% (kenaf) [7]. The cell wall structure and composition is very important in these species because they determine the properties of the extracted raw materials. Flax, for instance, has been used for several thousands of years to make ropes and linen tissues and more recently, for the production of environmentally friendly fiber-based composite materials [8, 9]. To optimize the use of these materials in the future, targeted engineering of the cell wall to improve product quality will require a thorough understanding of the biosynthetic pathways leading to the production of the main cell wall polymers. Over the past years, significant efforts including genome sequencing [10], development of molecular tools [11–13] and databases [14, 15] have allowed us not only to improve our understanding of flax fiber cell wall biology, but also to identify the different gene models potentially associated with the biosynthesis of different cell wall polymers. Nevertheless, many cell wall genes are part of multigenic families and a major challenge to engineering is the identification of the individual family members involved in lignification. For non-model plants such as flax, it is important to target the correct genes from the beginning since rapid and easy transformation to obtain high amounts of transgenic material is particularly difficult [16].

To clearly identify the genes that are indeed involved in flax lignin production, we first conducted a high throughput reverse-transcriptase quantitative PCR (HT-RT-qPCR) approach to establish the expression profile of in silico-identified genes using organs/tissues with contrasted lignin contents. To further characterize these genes and because it was not possible to use large-scale reporter gene transformation, we used in situ hybridization to confirm the expression of the identified genes in lignified tissues. This type of approach has already been used to detect some lignin gene transcripts in woody plants [17–19] but never on such a large scale. We also benefited from the development of this technique in flax to show that a microRNA, *miR397*, already shown to be involved in the cleavage of several laccase transcripts [20, 21] is co-expressed in the same cells as the lignin biosynthetic genes confirming the role of miRNAs on laccase regulation in flax. In the context of a possible impact of global climate changes on the quality of plant products, and because lignin is an important component in stress responses [5], the expression of the identified genes was also analysed under a range of different environmental conditions. The high amount of data obtained by HT-RT-qPCR has enabled us to show that the expression of most of these genes is coordinated, with a probable implication of MYB transcription factors.

Results

Identification of genes related to phenylpropanoid biosynthesis in flax

The whole genome shotgun assembly of flax [10] was first surveyed to identify the genes potentially related to phenylpropanoid biosynthesis. BLAST analyses were carried out to search against the 88,420 scaffolds including 43,471 gene models. A total of 69 genes were identified and organized in gene families containing between 3 and 15 members (Table 1) and the presence of the corresponding transcripts for each gene was checked among the different public EST databases. Only *Lus4CL8*, *Lus4CL9* and *LusCCoAOMT5* have no corresponding ESTs (E value <e-50) and the expression of *LusHCT1*, *LusCCR3*, *LusCCR8*, *LusCAD8* and *LusCAD11* was also undetectable when using whole genome microarrays [22]. In addition to these 8 genes, no expression data were obtained for *LusCCR9*, *LusCCR11*, *LusCCR12* and *LusCAD9* using EST-based microarrays [11, 23]. The intron/exon structure of the genes is graphically represented in Additional file 1: Figure S1.

In silico and expression approaches to identify monolignol biosynthetic genes

To identify the genes implicated in lignin precursor biosynthesis, a phylogenetic analysis was first performed by combining the protein sequences derived from the flax phenylpropanoid genes with those from *Arabidopsis thaliana*, *Vitis vinifera* and *Populus trichocarpa* (Additional file 2: Figure S2). Different *bona fide* proteins characterized at the biochemical level or by forward/reverse genetic approaches previously used to identify lignin genes in *Eucalyptus grandis* [24] were also added to the data. In addition to this in silico sequence comparison, we also performed HT-RT-qPCR on whole stems, roots and leaves as well as on inner stem xylem-rich tissues and outer stem bast fiber-rich tissues (Fig. 2). Gene expression was also determined in leaves and whole stems under different stress conditions (Additional file 3: Figure S3). Due to the high conservation of several gene sequences and intron/exon positions in some clades, but also the lack of annotation of the more specific 5′/3′-UTRs portions in the genome, we sometimes had to design primers targeting several close-related genes. Taken together, we performed amplifications on 35 single, 12 double, 2 triple and 4 quadruple gene groups (Additional file 4: Table S1). In our experimental conditions, we were unable to detect the expression of *LusPAL4*, *LusC4H4*, *Lus4CL3*, *Lus4CL6*, *LusCCR10*, *LusCCR12*, *LusF5H5*, *LusF5H6*, *LusF5H7*, *LusF5H8* and *LusCAD6* genes whatever the organs or environmental conditions. We then used both the phylogenetic analyses (Additional file 2: Figure S2) and the HT-RT-qPCR (Fig. 2) data to identify the potential lignin-specific genes in flax. The major conclusions are summarized in the following subchapters.

Table 1 Characteristics of the flax phenylpropanoid genes identified in this work

Gene name	Gene model	Genome	Microarray EST	GenBank EST	Ref
LusPAL1	Lus10040416	scaffold86	c247	JG065006	5
LusPAL2	Lus10023531	scaffold1216	c247	EH791565	4
LusPAL3	Lus10026518	scaffold617	c904	JG216766	5
LusPAL4	Lus10013805	scaffold618	c904	JG216766	5
LusC4H1	Lus10034449	scaffold310	c1220	JG081734	5
LusC4H2	Lus10019110	scaffold30	c32549	JG085936	1;5
LusC4H3	Lus10021671	scaffold208	c6570	GW864597	2
LusC4H4	Lus10035011	scaffold43	c4373	JG075414	1;5
LusC4H5	Lus10027598	scaffold2	c11079		
Lus4CL1	Lus10026143	scaffold319	c3640	CA483243	
Lus4CL2	Lus10008677	scaffold1635	c3640	CA483243	
Lus4CL3	Lus10024123	scaffold353	c680	JG220816	5
Lus4CL4	Lus10005390	scaffold547	c680	JG218659	5
Lus4CL5	Lus10002791	scaffold125	c15833	JG239276	5
Lus4CL6	Lus10016135	scaffold344	c3980	EH791222	4
Lus4CL7	Lus10021431	scaffold612	c3980	GW866203	2
Lus4CL8	Lus10025842	scaffold605			
Lus4CL9	Lus10038259	scaffold28			
LusHCT1	Lus10002321	scaffold120		JG213400	5
LusHCT2	Lus10026097	scaffold319	c3481	JG235206	5
LusHCT3	Lus10010786	scaffold18	c6233	JG202486	5
LusHCT4	Lus10026123	scaffold319	c8096	CV478884	2
LusHCT5	Lus10022163	scaffold342	c6233	JG202486	5
LusC3'H1	Lus10033524	scaffold701	c581	JG230193	5
LusC3'H2	Lus10020847	scaffold711	c5820	JG226878	5
LusC3'H3	Lus10020850	scaffold711	c5820	CA483318	
LusCCoAOMT1	Lus10027888	scaffold1143	c395	JG228418	5
LusCCoAOMT2	Lus10002837	scaffold810	c395	JG228418	5
LusCCoAOMT3	Lus10019841	scaffold1491	c2503	JG094269	5
LusCCoAOMT4	Lus10014074	scaffold1247	c2503	JG231505	3;5
LusCCoAOMT5	Lus10034584	scaffold9			
LusCCR1	Lus10041651	scaffold272	c5445	JG227337	5
LusCCR2	Lus10024068	scaffold353	c2262	GW866594	2
LusCCR3	Lus10008774	scaffold729		JG203738	5
LusCCR4	Lus10022239	scaffold225	c11788	GW865040	2
LusCCR5	Lus10024138	scaffold353	c13273	JG149727	5
LusCCR6	Lus10042399	scaffold123	c20639	JG103050	5
LusCCR7	Lus10026273	scaffold898	c20639	JG103050	5
LusCCR8	Lus10006885	scaffold329		JG097906	5
LusCCR9	Lus10003780	scaffold806		JG088940	5
LusCCR10	Lus10012930	scaffold434	c6761	JG215733	5
LusCCR11	Lus10030973	scaffold261		JG240727	5
LusCCR12	Lus10035369	scaffold151		JG240727	5
LusF5H1	Lus10028361	scaffold413	c772	JG214534	5

Table 1 Characteristics of the flax phenylpropanoid genes identified in this work *(Continued)*

LusF5H2	Lus10041811	scaffold272	c772	JG214534	5
LusF5H3	Lus10014273	scaffold275	c2179	GW865908	2
LusF5H4	Lus10025975	scaffold319	c2179	GW867449	2
LusF5H5	Lus10034300	scaffold310	c56836		
LusF5H6	Lus10012582	scaffold6	c56836		
LusF5H7	Lus10022303	scaffold225	c56836		
LusF5H8	Lus10041511	scaffold272	c56836		
LusCOMT1	Lus10015576	scaffold233	c2253	JG229091	5
LusCOMT2	Lus10032929	scaffold51	c629	JG229091	5
LusCOMT3	Lus10009442	scaffold981	c4476	GW865137	2
LusCAD1	Lus10027864	scaffold1143	c2456	JG020400	5
LusCAD2	Lus10002812	scaffold810	c2456	JG020400	5
LusCAD3	Lus10019811	scaffold1491	c3049	GW865020	2
LusCAD4	Lus10014104	scaffold1247	c3049	JG215109	5;6
LusCAD5	Lus10014363	scaffold275	c4852	JG019640	5
LusCAD6	Lus10010149	scaffold587	c6705	GW867690	2
LusCAD7	Lus10026070	scaffold319	c4852	JG102096	5
LusCAD8	Lus10023268	scaffold98		JG062652	5
LusCAD9	Lus10035956	scaffold76		EH792205	4
LusCAD10	Lus10002089	scaffold575	c4394	JG255907	5
LusCAD11	Lus10025706	scaffold605		GW866334	2
LusCAD12	Lus10002302	scaffold120	c5057	JG282203	5
LusCAD13	Lus10002300	scaffold120	c6631	JG216469	5
LusCAD14	Lus10009955	scaffold200	c3693	JG140771	5
LusCAD15	Lus10039595	scaffold15	c3758	JG217534	5

1: PR Babu, KV Rao and VD Reddy [86]); 2: A day, M Addi, W Kim, H David, F Bert, P Mesnage, C Rolando, B Chabbert, G Neutelings and S Hawkins [87]); 3: A day, G Neutelings, F Nolin, S Grec, a Habrant, D Cronier, B Maher, C Rolando, H David, B Chabbert, et al. [30]); 4: MJ roach and MK Deyholos [23]); 5: P Venglat, D Xiang, S Qiu, SL stone, C Tibiche, D cram, M Alting-Mees, J Nowak, S Cloutier, M Deyholos, et al. [88]); 6: M Wrobel-Kwiatkowska, M Starzycki, J Zebrowski, J Oszmianski and J Szopa [60])

PAL

The deamination step catalysed by the PAL enzymes leads to the formation of cinnamic acid, which may further undergo a series of esterification, hydroxylation and methylation steps. There are no clear reports on a single "lignin-specific" *PAL* gene in the usual plant models such as *Arabidopsis* or *Populus* because of a probable functional redundancy between the members of this gene family. For this reason, it was not easy to highlight *PAL* genes that are essential for lignin biosynthesis in flax when using only a phylogenetic approach. The 4 flax *PAL* genes identified in this work were all located among the *bona fide* genes. Although the bootstrap values were sometimes low among the subdivisions in this part of the tree, *LusPAL1* and *LusPAL2* were closely related to the 3 xylem-specific poplar genes *PtrPAL2*, *PtrPAL4* and *PtrPAL5* [25] while *LusPAL3* and *LusPAL4* were closer to the *Arabidopsis* orthologs *AthPAL1* and *AthPAL2* involved in lignin metabolism [26]. *LusPAL1* and *LusPAL2*

are highly expressed in stems and roots compared to leaves. Their respective expression levels in the inner stem (xylem tissues) are close to five- and nine-fold higher than that observed in the hypolignified external stem tissues. Consequently, *LusPAL1* and *LusPAL2* are the most likely candidates for lignin biosynthesis in flax.

C4H

C4H (CYP73) is the first of the three cytochrome P450 monooxygenases involved in the hydroxylation of phenylpropanoid precursors. In flax, five distinct genes located on different scaffolds were annotated by KEGG as trans-cinnamate 4-monooxygenases (K00487). In the neighbour joining tree, 2 distinct clades were present, one contained all of the *bona fide* functionally characterized orthologs, together with an additional *Vitis* sequence and the flax proteins LusC4H1–4. In the other clade, LusC4H5 was located with PtrC4H3 and the eucalyptus EgrC4H1, suggested to be the main C4H involved in

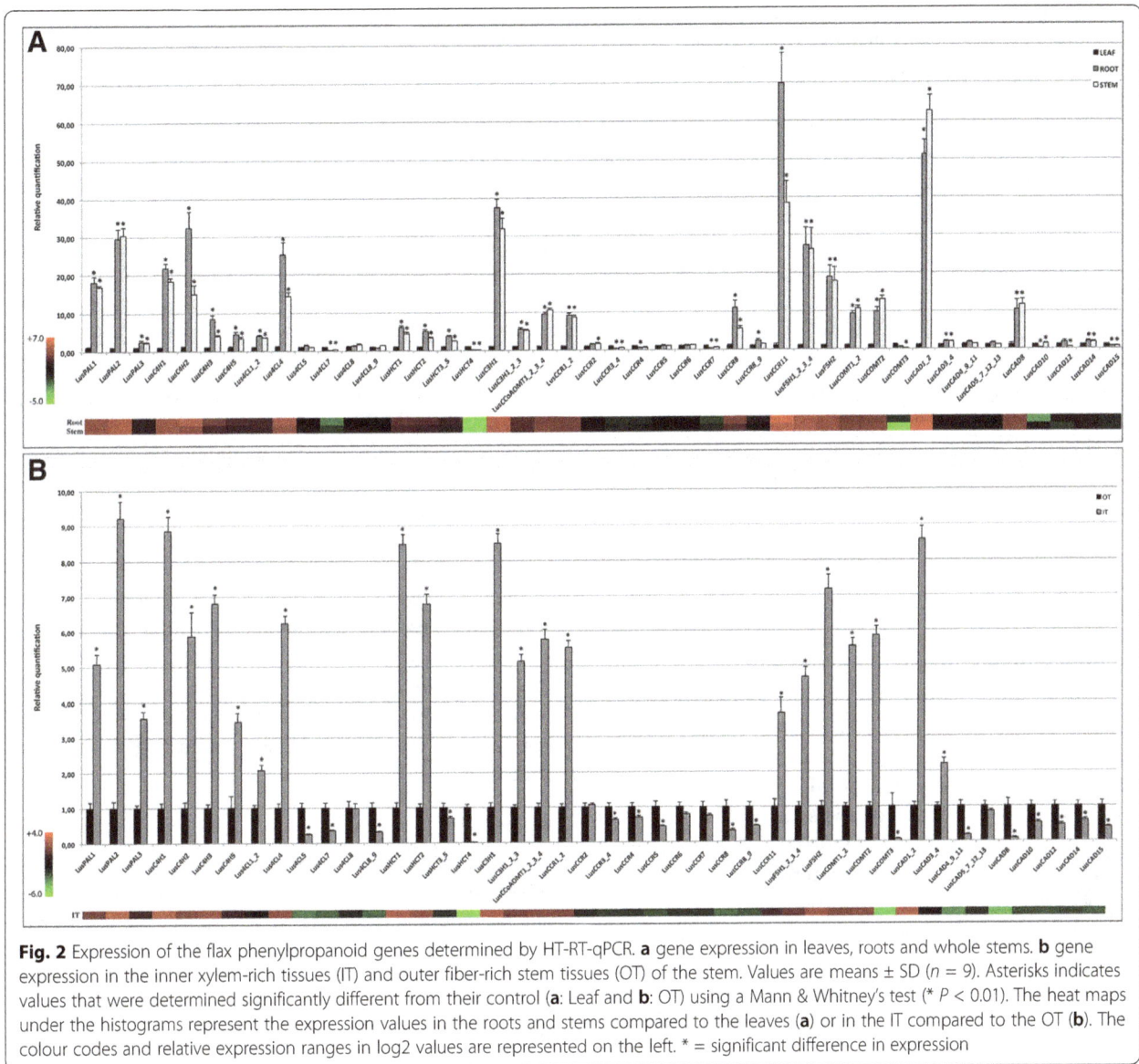

Fig. 2 Expression of the flax phenylpropanoid genes determined by HT-RT-qPCR. **a** gene expression in leaves, roots and whole stems. **b** gene expression in the inner xylem-rich tissues (IT) and outer fiber-rich stem tissues (OT) of the stem. Values are means ± SD ($n = 9$). Asterisks indicates values that were determined significantly different from their control (**a**: Leaf and **b**: OT) using a Mann & Whitney's test (* $P < 0.01$). The heat maps under the histograms represent the expression values in the roots and stems compared to the leaves (**a**) or in the IT compared to the OT (**b**). The colour codes and relative expression ranges in log2 values are represented on the left. * = significant difference in expression

lignin biosynthesis [24] even though *EgrC4H2* was also preferentially expressed in the xylem. The HT-RT-qPCR profiles of the flax genes showed that *LusC4H1*, *LusC4H2, LusC4H3* and *LusC4H5* were expressed in lignified tissues with highest levels for *LusC4H1–2* making them the best candidates for a major role in lignification.

4CL

The 4CL enzymes produce specific CoA thioesters of 4-hydroxycinnamic acids at an important crossroad in the phenylpropanoid pathway. Nine *4CL* gene models were identified in the genome of *Linum usitatissimum*. When the major known sequences were assembled in the phylogenetic tree, a delimited *bona fide* clade containing two separated classes was identified as shown in previous studies [27, 28]. The class I, predicted to be involved in monolignol biosynthesis, contained four flax proteins Lus4CL1–4. The *Lus4CL3* gene remained undetectable in our conditions whereas the other three genes had higher expression in roots and in stems. The relative expression of *Lus4CL4* was 6 fold higher in the xylem as compared to the hypolignified external stem tissues and 15- and 25- fold higher in the stems and roots as compared to the leaves and so is likely involved in lignin biosynthesis.

HCT

In the lignin biosynthetic pathway, HCTs first catalyse the formation of *p*-coumaroyl shikimate from *p*-coumaroyl CoA and shikimic acid. To produce caffeoyl CoA, the product is first hydroxylated by C3'H and then converted again by HCT acting in the reverse direction. The closely related acyltransferase hydroxyl-cinnamoyl CoA:quinate hydroxy-cinnamoyl transferase (HQT) can lead to the formation of chlorogenic acid by the transfer of quinic acid on the same *p*-coumaroyl CoA substrate, forming p-coumaroylquinic acid. The distinction between HCT and HQT genes without further biochemical characterization is not always very easy because of their very similar sequences. We identified five different genes in the flax genome closely related to the previous identified HCT/HQT genes. The expression profiles show that *LusHCT1* and *LusHCT2* are active in the roots and highly expressed in stem-internal tissues. Their position among the *bona fide* HCT sequences in the phylogenetic tree also suggests their implication in lignin biosynthesis. The *LusHCT4* gene is also among the true HCTs but the expression of this gene is approximately 50 fold times higher in the hypolignified external tissues of the stem.

C3'H

The C3'H genes generally belong to small hydroxylase subfamilies involved in monolignol biosynthesis. The enzyme (CYP98A3) catalyses the 3′-hydroxylation of 4-coumaroyl shikimate and 4-coumaroyl quinate into the corresponding caffeoyl-conjugated form. In flax, we identified only 3 gene models corresponding to C3'H. A gene duplication event possibly occurred because *LusC3'H2* and *LusC3'H3* are both located on the same scaffold in a 14 kb region. When considering the structure (Additional file 1: Figure S1) and homology between the 3 genes, it is possible to speculate that a first duplication event of *LusC3'H1* formed the *LusC3'H2* gene, which in turn was duplicated forming *LusC3'H3*. This gene was shorter at the 3′-end but still had between 92 and 98% amino acid sequence identity with the 2 former corresponding proteins. In *Arabidopsis* only one *C3'H* gene was identified whereas an expansion of this family by lineage-specific tandem duplications in *Populus* [29] and *Eucalyptus* [24] was described. The *LusC3'H1* gene is more strongly expressed in the stem and root tissues and therefore probably involved in lignin biosynthesis.

CCoAOMT

Among both methylation enzyme families, CCoAOMTs can transfer a −CH3 group from a donor to a hydroxycinnamoyl CoA ester, namely caffeoyl CoA. Among the 5 genes identified in this work, *LusCCoAOMT4* was previously characterized in knockdown flax plants [30]. The four genes *LusCCoAOMT1–4* have high sequence identity, similar gene structures and could not be individualized by the HT-RT-qPCR approach. The amplicon was detected at high levels in lignin-containing organs such as stems and roots compared to leaves and preferentially in the internal stem tissues. In the phylogenetic tree, they collocate

with the sequences identified as *bona fide* enzymes in *Arabidopsis*, *Eucalyptus*, *Nicotiana* and *Populus*.

CCR

This enzyme is important because it catalyses the first specific step of the monolignol production by reducing hydroxycinnamoyl CoA esters. Taking into account the large size of this gene family as determined by genome sequencing, they are likely to be involved in numerous other metabolic pathways. In flax, we identified 12 genes with very different intron/exon patterns. The *LusCCR3*, *LusCCR5* and *LusCCR6* genes do not fit the 3 patterns described for *Arabidopsis* [31]. Although most of the genes encode proteins with similar sizes, *LusCCR3* appears smaller and thus may be truncated. In *Arabidopsis*, 12 *CCR* genes were present in duplicated chromosome regions including 4 genes distributed in tandem [31]. No similar situation exists in flax although *LusCCR2* and *LusCCR5* were located on the same scaffold but at a distance of 360 kb. The differences in the structure of these 2 genes are however not in favour of a recent duplication event. The *bona fide* lignin biosynthesis genes were grouped in a common clade also containing *LusCCR1*, *LusCCR2*, *LusCCR11* and *LusCCR12*. The expression profiles of these four genes in the lignified tissues showed that *LusCCR1* and *LusCCR11* are the best lignin-associated candidates.

COMT

As indicated by the enzyme name, the caffeate/5 hydroxyferulate *O*-methyl-transferase was first thought to act as a bifunctional enzyme able to transfer a methyl group from S-adenosyl methionine to the C₃ and C₅ positions of the aromatic rings of caffeic and 5-hydroxyferulic acids [32]. In fact, the COMT enzyme promotes the biochemical pathway leading to the formation of sinapyl alcohol by catalysing the methylation of 5-hydroxyconiferyl alcohol [33]. Only 3 gene models were identified within the flax *COMT* gene family, which is low compared to many other species [34]. *LusCOMT1* and *LusCOMT2* have close exon/intron patterns but are present on different scaffolds. They both are expressed at higher rates in lignified tissues. The position of the corresponding proteins within the *bona fide* group in the phylogenetic tree seems to support the hypothesis of their implication in the lignin biosynthetic pathway.

F5H

The F5H cytochrome P450-dependent monooxygenase is required for the hydroxylation of coniferaldehyde and coniferyl alcohol leading to the formation of S lignin through the synthesis of sinapyl alcohol [35]. In flax, 8 gene models were identified and compared to orthologous models expressed in lignin-rich tissues. In the phylogenetic tree, these proteins are located on a branch separated from the *bona fide* enzymes by a node predicted by a low bootstrap value. The flax proteins were further subdivided into 2 groups of sequences with similar exon/intron patterns. The first group contained *F5H1*, *F5H2*, *F5H3* and *F5H4*, which could not clearly be separated by HT-RT-qPCR during the design of the primers except for *F5H2*. The common *F5H1_2_3_4* and specific *F5H2* amplicons were detected at high levels in lignin-containing organs such as roots and whole stems and also in xylem-rich outer tissues. The second group with *F5H5*, *F5H6*, *F5H7* and *F5H8* genes had undetectable expression.

CAD

The NADPH-dependent CAD enzymes play an important role in monolignol biosynthesis by catalyzing the reduction of hydroxycinnamyl aldehydes into their corresponding monolignols which is the last step in the biosynthetic pathway [36]. They are encoded by a multigenic family and their homologs have been detected widely in bacteria and eukaryota but not in animals [37]. Classification by several authors highlighted 3 to 7 groups/classes based on phylogenetic approaches [38–41]. Recently an accurate classification proposed a separation in 3 functional clades for both dicots and monocots [42] with a lignin specific class I and a possible lignin-associated class II. In the present study we identified 15 flax gene models predicted in the published genome. Both *LusCAD12* and *LusCAD13* were located on scaffold120 in a 5.5 kb fragment and have the same exon/intron pattern associated with a high (84%) amino acid sequence identity, reflecting a possible duplication event. *LusCAD1*, *LusCAD2*, *LusCAD3* and *LusCAD4* segregate with the known *bona fide* genes within the phylogenetic tree. Because of their very high sequence identity (98,6% at the protein level), it was not possible to separate *LusCAD1* and *LusCAD2* when designing the primers for qRT-PCR. Their expression profiles show that the amplicons were mostly present in the lignified tissues such as roots and whole stems and also in xylem-rich outer tissues. To a lesser extent, *LusCAD8* showed the same expression profile in these organs but revealed an opposite behaviour when comparing outer and inner stem tissues. This gene is located outside the *bona fide* clade.

Cell–/tissue-specific expression of lignin and selected lignin biosynthetic genes

The likely implication of identified candidate genes in the flax lignification process was further confirmed using in situ hybridization that established a close spatial correlation between lignifying cells determined by the Wiesner reaction and gene expression in stems, roots and leaves (Fig. 3 and summarized in Additional file 5: Table S2). These genes were *LusPAL1*, *LusC4H2*, *Lus4CL*, *LusHCT1*,

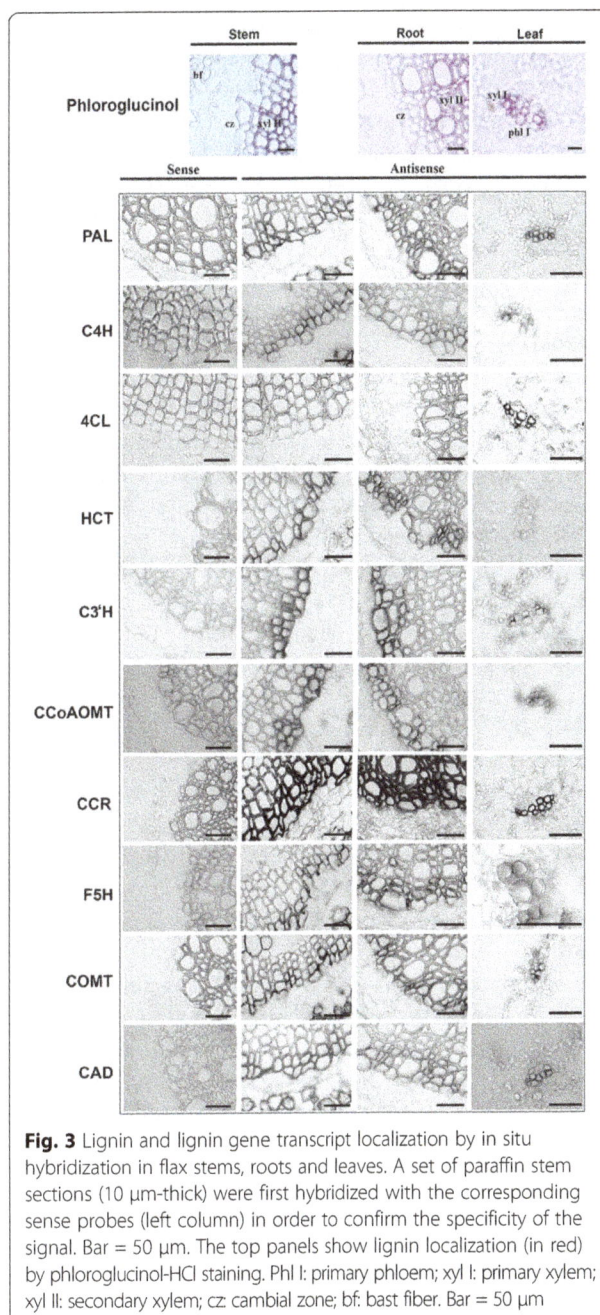

Fig. 3 Lignin and lignin gene transcript localization by in situ hybridization in flax stems, roots and leaves. A set of paraffin stem sections (10 µm-thick) were first hybridized with the corresponding sense probes (left column) in order to confirm the specificity of the signal. Bar = 50 µm. The top panels show lignin localization (in red) by phloroglucinol-HCl staining. Phl I: primary phloem; xyl I: primary xylem; xyl II: secondary xylem; cz: cambial zone; bf: bast fiber. Bar = 50 µm

LusC3'H1_2_3, *LusCCoAOMT1_2_3_4*, *LusCCR1_2*, *LusF5H1_2_3_4*, *LusCOMT1_2* and *LusCAD1_2*. For most genes, the strongest expression was observed in the stems and the roots in the first 2–5 secondary xylem cell layers from the cambial zone towards the inner part of the organs. In the leaves, the stain was in each case restricted to the primary xylem. For all genes, signal intensity was generally higher in stems and roots compared to leaves except for *CCR* for which equal intensity was observed in all 3 organs. In addition, the *4CL4* probe failed to reveal the presence of the corresponding transcripts in the stems and roots despite the clear organ specificity revealed by the HT-RT-qPCR results. The observed lack of signal might be due to poor probe efficiency since several fragments of the *4CL4* gene were tested but did not produce a positive signal in roots or stems. Overall, these results confirm that the expression of the selected genes is intimately associated with the actively differentiating secondary xylem zone thereby confirming their probable implication in the lignification process.

Identification of laccase genes and their regulation by microRNAs

Laccases play an important role in controlling lignin polymerization by catalysing the oxidation of the precursors [43]. Both in silico prediction and high throughput sequencing in different species [20, 21] have shown that laccase transcript levels are regulated by microRNAs. We identified 45 corresponding gene models in flax and used them to construct a phylogenetic tree (Fig. 4a). In comparison to previous published data reporting the classification of laccases in 6 clades [21], our tree contained an additional clade with 7 flax sequences. The expression profiles (Fig. 4b) of these laccase genes were extracted from public high throughput transcriptomic data [11, 15]. Of the 45 genes, 21 showed higher expression in 'high lignin' inner tissues compared to 'low lignin' outer tissues and are preferentially expressed on the top part of the stem where the development of the SCW takes place.

Based on previous studies reporting predictive and experimental data on the cleavage of *LAC4* and *LAC17* transcripts in *Arabidopsis* [20] and potentially 29 laccase transcripts in *Populus* [21] by *miR397*, we first searched for a flax orthologous miRNA precursor sequence in the EST and genome databases as previously described [44]. The c3244 EST from the genolin database [11] and scaffold1999 extracted from the phytozome genomic database [45] both contained the pre-miRNA sequence (Fig. 5). The deduced mature *lus-miR397* sequence was first used to search for targets among the flax laccases and among these, 11 genes were predicted with penalty mismatches included between 0 and 2.5 within the duplex. The amplification and sequencing of the 5′-end of the degraded transcripts by RLM-RACE allowed us to experimentally validate the cleavage of *LusLAC1*, *LusLAC5*, *LusLAC9*, *LusLAC38*, *LusLAC39* and *LusLAC44* among the predicted targets (Fig. 5). These 6 genes were present among the 21 genes highly expressed in the xylem-rich tissues. The five first laccase transcripts are close to the *Arabidopsis* lignin-related *AthLAC4* gene [43] and all belong to the clade 1 (Fig. 4A). *LusLAC44* belongs to clade 4 including *PtrLAC13* which was shown to be cleaved by *Ptr-miR397* [46]. These results demonstrate that flax laccase genes are regulated by microRNAs suggesting that this mechanism is implicated in controlling lignification in this species. Further evidence supporting this idea was provided by in

situ hybridisation data (Fig. 6) indicating a close correlation between tissue–/cell-specific expression of the *lus-miR397* precursor, monolignol specific genes and lignified tissues in stems and roots.

Stress-related flax phenylpropanoid gene expression

Since lignin is an important component in stress responses, we analysed the expression of the phenylpropanoid genes with a special focus on the previous identified lignin-associated genes to search for those regulated by environmental modifications. Flax plants and isolated organs were submitted to a range of different stress conditions and gene expression determined by HT-RT-qPCR. The expression profiles were shown in Additional file 3: Fig. S3 and summarized in Table 2. Salicylic acid (SA) plays an important role in plant defense signaling and accumulates in response to pathogen infection [47]. In flax stems, SA significantly down-regulated the lignin-related genes *LusC4H1* and *LusCCoAOMT1_2_3_4* but had an opposite effect on *LusPAL2* and the *LusF5H1_2_3_4* group. Higher expression levels were also observed for *Lus4CL8* and *LusCAD10*. In parallel, SA had a moderate effect on the expression of lignin genes in treated leaves. Only the up-regulation of the *LusCAD4_9_11* group and *LusCAD8* were evident. Methyljasmonate is a precursor of the active jasmonate form in plants [48] and mimics biotic as well as abiotic stresses [49]. MeJA application led to a very strong down-regulation of most of the phenylpropanoid genes including all of the identified *bona fide* lignin genes within the leaves except for *LusCCR11*. On the contrary, this signal regulator has a low overall effect on gene expression in flax stems. It is also interesting to note that some genes show no detectable expression levels in either the presence or absence of SA and MeJA, possibly due to the in vitro conditions used for organ incubation.

The effect of abiotic stresses was further tested by inducing water deficit on whole flax plants grown in soil. Under these conditions, *LusC4H1* and *LusCCR11* were the only lignin-associated genes showing higher expression in leaves while most of the other *bona fide* genes were slightly upregulated in the stems. The effect of wounding was also tested on flax stems and leaves. In these organs, the effect on the lignin-associated genes was much stronger in the stems compared to the leaves. The modulation of phenylpropanoid gene expression was also evaluated after 48 h of continuous light illumination. When compared to 16 h/8 h light/dark conditions, only the *LusPAL1* and *Lus4CL4* genes were downregulated and 7 genes or gene-groups were significantly upregulated. When taken together, our results showed that the strongest impact on the previously genes identified as involved in lignin biosynthesis, occurred in the stem during dehydration, wounding and in the presence of SA.

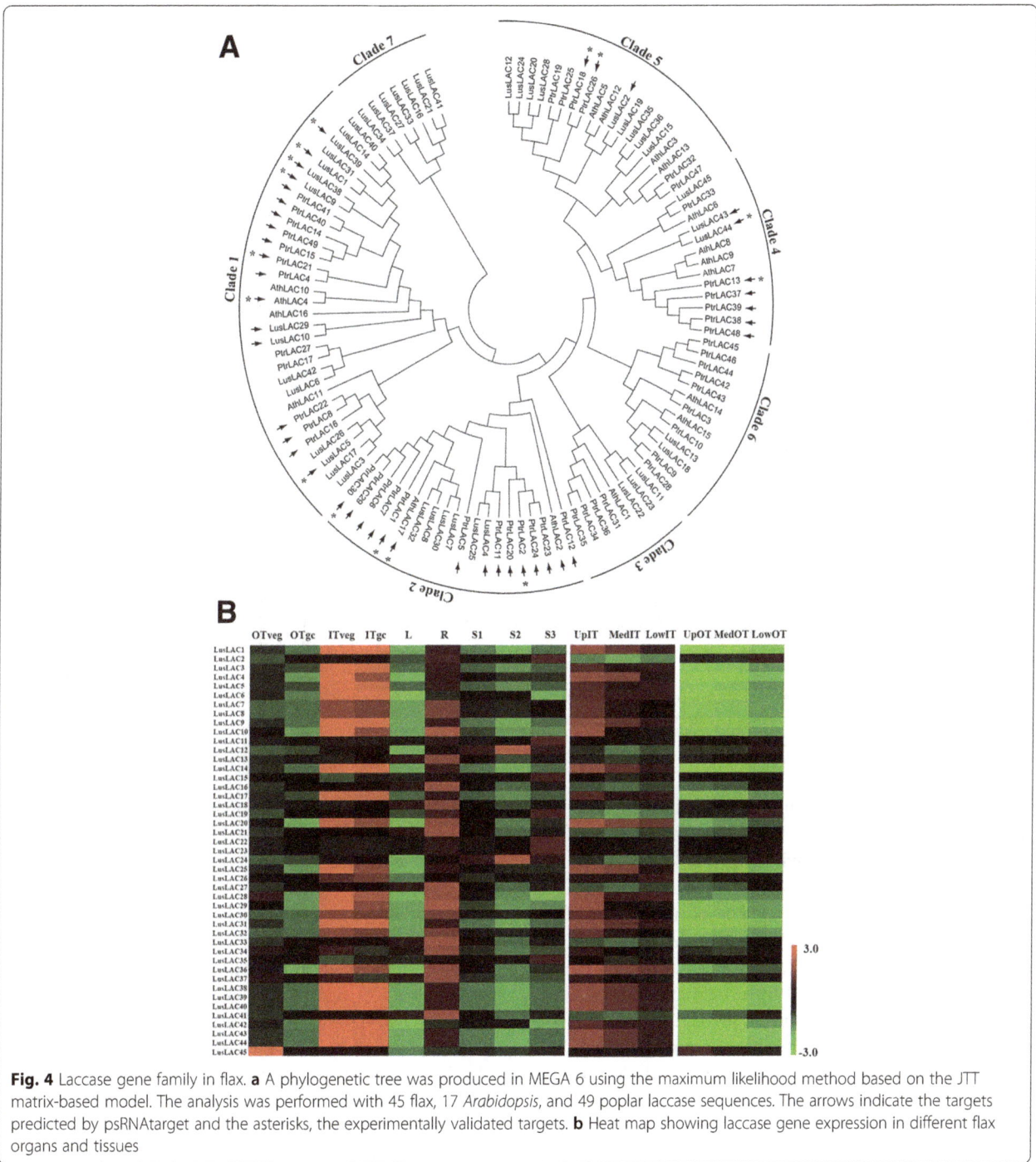

Fig. 4 Laccase gene family in flax. **a** A phylogenetic tree was produced in MEGA 6 using the maximum likelihood method based on the JTT matrix-based model. The analysis was performed with 45 flax, 17 *Arabidopsis*, and 49 poplar laccase sequences. The arrows indicate the targets predicted by psRNAtarget and the asterisks, the experimentally validated targets. **b** Heat map showing laccase gene expression in different flax organs and tissues

Gene clustering

Genes involved in the same biological process are often co-regulated. To further confirm the implication of the former identified genes in the lignification pathway, we searched for possible common expression profiles among the gene set. A k-means clustering approach was used to identify correlated groups exhibiting similar expression profiles in the different tissues and/or stress conditions. The number of flax genes in each cluster ranged between 2 and

14 (Fig. 7a). Among the flax phenylpropanoid associated genes, all the potential *bona fide* genes identified previously, except for *LusCAD1_2* and *LusCCR11*, were included in cluster 9 (Fig. 7b) showing that the expression of these genes were closely co-regulated in flax. Altogether, the clustering results provide further support for the involvement of the identified genes in lignification.

Gene co-regulation requires the activity of common transcription factors and since MYB proteins are known

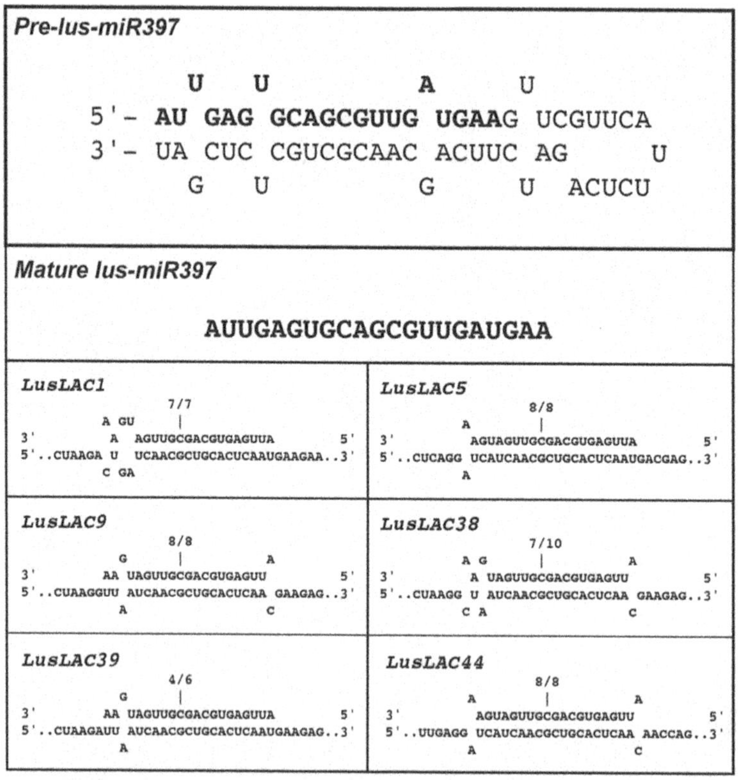

Fig. 5 *Lus-miR397* structure and validation of predicted targets. The sequences of the *pre-lus-miR397* and the mature *lus-miR397* are indicated. The association with laccase targets are shown only for the six transcripts validated by RLM 5'-RACE. The cleavage sites are indicated above the corresponding position and the number of cloned RACE products sequenced is shown above each sequence (number of sequenced fragments at this position/total number of sequenced clones)

to function as master switches of secondary cell wall and lignin gene transcription [50, 51] we examined the promoters of our identified genes for MYB consensus motifs named MBS (MYB binding site) (C/T)AAC(A/T)A(A/C)C and MBSIIG (or SMRE: secondary wall MYB-responsive element) (C/T)ACC(A/T)A(A/C)C. Our results (Additional file 6: Table S3) showed that the MBSIIG site was present in the 500 bp proximal promoter fragments of all the *bona fide* genes present in the cluster 9 whereas both MBS sites are absent from the *LusCCR11* promoter in cluster 1. Interestingly, both MYB sites were absent from the proximal promoters of the *LusF5H1_2_3_4* genes which is in accordance with previous results showing that *F5H* is directly regulated by a NAC SND1 transcription factor [52].

Discussion
Identification of genes involved in the lignin biosynthetic pathway in flax
In the past 25 years, a tremendous amount of information about lignin biosynthesis has been obtained using model plants such as *Arabidopsis* and *Populus* species. Currently, the possibility to obtain comprehensive genomic sequences in other species is allowing the scientific community to gather additional data on other species. Secondary cell walls (SCWs), whether they are specialized in water conduction (xylem) or mechanical support (sclerenchyma), usually contain around 25% lignin. However, in some specialized tissues such as bast fibers, the cell wall contains very low amounts of lignin together with high cellulose levels [53, 54]. Although these cells only contain low amounts of lignins, the presence of these phenolic molecules can have an important impact on the quality of the final products such as linen tissues and biocomposites.

In this context, we carried out a comprehensive genome-wide analysis of the gene set involved in phenylpropanoids, and more specifically in lignin biosynthesis in flax. The size of individual phenylpropanoid gene families is difficult to compare between species due to differences in the stringency of the sequence annotations. This is especially true for widespread enzymatic activities such as hydroxylases or methyltransferases. On the other hand, gene expression profiling often gives clear information about the function of the more specific *bona fide* lignin genes [24]. Usually, individual qRT-PCR experiments are performed to validate lignin transcript level pattern analysis [55] but HT-RT-qPCR approaches are increasingly used as they are more appropriately sized for characterizing the members of multigene

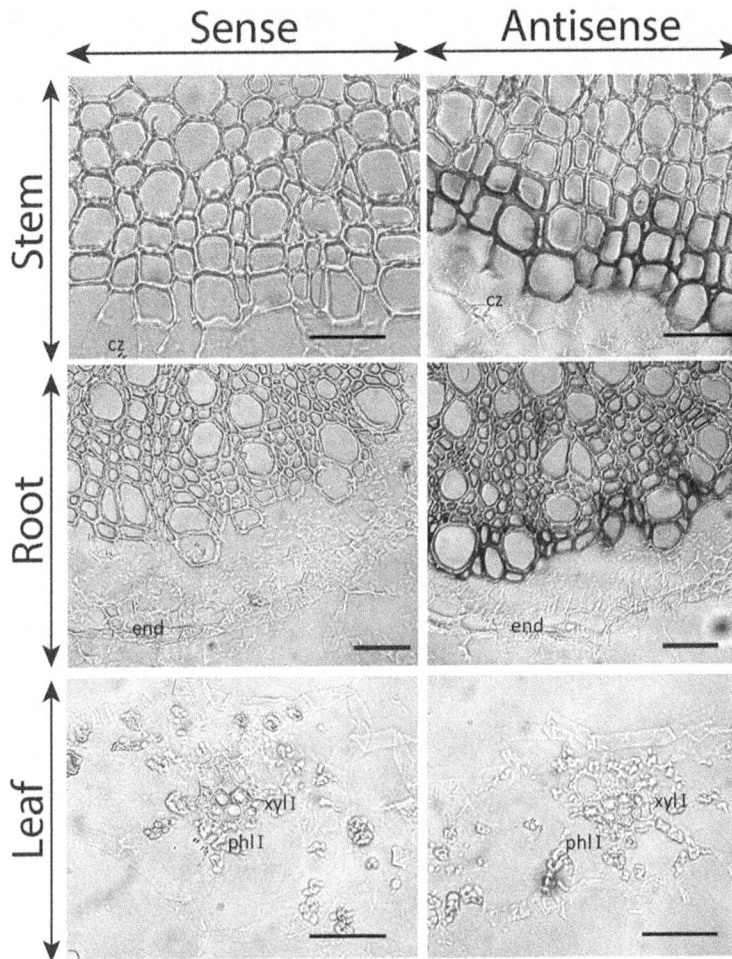

Fig. 6 Spatial localization of *lus-mir397* transcripts in flax organs. Paraffin sections (10 μm) of the stems, roots and leaves were hybridized with the sense controls (left) and the antisense specific probes (right). cz: cambial zone; end: endodermis; phl I: primary phloem; xyl I: primary xylem. Bar = 20 μm

families in a single step [56]. In flax, we first identified the genes potentially involved in lignification by examining the expression profiles in roots and stems compared to the leaves and also by comparison between the internal xylem stem tissues (high lignin) and the outer fiber-bearing tissues (low lignin).

Further confirmation of the likely implication of selected phenylpropanoid genes in lignification was provided by in situ hybridization, allowing the precise localization of the lignin biosynthetic gene expression. Both in situ hybridization and also promoter-reporter gene studies conducted on a low number of genes previously showed that *bona fide* lignin genes were expressed in the first few cell layers of differentiating secondary xylem [17–19, 57, 58]. To our knowledge, our work reports for the first time an in situ hybridization approach on a large number of lignin genes in a higher plant species. As was shown by the results of the high throughput expression analyses, the very high expression levels in the xylem-rich tissues compared to the bast fiber cells is in accordance with our previous results

suggesting that lignification of flax fibers is at least in part regulated at the transcriptional level [15].

Finally, when the expression patterns of all the phenylpropanoid genes were considered, most of these lignin-associated genes were found to be co-expressed within a same cluster confirming that these genes were indeed involved in a common metabolic pathway. The presence of common MYB *cis*-element motifs in the promoters of these genes associated to their co-expression was already described for the orthologous genes in *Arabidopsis* [59]. In this species, MYB58 can directly activate most lignin biosynthesis genes but not *F5H* showing that the syringyl lignin biosynthesis is activated by a different regulatory pathway probably controlled by NAC transcription factors [52].

Selection of candidates for lignin engineering in flax

The success of targeted engineering of cell wall lignin genes to improve the quality of different flax-derived products depends upon the correct identification of individual family

Table 2 Summary of the stress response of flax phenylpropanoid genes

Gene	Stress								
	SA		MeJA		Drought		Wound		Light
	Stem	Leaf	Stem	Leaf	Stem	Leaf	Stem	Leaf	Stem
LusPAL1*				−	−		+		−
LusPAL2*	+			−	−		+	−	
LusPAL3	−			−	−			+	
LusPAL4									
LusC4H1*	−		−	−		+	+	+	
LusC4H2*				−	+		+		
LusC4H3	−	+					+		
LusC4H4									
LusC4H5									−
Lus4CL1_2				−		+	+	+	
Lus4CL3									
Lus4CL4*				−	−		+		−
Lus4CL5	+	−	−	−	+		+	+	+
Lus4CL6									
Lus4CL7				−	−		−		
Lus4CL8	+			−					
Lus4CL8_9	+	−		−				+	
LusHCT1*				−	−		+		
LusHCT2*				−	−		+	−	
LusHCT3_5	+	−		−	+	+	+	+	+
LusHCT4			−	−			−	−	
LusHCT5									
LusC3'H1*				−	−		+		+
LusC3'H1_2_3				−	−		+		
LusCCoAOMT1_2_3_4*	−			−	−				
LusCCoAOMT5									
LusCCR1_2*				−	−				
LusCCR2		−		−					+
LusCCR3_4						+			
LusCCR4						+			
LusCCR5	+		−	−	+	+	+		+
LusCCR6			−	−					
LusCCR7			−	−					
LusCCR8				−					
LusCCR8_9				−				+	+
LusCCR10									
LusCCR11*					+	+	+		
LusCCR12									
LusF5H1_2_3_4*	+		+	−	+		+	+	
LusF5H2	+		−	−	+		+	+	
LusF5H5									
LusF5H6									

Table 2 Summary of the stress response of flax phenylpropanoid genes *(Continued)*

LusF5H7									
LusF5H8									
LusCOMT1_2*				–	+		+	+	
LusCOMT2*	+			–				+	
LusCOMT3	–			–	+				
LusCAD1_2*				–	–		+	+	
LusCAD3_4		+	–	–	+		+		
LusCAD4_9_11	–	+	–	+	+		+		
LusCAD5_7_12_13				–					+
LusCAD6									
LusCAD8	–	+	–		+	+		+	
LusCAD10	+			–	+	+	+		
LusCAD12		–		–	+		+	+	
LusCAD14					+		+	+	+
LusCAD15	+			–	+		+		+

Asterisks indicate potential lignin-specific flax genes

gene members involved in the trait to be modified. Some genes potentially involved in lignin biosynthesis have already been partially characterized in flax. The first flax lignin cDNA sequence was deposited in the GenBank database in 2005 and the corresponding *CCoAOMT* gene functionally characterized [30]. It was chosen because it is responsible for the synthesis of feruloyl CoA used for the formation of both G- and S-lignin units and, as confirmed by down-regulation, because it plays a central role in maintaining SCW integrity by regulating the quantity and the S/G proportion in the non-condensed lignin fraction. This gene corresponds to the *LusCCoAOMT4* identified in our study and is included in the *bona fide* group of genes expressed in lignified tissues. In another study, a *CAD* gene fragment sharing 100% homology with *LusCAD4* was targeted by a downregulation strategy [60]. This gene associated with *LusCAD3* was not retained in our work as a major lignin gene because of its much lower differential expression between lignified and poorly-lignified organs or tissues when compared to *LusCAD1_2*. Although the RNAi transformed plants contained lower amounts of lignins compared to the control, it is interesting to note that M Wrobel-Kwiatkowska, M Starzycki, J Zebrowski, J Oszmianski and J Szopa [60]) did not observe the typical brown-midrib phenotype of *CAD* mutants observed in other species. In contrast, TILLed flax chemical mutants of *LusCAD1* possess the characteristic orange-coloured xylem phenotype [16] confirming the likely involvement of this gene in the lignification process. Flax lignin has also received interest because of the very low amounts of S-units [53] so we were therefore interested to examine the *F5H* gene family. This enzyme is specifically required for the synthesis of S lignin since it

catalyses the 5-hydroxylation of coniferaldehyde and/or coniferyl alcohol [35]. Its down-regulation in *Arabidopsis* leads to a reduced S/G value [61]. The expression ratio of *LusF5H2* in the lignified organs and tissues is comparable to those of other identified genes so the biosynthesis of low amounts of S units in flax does not seem to be controlled by *F5H* gene expression. However, further computational modelling and biochemical characterization of the members of the F5H family have to be done in order to provide more clues on the specificity of these enzymes towards various substrates.

lus-miR397 controls several laccase genes in flax

Laccases are mono- or multimeric copper-oxidases in eukaryotes and procaryotes with a broad range of aromatic or non-aromatic substrates [62] and have been implicated in the oxidative polymerization of monolignols. Among the 17 gene models described in *Arabidopsis*, *AthLAC4*, *AthLAC11* and *AthLAC17* were clearly shown to contribute to the constitutive lignification of the floral stems [43, 63]. In *Populus trichocarpa*, 49 gene models have been identified but it was not yet possible to clearly identify those involved in lignification [21]. In flax, we identified 45 laccase gene models of which 23 were highly expressed in lignified tissues. MicroRNA targeting (*miR397*) of some laccase transcripts, including *AthLAC4* and *AthLAC17*, contribute to the overall gene regulation [20, 21]. This microRNA has been identified in many gymnosperms, monocots and dicots with 1 to 3 copies per species [64]. In this study, we experimentally demonstrated that transcripts corresponding to six flax laccase genes were targeted by *lus-miR397*. On the basis of our phylogenetic analysis, five of these

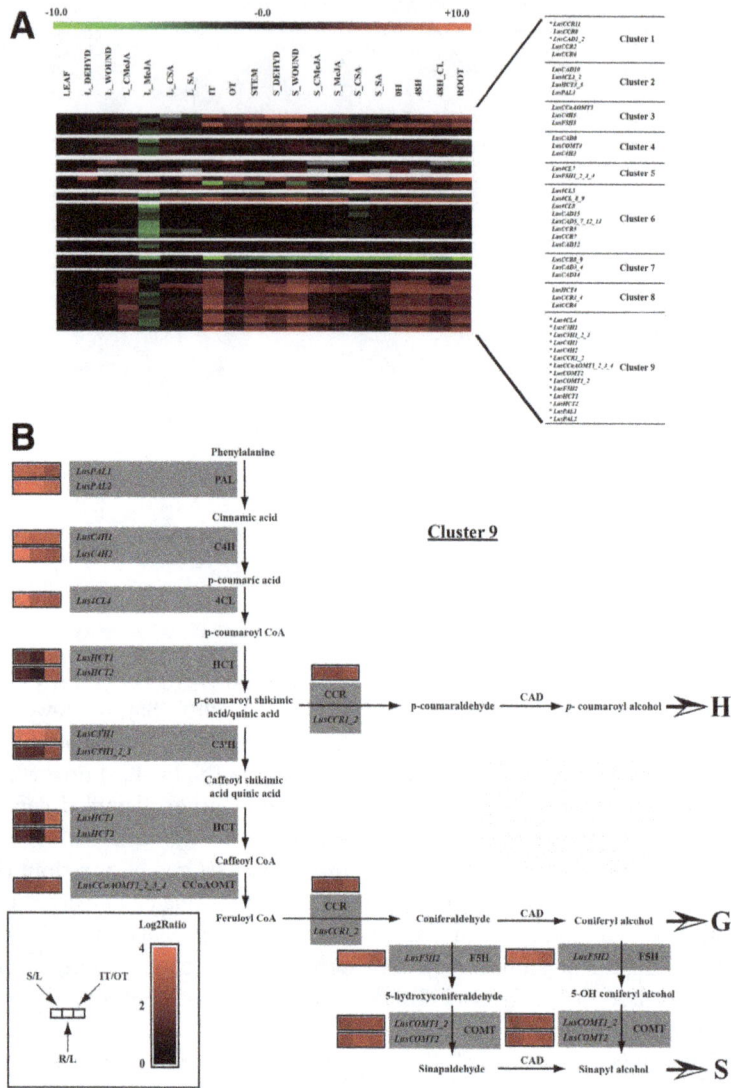

Fig. 7 Clustering of flax phenylpropanoid genes according to their expression profile in different organs and stress conditions. **a** the k-means medians clustering function from the TIGR MultiExperiment Viewer (TM4 Mev v4.8.1) was used to cluster the genes according to their expression profiles. The distance metric was Pearson correlation and the default parameters were used for the k-means calculation. The expression profiles are represented as heat maps. The colour code and expression range in log2 values are represented on top of the figure. **b** the biosynthetic pathway leading to monolignol production and further to H, G and S units is summarized. Three expression ratios are indicted as heat maps beside the genes. C: control; CL: continuous light; DEHYD: drought stressed stems; IT: internal stem tissues; L: leaf; MeJA: methyljasmonate; OT: outer stem tissues; R: root; S; stem; SA: salicylic acid

genes were found in the clade 1, which also contained *AthLAC4* whereas six non-targeted genes were present in clade 2 along with *AthLAC17* showing that either no functional ortholog of this gene is present in flax or that these genes have lost their ability to be cleaved by *lus-miR397*. Strong evidence for a likely role of microRNA regulation of laccase-mediated lignification in flax was provided by the cell-/tissue-specific localization of the corresponding transcript precursors. Our results are in agreement with those obtained with a *GUS-GFP* fusion driven by the *miR397b* promoter, showing that this gene is expressed in the stem vascular tissues of *Arabidopsis* [65]. Further

analyses should allow a better understanding of whether *miR397* plays a role in maintaining constant transcript levels or is implicated in a finer control of different laccase paralogs.

Stress related phenylpropanoid genes

Flax plants selected for fiber production are mostly cultivated around the Mediterranean and temperate climate zones. In the context of climate change, cell wall metabolism, structure and fiber quality are likely to be affected in the future and it is therefore informative to know which genes are affected by biotic or abiotic stresses. We took advantage of the high throughput gene expression approach to try and distinguish between developmental lignin genes and stress-related phenylpropanoid gene expression.

Defence-related hormones such as SA and MeJA can control lignin gene activation. SA is an important signalling molecule for systemic acquired resistance [66] and triggers the expression of lignin genes in many different species [67]. In flax, several lignin genes showed opposite expression profiles in the presence of SA suggesting that the composition of lignin may change during the defence mechanism or that more specific metabolic pathways including these genes are activated. The effect of MeJA on this metabolic pathway seems more subtle. This component has been shown to participate in the signal transduction pathway in response to different forms of stresses [68]. A dose effect was described in *Fragaria* fruits [69] as well as an opposite *CAD* gene expression pattern in tea plants [70]. In our experimental conditions, most of the flax genes were underexpressed indicating that the synthesis of lignin is probably not part of the response to this elicitor. Abiotic stresses can also have an influence on lignification. Mechanical stresses can be responsible for the deposition of various polymers including lignin that may function as a physical barrier around the injury zone [71]. In this context, lignin genes are activated in different species including sweet potato [72], maize [73] or *Arabidopsis* [74]. Most of the flax lignin genes were activated in the stem but less in the leaves showing that in this species, lignin is also synthesized in reaction to wounding. Outsides these *bona fide* genes, several common phenylpropanoid genes including *CADs* were also activated in both organs showing that they may potentially be involved in the production of other abiotic stress responsive metabolites as was already proposed in poplar [38]. The effect of drought stress on plants is more difficult to understand because it can have positive or negative roles on lignification depending on the species, the organs/tissues and the intensity and duration of the stress period [75]. In our experimental conditions (12 days after watering stop) most of the lignin genes were upregulated in the stems and the leaves. The increase in lignification has been described as one of the reactions included in a general adaptation strategy of plants faced with water loss and may result in an increase of mechanical strength and/or water impermeability [76].

Conclusion

We have identified, in this study, the individual members of the multigenic families most likely involved in lignin biosynthesis (and polymerisation) in flax. These data provide the knowledge base necessary to undertake targeted engineering and/or marker-based selection of lignin biosynthesis in flax, as well as important information

on the behaviour of these genes in a number of different stress conditions. The latter information will be particularly useful for farmers and breeders in the current context of increasing variability in weather conditions associated with climate change.

Methods
Plant material

For HT-RT-qPCR expression analysis, flax plants (*Linum usitatissimum* L. cv Diane) were grown in 12 × 12 cm soil-containing pots in a greenhouse under 16 h/20 °C day and 8 h/18 °C night conditions for 35 days. For developmental expression studies, (i) leaves, ii) 2-cm root fragments taken under the stem base, (iii) top 4 cm of stem, (iv) fiber-rich outer-tissues and (v) xylem-rich inner-tissues from stem base were harvested. For stress expression studies, the following samples were taken from these 35-day-old plants: (i) leaves and stems harvested 24 h after wounding with a scalpel, (ii) leaves and stems incubated for 24 h in Petri dishes containing 200 μM salicylic acid (SA) or methyljasmonate (MeJA) dissolved in ½ MS medium [77], (iii) leaves and stems from plants submitted to a drought stress imposed by withholding water for 12 days, (iv) stems from plants first placed in phytotrons during 7 days for acclimation and then submitted to continuous light for 48 h at 20 °C. All the samples were triplicates and always frozen immediately in liquid nitrogen before storage at –80 °C prior to RNA extraction. For in situ hybridization, the median stem, leaves and 2-cm root fragments were collected and fixed as described below.

In silico identification and analysis of phenylpropanoid genes in *L. usitatissimum*

The flax phenylpropanoid protein and gene sequences were identified in the annotated v1.0 of the *Linum usitatissimum* genome hosted on the Phytozome website [45]. The entire database derived from the published genome sequences [10] was queried with the names of the gene families used as keywords and also by BLASTp interrogation with previously lignin-related identified sequences from *Arabidopsis thaliana* [78] and *Populus trichocarpa* [25]. When it was necessary, unknown portions of some sequences were replaced by the corresponding sequences found in public EST databases and truncated sequences were discarded. For each gene model, the presence of a corresponding EST was searched in the dbEST [79] and genolin [11] databases. The intron/exon structures were also retrieved and the identities of the protein sequences compared. The k-means medians clustering function from the TIGR MultiExperiment Viewer (TM4 Mev v4.8.1) was used to cluster the genes according to their expression profiles. The distance metric was Pearson correlation and the default parameters were used for the k-means calculation.

Phylogenetic analyses

For each gene family, the protein sequences were aligned with ClustalW and the phylogenetic analyses obtained with the MEGA 6.06 [80] software for performed by maximum likelihood method based on the JTT matrix-based model [81]. Consensus trees were generated with 1000 bootstrap replicates.

HT-RT-qPCR using a 96.96 dynamic array

Total RNA was extracted from the different organs and tissues using the TriReagent method (Molecular Research Center). RNA integrity and concentration were evaluated with RNA Standard Sens Chips in the Experion automated capillary electrophoresis system (Bio-Rad). cDNA was then synthesized using the High Capacity RNA-to-cDNA Kit (Applied Biosystems) according to the manufacturer's instructions. The large scale quantitative RT-PCR was performed with a BioMark HD System using a Fluidigm 96.96 dynamic array (IntegraGen, Evry France) according to the Fluidigm Advanced Development Protocol with EvaGreen using primer pairs listed in Additional file 4: Table S1. A first preamplification reaction (1 cycle: 95 °C 10 min; 14 - cycles: 95 °C 15 s, 60 °C 4 min) was performed for each sample in 10 µl by pooling primer pairs (final concentration, 50 nM), 3.3 µl cDNA, and 5 µl 2X PreAmp Master Mix (Applied Biosystems). For each assay, 5 µl 10X Assay Mix containing 9 µM forward primer, 9 µM reverse primer, and 1X Assay Loading Reagent was loaded into the chip. The following thermal cycles were executed: 1 cycle: 95 °C 10 min; 40 cycles: 95 °C 15 s, 60 °C 1 min. The amplifications on the three biological samples were always conducted in triplicate and were also performed on a mix of cDNAs to determine the amplification efficiencies. The Ct values were analysed with the qBase + software (Biogazelle, Belgium). The data normalization was performed using the previous experimentally validated reference genes *LusE-TIF5A1*, *LusUBI1* and *LusEF1A* [82].

Histochemical analysis

Flax stems, roots and leaves were fixed with 4% para-formaldehyde in phosphate buffer (0.2 M, pH 7) for 16 h. They were then washed in phosphate buffer containing 0.1 M glycine and dehydrated through a series of ethanol solutions and progressively embedded in paraffin (ParaplastPlus; Oxford Labware). Tissue sections were obtained with a Leica RM2065 microtome and placed on silanised-coated slides. Before staining, tissue sections were deparaffinised with Histoclear (Labonord) and rehydrated with decreasing concentrations of ethanol. The presence of lignin was determined by staining with the Wiesner reagent (phloroglucinol–HCl) [83].

Spatial gene expression determination by in situ hybridization

To obtain DNA templates for RNA probe synthesis, PCR amplifications were performed with gene-specific primers tailed with a T7 RNA polymerase binding site Additional file 4: Table S1. The amplicons (1 µg) were used as templates to synthesize sense and antisense probes for each gene, with the incorporation of UTP–digoxigenin as the label using the DIG labelling mix (Roche). Deparaffinised sections were treated as previously described [84], hydridized and probe detected with NBT (nitro-blue tetrazolium chloride)/BCIP (5-bromo-4-3'-indolylphosphate p-toluidine salts) solutions according to the manufacturer's instructions. Controls without probe or with sense probe were performed to check for endogenous alkaline phosphate activity.

miRNA target prediction and experimental validation

The sequence of the *lus-miR397* precursor was identified among the known flax ESTs using stringent rules described elsewhere [44]. The laccase transcript models were analysed with psRNATarget [85] to predict the targets that can potentially match with the microRNA without any gaps. When referring to the miRNA sequence, there were not more than one mismatch between the nucleotides number 1 and 9, none between 10 and 11, and no more than 2 consecutive mismatches after position 11. The experimental validation of the laccase targets was carried out by using a modified protocol of the RNA ligation mediated (RLM) RACE included in the GeneRacer kit (Invitrogen). RNA was extracted from a mix of flax organs and the cleaved transcripts (without 5'-cap) were selectively ligated to a 5'-RNA adaptor. After reverse transcription, PCR amplification was performed with a 5' adaptor-specific primer and a reverse gene specific primer Additional file 4: Table S1 located downstream of the predicted cleavage site. The fragments were then gel purified and sequenced.

Additional files

Additional file 1: Figure S1. Flax phenylpropanoid gene structures. The name of the gene is indicated followed by the length in bp between the start and stop codons in parentheses. The exons are shown as boxes and the introns as lines. The scale in bp is shown on the top of the figure.

Additional file 2: Figure S2. Molecular phylogenetic analysis of the phenylpropanoid genes.

Additional file 3: Figure S3. Expression of the flax phenylpropanoid genes under stress conditions as determined by HT-RT-qPCR. 0H: start of the light period on the first day; 48H: end of the night on the second day; 48H_CL: 48 h after continuous light; CONT: control; Dehyd: dehydration; L: leaf; MeJA: methyljasmonate; S: stem; SA: salicylic acid.

Additional file 4: Table S1. Primers used for HT-RT-qPCR, RLM RACE and in situ hybridization.

Additional file 5: Table S2. Description of the gene expression localization in the tissues of flax roots, stems and leaves.

Additional file 6: Table S3. Position of the MBSII and MBSIIG specific *cis* elements on both strands (+/−) of the flax phenylpropanoid gene promoters.

Abbreviations
4CL: 4-coumarate:CoA ligase; C3'H: *p*-coumarate 3-hydroxylase; C4H: cinammate 4-hydroxylase; CAD: cinnamyl alcohol dehydrogenase; CCoAOMT: caffeoyl CoA 3-O-methyltransferase; CCR: cinnamoyl CoA reductase; COMT: caffeate/5-hydroxyferulate O-methyl-transferase; F5H: ferulate 5-hydroxylase; HCT: hydroxycinnamoyl-CoA:shikimate hydroxycinnamoyl transferase; MeJA: methyljasmonate; PAL: Phenylalanine ammonia-lyase; SA: salicylic acid

Acknowledgements
We are indebted to the Research Federation FRABio (Univ. Lille, CNRS, FR 3688, FRABio, Biochimie Structurale et Fonctionnelle des Assemblages Biomoléculaires) for providing the scientific and technical environment conducive to achieving this work. JLR gratefully acknowledges the french Ministère de l'Enseignement Supérieur et de la Recherche for financial support.

Funding
This research was supported by the french Ministère de l'Enseignement Supérieur et de la Recherche.

Authors' contributions
JLR and ASB performed the sample preparations and experimental procedures, GN participated in the experiments, AC performed data analysis, BH and SH critically revised the manuscript and GN designed the study, supervised the work and wrote the manuscript. All authors read and approved the final manuscript.

Competing interests
The authors declare that they have no competing interests.

References
1. Le Roy J, Huss B, Creach A, Hawkins S, Neutelings G. Glycosylation is a major regulator of phenylpropanoid availability and biological activity in plants. Front Plant Sci. 2016;7
2. Boerjan W, Ralph J, Baucher M. Lignin biosynthesis. Annu Rev Plant Biol. 2003;54:519–46.
3. Freudenberg K. Lignin biosynthesis. Berlin, Heidelberg, New York: Springer; 1968.
4. Sterjiades R, Dean JF, Eriksson KE. Laccase from sycamore maple (*Acer pseudoplatanus*) polymerizes Monolignols. Plant Physiol. 1992;99(3): 1162–8.
5. Moura JCMS, Bonine CAV. De Oliveira Fernandes Viana J, Dornelas MC, Mazzafera P. Abiotic and biotic stresses and changes in the lignin content and composition in plants. J Integr Plant Biol. 2010;52(4):360–76.
6. Neutelings G. Lignin variability in plant cell walls: contribution of new models. Plant Sci. 2011;181(4):379–86.
7. Mohanty AK, Misra M, Hinrichsen G. Biofibres, biodegradable polymers and biocomposites: an overview. Macromol Mater Eng. 2000;276-277(1):1–24.
8. Zohary D, Hopf M. Domestication of plants in the old world, third edition edn. New York: Oxford University Press; 2004.
9. Summerscales J, Dissanayake NPJ, Virk AS, Hall W. A review of bast fibres and their composites. Part 1 – Fibres as reinforcements. Compos A: Appl Sci Manuf. 2010;41(10):1329–35.
10. Wang Z, Hobson N, Galindo L, Zhu S, Shi D, McDill J, Yang L, Hawkins S, Neutelings G, Datla R, et al. The genome of flax (*Linum usitatissimum*) assembled de novo from short shotgun sequence reads. Plant J. 2012;72(3):461–73.
11. Fenart S, Ndong Y-P, Duarte J, Riviere N, Wilmer J, van Wuytswinkel O, Lucau A, Cariou E, Neutelings G, Gutierrez L. Development and validation of a flax (*Linum usitatissimum* L.) gene expression oligo microarray. BMC Genomics. 2010;11(1):592.
12. Chantreau M, Chabbert B, Billiard S, Hawkins S, Neutelings G. Functional analyses of cellulose synthase genes in flax (*Linum usitatissimum*) by virus-induced gene silencing. Plant Biotechnol J. 2015;13(9):1312–24.
13. Ragupathy R, Rathinavelu R, Cloutier S. Physical mapping and BAC-end sequence analysis provide initial insights into the flax (*Linum usitatissimum* L.) genome. BMC Genomics. 2011;12:217.
14. Day A, Fenart S, Neutelings G, Hawkins S, Rolando C, Tokarski C. Identification of cell wall proteins in the flax (*Linum usitatissimum*) stem. Proteomics. 2013;13(5):812–25.
15. Huis R, Morreel K, Fliniaux O, Lucau-Danila A, Fenart S, Grec S, Neutelings G, Chabbert B, Mesnard F, Boerjan WA. Natural hypolignification is associated with extensive oligolignol accumulation in flax stems. Plant Physiol. 2012; 158(4):1893–915.
16. Chantreau M, Grec S, Gutierrez L, Dalmais M, Pineau C, Demailly H, Paysant-Leroux C, Tavernier R, Trouve JP, Chatterjee M, et al. PT-flax (phenotyping and TILLinG of flax): development of a flax (*Linum usitatissimum* L.) mutant population and TILLinG platform for forward and reverse genetics. BMC Plant Biol. 2013;13:159.
17. Harding SA, Leshkevich J, Chiang VL, Tsai CJ. Differential substrate inhibition couples kinetically distinct 4-Coumarate:coenzyme a ligases with spatially distinct metabolic roles in quaking aspen. Plant Physiol. 2002;128(2):428–38.
18. Hawkins S, Boudet A, Grima Pettenati J. Characterisation of caffeic acid O-methyltransferase and cinnamyl alcohol dehydrogenase gene expression patterns by in situ hybridisation in *Eucalyptus gunnii* hook. Plantlets. Plant Sci. 2003;164(2):165–73.
19. Lacombe E, Hawkins S, Van Doorsselaere J, Piquemal J, Goffner D, Poeydomenge O, Boudet AM, Grima-Pettenati J. Cinnamoyl CoA reductase, the first committed enzyme of the lignin branch biosynthetic pathway: cloning, expression and phylogenetic relationships. Plant J. 1997;11(3):429–41.
20. Abdel Ghany SE, Pilon M. MicroRNA-mediated systemic down-regulation of copper protein expression in response to low copper availability in Arabidopsis. J Biol Chem. 2008;283(23):15932–45.
21. Lu S, Li Q, Wei H, Chang MJ, Tunlaya Anukit S, Kim H, Liu J, Song J, Sun YH, Yuan L, et al. Ptr-miR397a is a negative regulator of laccase genes affecting lignin content in *Populus trichocarpa*. Proceedings of the Natlional Academy of Science USA. 2013;110(26):10848–53.
22. Chantreau M, Portelette A, Dauwe R, Kiyoto S, Cronier D, Morreel K, Arribat S, Neutelings G, Chabi M, Boerjan W, et al. Ectopic lignification in the flax lignified bast fiber1 mutant stem is associated with tissue-specific modifications in gene expression and Cell Wall composition. Plant Cell. 2014;26(11):4462–82.
23. Roach MJ, Deyholos MK. Microarray analysis of flax (*Linum usitatissimum* L.) stems identifies transcripts enriched in fibre-bearing phloem tissues. Mol Gen Genomics. 2007;278(2):149–65.
24. Carocha V, Soler M, Hefer C, Cassan-Wang H, Fevereiro P, Myburg AA, Paiva JA, Grima-Pettenati J. Genome-wide analysis of the lignin toolbox of *Eucalyptus grandis*. New Phytologist. 2015;206(4):1297–313.
25. Shi R, Sun YH, Li Q, Heber S, Sederoff R, Chiang VL. Towards a systems approach for lignin biosynthesis in *Populus trichocarpa*: transcript abundance and specificity of the monolignol biosynthetic genes. Plant Cell Physiol. 2010;51(1):144–63.
26. Rohde A, Morreel K, Ralph J, Goeminne G, Hostyn V, De Rycke R, Kushnir S, Van Doorsselaere J, Joseleau JP, Vuylsteke M, et al. Molecular phenotyping of the pal1 and pal2 mutants of *Arabidopsis thaliana* reveals far-reaching consequences on phenylpropanoid, amino acid, and carbohydrate metabolism. Plant Cell. 2004;16(10):2749–71.

27. De Azevedo SC, Barbazuk B, Ralph SG, Bohlmann J, Hamberger B, Douglas CJ. Genome-wide analysis of a land plant-specific acyl:coenzymeA synthetase (ACS) gene family in Arabidopsis, poplar, rice and Physcomitrella. New Phytol. 2008;179(4):987–1003.

28. Ehlting J, Buttner D, Wang Q, Douglas CJ, Somssich IE, Kombrink E. Three 4-coumarate:coenzyme a ligases in Arabidopsis thaliana represent two evolutionarily divergent classes in angiosperms. Plant J. 1999;19(1):9–20.

29. Hamberger B, Ellis M, Friedmann M, de Azevedo SC, Barbazuk B, Douglas CJ. Genome-wide analyses of phenylpropanoid-related genes in Populus trichocarpa, Arabidopsis thaliana, and Oryza sativa: the Populus lignin toolbox and conservation and diversification of angiosperm gene families. Can J Bot. 2007;85(12):1182–201.

30. Day A, Neutelings G, Nolin F, Grec S, Habrant A, Cronier D, Maher B, Rolando C, David H, Chabbert B, et al. Caffeoyl coenzyme A O-methyltransferase down-regulation is associated with modifications in lignin and cell-wall architecture in flax secondary xylem. Plant Physiol Biochem. 2009;47(1):9–19.

31. Barakat A, Yassin NB, Park JS, Choi A, Herr J, Carlson JE. Comparative and phylogenomic analyses of cinnamoyl-CoA reductase and cinnamoyl-CoA-reductase-like gene family in land plants. Plant Sci. 2011;181(3):249–57.

32. Bugos RC, Chiang VL, Campbell WH. cDNA cloning, sequence analysis and seasonal expression of lignin-bispecific caffeic acid/5-hydroxyferulic acid O-methyltransferase of aspen. Plant Mol Biol. 1991;17(6):1203–15.

33. Osakabe K, Tsao CC, Li L, Popko JL, Umezawa T, Carraway DT, Smeltzer RH, Joshi CP, Chiang VL. Coniferyl aldehyde 5-hydroxylation and methylation direct syringyl lignin biosynthesis in angiosperms. Proceedings of the National Academy of Science USA. 1999;96(16):8955–60.

34. Xu Z, Zhang D, Hu J, Zhou X, Ye X, Reichel KL, Stewart NR, Syrenne RD, Yang X, Gao P et al. Comparative genome analysis of lignin biosynthesis gene families across the plant kingdom. BMC Bioinformatics 2009; 10 Suppl 11:S3.

35. Humphreys JM, Hemm MR, Chapple C. New routes for lignin biosynthesis defined by biochemical characterization of recombinant ferulate 5-hydroxylase, a multifunctional cytochrome P450-dependent monooxygenase. Proceedings of the National Academy of Science USA. 1999;96(18):10045–50.

36. Mansell RE, Stöckigt J, Zenk H. Reduction of Ferulic acid to Coniferyl alcohol in a cell free system from a higher plant. Z Pflanzenphysiol. 1972;68:286–8.

37. Guo DM, Ran JH, Wang XQ. Evolution of the Cinnamyl/Sinapyl alcohol dehydrogenase (CAD/SAD) gene family: the emergence of real lignin is associated with the origin of Bona fide CAD. J Mol Evol. 2010;71(3):202–18.

38. Bagniewska Zadworna A, Barakat A, Lakomy P, Smolinski DJ, Zadworny M. Lignin and lignans in plant defence: insight from expression profiling of cinnamyl alcohol dehydrogenase genes during development and following fungal infection in Populus. Plant Sci. 2014;229:111–21.

39. Chao N, Liu SX, Liu BM, Li N, Jiang XN, Gai Y. Molecular cloning and functional analysis of nine cinnamyl alcohol dehydrogenase family members in Populus tomentosa. Planta. 2014;240(5):1097–112.

40. Jin Y, Zhang C, Liu W, Qi H, Chen H, Cao S. The cinnamyl alcohol dehydrogenase gene family in melon (Cucumis melo L.): bioinformatic analysis and expression patterns. PLoS One. 2014;9(7):e101730.

41. Saballos A, Ejeta G, Sanchez E, Kang C, Vermerris W. A genomewide analysis of the cinnamyl alcohol dehydrogenase family in sorghum [Sorghum bicolor (L.) Moench] identifies SbCAD2 as the brown midrib6 gene. Genetics. 2009; 181(2):783–95.

42. van Parijs FR, Ruttink T, Boerjan W, Haesaert G, Byrne SL, Asp T, Roldan-Ruiz I, Muylle H. Clade classification of monolignol biosynthesis gene family members reveals target genes to decrease lignin in Lolium perenne. Plant Biology (Stuttg). 2015;17(4):877–92.

43. Berthet S, Demont-Caulet N, Pollet B, Bidzinski P, Cezard L, Le Bris P, Borrega N, Herve J, Blondet E, Balzergue S, et al. Disruption of LACCASE4 and 17 results in tissue-specific alterations to lignification of Arabidopsis thaliana stems. Plant Cell. 2011;23(3):1124–37.

44. Neutelings G, Fenart S, Lucau-Danila A, Hawkins S. Identification and characterization of miRNAs and their potential targets in flax. J Plant Physiol. 2012;169(17):1754–66.

45. Goodstein DM, Shu S, Howson R, Neupane R, Hayes RD, Fazo J, Mitros T, Dirks W, Hellsten U, Putnam N, et al. Phytozome: a comparative platform for green plant genomics. Nucleic Acids Res. 2012;40(Database issue):D1178–86.

46. Lu S, Yang C, Chiang VL. Conservation and diversity of microRNA-associated copper-regulatory networks in Populus trichocarpa. J Integr Plant Biol. 2011; 53(11):879–91.

47. Lu H. Dissection of salicylic acid-mediated defense signaling networks. Plant Signal Behav. 2009;4(8):713–7.

48. Denness L, McKenna JF, Segonzac C, Wormit A, Madhou P, Bennett M, Mansfield J, Zipfel C, Hamann T. Cell Wall damage-induced lignin biosynthesis is regulated by a reactive oxygen species- and Jasmonic acid-dependent process in Arabidopsis. Plant Physiol. 2011;156(3):1364–74.

49. Kim EH, Kim YS, Park SH, Koo YJ, Choi YD, Chung YY, Lee IJ, Kim JK. Methyl Jasmonate reduces grain yield by mediating stress signals to Alter spikelet development in Rice. Plant Physiol. 2009;149(4):1751–60.

50. Rahantamalala A, Rech P, Martinez Y, Chaubet-Gigot N, Grima-Pettenati J, Pacquit V. Coordinated transcriptional regulation of two key genes in the lignin branch pathway–CAD and CCR–is mediated through MYB- binding sites. BMC Plant Biol. 2010;10:130.

51. Zhong R, McCarthy RL, Haghighat M, Ye ZH. The poplar MYB master switches bind to the SMRE site and activate the secondary wall biosynthetic program during wood formation. PLoS One. 2013;8(7):e69219.

52. Zhao Q, Wang H, Yin Y, Xu Y, Chen F, Dixon RA. Syringyl lignin biosynthesis is directly regulated by a secondary cell wall master switch. Proceedings of the National Academy of Science USA. 2010;107(32):14496–501.

53. Day A, Ruel K, Neutelings G, Crônier D, David H, Hawkins S, Chabbert B. Lignification in the flax stem: evidence for an unusual lignin in bast fibers. Planta 2005; 222(2):234–45.

54. Lapierre C, Pollet B, Petit Conil M, Toval G, Romero J, Pilate G, Leplé JC, Boerjan W, Ferret V, De Nadai V, et al. Structural alterations of Lignins in transgenic poplars with depressed Cinnamyl alcohol dehydrogenase or Caffeic acid O-Methyltransferase activity have an opposite impact on the efficiency of industrial Kraft pulping. Plant Physiol. 1999;119(1):153–64.

55. Shen H, Mazarei M, Hisano H, Escamilla-Trevino L, Fu C, Pu Y, Rudis MR, Tang Y, Xiao X, Jackson L, et al. A genomics approach to deciphering lignin biosynthesis in switchgrass. Plant Cell. 2013;25(11):4342–61.

56. Pinzon-Latorre D, Deyholos MK. Characterization and transcript profiling of the pectin methylesterase (PME) and pectin methylesterase inhibitor (PMEI) gene families in flax (Linum usitatissimum). BMC Genomics. 2013;14:742.

57. Feuillet C, Lauvergeat V, Deswarte C, Pilate G, Boudet A, Grima-Pettenati J. Tissue- and cell-specific expression of a cinnamyl alcohol dehydrogenase promoter in transgenic poplar plants. Plant Mol Biol. 1995;27(4):651–67.

58. Hawkins S, Samaj J, Lauvergeat V, Boudet A, Grima-Pettenati J. Cinnamyl alcohol dehydrogenase: identification of new sites of promoter activity in transgenic poplar. Plant Physiol. 1997;113(2):321–5.

59. Ma S, Shah S, Bohnert HJ, Snyder M, Dinesh Kumar SP. Incorporating motif analysis into gene co-expression networks reveals novel modular expression pattern and new signaling pathways. PLoS Genet. 2013;9(10):e1003840.

60. Wrobel-Kwiatkowska M, Starzycki M, Zebrowski J, Oszmianski J, Szopa J. Lignin deficiency in transgenic flax resulted in plants with improved mechanical properties. J Biotechnol. 2007;128(4):919–34.

61. Ohman D, Demedts B, Kumar M, Gerber L, Gorzsas A, Goeminne G, Hedenstrom M, Ellis B, Boerjan W, Sundberg B. MYB103 is required for FERULATE-5-HYDROXYLASE expression and syringyl lignin biosynthesis in Arabidopsis stems. Plant J. 2013;73(1):63–76.

62. Berthet S, Thevenin J, Baratiny D, Demont-Caulet N, Debeaujon I, Przemyslaw B, Leplé J, Huis R, Hawkins S, Gomez LD, et al. Role of plant Laccases in lignin polymerization. Chapter 5. Adv Bot Res. 2012; 61:145–72.

63. Zhao Q, Nakashima J, Chen F, Yin Y, Fu C, Yun J, Shao H, Wang X, Wang ZY, Dixon RA. Laccase is necessary and nonredundant with peroxidase for lignin polymerization during vascular development in Arabidopsis. Plant Cell. 2013;25(10):3976–87.

64. Kozomara A, Griffiths-Jones S. miRBase: annotating high confidence microRNAs using deep sequencing data. Nucleic Acids Res. 2013;42(D1):D68–73.

65. Wang CY, Zhang S, Yu Y, Luo YC, Liu Q, Ju C, Zhang YC, Qu LH, Lucas WJ, Wang X, et al. MiR397b regulates both lignin content and seed number in Arabidopsis via modulating a laccase involved in lignin biosynthesis. Plant Biotechnol J. 2014;12(8):1132–42.

66. Yalpani N, Raskin I. Salicylic acid: a systemic signal in induced plant disease resistance. Trends Microbiology. 1993;1(3):88–92.

67. Miedes E, Boerjan W, Vanholme R, Molina A. The role of the secondary cell walls in plant resistance to pathogens. Front Plant Sci. 2014;5

68. Sharma M, Laxmi A. Jasmonates: emerging players in controlling temperature stress tolerance. Front Plant Sci. 2015;6:1129.

69. Concha CM, Figueroa NE, Poblete LA, Onate FA, Schwab W, Figueroa CR. Methyl jasmonate treatment induces changes in fruit ripening by modifying the expression of several ripening genes in Fragaria chiloensis fruit. Plant Physiol Biochem. 2013;70:433–44.

70. Deng WW, Zhang M, Wu JQ, Jiang ZZ, Tang L, Li YY, Wei CL, Jiang CJ, Wan XC. Molecular cloning, functional analysis of three cinnamyl alcohol dehydrogenase (CAD) genes in the leaves of tea plant. *Camellia sinensis* Journal of Plant Physiology. 2013;170(3):272–82.

71. Hawkins S, Boudet A. Wound-induced lignin and suberin deposition in a woody angiosperm *(Eucalyptus gunnii* hook.): Histochemistry of early changes in young plants. Protoplasma. 1996;191(1):96–104.

72. Kim YH, Bae JM, Huh GH. Transcriptional regulation of the cinnamyl alcohol dehydrogenase gene from sweet potato in response to plant developmental stage and environmental stress. Plant Cell Rep. 2010;29(7):779–91.

73. de Obeso M, Caparros-Ruiz D, Vignols F, Puigdomenech P, Rigau J. Characterisation of maize peroxidases having differential patterns of mRNA accumulation in relation to lignifying tissues. Gene. 2003;309(1):23–33.

74. Delessert C, Wilson IW, Van Der Straeten D, Dennis ES, Dolferus R. Spatial and temporal analysis of the local response to wounding in Arabidopsis leaves. Plant Mol Biol. 2004;55(2):165–81.

75. Cabané M, Afif D, Hawkins S. Advances in Botanical Research. In: Lignins and Abiotic Stresses, vol. 61: London: Academic Press Elsevier; 2012. p. 220–62.

76. Hu Y, Li WC, Xu YQ, Li GJ, Liao Y, Fu FL. Differential expression of candidate genes for lignin biosynthesis under drought stress in maize leaves. J Appl Genet. 2009;50(3):213–23.

77. Murashige T, Skoog F. A revised medium for rapid growth and bio assays with tobacco tissue cultures. Physiol Plant. 1962;15.

78. Raes J, Rohde A, Christensen JH, Van de Peer Y, Boerjan W. Genome-wide characterization of the lignification toolbox in Arabidopsis. Plant Physiol. 2003;133(3):1051–71.

79. Boguski M, Lowe T, Tolstoshev C. dbEST database for expressed sequence tags. Nat Genet. 1993;4(4):332–3.

80. Tamura K, Stecher G, Peterson D, Filipski A, Kumar S. MEGA6: molecular evolutionary genetics analysis version 6.0. Mol Biol Evol. 2013;30(12):2725–9.

81. Jones DT, Taylor WR, Thornton JM. The rapid generation of mutation data matrices from protein sequences. Bioinformatics. 1992;8(3):275–82.

82. Huis R, Hawkins S, Neutelings G. Selection of reference genes for quantitative gene expression normalization in flax (*Linum usitatissimum* L.). BMC Plant Biol. 2010;10:71.

83. Clifford MN. Specificity of acidic phloroglucinol reagents. J Chromatogr A. 1974;94:321–4.

84. Roongsattham P, Morcillo F, Jantasuriyarat C, Pizot M, Moussu S, Jayaweera D, Collin M, Gonzalez-Carranza ZH, Amblard P, Tregear JW, et al. Temporal and spatial expression of polygalacturonase gene family members reveals divergent regulation during fleshy fruit ripening and abscission in the monocot species oil palm. BMC Plant Biol. 2012;12:150.

85. Dai X, Zhao PX. psRNATarget: a plant small RNA target analysis server. Nucleic Acids Research 2011; 39(Web Server issue):W155-W159.

86. Babu PR, Rao KV, Reddy VD. Structural organization and classification of cytochrome P450 genes in flax (*Linum usitatissimum* L.). Gene. 2013;513(1): 156–62.

87. Day A, Addi M, Kim W, David H, Bert F, Mesnage P, Rolando C, Chabbert B, Neutelings G, Hawkins S. ESTs from the fibre bearing stem tissues of flax (*Linum usitatissimum* L.): expression analyses of sequences related to Cell Wall development. Plant Biol. 2005;7(1):23–32.

88. Venglat P, Xiang D, Qiu S, Stone SL, Tibiche C, Cram D, Alting-Mees M, Nowak J, Cloutier S, Deyholos M, et al. Gene expression analysis of flax seed development. BMC Plant Biol. 2011;11:74.

4

Characterization of molecular diversity and genome-wide mapping of loci associated with resistance to stripe rust and stem rust in Ethiopian bread wheat accessions

Kebede T. Muleta[1], Matthew N. Rouse[2], Sheri Rynearson[1], Xianming Chen[3], Bedada G. Buta[4] and Michael O. Pumphrey[1*]

Abstract

Background: The narrow genetic basis of resistance in modern wheat cultivars and the strong selection response of pathogen populations have been responsible for periodic and devastating epidemics of the wheat rust diseases. Characterizing new sources of resistance and incorporating multiple genes into elite cultivars is the most widely accepted current mechanism to achieve durable varietal performance against changes in pathogen virulence. Here, we report a high-density molecular characterization and genome-wide association study (GWAS) of stripe rust and stem rust resistance in 190 Ethiopian bread wheat lines based on phenotypic data from multi-environment field trials and seedling resistance screening experiments. A total of 24,281 single nucleotide polymorphism (SNP) markers filtered from the wheat 90 K iSelect genotyping assay was used to survey Ethiopian germplasm for population structure, genetic diversity and marker-trait associations.

Results: Upon screening for field resistance to stripe rust in the Pacific Northwest of the United States and Ethiopia over multiple growing seasons, and against multiple races of stripe rust and stem rust at seedling stage, eight accessions displayed resistance to all tested races of stem rust and field resistance to stripe rust in all environments. Our GWAS results show 15 loci were significantly associated with seedling and adult plant resistance to stripe rust at false discovery rate (FDR)-adjusted probability (P) <0.10. GWAS also detected 9 additional genomic regions significantly associated (FDR-adjusted $P < 0.10$) with seedling resistance to stem rust in the Ethiopian wheat accessions. Many of the identified resistance loci were mapped close to previously identified rust resistance genes; however, three loci on the short arms of chromosomes 5A and 7B for stripe rust resistance and two on chromosomes 3B and 7B for stem rust resistance may be novel.

Conclusion: Our results demonstrate that considerable genetic variation resides within the landrace accessions that can be utilized to broaden the genetic base of rust resistance in wheat breeding germplasm. The molecular markers identified in this study should be useful in efficiently targeting the associated resistance loci in marker-assisted breeding for rust resistance in Ethiopia and other countries.

Keywords: Bread wheat, Stripe rust, Stem rust, Genetic resistance, Genetic diversity, Association mapping

* Correspondence: m.pumphrey@wsu.edu
[1]Department of Crop and Soil Sciences, Washington State University, Pullman, WA 99164-6420, USA
Full list of author information is available at the end of the article

Background

Stripe rust [*Puccinia striiformis* Westend. f. sp. *tritici* Erikss. (*Pst*)] and stem rust [*Puccinia graminis* Pers.:Pers. f. sp. *tritici* Erikss. & E. Henn. (*Pgt*)] are two of the most damaging diseases of wheat worldwide [1–3]. The ability of wheat rust pathogen populations to quickly evolve new pathotypes that overcome deployed resistance genes and produce multiple cycles of urediniospores in a single season, along with their capacity for long distance dispersal, create potential for destructive epidemics in susceptible varieties under favorable conditions. At various times in history, epidemics of both diseases have caused massive losses to wheat production globally. Stripe rust has been a disease of wheat mainly in areas with cooler climates [1]. More recently, however, the emergence of aggressive races of *Pst* tolerant to higher temperatures has resulted in yield loss in areas normally considered too warm for serious epidemic development. These new strains of *Pst* are currently widespread and threatening wheat production on a global scale [1, 4, 5]. In recent years, wheat growing regions in East Africa, Central and West Asia, and the Caucasus countries experienced one of the largest stripe rust epidemics in the recent history [6]. In Ethiopia alone, this epidemic affected more than 600,000 ha of wheat and led to an expenditure of more than $US 3.2 million on fungicides, while significant widespread losses were still realized [7].

During the last 40 years, introgression of different combinations of stem rust resistance genes and other epidemic mitigation strategies have reduced global stem rust epidemics [8, 9]. However, the discovery of race TTKSK (isolate Ug99) of *Pgt* in Uganda in 1998 has raised a particular concern to global wheat production and food security due to its wide virulence spectrum [10]. Currently, at least eight variants of TTKSK (the Ug99 lineage races) have been described with virulence for additional resistance genes, including *Sr24*, *Sr36*, *Sr9h* and *SrTmp* [11–14]. These races are currently spreading across eastern, southern and northern Africa as well as the Middle East [8, 14, 15]. In 2013–2014, a new stem rust race designated as TKTTF caused a severe epidemic on the variety 'Digalu' carrying *SrTmp* in Ethiopia [16]. The continued emergence of new virulent races of stem rust emphasizes the dynamic challenges of breeding for stem rust resistance.

To mitigate losses due to rust diseases in wheat, world wheat production has largely depended on the use of resistant wheat varieties [1, 17, 18]. The emergence of pathogen populations virulent against deployed resistance genes has caused not only tremendous yield and quality losses, but has also led to frequent replacement of the otherwise agronomically superior cultivars, as well as interference with progress in improving other important traits [17]. Achieving durable varietal performance

has therefore been the primary focus of many breeding programs in wheat. Two categories of resistance genes have been widely recognized in wheat breeding for rust resistance; all-stage resistance (also called seedling resistance) and adult-plant resistance (APR) [1]. Developing cultivars carrying effective seedling resistance in combination with APR genes is more desirable to minimize the damage caused by new mutants of the pathogen [3, 17]. The deployment of cultivars with durable rust resistance in wheat is particularly desirable in regions such as Ethiopia and the neighboring countries in East Africa that are characterized by the presence of year-round inoculum due to nearly constant wheat cropping seasons and a suitable environment. These conditions provide the pathogen not only with a continuous substrate both in area and in time due to a green bridge between the seasons, but may also result in rapid selection for virulent races of the pathogen.

Limited variation in elite germplasm may constrain deployment of diverse resistance genes in released cultivars and the capacity for countering new virulence in pathogen populations [19, 20]. Landraces and other germplasm collections maintained in germplasm banks may provide access to diverse alleles for disease resistance and reverse the trend of genetic diversity erosion in established, elite cultivars [21, 22]. Such genetic resources are potentially untapped sources of useful genetic diversity owing to their limited use in modern plant breeding programs. The usefulness of wheat landraces as a good source of resistance to diseases in wheat have been demonstrated [23–27]. The objectives of this study were: 1) to evaluate Ethiopian bread wheat landraces and cultivars for their field and seedling resistance to stripe rust and stem rust across a range of environments and multiple races of the pathogens, 2) to assess the genetic diversity of Ethiopian bread wheat landraces and cultivars based on high-density genotyping, and 3) to conduct a genome-wide search for molecular markers associated with loci underpinning seedling and field resistance to stripe rust and stem rust diseases of wheat.

Methods

Plant materials

One hundred and ninety bread wheat accessions were used in this study. The panel was comprised of 124 landraces and 66 commercial cultivars and advanced breeding lines from the Ethiopian Institute of Agricultural Research (EIAR). Most of the commercial cultivars and advanced breeding lines were originally developed by the International Maize and Wheat Improvement Center (CIMMYT), some by the International Center for Agricultural Research in the Dry Areas (ICARDA), but evaluated and released as cultivars by the Ethiopian wheat improvement program. One hundred and ten accessions were landraces of hexaploid wheat originally from the Ethiopian Institute of Biodiversity, characterized and maintained by EIAR at Debre-Zeit Agricultural Research Center (DZARC), from where they were obtained. Fourteen landraces were obtained from the United States Department of Agriculture, Agricultural Research Service (USDA-ARS) National Small Grains Collection (NSGC) in Aberdeen, ID, USA. Two susceptible cultivars, Avocet S and Morocco, used in this study as checks were obtained from the USDA-ARS, Wheat Health, Genetics, and Quality Research Unit, Pullman, WA; and DZARC, respectively.

Stripe rust response evaluation under field conditions
The panel was tested for response to stripe rust infection under field conditions in four nurseries in the Pacific Northwest (PNW) of the US and in Ethiopia for three consecutive growing seasons (2012–2014). The locations in the PNW were Mount Vernon (48° 25′ 12″N; 122° 19′ 34″W), a high rainfall area located west of the Cascade Mountain range; and Pullman (46° 43′ 59″ N; 117° 10′ 00″W), a semi-arid wheat belt area located east of the Cascade Mountain range. Locations in Ethiopia included Meraro (07°24′27″ N; 39°14′56″ E) and Arsi Robe (07° 53′02″ N; 39°37′40″ E), sub-stations of Kulumsa Agricultural Research Center of the Ethiopian Institute of Agricultural Research, representing relatively high annual rainfall and cool temperatures. Each nursery location is subject to high disease pressure on an annual basis, but has variable Pst races and environmental conditions. The accessions were planted as single rows of non-replicated trials with repeating checks. The susceptible cultivars Avocet S (at Mount Vernon and Pullman) and Morocco (at Meraro and Arsi Robe) were planted every 20 rows and on each side of the plot to ensure uniform disease pressure across the experimental plots. Accessions were evaluated for response Pst under natural disease epidemics. Host response to infection [i.e. infection types (IT)] to stripe rust was estimated using a 0–9 scale as described by Line and Qayoum [28]. Stripe rust disease severity (DS) was recorded as percent leaf area showing disease symptoms. No permits were required to carry out field experiments. In addition, the field studies did not involve endangered or protected species.

Seedling resistance screening to Pst
Seedlings of the 190 accessions were evaluated for response to five isolates representing five races of Pst under greenhouse conditions at Washington State University, Pullman, WA. Three of the races (PSTv-14, PSTv-37 and PSTv-40) were of US origin [29], one race from Ethiopia, but also detected in the USA (PSTv-41) and one race predominant only in Ethiopia (PSTv-106) were used in the study [30].

The virulence/avirulence formulae of the isolates are presented in Additional file 1. All Pst races used in this study were obtained from the USDA-ARS, Wheat Health, Genetics, and Quality Research Unit, Pullman, WA.

For the seedling test, three to five seeds of each accession were planted into pots filled with the Sunshine® mix growing medium (Sun Gro Horticulture, Agawam, MA, and USA). The susceptible check, 'Avocet S' and the stripe rust single-gene differential lines were also included in each experiment. Ten-day old seedlings were inoculated with urediniospores of each race. The urediniospores were mixed with talcum to ensure uniform coverage of the spores over the leaves of the seedlings. Inoculated seedlings were kept in a dark dew chamber for 24 h at 10 °C, and then transferred into a growth chamber with temperature programmed to change gradually between 4 °C at 2:00 am during the 8 h dark period and 20 °C at 2:00 pm during the 16-h light period [29]. Stripe rust infection types (IT) based on the 0–9 scale were recorded at 18–20 days after inoculation as described by Line and Qayoum [28].

Seedling resistance screening to Pgt
The Ethiopian bread wheat cultivars and landraces were evaluated for reaction to four Pgt races at the USDA-ARS Cereal Disease Laboratory (CDL), St. Paul, MN. The four stem rust races were TTKSK (isolate 04KEN156/04), TRTTF (06YEM34–1), TTTTF (01MN84A-1-2) and TKTTF (13ETH18–1) that were selected based on their differential virulence patterns and potential impact on wheat production. TRTTF is broadly virulent to stem rust resistance genes including $Sr1RS^{Amigo}$ and was detected in Yemen and Ethiopia [31]. Race TTKSK (Ug99) has a wide virulence spectrum and is rapidly evolving in East Africa. Race TTTTF is the most virulent race from the United States, producing high infection types (ITs) on the majority of stem rust differential lines [32]. TKTTF is a new race that was responsible for the localized severe stem rust epidemic in Ethiopia in 2013–2014 due to its virulence to the stem rust resistance gene SrTmp in cultivar 'Digalu' [16]. Seedling resistance evaluations were performed as described previously by Rouse et al., 2011 [33]. Infection types (ITs) were recorded on a 0 to 4 scale according to Stakman et al., 1962 [34]. Stakman ITs were converted to a linear scale using a conversion algorithm [35]. All the Pgt races used in this study were obtained from the USDA-ARS Cereal Disease Laboratory (CDL), St. Paul, MN.

Genotyping
The 190 Ethiopian bread wheat accessions were genotyped using the 90,000 SNP Illumina Infinium assay at USDA-ARS Fargo, ND, USA. Illumina genotypic data were analyzed using Illumina GenomeStudio v2011.1

software to optimize the SNP call rates. After removing SNPs with low-quality clustering and those with minor allele frequency (MAF) less than 5%, 24,281 high quality SNP markers with genetic map information were used for GWAS analyses. The genetic positions of the SNP markers were based on the wheat 90 K SNP consensus map [36]. The dataset was also filtered using a 10% cut-off for missing data in the accessions. Accordingly, a total of 179 accessions were retained in the GWAS and population structure analyses.

For internal controls, the accessions were screened with molecular markers for known stripe rust and stem rust resistance genes, including *Lr34/Yr18/Pm38* (KASP marker wMAS000003) [37], *Lr37/Yr17/Sr38* (CAPS marker Ventriup-LN2) [37], *Lr67/Yr46* (Kasp856) [38], *Sr31/Yr9/Lr26* (KASP 1RS:1BL_6110) [39], *Sr24* (barc71) [40], *Sr36* (KASP wMAS000015), [37], *Sr2/Yr30* (KASP wMAS000005) [37]. Markers with a MAF greater than 5% were included in the GWAS and linkage disequilibrium analyses.

Analyses of molecular diversity and population structure

Genetic diversity was analyzed based on Nei's gene diversity, which is an estimate of the probability that two randomly chosen alleles from the population are different, and polymorphism information content (PIC), which reflects the probability of polymorphism between two random samples in the germplasm, using POWERMARKER v3.25 [41, 42]. The two indices of genetic diversity were used to compare the extent of molecular diversity among the landrace collection and the contemporary Ethiopian bread wheat cultivars. A *t*-test was performed using the software package JMP Genomics v6.0 (SAS Institute, Cary, NC) to compare the extent of molecular diversity between landraces and elite Ethiopian cultivars.

Cluster analysis based on the neighbor joining (NJ) tree algorithm according to shared-allele distance was also used to determine the genetic structure of the accessions using the phylogenetic tree analysis package in POWER-MARKER v3.25. The assessment of the branching pattern in the NJ tree was based on bootstrapping over loci with 1000 replications. FigTree program v1.4 (http://tree.bio.ed.ac.uk/software/figtree/) was used to display the consensus bootstrap value generated by PHYLIP v3.66 [43].

Population structure was analyzed using a set of 355 random SNP markers distributed across the 21 wheat chromosomes (12–23 markers per chromosome) spaced at >10 cM apart. The admixture model with correlated allele frequency implemented in STRUCTURE software version 2.2.3 was applied to detect population structure in the germplasm panel [44]. The parameters were set to a burn-in of 20,000 iterations and 50,000 Monte Carlo Markov Chain (MCMC) replicates to determine K values

in the range of 1 to 10. For each K, five independent runs were carried out. The Evanno method [45] was used to determine the likely number of subpopulations. Principal components (PC) were inferred using the software package JMP Genomics v6.0 (SAS Institute, Cary, NC) to further analyze population sub-structuring and compare the results with results from analysis with STRUCTURE software.

Linkage disequilibrium analysis

Pairwise measures of linkage disequilibrium (LD) between SNP markers were estimated using the software package JMP Genomics v6.0 (SAS Institute, Cary, NC). LD was estimated as squared allele frequency correlation (r^2) between pairs of intra-chromosomal SNPs with known chromosomal position. To determine the average pattern of genome-wide LD decay over genetic distance, a scatter-plot of r^2 values against the corresponding genetic distance between markers was constructed. The second-degree locally weighted polynomial regression (LOESS)-based curve was fitted to estimate the extent of LD decay [46]. The critical r^2 value that indicates the demarcation beyond which LD is due to true physical linkage was determined by taking the 95th percentile of the square root of transformed r^2 data of unlinked markers [47]. The genetic distance at which the LD decay curve intersects with the critical r^2 value was used as a threshold to determine the confidence interval of significant QTL.

Analyses of variance and heritability

Variance components were estimated for stripe rust IT and DS from the field experiments with a mixed linear model using the Restricted Maximum Likelihood (REML) method [48]. Genotype, location, genotype x location interaction and replication (season) were considered to have random effects whereas the overall mean was considered to have fixed effect. Variance components were estimated for each location and across all locations according to the following model:

$$y_{ijk} = \mu + g_i + k_j + e_j + ge_{ij} + e_{ij}$$

Where y_{ij} is the observation for genotype i at environment j in season k, μ is the overall mean, g_i the effect of the accession i, e_j the effect of environment j, k_j the effect of season k, ge_{ij} the interaction between accession i within environment j, and e_{ij} the residual.

Heritability (H^2) estimates were calculated for each location and across all locations as:

$$H^2 = \sigma^2{}_G / \left(\sigma^2{}_G + \left(\sigma^2{}_E / y \right) + \left(\sigma^2{}_{GXE} / y \right) + \sigma^2{}_{error} / y \right)$$

Where σ^2_G is the genotypic variance, σ^2_E is the environment variance, σ^2_{GXE} is the genotype by environment interaction variance, and σ^2_{error} is the residual error variance and y is the number years within each location or location by year for the estimates of heritability across all environments. Genotype adjusted means were computed based on best linear unbiased predictions (BLUPs). Pearson correlation coefficient between locations and seasons were calculated to determine the consistency of IT and DS across the environments.

Marker-trait association (MTA)

To identify loci associated with response to *Pst* and *Pgt*, genome-wide association analyses were conducted using a total of 24,281 high quality SNP markers and phenotypic traits values (IT and DS) from the field and greenhouse experiments. Marker-trait associations (MTAs) were identified using the compressed mixed linear model (CMLM) [49, 50] implemented in GAPIT (Genomic Association and Prediction Integrated Tool) R package [51]. The first three principal components (PC3) and a compressed relationship matrix (K) were included as fixed and random effect, respectively. Three additional association test models were compared to the PC3 + K CMLM model for correcting population structure and kinship based on the deviances of observed probability from expected distribution in Q-Q plot were used to compare the models. The following three additional models were tested: a fixed general linear model with no correction for population structure (GLM), general linear model with the first three principal components included as a fixed covariate (PC3 GLM) and mixed linear model with compressed relationship matrix included as a random covariate (K CMLM). Association analyses were conducted for IT and DS values from each location separately and estimates of BLUPs for each location and across all environments.

Results

Phenotypic variability and estimates of heritability

Analysis of variance revealed a highly significant ($P < 0.0001$) difference among the genotypes both for individual locations and data combined across locations (Table 1). Genotype by environment interactions were significant ($P < 0.01$) for disease severity at Pullman, Mount Vernon and the combined analysis across all locations. The variance components for locations were not significant across all analyses. Based on the IT data, 21 accessions (11%) were highly resistant to all races of *Pst* as seedlings and across all locations and seasons at adult plant stages. Eleven of the 21 resistant accessions were landraces, while the remaining 10 accessions were cultivars. Twenty-three accessions (13%) showed highly susceptible reactions to all races as seedlings and across all locations and seasons at adult plant stage (Fig. 1). Heritability (H^2) values for stripe rust IT and DS ranged from 73% to 93% (Table 1). Correlation coefficients between locations and seasons for stripe rust IT and DS are summarized in Additional file 2. The Pearson correlation coefficients for stripe rust IT and DS between the multiple locations over multiple growing seasons averaged 0.64 and 0.72, respectively. Average correlations between seasons within locations were 0.79, 0.76 and 0.64 for IT, and 0.80, 0.86 0.68 for DS at Pullman, Mount Vernon and Ethiopia, respectively. Correlation between Pullman and Mount Vernon (over multiple seasons) (averaged 0.72 and 0.78 for IT and DS, respectively), were significantly higher than the correlation between Ethiopia and the two locations in PNW; the respective average correlation coefficients for IT and DS were 0.60 and 0.58 between Pullman and Ethiopia, and 0.54 and 0.55 between Mount Vernon and Ethiopia.

Table 1 Mean response to *Puccinia striiformis* f. sp. *tritici* infection, estimates of variance components and heritability

Parameter	Pullman		Mount Vernon		Ethiopia	Across locations	
	IT (0–9)	Severity (%)	IT (0–9)	Severity (%)	IT (0–9)	IT (0–9)	Severity (%)
Minimum	0.0	0.0	0.0	0.0	0.0	0.0	0.0
Mean	4.2	41.0	4.2	43.0	4.8	4.4	42.0
Maximum	9.0	100.0	9.0	100.0	9.0	0.0	100.0
σ^2_G	2.7***	360.0****	4.5***	612.0****	4.1***	2.9***	451.2****
σ^2_E	0.9ns	141.0ns	0.1ns	4.4ns	0.1ns	0.3ns	48.1
σ^2_{GXE}	0.1ns	117.0***	0.0	90.9**	0.3*	0.2*	136.9***
σ^2_{error}	1.0	1.0	0.7***	1.0	1.0	1.8	1.0
Heritability	0.73	0.74	0.92	0.93	0.85	0.89	0.91

σ^2_G = estimate of genotypic variance
σ^2_E = estimate of environmental variance
σ^2_{GE} = estimate of genotype x environment variance
σ^2_e = estimate of residual variance; H^2 = heritability
IT infection type; *DS* disease severity; *ns* not significant
*$P < 0.05$
**$P < 0.01$
***$P < 0.001$
****$P < 0.0001$

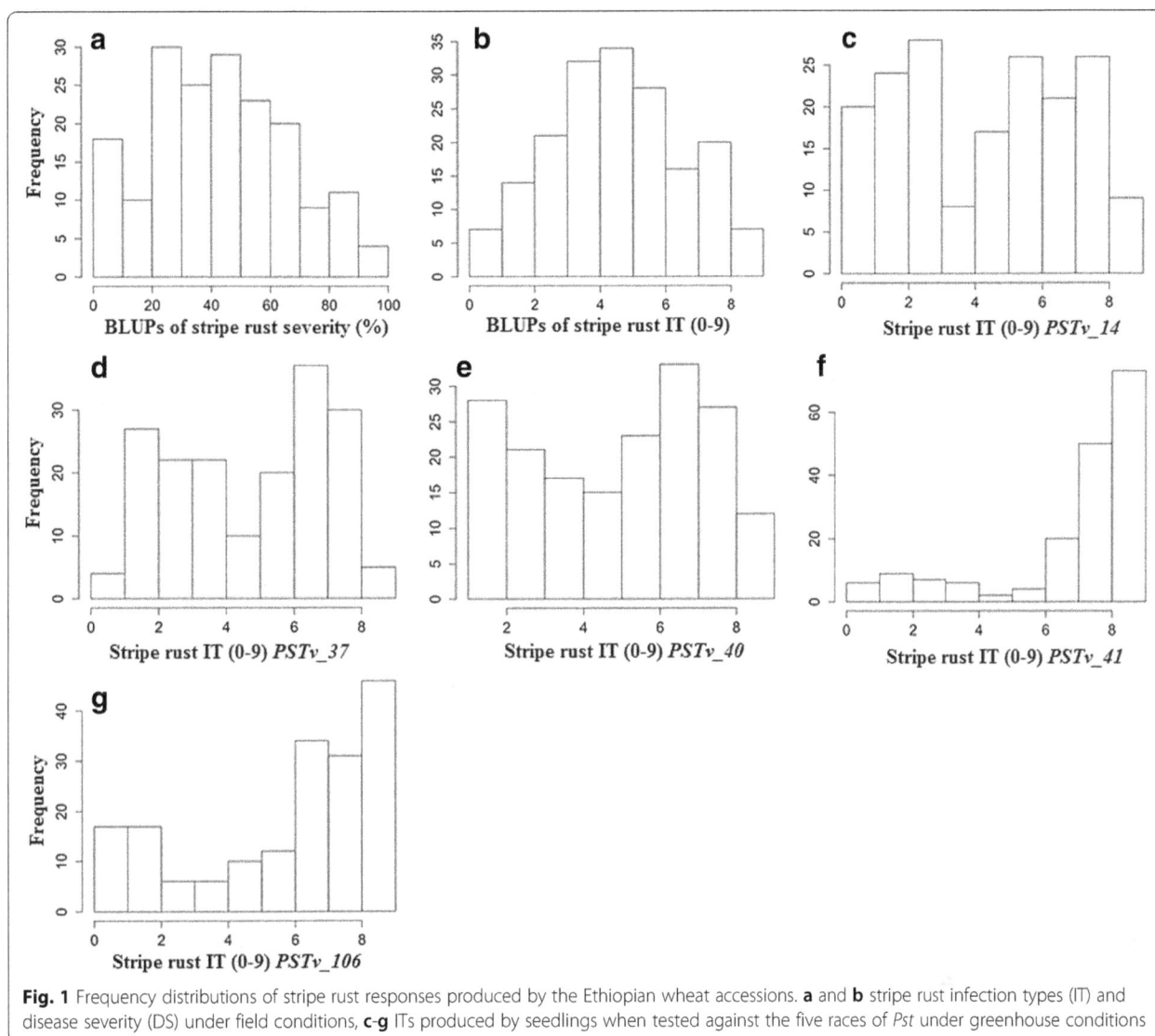

Fig. 1 Frequency distributions of stripe rust responses produced by the Ethiopian wheat accessions. **a** and **b** stripe rust infection types (IT) and disease severity (DS) under field conditions, **c-g** ITs produced by seedlings when tested against the five races of *Pst* under greenhouse conditions

The frequency distribution of the linearized ITs produced by the four *Pgt* races on the Ethiopian cultivars and landraces is shown in Fig. 2. Among all accessions, 75 (41%), 84 (46%), 61 (34%) and 69 (38%) showed seedling resistance to races TTKSK, TRTTF, TTTTF and TKTTF, respectively. Fifteen (8%) of the accessions were resistant (IT = 0 to 23) to all four races, of which the majority were landraces. Eight accessions (5%) shared resistance to the four races of *Pgt* and field resistance to *Pst* at all environments. Five of the 8 accessions resistant to the four races of *Pgt* and with field resistance to *Pst* at all environments were cultivars, while the remaining 3 were landraces. Additional file 3 summarizes the responses of the accessions to *Pst* and *Pgt* infection in the seedling and field resistance screenings.

Molecular diversity and population structure
Genome-specific analyses of Nei's gene diversity and PIC values were used to compare the extent of genetic variation in modern Ethiopian cultivars and accessions maintained in the germplasm collection. The analyses revealed a highly significant difference ($P < 0.001$) between cultivars and landraces in terms of both Nei's gene diversity index and PIC values. Both diversity indices were significantly higher in the landraces in 13 of the 21 wheat chromosomes (including 1A, 2D, 3A, 3B, 3D, 4A, 5A, 5B, 6A, 6D, 7A, 7B and 7D). The cultivars showed higher values of gene diversity and PIC values over the landraces only in chromosome 1B and 2A. Genome-wide average Nei's gene diversity index values were 0.36 and 0.32 for the landraces and cultivars, respectively, while the PIC values were 0.29 and 0.25 for the landraces and cultivars, respectively (Fig. 3).

A neighbor-joining (NJ) phylogeny analysis based on shared allele distance showed that the Ethiopian improved varieties exhibited a high degree of genetic relatedness and were clearly distinct from the majority of landraces (Fig. 4). Based on pairwise kinship analysis

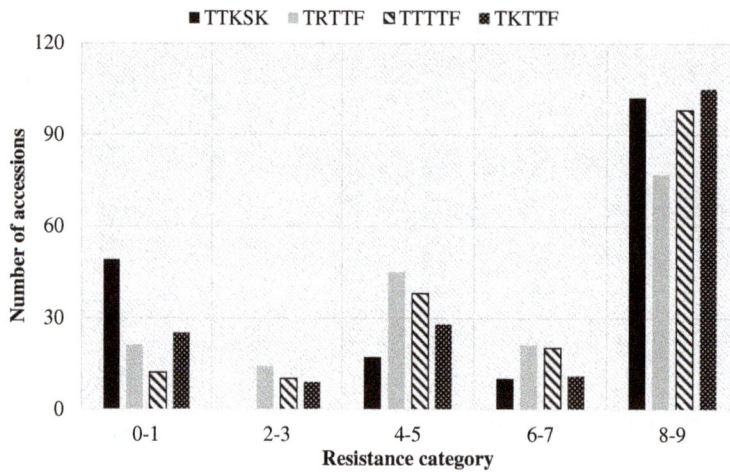

Fig. 2 Frequency distributions of stem rust infection types produced by seedlings of the Ethiopian wheat accessions

using the Fast Ward clustering algorithms, about 31% of the released commercial cultivars showed coefficient of kinship greater than 0.9 (1.0 is exact resemblance) with each other.

Bayesian analysis of population structure yielded a strong signal for the existence of two distinct genetic clusters in the germplasm panel. With K = 2, more than 91% of the accessions were assigned to a specific cluster with membership probability higher than 75%, with the remaining accessions (9%) being classified as admixed. Kinship and principal component analysis (PCA) tended to group the accessions into three subgroups, in which all of the improved varieties were categorized into a distinct third cluster (Fig. 5). The structure analysis yielded

similar clustering patterns as PCA when the second subpopulation was further analyzed separately. PCA revealed that the first and second principal component PCs accounted for 11.5% and 6.6% of the variance, respectively.

Linkage disequilibrium estimation

The extent of LD and average rate of LD decay were estimated by squared correlation coefficient (r^2) for all pairs of SNPs along each chromosome. LD due to physical linkage, estimated as the 95th percentile of the distribution of LD r^2 between unlinked SNPs, was 0.18. On a genome-wide level, 33.3% of all pairwise SNP loci showed LD greater than the critical value

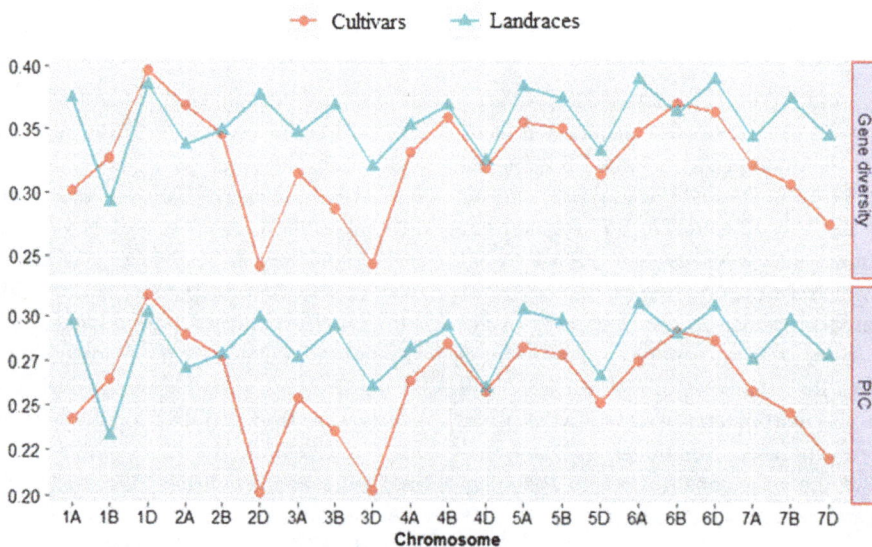

Fig. 3 Genome specific comparisons of molecular diversity between elite cultivars and gene bank conserved landrace accessions. Nei's gene diversity and polymorphism information content (PIC) values were used to compare the extent of genetic variation in modern Ethiopian cultivars and landrace accessions

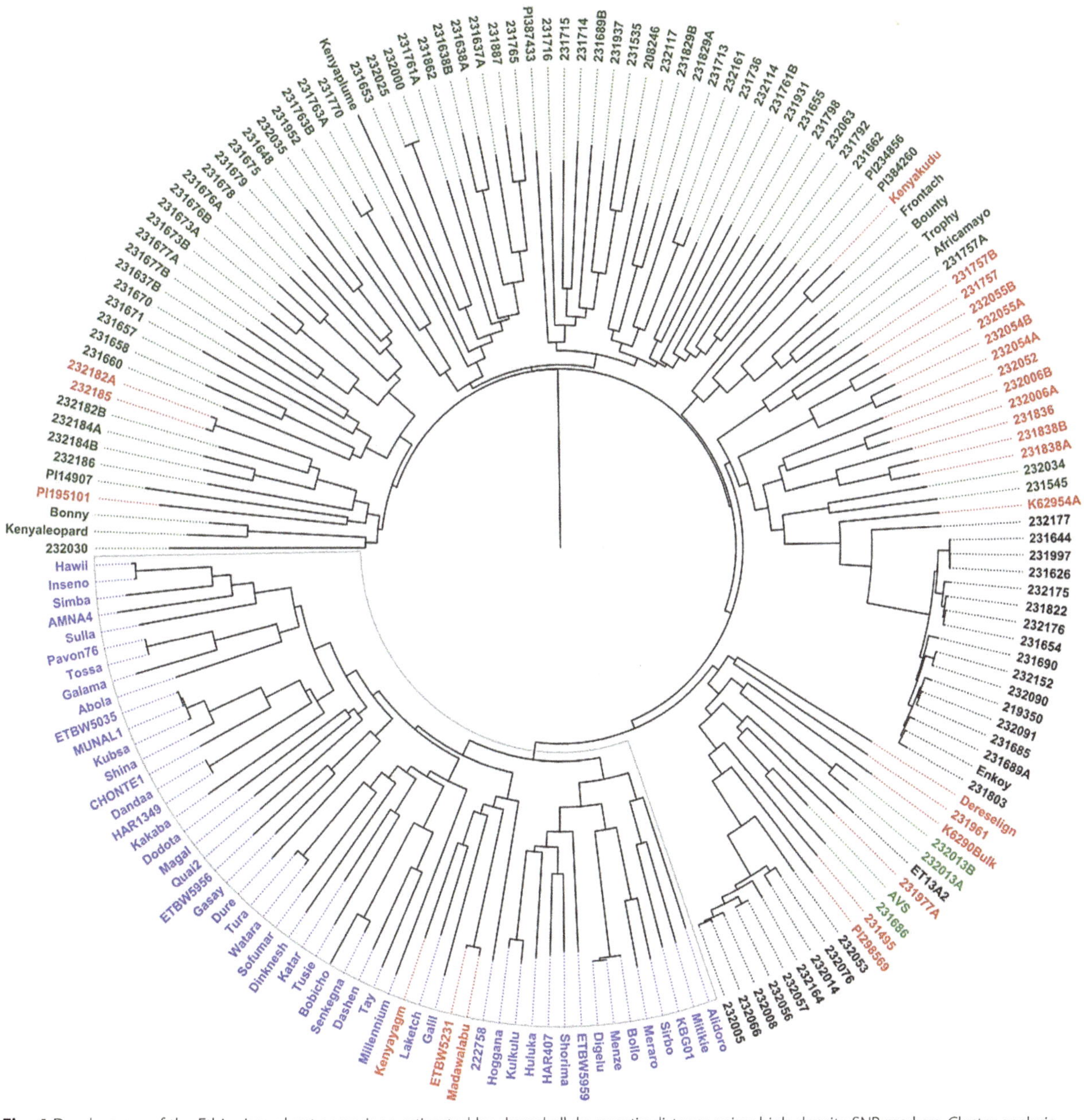

Fig. 4 Dendrogram of the Ethiopian wheat accessions estimated by shared-allele genetic distance using high-density SNP markers. Cluster analysis was based on the neighbor-joining algorithm. Accessions have been assigned colors based on STRUCTURE analysis at K = 3. *Black* = sub-population 1 (SP_I), *Green* = sub-population 2 (SP_II (i)) and *Blue* = subpopulation 3 (SP_II (ii)). Structure analysis indicated the likely number of sub-population to be two (SP_I and SP_II). Separate analysis of population structure within SP_I showed no sign of further sub-division, while SP_II subdivided into SP_II (i) and SP_II (ii). The blue colored accessions (SP_II (ii) in the structure analysis) are Ethiopian cultivars, *while the rest are landraces*

($r^2 > 0.18$). In the A genome, the percentage of marker pairs that were in LD greater than 0.18 for chromosome 1A was the highest (6.8%), while in the B genome chromosome 1B contained the highest percentage of marker pairs (11.0%) that showed LD greater than the critical r^2 value. For the D genome, the percentage of SNP pairs that showed LD greater than 0.18 for chromosome 2D was the highest (4.2%).

LD decayed to the critical r^2 value (0.18) at about 2.5 cM, which was used to determine the confidence interval for declaring distinct MTAs (Fig. 6).

Association analysis for field-based and seedling resistance to Pst

Genome-wide association analyses were performed for stripe rust IT and DS within each of the eight

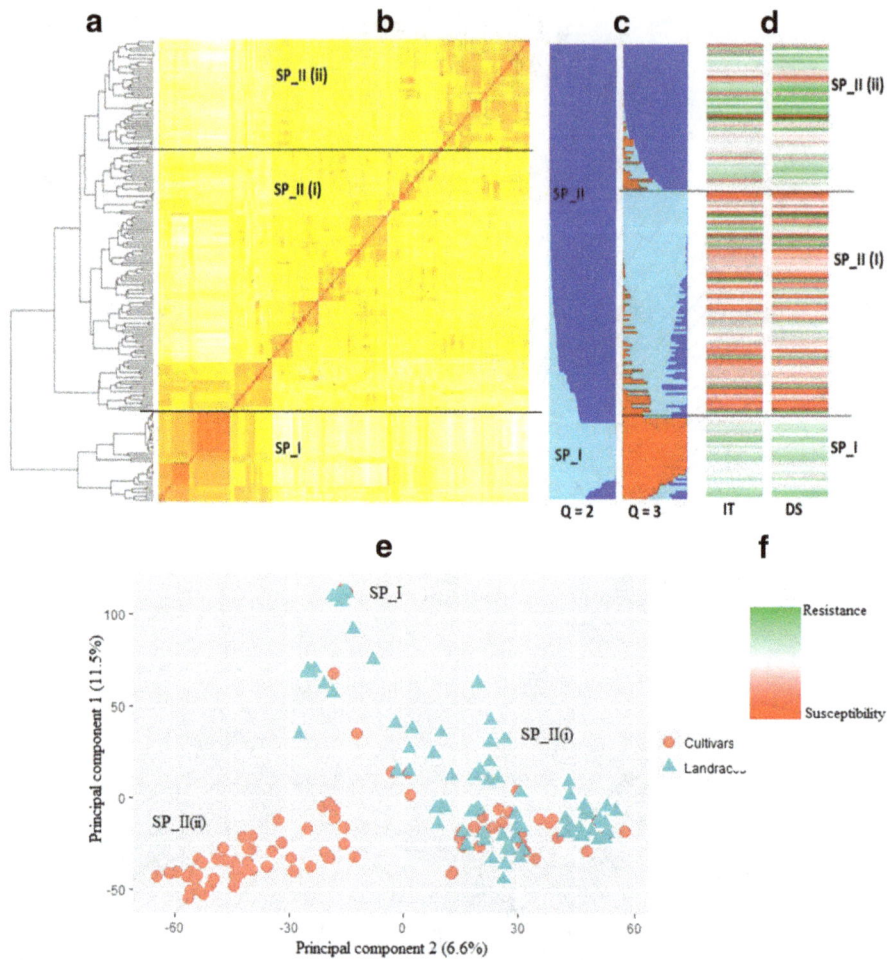

Fig. 5 Genetic relatedness, population structure and the relationship between population sub-clustering and stripe rust resistance. **a** The distance based hierarchical clustering of the accessions based on the Fast Ward grouping algorithm. **b** Heat map of identity-by-decent (IBD) kinship matrix. **c** Clustering diagrams of population structure based on the model based quantitative assessment of subpopulation membership. **d** Heat map of reactions of the accessions to stripe rust IT and DS based on BLUP values across all environments. **e** Principal component analysis (PCA). **f** Color key for the heat map of the phenotypes. SP_I and SP_II are sub-populations defined according to the optimum number of clusters determined by STRUCTURE analysis. SP-II was further sub-grouped into two (SP-II (i) and SP-II (ii)) based on kinship analysis that agrees with the STRUCTURE analysis when K = 3 was considered. Accessions in SP-II (ii) are all recent and historical Ethiopian bread wheat cultivars

environments. Based on the assessment of the performance of the GWAS models for controlling population structure and kinship, the K + PC3 CMLM model (containing the first three principal components and a compressed kinship) gave a minimum deviance of observed P values from expected distribution in Q-Q plot (Additional file 4) and was used for all GWAS analyses. The analyses revealed 67 genomic regions significantly associated with adult plant IT or DS at nominal probability (P) < 0.005 in at least four test environments. A confidence interval for linkage blocks of the 67 regions was determined based on average LD decay rate (±2.5 cM) and adjacent markers were assigned to the linkage blocks based on the significance and consistency of associations across environments. Markers with the most significant associations to

each trait were used as the MTA-tagging marker. Altogether, the 67 genomic regions explained 54.3% and 52.1% of the total variation in IT and DS, respectively (excluding variation explained by the population structure). Additional file 5 summarizes detailed information of the 67 putative resistance loci.

Eleven of the 67 genomic regions were significant at False Discovery Rate (FDR) [52] adjusted P < 0.1 (Table 2). The 11 high-probability QTL were located on chromosomes 1B, 2A, 4A, 5A, 6A and 7B. The phenotypic variation (R^2) explained by these genomic regions were in the range of 3–11%. When combined, these loci explained 27.5 and 26.9% of the total variation in *Pst* IT and DS responses, respectively. Results of the GWAS tests are discussed in detail focusing on

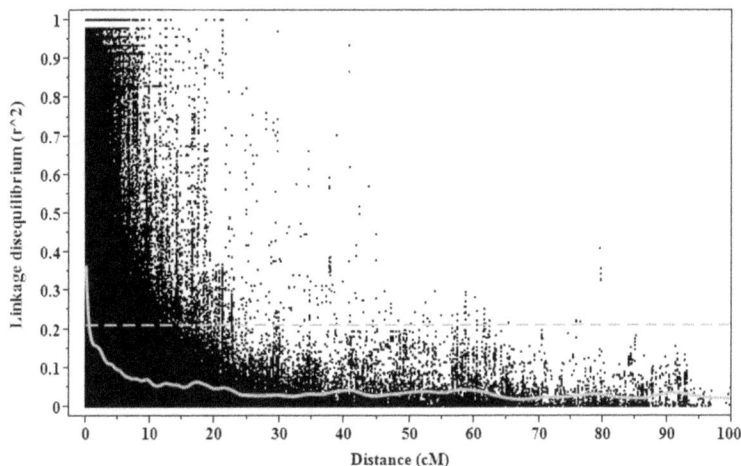

Fig. 6 Genome-wide linkage disequilibrium decay plot for the Ethiopian wheat accessions based on high-density SNP markers. LD, measured as r^2 between pairs of polymorphic marker loci is plotted against the genetic distance (cM) between the markers. LOESS smoothening curve (*solid line*) and mean LD (*broken line*) were fitted to the LD decay

the chromosome regions with significance at FDR-adjusted $P < 0.1$.

Among the 11 highly significant genomic regions, a block of haplotypes covering a genetic distance from 65 to 74 cM on chromosome 1B was identified to harbor major *Pst* resistance QTL/genes. SNPs *IWB11553* (mapped at 65 cM) and *IWB9661* (mapped at 74 cM) represented loci significantly associated with stripe rust IT and DS on 1BS (Table 2). The phenotypic variation explained by *IWB11553* and *IWB9661* loci were in the range of 3.1–9.6% and 3.4–6.7%, respectively.

On the short arm of chromosome 2A, a cluster of significant SNPs covering a genetic distance of 9.4–47.0 cM was detected. Among these, *IWB11136* (mapped at 9.4 cM), *IWB6584* (mapped at 26 cM) and *IWB57199* (mapped at 47.0 cM) showed strong associations (FDR $P < 0.1$) across multiple locations and growing seasons. The phenotypic variation explained by *IWB11136*, *IWB6584* and *IWB57199* loci ranged from 3.5–8.3%, 3.1–11.2% and 3.4–9.5%, respectively. Each SNP was in strong LD with each other ($r^2 = 0.60$, 0.68 and 0.88 between *IWB11136* and *IWB6584*, *IWB11136* and *IWB57199*; and *IWB6584* and *IWB57199*, respectively), in spite of extended genetic distance in the consensus map.

On chromosome 4A, SNP *IWB24187* (located 114.5 cM) was among the 11 strongly associated markers representing effective stripe rust resistance loci. The amount of variation explained by *IWB24187* ranged 2.8–6.1%. On chromosomes 5A, SNP markers *IWA1280* and *IWB28837*, mapped at 15.7 cM and 123 cM, were significant with R^2 values in the range of 2.7–6.1 and 3.0–7.1%, respectively. Several other SNP markers that were mapped close to *IW1280* and *IWB28837* were also significant at nominal probability of association ($P < 0.005$ across

multiple locations and growing seasons), but unable to pass the stringent criteria based on FDR-adjusted $P < 0.1$.

On the long arm of chromosome 6A, SNP *IWB69846* (mapped at 136.9 cM) also showed strong association with *Pst* IT and DS. The amount of variation explained by the resistance loci associated with *IWB69846* ranged from 3.1–8.3%. Two resistance loci were identified on chromosome 7B, associated with SNP markers *IWA3506* (mapped on the short arm at 57 cM) and *IWB58601* (mapped on the long arm at 155.4 cM). The resistance locus tagged by *IWA3506* was detected as significant both in the seedling and adult plant tests with an average R^2 value of 6% at adult stage and 11.4% at seedling stage. The resistance locus tagged by *IWB58601* was effective across multiple locations and seasons at adult plant stage and showed average R^2 value of 4.8%.

Seedling resistance to *Pst* detected by GWAS among the Ethiopian wheat accessions is summarized in Table 2. Associations significant at FDR-adjusted P value <0.1 were detected for eight genomic regions on chromosomes 1B, 2A, 2B, 2D, 5B and 7B. The phenotypic variation explained by each of the eight regions ranged from 9.4 to 13.8%. Among these, *IWB9661* (mapped to chromosome 1B at 74.4 cM), *IWB6584* and *IWB57199* (mapped to chromosome 2A at 26 cM) and *IWA3506* (mapped to chromosome 7B at 56.9 cM) were commonly detected in the seedling and adult plant tests. On chromosome 2B, SNP *IWB43797* (located at 134.5 cM) was significantly associated with seedling resistance to PSTv-41 and PSTv-106. On chromosome 2D, SNP *IWB10441* showed strong association with resistance to PSTv-106. Two FDR $P < 0.1$ significant genomic regions were detected on chromosome 5B for seedling resistance to *Pst*, which were represented by SNPs *IWB14332* (mapped at 29.1 cM) and

Table 2 Chromosomal location, probability of association (P) and R^2 values of SNPs markers representing genomic regions significantly associated (FDR adjusted $P < 0.1$) with stripe rust infection type (IT) and disease severity (DS) from field experiments and seedling reactions against five races of the stripe rust pathogen in the Ethiopian bread wheat accessions

QTL-tagging SNP index[a]	SNP[b]	Associated SNP index[c]	Chr.	Position (cM)[d]	RAF (%)	P values (-log)[e]	R^2 (%)	Traits/Pst races[f]	Previously mapped Yr genes/QTL[g]
Field experiments									
IWB11553	T/G	IWB63258, IWB42176	1B	64.9	8.4	2.6-6.3	3.1-9.6	PLM_IT_12, MTV_DS_13, MTV_IT_13, MTV_IT_14	Yr15, Yr24, YrAlp, Yr64, YrCH52, Yr65, QYr.caas-1BL.1RS_SHA3/CBRD
IWB9661	A/G	IWB20091	1B	74.4	7.0	2.7-4.6	3.4-6.7	PSTv-37, PLM_IT_12	QYr.cim-1BS_Pastor
IWB11136	T/G	IWB22615, IWB71860	2A	9.4	6.7	2.5-5.2	3.5-8.3	ETH_IT_12, ETH_IT_14, ETH_DS_14, PLM_IT_12, PLM_DS_12	Yr17, Yr56, QYr.tam-2AS_TAM111, QYr.uga-2AS_26R61, QYr.ufs-2AS_Cappelle Desprez
IWB6584	A/G	IWB42693	2A	26.0	10.6	2.4-7.3	3.1-11.2	PSTv-41, ETH_IT_12, ETH_IT_14, PLM_IT_12, PLM_DS_12	QYrva.vt-2AS_VA00W-38, QYr.inra-2AL_CampRemy
IWB57199	T/C	IWB63394	2A	47.2	9.5	2.6-6.8	3.4-9.5	PSTv-41, ETH_IT_12, ETH_IT_14, ETH_DS_14, PLM_IT_12, PLM_DS_12	QYrva.vt-2AS_VA00W-38, QYr.inra-2AL_CampRemy
IWB24187	T/C	–	4A	114.5	19.0	2.3-4.3	2.8-6.1	MTV_DS_12, MTV_IT_14	Yr51, QYr-4A_Sachem, QYrst.orr-4AL_Stephens
IWA1280	A/G	IWA7226, IWA8119, IWB11086, IWB58658	5A	15.7	17.7	2.1-4.6	2.7-6.1	PLM_DS_12, PLM_IT_14, ETH_IT_14	–
IWB28837	T/C	IWB79251	5A	89.6	44.7	2.6-4.5	3.0-7.1	MTV_DS_12, MTV_IT_13	–
IWB69846	A/G	–	6A	136.9	10.6	2.4-5.5	3.1-8.3	PLM_DS_12, PLM_IT_12	QYr-6A_Avocet, QYr-6A_Saar, QYr.ufs-6A_Kariega, YrLM168, QYr.cim-6AL_Francolin
IWA3506	A/G	IWB33903, IWB39783	7B	56.9	64.8	2.6-5.2	3.1-8.1	PSTv-41, PLM_DS_14, MTV_IT_14, PLM_IT_12	–
IWB58601	A/G	–	7B	155.4	5.0	2.3-5.3	3.1-8.4	PLM_IT_14, MTV_DS_14	Yr59, YrC591, Yr52, Yr67, YrZH84, QYr7BL_Strongfield, QYr.cim-7BL_Pastor
Seedling resistance									
IWB9661	A/G	IWB20091	1B	74.4	7.4	5.9	13.2	PSTv-37, PLM_IT_12	QYr.cim-1BS_Pastor
IWB6584	A/G	IWB42693	2A	26.0	10.4	4.4	9.4	PSTv-41, ETH_IT_12, ETH_IT_14, ETH_DS_14, PLM_IT_12, PLM_DS_12	Yr17, Yr56, QYr.tam-2AS_TAM111, QYr.uga-2AS_26R61, QYr.ufs-2AS_Cappelle Desprez
IWB57199	T/C	IWB57200	2A	47.2	11.0	4.9	10.6	PSTv-41, ETH_IT_12, ETH_IT_14, ETH_DS_14, PLM_IT_12, PLM_DS_12	QYrva.vt-2AS_VA00W-38, QYr.inra-2AL_CampRemy
IWB43797	T/C	IWB29589, IWB81562, IWB32481	2B	134.5	33.5	6.0	13.7	PSTv-41, PSTv-106	Yr43

Table 2 Chromosomal location, probability of association (P) and R[2] values of SNPs markers representing genomic regions significantly associated (FDR adjusted P < 0.1) with stripe rust infection type (IT) and disease severity (DS) from field experiments and seedling reactions against five races of the stripe rust pathogen in the Ethiopian bread wheat accessions *(Continued)*

IWB10441	T/**G**	–	2D	82.8	12.5	6.6	11.9	PSTv-106	QPst.jic-2D_Guardian
IWB14332	T/**C**	–	5B	29.1	45.4	4.6	9.7	PSTv-40	Yr47, QYr.uga-5B_AGS2000
IWB48375	**T**/C	IWB81061, IWB78446	5B	71.9	38.4	6.2	11.7	PSTv-14	YrEXP2, QYr.caas-5BL.1_Libellula, QYr-5B_Oligoculm
IWA3506	A/**G**	IWB33903, IWB39783	7B	56.9	64.8	2.6–5.2	3.1–8.1	PSTv-41, PLM_DS_14, MTV_IT_14, PLM_IT_12	–

[a]SNP index from the wheat 90 K iSelect assay. Within the confidence interval of the QTL, marker with the most significant associations were used as the QTL-tagging marker

[b]Underline indicates favorable allele

[c]Other significant SNPs identified within the confidence interval of the QTL other than the QTL-tagging marker

[d]Scaled position from hexaploid wheat consensus map (Wang et al. 2014)

[e]Marker-wise P value ranges - all of the 11 QTL were significant at FDR P value <0.1 in at least one environment

[f]Stripe rust IT and severity from the field experiments and response to Pst races to which the marker showed significant association at FDR-adjusted probabilities

[g]References given in the text, RAF = Resistance allele frequency, Chr. = chromosome

PLM_IT_12 = Pullman, infection types, 2012
PLM_IT_14 = Pullman, infection types, 2014
PLM_DS_12 = Pullman, disease severity, 2012
PLM_DS_14 = Pullman, disease severity, 2014
MTV_IT_12 = Mount Vernon, infection types, 2012
MTV_IT_13 = Mount Vernon, infection types, 2013
MTV_IT_14 = Mount Vernon, infection types, 2014
MTV_DS_12 = Mount Vernon, disease severity, 2012
MTV_DS_13 = Mount Vernon, disease severity, 2013
MTV_DS_14 = Mount Vernon, disease severity, 2014
ETH_IT_12 = Ethiopia, infection types, 2012
ETH_IT_14 = Ethiopia, infection types, 2014
ETH_DS_14 = Ethiopia, disease severity, 2014

IWB48375 (mapped at 72 cM). The LD between the two genomic regions was non-significant ($r^2 < 0.001$), indicating that they represent two different resistance loci.

Mapping of seedling resistance to Pgt

GWAS identified nine genomic regions significantly associated with seedling response to the four tested races of *Pgt* (Table 3). Among these, a resistance locus mapped to the short arm of chromosome 2B was significantly associated with response to race TTKSK. This locus was represented by SNP *IWB26389*, which was mapped at 99.9 cM on 2BS. *IWB26389* explained 8.0% of the phenotypic variance in the germplasm panel. One additional genomic region on 7B, *IWB64169* at 171.1 cM, was significantly associated with race TTKSK. This genomic region accounted for about 2.4% of the total variation in IT response against TTKSK. GWAS of seedling resistance to race TTTTF identified a single genomic region (represented by *IWB34733*) that was mapped to the long arm of chromosome 4A at 144.4 cM. The phenotypic variance explained by *IWB34733* was 15%. The locus tagged by *IWB34733* was also effective against race TKTTF. Five genomic regions were significantly associated with seedling resistance to race TRTTF. These putative resistance loci were mapped to chromosomes 1B, 2A, 3B, 6A and 7B. The five genomic regions were tagged by *IWB39306* (chromosome 1B at 60.6 cM), *IWB12320* (chromosome 2A at 106.3 cM), *IWB19479* (chromosome 3B at 136.3 cM), *IWA5416* (chromosome 6A at 5.6 cM) and *IWB2378* (chromosome 7B at 92.5 cM). The phenotypic variance explained by the genomic regions significant for race TRTTF ranged 5.1% to 6.4%. On chromosome 6B, a

genomic region tagged by SNP marker *IWB14375* showed highly significant association with response to race TKTTF. This genomic region was responsible for 9% of the phenotypic variation. *IWB14375* was significantly associated also with resistance to races TTTTF and TRTTF.

Validation of previously mapped rust resistance genes

Genotyping of the accessions with molecular markers linked to previously mapped stripe rust and stem rust resistance genes/QTL revealed the presence of *Lr34/ Yr18/Sr57*, *Yr17/Sr38*, *Sr31/Yr9/Pm8/Lr26*, *Sr2/Yr30*, *Sr36* and *Sr24* at a frequency of 12.0%, 6.2%, 9.0%, 13%, 30% and 2.6%, respectively (Additional files 3 and 6). A single accession (Alidoro) was positive for the resistance-associated alleles of both *Lr34/Yr18/Sr57* and *Yr17/Sr38*, while four lines (Sofumer, Tusie, Bobicho and AMNA-4) carry both *Lr34/Yr18/Sr57* and *Sr31/Yr9/Lr26*. Three accessions (Sofumer, Bonny, and 231,658) carried both *Lr34/Yr18/Sr57* and *Yr30/Sr2*, while only one accession (KBG-01) was positive for the resistance-associated alleles of both *Yr17/Sr38* and *Yr30/Sr2*. Four accessions identified as landraces (231,658, 232,034 and 232,014) carried the resistance alleles of both *Yr30/Sr2* and *Sr36*. Accession 231,658 carried *Lr34/Yr18/Sr57*, *Yr30/Sr2* and *Sr36*. None of the accessions were positive for the *Lr67/Yr46* gene. Association analysis revealed a significant association of the *Lr34/Yr18* KASP marker wMAS000003 with field resistance to stripe rust at marker wise *P* value <0.005 in at least half of the test environments. However, no significant SNPs were detected on this region of chromosome

Table 3 Genomic regions significantly associated with seedling responses to the four races of *Puccinia graminis* f. sp. *tritici* in the Ethiopian bread wheat accessions

QTL-tagging SNP index	SNP	Associated SNPs index	Chr.	Position (cM)	MAF	P value	R^2 (%)	FDR P values
TRTTF								
IWB39306	**A**/G	–	1B	60.6	0.1	1.30E-04	5.08	*
IWB12320	**T**/C	–	2A	106.3	0.15	3.40E-05	5.98	*
IWB19479	**T**/G	IWB1595	3B	136.3	0.07	7.20E-05	5.47	*
IWA5416	**T**/C	IWB11315, IWB22036, IWB60233, IWB67415	6A	5.6	0.08	2.00E-05	6.35	*
IWB2378	T/**C**	IWB4726, IWB4725, IWB7743, IWB65791	7B	92.5	0.19	2.00E-05	6.35	*
TTKSK								
IWB26389	**A**/G	IWB42742, IWB1518, IWB24614, IWB32327	2B	99.9	0.31	2.40E-10	8.02	****
IWB64169	**A**/G	IWB63221	7B	171.1	0.13	3.00E-04	2.41	**
TTTTF								
IWB34733	A/**G**	IWB63979, IWB61381, IWB59018	4A	144.4	0.08	8.90E-10	14.62	****
TKTTF								
IWB14375	T/**G**	IWB59006, IWB72471, IWB59005	6B	113.7	0.096	3.60E-07	0.09	**

****= <0.0001
***< 0.001
**< 0.05
*< 0.1

7D and none of the SNPs on the 7D chromosome were in significant LD with the KASP marker. Significant associations were also detected for *Yr17/Lr37/Sr38* KASP marker together with several other SNP markers including *IWB11136, IWB6584, IWB42693, IWB30196, IWB57199* and *IWB63394* that were in LD with the *Yr17* marker *Ventriup-LN2*.

Although there was no significant MTA detected for *Yr9* KASP marker 1RS.1BL_6110 in association with *Pst* resistance, the associated *Sr* gene (*Sr31*) showed an effect of reducing IT responses to races TRTTF, TTTTF and TKTTF by 44%, 50% and 45%, respectively. As expected, accessions postulated to carry the resistance allele of *Sr31* were highly susceptible to TTKSK. The resistance-associated allele of the marker for *Sr24* was carried by cultivars Huluka, Shorima, Hoggana, Millennium, and Kulkullu in this panel, and showed highly resistant responses to all races of the stem rust pathogen. Accessions carrying *Sr36* showed highly effective responses to TTKSK, but were mostly susceptible to TKTTK and TTTTF.

Discussion

Phenotypic variability and molecular diversity of the Ethiopian bread wheat accessions

A better understanding of the extent of genetic diversity and genetic basis of the responses to rust diseases in wheat may offer the prospect of effective exploitation of genetic resources and enhanced breeding for durable rust resistance in wheat. To this end, the current study surveyed Ethiopian bread wheat landraces and cultivars for resistance to *Pst* and *Pgt* under diverse environments and variable pathogen populations. Our data revealed that the accessions, most notably the landraces, possess considerable variation for resistance to both diseases. The sources of resistance identified with high levels of combined resistance to stripe rust and stem rust may be used as parental breeding lines for improving wheat resistance to these diseases. Profiling of the accessions with high-density SNP markers facilitated analyses of genome-specific molecular diversity and phylogeny between the landraces and elite cultivars as well as a genome-wide survey of marker-trait associations. It was evident that the landrace accessions were characterized by high levels of molecular diversity as measured by Nei's gene diversity index, PIC values, NJ clustering algorithms and kinship analyses. By contrast, the Ethiopian bread wheat elite cultivars show high genetic similarity, with some of them nearly genetically redundant. Some of these nearly genetically redundant cultivars were released by different breeding programs and are widely grown across Ethiopia, which demonstrates one of the likely reasons for the frequent failures of deployed resistance genes and highlights the continued risk of disease epidemics due to minimal diversity. The loss of genetic diversity in modern

wheat varieties compared to landraces has been reported by previous studies [19, 20, 53]. Such erosion of genetic variation in the elite wheat varieties may constrain the capacity to counter threats from changing pathogen populations. A particular concern in this regard is the fact that the highlands of Ethiopia are 'hot spot' areas for the *Triticum-Puccinia* pathosystem [16, 54, 55] and therefore the presence of high inoculum pressure and rapid changes in pathogen virulence. Cultivation of genetically uniform wheat cultivars and the practice of frequent double cropping in some wheat growing areas of the country may result in rapid selection for virulent races of the pathogen. The results of our analyses of genetic diversity in Ethiopian bread wheat cultivars highlight the continued risk of wheat rust epidemics due to the genetic vulnerability of the current wheat varieties and suggest the utilization of more genetic diversity to broaden the genetic base of elite wheat cultivars. We also believe that Ethiopian landraces are a unique source of unexploited diversity for wheat improvement [27].

MTAs for Pst resistance and co-localization of the putative QTL with previously identified Yr genes/QTL

The mixed model analyses of association between the high-density genome-wide SNP markers and response to *Pst* in the Ethiopian bread wheat accessions highlighted several chromosome regions harboring putative resistance loci. Eleven FDR $P < 0.1$-significant regions, perhaps of large effect, were identified when employing the most stringent criteria. Four additional genomic regions were significantly associated (FDR-adjusted $P < 0.10$) with seedling resistance to stripe rust. Aiming to keep a reasonable power of statistical association and identify loci conferring intermediate effects, less stringent GWAS tests based on marker-wise significance tests were also considered. The latter approach was supported by associations detected across multiple location and seasons. Fifty-six additional *Pst* resistance loci, at a marker-wise $P < 0.005$ in at least half of the test environments, were detected. The amount of variation explained by the 67 genomic regions ranged from 54.3 to 52.1%. The reason for the partial account of the variation in the field response to stripe rust by the significant markers could be attributed to the fact that the current GWAS methods are more suited to the identification of common variants of large to intermediate effects [56]. In addition, the 90 K iSelect chip did not provide enough marker coverage in the D genome to identify all *Pst* resistance loci present in the germplasm panel. It was evident that none of the 90 K SNP markers mapped to the short arm of chromosome 7D were linked to the marker for known *Pst* resistance gene *Yr18* (*Lr34/Yr18/Pm38*) included as a control in a separate

GWAS that showed experiment-wise significant associations with *Pst* resistance at multiple locations.

The percent of variation explained by the 56 MTAs from the field experiment support that many are unlikely false-positives. Nonetheless, the details of only the 11 highest-confidence resistance-associated MTAs from the field experiment and four MTAs from the greenhouse seedling resistance screening are further discussed.

To determine whether any known resistance genes coincided with the 15 putative resistance regions identified in this study, the current results were compared with the integrated genetic maps of *Pst* resistance genes and SNP markers constructed by Maccaferri et al. [24] and Bulli et al. [26]. The genomic regions harboring eight of the eleven loci identified from our field studies of stripe rust resistance overlap with positions of previously reported QTL. Among the three potentially novel resistance loci, *IWB28837* and *IWA1280* were associated with resistance on chromosomes 5A at 15.7 and 89.6 cM positions, respectively. In line with their distant map positions, *IWB28837* and *IWA1280* are not in LD with each other, and thus tag distinct genomic regions. The confidence intervals of MTAs tagged by these two SNPs do not overlap with positions of any of the previously reported QTL or *Yr* genes. Thus, *IWA1280* and *IWB28837* likely represent novel resistance loci for *Pst*. On the short arm of 7B, *IWA3506* (at 57.0 cM) was significant in both seedling and adult plant tests. *Yr63* has been mapped to the short arm of 7B chromosome (McIntosh et al., 2013). Based on the relative genetic position of *Yr63* in the integrated genetic map constructed by [24], *IWA3506* was mapped at an average genetic distance of 30.0 cM from *Yr63*, which indicates that *IWA3506* is likely different from *Yr63*, and thus represents a newly documented stripe rust resistance locus.

The haplotype blocks of significant stripe rust resistance loci detected on chromosome 1B spanning a segment of 63–74 cM likely harbor more than one resistance locus. Figure 7a illustrates LD relationships between the two loci on 1BS (*IWB11553* at 64.9 cM and *IWB9661* at 74.4 cM), which were all less than $r^2 = 0.03$. This indicates that *IWB11553* and *IWB9661* likely represent distinct genomic regions. Chromosome 1B is rich with *Yr* genes/QTL, and a number of *Yr* genes have been mapped to the proximity of the genomic regions tagged by *IWB9314*, *IWB11553* and *IWB9661*, including *Yr15*, *Yr24/Yr26/YrCh42*, *YrAlp*, *YrCH52*, *Yr52*, *Yr64* and *Yr65*. Other *Pst* resistance QTL were also previously mapped in this region of chromosome 1B, including *QYr.caas-1BL.1RS_SHA3/CBRD* and *QYr.cim-1BS_Pastor* [57–61]. Additional research is needed to fully characterize chromosome 1B resistance.

We detected significant associations between stripe rust resistance and SNP markers mapped close to the translocation segment harboring *Yr17* including the *Yr17* marker *Ventriup-LN2*. Since virulence to *Yr17* is widespread [29, 30], the significant association of *Yr17* and other SNP markers may indicate the presence of another effective resistance gene tightly linked to *Yr17*. Among the three significant SNPs detected, *IWB11136* (located at 9.4 cM) was in perfect LD ($r^2 = 1.0$) with the *Yr17* KASP marker. *IWB6584* and *IWB57199* were mapped at the 26 and 47 cM positions, respectively and in LD with the *Yr17* KASP marker with r^2 value of 0.64 and 0.68, respectively (Fig. 7b). Hence, *IWB6584* and *IWB57199* may not be located on the 2NS.2AS translocation segment carrying *Yr17*, but tightly linked and likely represent a gene responsible for the high level of resistance exhibited by lines positive for the favorable alleles of the SNPs. Several other temporarily designated QTL have been identified on the short arm of chromosome 2A, including *QYr.sun-2A_Wollaroi* [62], *QYr.tam-2AS_TAM111* [63], *QYrva.vt-2AS_VA00W-38* [64], *QYr.-inra_2A1_Recital* [65] and *QYr.ucw-2A_PI610750* [66]. *IWB6584* and *IWB57199* could be related to these QTL. Tan et al. (2013) reported mapping of a QTL on 2AS encoding 12-oxo-phytodienoic acid reductase (OPR) that is collinear to an *Yr17*-containing-fragment of the chromosome arm. The genes encoding OPR are known to play critical roles in insect and disease resistance pathways in higher plants [67, 68]. Further genetic analysis is required to determine the relationship of MTAs detected by *IWB6584* and *IWB57199* with previously mapped genes/QTL.

On chromosome 6A, an MTA associated with field resistance to *Pst* was identified at FDR-adjusted probability *(P)* <0.1 at multiple locations and seasons. This locus was tagged by the SNP *IWB69846* at the map position of 136.9 cM. Previously, *YrLM168* [69] and other temporarily designated QTL [*QYr-6A_Avocet* [70], *QYr-6A_Saar* [71], *QYr.ufs-6A_Kariega* [72] and *QYr.cim-6AL_Francolin* [73] have been mapped to the vicinity of *IWB69846*, and could be related to the locus tagged by *IWB69846*. On chromosome 4A, *IWB24187* represents an MTA mapped closely to *Yr51* [74] and two other QTL, *QYr-4A_Sachem* [75] and *QYrst.orr-4AL_Stephens* [76]. Further genetic analysis will be required to determine the relationship between the resistance locus tagged by *IWB24187* and the previously mapped QTL/genes.

On the long arm of chromosome 7B, SNP marker *IWB58601* was associated with *Pst* resistance across multiple field tests. The confidence interval of the genomic region tagged by *IWB58601* overlaps with the map positions of *Yr59*, *YrC591*, *Yr52*, *Yr67*, *YrZH84* and QTL, *QYr.cim-7BL_Pastor*, *QYr7BL_Strongfield* [58, 77, 78]. *Yr52* and *Yr59* confer high-temperature adult-plant (HTAP) resistance to stripe rust, while *YrC591*, *Yr67* and *YrZH84* are dominant genes conferring seedling resistance

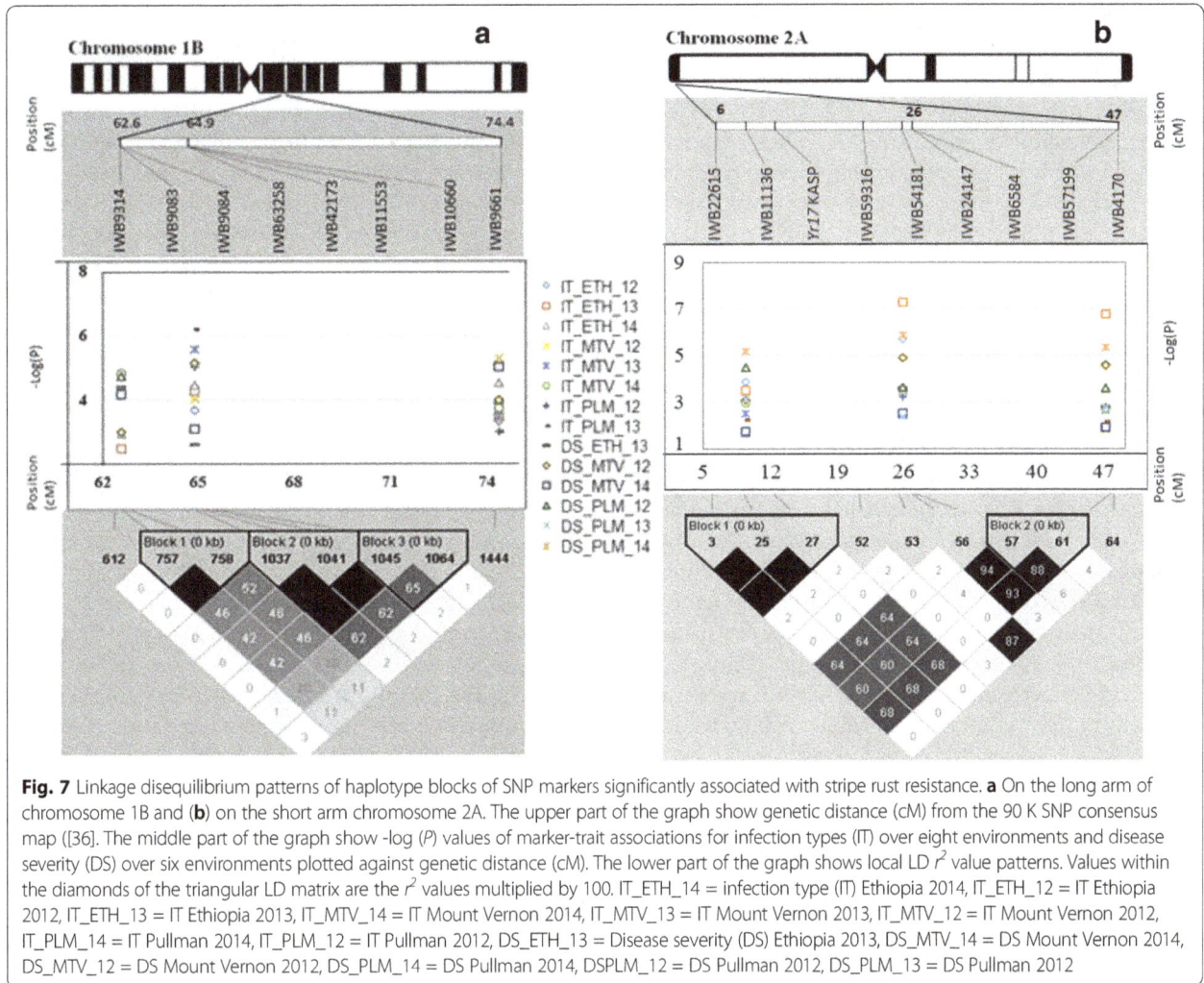

Fig. 7 Linkage disequilibrium patterns of haplotype blocks of SNP markers significantly associated with stripe rust resistance. **a** On the long arm of chromosome 1B and (**b**) on the short arm chromosome 2A. The upper part of the graph show genetic distance (cM) from the 90 K SNP consensus map ([36]. The middle part of the graph show -log (*P*) values of marker-trait associations for infection types (IT) over eight environments and disease severity (DS) over six environments plotted against genetic distance (cM). The lower part of the graph shows local LD *r*² value patterns. Values within the diamonds of the triangular LD matrix are the *r*² values multiplied by 100. IT_ETH_14 = infection type (IT) Ethiopia 2014, IT_ETH_12 = IT Ethiopia 2012, IT_ETH_13 = IT Ethiopia 2013, IT_MTV_14 = IT Mount Vernon 2014, IT_MTV_13 = IT Mount Vernon 2013, IT_MTV_12 = IT Mount Vernon 2012, IT_PLM_14 = IT Pullman 2014, IT_PLM_12 = IT Pullman 2012, DS_ETH_13 = Disease severity (DS) Ethiopia 2013, DS_MTV_14 = DS Mount Vernon 2014, DS_MTV_12 = DS Mount Vernon 2012, DS_PLM_14 = DS Pullman 2014, DSPLM_12 = DS Pullman 2012, DS_PLM_13 = DS Pullman 2012

to stripe rust. It has been reported that *IWB37096* and *IWB71995*, both mapped at 148.7 cM on 7BL [36], are flanking *Yr67* with a respective genetic distance of 1.1 and 0.6 cM [79]. In the present study, these two SNPs (*IWB37096* and *IWB71995*) were not among the 67 significant MTAs and in non-significant LD with *IWB58601*, suggesting that the QTL tagged by *IWB58601* may be different from those of *IWB37096* and *IWB71995* as well as *Yr63*. However, since these SNPs and the previously mapped resistance genes are located close to each other, further characterization and genetic analysis is required to determine whether they are distinct from those previously mapped resistance genes.

MTAs for Pgt resistance and co-localization of the resistance loci with previously identified *Sr* genes/QTL

Five genomic regions were significantly associated with seedling resistance to race TKTTF. Among these, *IWB39306* was mapped close to the segment of 1BS that harbors the 1BL.1RS rye translocation carrying

Sr31 [80]. *Sr31* is ineffective to TTKSK, but confers effective resistance to TRTTF and TKTTF [16, 33]. To determine if *IWB39306* is related to *Sr31*, we genotyped the accessions with a KASP marker for 1BL.1RS and performed LD analysis. *IWB39306* and the KASP marker for 1BL.1RS were in complete LD, indicating that *IWB39306* is detecting allelic variation based on the presence or absence of 1BL.1RS. The accessions carrying the resistance-associated allele of *IWB39306* are all elite Ethiopian cultivars. Working on the North American spring wheat breeding germplasm, Bajgain et al. [81] identified 13 significant SNPs on the short arm of chromosome 1B (position range 44–65 cM) that also showed various levels of significant associations with resistance to race TRTTF in our study, but did not pass the FDR threshold of <0.1. We found strong LD between the SNP markers identified by Bajgain et al. [81], *Sr31* and *IWB39306* (*r*² ranging from 0.41 to 1.00). This further clarifies the uncertainty whether the 13 significant markers are linked to *Sr31*, or to a novel gene of resistance to TRTTF.

The SNP locus *IWB12320* on chromosome 2A at 106.3 cM was significantly associated with response to race TRTTF. Eighteen of the 25 accessions carrying the resistance-associated allele of *IWB12320* are landraces. Stem rust resistance gene *Sr21* was mapped approximately 50 cM from the centromere on the long arm of 2A [82, 83]. *Sr21* confers resistance to several races of *Pgt*, including those in the Ug99 group. However, *Sr21* is derived from *T. monococcum* and was later transferred to hexaploid wheat [82]. Hence, *Sr21* is not common in landrace accessions, and suggests that *IWB12320* is tagging a resistance locus or allele distinct from *Sr21*. Stem rust resistance genes *Sr48* and *SrTm4* have also been previously mapped to the long arm of chromosome 2A [84, 85]. However, *IWB12320* may not be related to *SrTm4*, as the latter was identified from the diploid wheat, *Triticum monococcum*, and is not common in hexaploid wheat. *Sr48* (first designated as *SrAnl*) is a seedling resistance gene that was mapped to chromosome arm 2AL of the bread wheat cultivar Arina. Based on the genetic map position of *Sr48* (~55 cM from the centromere), it is likely that the genomic region tagged by *IWB12320* corresponds to *Sr48*. Allelism tests will be required to test the relationships between *IWB12320* and *Sr48*.

On 3BL, SNP *IWB19479* (mapped at 136.3 cM) identified a genomic region significantly associated with the response to race TRTTF. This locus may represent a new resistance gene since no race TRTTF-effective seedling resistance genes have been mapped to chromosome 3B so far. Similarly, no TTKSK-effective seedling resistance genes have been identified on chromosome 7B [86]. Therefore, the significant genomic region detected on chromosome 7B in the present study (*IWB2378* mapped at 92.5 cM) is likely tagging a new resistance gene effective to TTKSK. On the short arm of chromosome 6A, SNP *IWA5416* (mapped at 5.6 cM) tagged a genomic region significantly associated with resistance to TRTTF. This 6AS locus is most likely linked to the gene *Sr8a*, which is effective to race TRTTF [87, 31]. Bajgain et al. [81] identified a significant association of *IWA5416* and 56 other SNP markers with race TRTTF, validating the association of this genomic region with resistance to *Pgt*.

Analysis of response to TTTTF identified a significant genomic region that was mapped to long arm of chromosome 4A. This locus was identified by SNP *IWB34733*, mapped at 144.4 cM, and was also effective to TTKSK and TRTTF, but did not pass the FDR threshold of significant associations for these races. It is likely that *IWB34733* is related to *Sr7a*. Bajgain et al. [81] also identified 51 significant SNPs mapped on 4AL, spanning 142–164 cM, in North American spring wheat breeding germplasm, one of which was *IWB34733*.

A single genomic region was detected that was significantly associated with race TKTTF, a newly detected Ethiopian *Pgt* race that defeated resistance in the popular cultivar 'Digalu' [16]. SNP marker *IWB14375*, mapped at 113.7 cM on 6B, identified this genomic region. This region of chromosome 6B is known to harbor *Sr11*, which is effective against TKTTF [86]. Bajgain et al. [81] also detected significant association of *IWB14375* with response to race TKTTF in North American spring wheat germplasm. Nirmala et al. [88] identified seven SNP markers linked to *Sr11*. Among the seven SNP markers, *IWB59006* and *IWB72471* also showed highly significant association with the response to TKTTF in the present study. Both *IWB59006* and *IWB72471* were in strong LD (r^2 = 0.8–1.0) with *IWB14375*, indicating that the locus tagged by *IWB14375* corresponds to *Sr11*.

A genomic region on the short arm of chromosome 2B, tagged by *IWB26389*, was significantly associated with IT response to race TTKSK. Although the short arms of 2A, 2B and 2D are known to carry *Sr32* [79], it is unlikely that the three markers represent *Sr32* as it is not common in landrace accessions due to its origin from *Ae. speltoides* and relatively recent translocations to hexaploid wheat. The stem rust resistance gene *Sr36* derived from *Triticum timopheevi* is also located on 2BS and confers resistance against TTKSK [89]. To determine if *IWB26389* is associated with *Sr36*, we determined LD between the KASP marker for *Sr36* (wMAS000015) and *IWB26389*. The result showed that *IWB26389* and *Sr36* KASP marker are in complete LD, which indicates that *IWB26389* is also tagging the *T. timopheevi* segment that harbors *Sr36*.

We identified a genomic region on 7BL, tagged by *IWB64169* at 171 cM, significantly associated with response to race TTKSK. Although there are several QTL effective to Ug99 on chromosome 7B, no effective seedling resistance gene has been identified so far [86]. Therefore, the significant genomic region detected on 7BL in the present study may be a new resistance gene effective to TTKSK. Based on LD analysis, *IWB64169* represents a different genomic region from *IWB2378*, which was significant to race TRTTF.

Conclusion

The limited pool of alleles present in the Ethiopian bread wheat cultivars may offer narrow perspective for germplasm improvement and consequently limit the capacity to deal with threats from wheat rust epidemics. Notwithstanding, enough genetic variation resides within the landrace accessions that can be utilized to broaden the genetic base of rust resistance in wheat breeding germplasm. In addition to characterizing the genetic diversity of the Ethiopian bread wheat cultivars and landraces, this study highlights the potential of GWAS to accelerate the translation of landrace germplasm diversity toward applied wheat improvement. Several newly documented resistance loci were discovered from Ethiopian landraces, along with detection of previously reported resistance genes. The molecular markers linked to resistance loci identified in the current study can be used to efficiently select for resistance to diversify the genetic basis of elite cultivars.

Additional files

Additional file 1: Virulence/avirulence formula of *Puccinia striiformis* f. sp. *tritici* isolates used for seedling resistance screening of the Ethiopian wheat accessions.

Additional file 2: Pearson's correlation coefficients among stripe rust infection type (IT) and disease severity (DS) in each location and season.

Additional file 3: Infection types and disease severity observed on Ethiopian wheat cultivars and landraces tested with races of *Puccinia striiformis* f. sp. *tritici* and *Puccinia graminis* f. sp. *tritici* at seedling stage and evaluated under field conditions. Allele report for markers linked to previously known stripe rust and stem rust resistance genes and SNP markers significantly associated with the resistance to both diseases (identified in this study) are summarized.

Additional file 4: Comparison of different association test models using Quantile-Quantile (Q-Q) plots using BLUP values of stripe rut IT and DS across all environments. The Q-Q plot determines the magnitude of the deviation of the observed association between the markers and response to stripe rust from the expected null hypothesis of no association. Under the assumption of no association between the SNPs and the traits, a large inflation of observed P values from the expected P values indicates spurious associations. Only few true SNP-trait associations are expected to deviate from the null hypothesis.

Additional file 5: Genomic regions significantly associated with field resistance of the Ethiopian bread wheat accessions to *Puccinia striiformis* f. sp. *tritici* at marker wise $P < 0.005$ in at least four of the eight environments.

Additional file 6: Frequency of favorable alleles of the molecular markers linked to previously mapped stripe rust and stem rust resistance genes/QTL in the Ethiopian wheat accessions.

Abbreviations

APR: Adult plant resistance; CMLM: Compressed mixed linear model; DS: Disease severity; GLM: General linear model; GWAS: Genome-wide association study; IT: Infection types; LD: Linkage disequilibrium; MAF: Minor allele frequency; MLM: Mixed linear model; MTA: Marker-trait association; PC: Principal component; PCA: Principal component analysis; Pgt: Puccinia graminis f. sp. tritici; PNW: Pacific Northwest; Pst: Puccinia striiformis f. sp. tritici; Q-Q: quantile-quantile; QTL: Quantitative trait locus; SNP: Single nucleotide polymorphism

Acknowledgments

We would thank our colleagues at the Ethiopian Institute of Agricultural Research (EIAR), Kulumsa Agricultural Research Center (KARC) and the International Maize and Wheat Improvement Center (CIMMYT), Ethiopia Office for providing the necessary resources and technical support during the field experiments in Ethiopia. We also thank John Kuehner, Victor DeMacon and Dr. Kent Evans for providing technical support in the field and greenhouse.

Competing of interests

The authors declare that they have no competing interests.

Funding

This project was partly funded by Monsanto Beachell-Borlaug International Scholars Program, by the National Research Initiative Competitive Grant 2011–68,002-30,029 (Triticeae-CAP) from the United States Department of Agriculture National Institute of Food and Agriculture, and by the Durable Rust Resistance in Wheat Project with funds through Cornell University originating from the Bill and Melinda Gates Foundation and the Department for International Development of the United Kingdom.

Authors' contributions

KTM MP and XC conceived and designed the experiments. KTM MP MR SR XC BGB performed the experiments. KTM carried out the statistical analysis, interpretation of the results, and drafted the manuscript. MP MR XC BGB revised the paper. All authors read and approved the final manuscript.

Author details

[1]Department of Crop and Soil Sciences, Washington State University, Pullman, WA 99164-6420, USA. [2]USDA-ARS Cereal Disease Laboratory, Department of Plant Pathology, University of Minnesota, St. Paul, MN 55108, USA. [3]USDA-ARS, Wheat Health, Genetics, and Quality Research Unit, and Department of Plant Pathology, Washington State University, Pullman, WA 99164-6430, Pullman, WA 99164-6430, USA. [4]Ethiopian Institute of Agricultural Research, Kulumsa Agricultural Research Center, P. O. Box 489, Assela, Ethiopia.

References

1. Chen XM. Epidemiology and control of stripe rust [*Puccinia striiformis* f. Sp. *tritici*] on wheat. Can. J. Bot. 2005;27:314–37.
2. Kolmer JA. Tracking wheat rust on a continental scale. Curr Opin Plant Biol. 2005;8:441–9.
3. Singh RP, Huerta-espino J, William HM. Genetics and breeding for durable resistance to leaf and stripe rusts in wheat. Turk J Agric. 2005;29:121–7.
4. Hovmøller MS, Yahyaoui AH, Milus EA, Justesen AF. Rapid global spread of two aggressive strains of a wheat rust fungus. Mol Ecol. 2008;17:3818–26.
5. Milus EA, Kristensen K, Hovmøller MS. Evidence for increased aggressiveness in a recent widespread strain of *Puccinia striiformis* f sp *tritici* causing stripe rust of wheat. Phytopathology. 2009;99:89–94.
6. Solh M, Nazari K, Tadesse W, Wellings CR. The growing threat of stripe rust worldwide. Borlaug Glob. Rust Initiat. Tech. Work. 1–4 Sept. 2012, Beijing, China. 2012;1–10.
7. Global Rust Reference Center. http://wheatrust.org/. Accessed 15 July 2016.
8. Mcintosh RA, Brown GN. Anticipatory breeding for resistance to rust diseases in wheat. Annu Rev Phytopathol. 1997;35:311–26.
9. Singh RP, Hodson DP, Jin Y, Lagudah ES, Ayliffe MA, Bhavani S, et al. Emergence and spread of new races of wheat stem rust fungus : continued threat to food security and prospects of genetic control. Phytopathology. 2015;105:872–84.
10. Pretorius ZA, Singh RP, Wagoire WW, Payne TS. Detection of virulence to wheat stem rust resistance gene Sr31 in *Puccinia graminis*. f. Sp. *tritici* in Uganda. Plant Dis. 2000;84:203.

11. Jin Y, Szabo LJ, States U, Agricultural A, Disease C. Detection of virulence to resistance gene Sr24 within race TTKS of *Puccinia graminis* f. Sp. *tritici*. Plant Dis. 2008;92:923–6.

12. Jin Y, Szabo LJ, States U, Agricultural A, Disease C, Paul S, et al. Detection of virulence to resistance gene Sr36 within the TTKS race lineage of *Puccinia graminis* f. Sp. *tritici*. Plant Dis. 2009;93:367–70.

13. Rouse MN, Nirmala J, Pretorius ZA, Hiebert CW. Characterization of *Sr9h*, a wheat stem rust resistance allele effective to Ug99. Theor Appl Genet. 2014; 127:1681–8.

14. Patpour M, Hovmøller MS, Justesen AF, Newcomb M, Olivera P, Jin Y, et al. Emergence of virulence to *SrTmp* in the Ug99 race Group of Wheat Stem Rust, *Puccinia graminis* f. Sp. *tritici*, in Africa. Plant Dis. 2016;100:522.

15. Patpour M, Hovmoller MS, Shahin AA, Newcomb M, Olivera P, Jin Y, et al. First report of the Ug99 race Group of Wheat Stem Rust, *Puccinia graminis* f. Sp. *tritici*, in Egypt in 2014. Plant Dis. 2016;100:863.

16. Olivera P, Newcomb M, Szabo LJ, Rouse M, Johnson J, Gale S, et al. Phenotypic and Genotypic Characterization of Race TKTTF of *Puccinia graminis* f. sp. *tritici* that caused a wheat stem rust epidemic in southern Ethiopia in 2013–14. Phytopathology. 2015;105:917–28.

17. Burdon JJ, Barrett LG, Rebetzke G, Thrall PH. Guiding deployment of resistance in cereals using evolutionary principles. Evol Appl. 2014;7:609–24.

18. Hulbert S, Pumphrey M. A time for more booms and fewer busts? Unraveling cereal-rust interactions. Mol Plant-Microbe Interact. 2014;27:207–14.

19. Haudry A, Cenci A, Ravel C, Bataillon T, Brunel D, Poncet C, et al. Grinding up wheat : a massive loss of nucleotide diversity since domestication. Mol Biol Evol. 2007;24:1506–17.

20. Peng JH, Sun D, Nevo E. Domestication evolution, genetics and genomics in wheat. Mol Breed. 2011;28:281–301.

21. Feuillet C, Langridge P, Waugh R. Cereal breeding takes a walk on the wild side. Trends Genet. 2007;24:24–32.

22. Vasudevan K, Vera Cruz CM, Gruissem W, Bhullar NK. Large scale germplasm screening for identification of novel rice blast resistance sources. Front Plant Sci. 2014;5:1–9.

23. Zurn JD, Newcomb M, Rouse MN, Jin Y, Chao S, Sthapit J, et al. High-density mapping of a resistance gene to Ug99 from the Iranian landrace PI 626573. Mol Breed. 2014:1–11.

24. Maccaferri M, Zhang J, Bulli P, Abate Z, Chao S, Cantu D, et al. A Genome-Wide Association Study of Resistance to Stripe Rust (*Puccinia striiformis* f. sp. *tritici*) in a Worldwide Collection of Hexaploid Spring Wheat (*Triticum aestivum* L.). G3; Genes|Genomes|Genetics. 2015;5:449–65.

25. Kertho A, Mamidi S, Bonman JM, McClean PE, Acevedo M. Genome-wide association mapping for resistance to leaf and stripe rust in winter-habit hexaploid wheat landraces. PLoS One. 2015;10:e0129580.

26. Bulli P, Zhang J, Chao S, Chen X, Pumphrey M, Health W. Genetic architecture of resistance to stripe rust in a global winter wheat germplasm collection. G3 Genes|Genomes|Genetics. 2016;6:2237–53.

27. Mengistu DK, Kidane YG, Catellani M, Frascaroli E, Fadda C, P ME. High-density molecular characterization and association mapping in Ethiopian durum wheat landraces reveals high diversity and potential for wheat breeding. Plant Biotechnol. 2016; doi:10.1111/pbi.12538.

28. Line RF, Qayoum A. Virulence, aggressiveness, evolution, and distribution of races of Puccinia striiformis (the cause of stripe rust of wheat) in North America. United States dep. Agric Tech Bull. 1968-1987;1992:44.

29. Wan A, Chen X. Virulence characterization of *Puccinia striiformis* f. Sp. *tritici* using a new set of *Yr* single-gene line differentials in the United States in 2010. Plant Dis. 2014;98:1534–42.

30. Wan A, Muleta KT, Zegeye H, Hundie B, Pumphrey M, Chen X. Virulence characterization of wheat stripe rust fungus *Puccinia striiformis* f. Sp. *tritici* in ethiopia and evaluation of ethiopian wheat germplasm for resistance to races of the pathogen from Ethiopia and the United States. Plant Dis. 2017;101:73–80.

31. Olivera PD, Jin Y, Rouse M, Badebo A, Fetch T, Singh RP, et al. Races of *Puccinia graminis* f. Sp. *tritici* with combined virulence to Sr13 and Sr9e in a field stem rust screening nursery in Ethiopia. Plant Dis. 2012;96:623–8.

32. Jin Y, Sigh RP, Ward RW, Wanyera R, Kinyua M, Njau P, et al. Characterization of seedling infection types and adult plant infection responses of monogenic *Sr* gene lines to race TTKS of *Puccinia graminis* f. Sp. *tritici*. Plant Dis. 2007;91:1096–9.

33. Rouse MN, Jin Y. Stem rust resistance in A-genome diploid relatives of wheat. Plant Dis. 2011;95:941–4.

34. Stakman EC, Steward DM, Loegering WQ. Identification of physiologic races of *Puccinia graminis var. tritici*. US Dep. Agric. Agric. Res. Serv. 1962;E-617.

35. Zhang D, Bowden R, Bai G. A method to linearize Stakman infection type ratings for statistical analysis. Proc. 2011 Borlaug Glob. Rust Initiat. Tech. Work. 17–20 June 2011, St. Paul, MN. 2011;P. 28.

36. Wang S, Wong D, Forrest K, Allen A, Chao S, Huang BE, et al. Characterization of polyploid wheat genomic diversity using a high-density 90 000 single nucleotide polymorphism array. Plant Biotechnol. 2014;12:787–96.

37. Marker Assisted Selection in Wheat (MAS Wheat). http://maswheat.ucdavis.edu/Index.htm. Accessed 20 May 2016. 2016;2016.

38. Forrest K, Pujol V, Bulli P, Pumphrey M, Wellings C, Herrera-Foessel S, et al. Development of a SNP marker assay for the *Lr67* gene of wheat using a genotyping by sequencing approach. Mol Breed. 2014;34:2109–18.

39. Rasheed A, Wen W, Gao F, Zhai S, Jin H, Liu J, et al. Development and validation of KASP assays for genes underpinning key economic traits in bread wheat. Theor. Appl. Genet. Springer. Berlin Heidelberg. 2016;129:1843–60.

40. Mago R, Bariana HS, Dundas IS, Spielmeyer W, Lawrence GJ, Pryor AJ, et al. Development of PCR markers for the selection of wheat stem rust resistance genes Sr24 and Sr26 in diverse wheat germplasm. Theor Appl Genet. 2005;111:496–504.

41. Liu K, Muse SV. PowerMarker : an integrated analysis environment for genetic marker analysis. Oxford Univ. Presstime. 2005;21:2128–9.

42. Weir BS. Genetic data analysis II methods for discrete population genetic data. Sundarland, MA: Sinauer Assoc; 1996.

43. Felsenstein J. PHYLIP (Phylogeny Inference Package) Version 3.66. Dep. Genome Sci. Univ. Washingt. Seattle, WA. 2006.

44. Pritchard JK, Stephens M, Donnelly P. Inference of population structure using multilocus genotype data. Genet Soc Am. 2000;155:945–59.

45. Evanno G, Regnaut S, Goudet J. Detecting the number of clusters of individuals using the software structure: a simulation study. Mol Ecol. 2005;14:2611–20.

46. Cleveland WS. Robust locally weighted regression and smoothing scatterplots. J Am Stat Assoc. 1979;74:829–36.

47. Breseghello, Sorrells FM. Association Mapping of Kernel Size and Milling Quality in Wheat (*Triticum aestivum* L.) Cultivars. Genetics. 2005;172:1165–1177.

48. Corbeil RR, Searle SR. Restricted maximum likelihood (REML) estimation of variance components in the mixed model. Technometrics. 1976;18:31–8.

49. Yu J, Pressoir G, Briggs WH, Vroh Bi I, Yamasaki M, Doebley JF, et al. A unified mixed-model method for association mapping that accounts for multiple levels of relatedness. Nat Genet. 2006;38:203–8.

50. Zhang Z, Ersoz E, Lai CQ, Todhunter RJ, Tiwari HK, Gore MA, et al. Mixed linear model approach adapted for genome-wide association studies. Nat Genet. 2010;42:355–60.

51. Lipka AE, Tian F, Wang Q, Peiffer J, Li M, Bradbury PJ, et al. GAPIT: genome association and prediction integrated tool. Bioinformatics. 2012;28:2397–9.

52. Benjamini Y, Hochberg Y. Controlling the false discovery rate: a practical and powerful approach to multiple testing. J R Stat Soc. 1995:289–300.

53. Ren J, Chen L, Sun D, You FM, Wang J, Peng Y, et al. SNP-revealed genetic diversity in wild emmer wheat correlates with ecological factors. BMC Evol Biol. 2013;13:1–15.

54. Negassa M. Estimates of phenotypic diversity and breeding potential of Ethiopian wheats. Hereditus. 1986;8:41–8.

55. Tadesse K, Ayalew A, Badebo A. Effect of fungicide on the development of wheat stem rust and yield of wheat varieties in highlands of Ethiopia. African Crop Sci J. 2010;18:23–33.

56. Korte A, Farlow A. The advantages and limitations of trait analysis with GWAS: a review. Plant Methods. 2013;9:29.

57. Ma J, Zhou R, Dong Y, Wang L, Wang X, Jia J. Molecular mapping and detection of the yellow rust resistance gene *Yr26* in wheat transferred from *Triticum turgidum* L. using microsatellite markers. Euphytica. 2001; 120:219–26.

58. Li ZF, Zheng TC, He ZH, Li GQ, Xu SC, Li XP, et al. Molecular tagging of stripe rust resistance gene YrZH84 in Chinese wheat line Zhou 8425B. TAG Theor Appl Genet. 2006;112:1098–103.

59. Ren RS, Wang MN, Chen XM, Zhang ZJ. Characterization and molecular mapping of Yr52 for high-temperature adult-plant resistance to stripe rust in spring wheat germplasm PI 183527. Theor Appl Genet. 2012;125: 847–57.

60. Rosewarne GM, Singh RP, Huerta-Espino J, Herrera-Foessel SA, Forrest KL, Hayden MJ, et al. Analysis of leaf and stripe rust severities reveals pathotype changes and multiple minor QTLs associated with resistance in an avocet × pastor wheat population. Theor Appl Genet. 2012;124:1283–94.

61. Cheng P, Xu LS, Wang MN, See DR, Chen XM. Molecular mapping of genes *Yr64* and *Yr65* for stripe rust resistance in hexaploid derivatives of durum wheat accessions PI 331260 and PI 480016. Theor Appl Genet. 2014;127: 2267–77.

62. Bansal UK, Kazi AG, Singh B, Hare RA, Bariana HS. Mapping of durable stripe rust resistance in a durum wheat cultivar Wollaroi. Mol Breed. 2014;33:51–9.

63. Basnet BR, Ibrahim AMH, Chen X, Singh RP, Mason ER, Bowden RL, et al. Molecular mapping of stripe rust resistance in hard red winter wheat TAM 111 adapted to the U.S. High Plains. Crop Sci. 2014;54:1361–73.

64. Christopher MD, Liu S, Hall MD, Marshal DS, Fountain MO, Johnson JW, et al. Identification and mapping of adult-plant stripe rust resistance in soft red winter wheat cultivar USG 3555. Plant Breed. 2013;132:53–60.

65. Dedryver F, Paillard S, Mallard S, Robert O, Trottet M, Nègre S, et al. Characterization of genetic components involved in durable resistance to stripe rust in the bread wheat "Renan". Phytopathology. 2009;99:968–73.

66. Lowe I, Cantu D, Dubcovsky J. Durable resistance to the wheat rusts: integrating systems biology and traditional phenotype-based research methods to guide the deployment of resistance genes. Euphytica. 2011;179:69–79.

67. Zhang J, Simmons C, Yalpani N, Crane V, Wilkinson H, Kolomiets M. Genomic analysis of the 12-oxo-phytodienoic acid reductase gene family of *Zea mays*. Plant Mol Biol. 2005;59:323–43.

68. Tan CT, Carver BF, Chen M, Gu Y, Yan L. Genetic association of OPR genes with resistance to hessian fly in hexaploid wheat. BMC Genomics. 2013;14:1–11.

69. Feng J, Chen G, Wei Y, Liu Y, Jiang Q, Li W, et al. Identification and genetic mapping of a recessive gene for resistance to stripe rust in wheat line LM168-1. Mol Breed. 2014;33:601–9.

70. William HM, Singh RP, Huerta-Espino J, Palacios G, Suenaga K. Characterization of genetic loci conferring adult plant resistance to leaf rust and stripe rust in spring wheat. Genome. 2006;49:977–90.

71. Lillemo M, Asalf B, Singh RP, Huerta-Espino J, Chen XM, He ZH, et al. The adult plant rust resistance loci *Lr34/Yr18* and *Lr46/Yr29* are important determinants of partial resistance to powdery mildew in bread wheat line Saar. Theor Appl Genet. 2008;116:1155–66.

72. Prins R, Pretorius ZA, Bender CM, Lehmensiek A. QTL mapping of stripe, leaf and stem rust resistance genes in a Kariega x avocet S doubled haploid wheat population. Mol Breed. 2011;27:259–70.

73. Suenaga K, Singh RP, Huerta-Espino J, William HM. Microsatellite markers for genes *Lr34/Yr18* and other quantitative trait loci for leaf rust and stripe rust resistance in bread wheat. Phytopathology. 2003;93:881–90.

74. Randhawa M, Bansal U, Valarik M, Klocova B, Dolezel J, Bariana H. Molecular mapping of stripe rust resistance gene *Yr51* in chromosome 4AL of wheat. Theor Appl Genet. 2014;127:317–24.

75. Singh A, Pandey MP, Singh AK, Knox RE, Ammar K, Clarke JM, et al. Identification and mapping of leaf, stem and stripe rust resistance quantitative trait loci and their interactions in durum wheat. Mol Breed. 2013;31:405–18.

76. Vazquez MD, Peterson CJ, Riera-Lizarazu O, Chen X, Heesacker A, Ammar K, et al. Genetic analysis of adult plant, quantitative resistance to stripe rust in wheat cultivar "Stephens" in multi-environment trials. Theor Appl Genet. 2012;124:1–11.

77. Xu H, Zhang J, Zhang P, Qie Y, Niu Y, Li H, et al. Development and validation of molecular markers closely linked to the wheat stripe rust resistance gene *YrC591* for marker-assisted selection. Euphytica. 2014; 198:317–23.

78. Zhou XL, Wang MN, Chen XM, Lu Y, Kang ZS, Jing JX. Identification of *Yr59* conferring high-temperature adult-plant resistance to stripe rust in wheat germplasm PI 178759. Theor Appl Genet. 2014;127:935–45.

79. McIntosh R, Yamazaki Y, Dubcovsky J, Rogers WJ, Morris C, Appels R, et al. Catalogue of gene symbols for wheat in 12th international wheat genetics symposium. McIntosh, Yokohama, Japan: Ed. by R.A; 2013.

80. Mago R, Spielmeyer W, Lawrence GJ, Lagudah ES, Ellis JG, Pryor A. Identification and mapping of molecular markers linked to rust resistance genes located on chromosome 1RS of rye using wheat-rye translocation lines. Theor Appl Genet. 2002;104:1317–24.

81. Bajgain P, Rouse MN, Bulli P, Bhavani S, Gordon T, Wanyera R, et al. Association mapping of north American spring wheat breeding germplasm reveals loci conferring resistance to Ug99 and other African stem rust races. BMC Plant Biol. 2015;15:1–19.

82. The TT. Chromosome location of genes conditioning stem rust resistance transferred from diploid to hexaploid wheat. Nat New Biol. 1973;241:256.

83. Chen S, Rouse MN, Zhang W, Jin Y, Akhunov E, Wei Y, et al. Fine mapping and characterization of *Sr21*, a temperature-sensitive diploid wheat resistance gene effective against the *Puccinia graminis* f. Sp. *tritici* Ug99 race group. Theor. Appl. Genet. 2015;128:645–56.

84. Bansal UK, Bossolini E, Miah H, Keller B, Park RF, Bariana HS. Genetic mapping of seedling and adult plant stem rust resistance in two European winter wheat cultivars. Euphytica. 2008;164:821–8.

85. Briggs J, Chen S, Zhang W, Nelson S, Dubcovsky J, Rouse MN. Mapping of *SrTm4*, a recessive stem rust resistance gene from diploid wheat effective to Ug99. Phytopathology. 2015;105:1347–54.

86. Yu LX, Barbier H, Rouse MN, Singh S, Singh RP, Bhavani S, et al. A consensus map for Ug99 stem rust resistance loci in wheat. Theor Appl Genet. 2014; 127:1561–81.

87. Letta T, Olivera P, Maccaferri M, Jin Y, Ammar K, Badebo A, et al. Association mapping reveals novel stem rust resistance loci in durum wheat at the seedling stage. Plant Genome. 2014;7:1–13.

88. Nirmala J, Chao S, Olivera P, Babiker EM, Abeyo B, Tadesse Z, et al. Markers Linked to Wheat Stem Rust Resistance Gene Sr11 Effective to *Puccinia graminis* f. sp. *tritici* Race TKTTF. Phytopathology. 2016;106:1352–8.

89. Rouse MN, Nava IC, Chao S, Anderson JA, Jin Y. Identification of markers linked to the race Ug99 effective stem rust resistance gene *Sr28* in wheat (*Triticum aestivum* L.). Theor. Appl. Genet. 2012;125:877–85.

5

Comparative studies on tolerance of rice genotypes differing in their tolerance to moderate salt stress

Qian Li[1,2†], An Yang[1†] and Wen-Hao Zhang[1,2,3*]

Abstract

Background: Moderate salt stress, which often occurs in most saline agriculture land, suppresses crop growth and reduces crop yield. Rice, as an important food crop, is sensitive to salt stress and rice genotypes differ in their tolerance to salt stress. Despite extensive studies on salt tolerance of rice, a few studies have specifically investigated the mechanism by which rice plants respond and tolerate to moderate salt stress. Two rice genotypes differing in their tolerance to saline-alkaline stress, Dongdao-4 and Jigeng-88, were used to explore physiological and molecular mechanisms underlying tolerance to moderate salt stress.

Results: Dongdao-4 plants displayed higher biomass, chlorophyll contents, and photosynthetic rates than Jigeng-88 under conditions of salt stress. No differences in K^+ concentrations, Na^+ concentrations and Na^+/K^+ ratio in shoots between Dongdao-4 and Jigeng-88 plants were detected when challenged by salt stress, suggesting that Na^+ toxicity may not underpin the greater tolerance of Dongdao-4 to salt stress than that of Jigeng-88. We further demonstrated that Dongdao-4 plants had greater capacity to accumulate soluble sugars and proline (Pro) than Jigeng-88, thus conferring greater tolerance of Dongdao-4 to osmotic stress than Jigeng-88. Moreover, Dongdao-4 suffered from less oxidative stress than Jigeng-88 under salt stress due to higher activities of catalase (CAT) in Dongdao-4 seedlings. Finally, RNA-seq revealed that Dongdao-4 and Jigeng-88 differed in their gene expression in response to salt stress, such that salt stress changed expression of 456 and 740 genes in Dongdao-4 and Jigeng-88, respectively.

Conclusion: Our results revealed that Dongdao-4 plants were capable of tolerating to salt stress by enhanced accumulation of Pro and soluble sugars to tolerate osmotic stress, increasing the activities of CAT to minimize oxidative stress, while Na^+ toxicity is not involved in the greater tolerance of Dongdao-4 to moderate salt stress.

Keywords: Dongdao-4, Osmotic regulation, ROS detoxifying mechanism, Jigeng-88, *Oryza sativa* L., Rice, Moderate salt stress

Background

Soil salinity is a global environmental challenge, limiting crop production over 800 million hectares worldwide [1]. The majority of the saline land has arisen from natural events and human intervention, such as release of soluble salts of various types during weathering of parental rocks, the deposition of oceanic salts by wind and rain as well as irrigation containing trace amounts of sodium chloride [2]. Rice is an important cereal that provides 50–80% of daily calorie intake for more than 3 billion people. Rice plants are sensitive to salt stress, particularly at the seedling and reproductive stages. However, rice genotypes differ in their sensitivity to salt stress, and some rice genotypes tolerant to salt stress have been reported, including those of genotypes Pokkali [3] and IR63731–1–1-4-3-2 [4]. Elucidating of the molecular and physiological mechanisms by which rice genotypes respond and adapt to salt stress are pivotal for selecting and breeding rice genotypes capable of growth in the saline soils.

* Correspondence: whzhang@ibcas.ac.cn

†Equal contributors

[1]State Key Laboratory of Vegetation and Environmental Change, Institute of Botany, the Chinese Academy of Sciences, Beijing 100093, People's Republic of China

[2]University of Chinese Academy of Sciences, Beijing 100049, People's Republic of China

Full list of author information is available at the end of the article

Plants suffering from high salt stress often display symptoms of Na^+ toxicity due to accumulation of Na^+, which in turn reduces nutrient acquisition, leading to nutritional imbalances, and oxidative damage [5]. Plants have evolved several mechanisms to cope with these problems. Minimizing Na^+ toxicity by compartmenting toxic Na^+ into vacuoles and/or restricting Na^+ uptake by plants are the most common strategy for plants to tolerate salt stress [2]. For example, greater tolerance to Na^+ confers barley greater tolerance to salt stress than wheat despite similar foliar Na^+ concentrations in barely to those of wheat [6].

In addition, plants exposed to salt stress can also suffer from osmotic stress. Therefore, plants have to equip with capacity to tolerate osmotic stress under saline conditions. Salt stress limits plant growth by increasing the osmotic potential of the soil and, thus, decreasing water uptake by the roots. Accumulation of compatible osmolytes in the cytosol, lowering osmotic potential to sustain water absorption from saline soil solutions, is an important salinity tolerance mechanism [7, 8]. Many attempts to molecular breeding plants tolerant to drought have been made by introducing genes that encode key enzymes for biosynthesis of compatible solutes [9].

Increases in activities of enzymes that detoxify reactive oxygen species also contribute to plant tolerance to salinity [10]. For example, Mishra et al. (2013) reported that salt tolerant rice seedlings have a better protection against reactive oxygen species (ROS) by increasing the activities of antioxidant enzymes under salt stress [11]. Transgenic plants overexpressing genes encoding antioxidant enzymes are more tolerant to salt stress than their wild-type counterparts [12, 13].

Several Na^+ transporters responsible for Na^+ uptake, translocation and compartmentation have been identified in rice plants, including those OsHKTs, OsHAKs, OsNHXs [14–16]. However, the molecular mechanisms underlying Na^+ transport from soil solution and within plants remain largely elusive. A comprehensive genome-wide analysis of gene expression in response to salt stress may shed some light on molecular mechanisms responsible for tolerance of rice plants to salt stress. RNA-seq, a high-throughput sequencing technology, was widely used to dissect transcriptomic information. These transcriptome-wide studies have provided new insights on genes and regulatory mechanisms involved in abiotic stresses.

Soil is often referred to as saline one when the electrical conductivity (equivalent to the concentration of salts in saturated soil or in a hydroponic solution) is greater than 4 dS m^{-1}, which is equivalent to approximately 40 mM NaCl and yields of most crops are suppressed when grown in such saline soils [2]. However, higher levels of concentration of NaCl (150–200 mM)

have been frequently used to study the physiological and molecular mechanisms to saline stress in the literature [3, 17, 18]. Therefore, elucidating the mechanisms underlying response and tolerance to moderate salt stress that is similar to natural saline soils will be of practical implications regarding for breeding crops capable of growing in saline soils. In our previous studies, we collected more than 100 rice genotypes and assessed their tolerance to saline-alkaline stress. Among these rice genotypes, Dongdao-4 is an elite rice genotype that is capable of growing in saline-alkaline soils in the northeast of China, while the genotype Jigeng-88 is a relatively saline-alkaline-sensitive genotype. In addition, we have indicated that Dongdao-4 rice genotype is more tolerant to saline-alkaline stress than Jigeng-88 by more efficient acquisition of iron under saline-alkaline conditions [19]. However, whether the two rice genotypes differ in their tolerance to moderate, neutral salts stress is unclear. In the present study, we made a comparative study on the effects of moderate salt stress on the two genotypes differing in their tolerance to saline-alkaline stress. We further investigated the physiological and molecular mechanisms responsible for salt tolerance in rice plants.

Results

Dongdao-4 seedlings are more tolerant to moderate salt stress than Jigeng-88 seedlings

To characterize the differences in salt tolerance between Dongdao-4 and Jigeng-88 plants, three-week-old rice seedlings of the two rice genotypes were exposed to solution supplemented with 20 mM NaCl for one day, and 40 mM NaCl for another day, and then exposed to 60 mM NaCl for one week. We monitored the effects of salt stress on plant growth (Additional file 1: Figure S1), photosynthetic rates (Additional file 1: Figure S2b), and Na^+, K^+ concentrations (Additional file 1: Figure S3). Our results showed that, similar to previous results of saline-alkaline stress [19], Dongdao-4 was more tolerant to moderate salt stress than Jigeng-88. Chlorophyll content in Dongdao-4 plants was little affected by saline stress, while a significant reduction in chlorophyll content in Jigeng-88 plants was observed by salt stress, leading to a significantly higher chlorophyll content in Dongdao-4 than in Jigeng-88 when exposed to salt stress (Additional file 1: Figure S2a).

Dongdao-4 plants accumulated more soluble sugars and Pro

In addition to toxic effect of Na^+, plants also suffer from osmotic stress when challenged by salt stress [2]. To cope with osmotic stress, maintenance of turgor pressure by synthesizing, transporting and accumulating low-molecular weight organic compounds is a common strategy [8]. To test whether the enhanced tolerance of Dongdao-4 seedlings to salt stress is related to the capacity to accumulate soluble sugars and Pro, the effect of salt stress on contents of soluble sugars and Pro in Dongdao-4 and Jigeng-88 plants was investigated. The two genotypes exhibited comparable soluble sugars and Pro contents in their shoots in non-salt, control medium (Fig. 1). There were significant increases in shoot soluble sugars and Pro contents of both Dongdao-4 and Jigeng-88 seedlings upon exposure to salt medium. However, the magnitude of salt-induced increases in shoot soluble sugars and Pro contents in Dongdao-4 plants was greater than in Jigeng-88 plants, leading to a significantly higher shoot soluble sugars and Pro contents in Dongdao-4 than that in Jigeng-88 seedlings under salt stress (Fig. 1).

To further elucidate the mechanisms by which Dongdao-4 accumulated greater amounts of Pro than Jigeng-88 under salt stress, the effects of salt stress on the expression of genes responsible for Pro biosynthesis (Δ^1-pyrroline-5-carboxylate synthase genes, $Os05g0455500$ and $Os01g0848200$), Pro transport ($Os03g0644400$ and $Os07g0100800$) were investigated. As shown in Fig. 1c, treatment with salt stress led to a great increase in transcripts of $Os05g0455500$ in Dongdao-4 seedlings, whereas expression levels of $Os05g0455500$ in Jigeng-88 seedlings remained relatively unchanged, leading to a significantly higher expression level of $Os05g0455500$ in Dongdao-4 than in Jigeng-88 plants. Exposure to salt stress led to a similar increase in transcripts of $Os01g0848200$ in the two genotypes (Fig. 1d). Treatment with salt stress led to a significant increase in transcripts levels of $Os03g0644400$ in

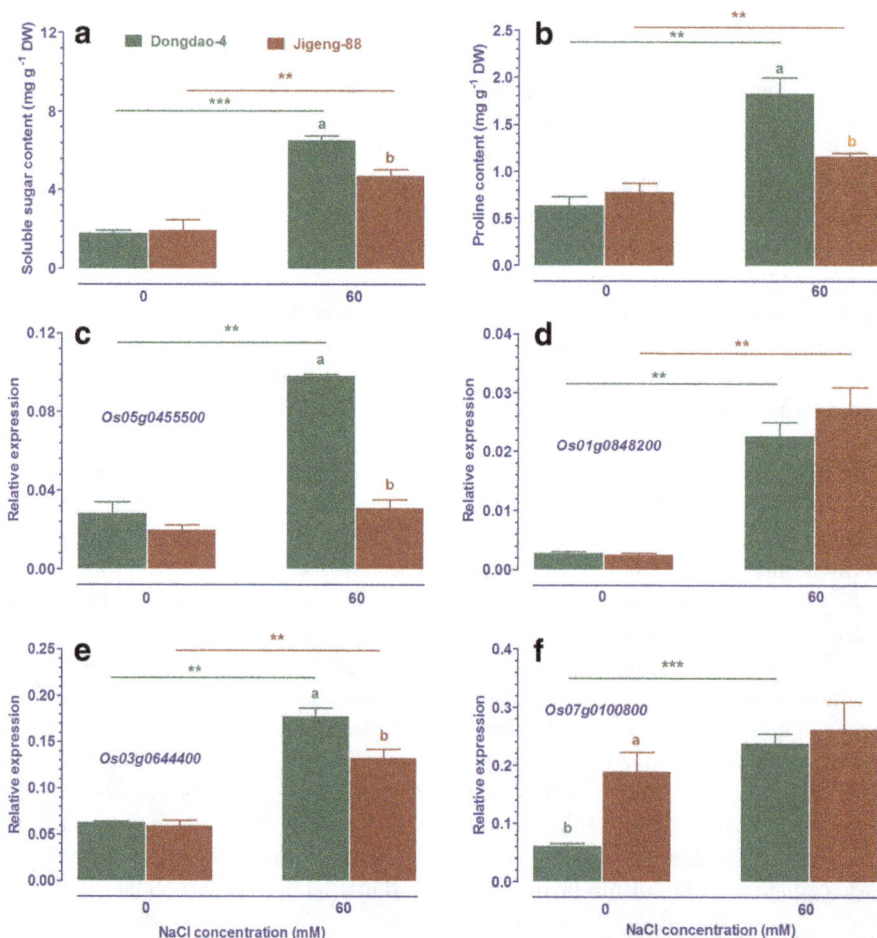

Fig. 1 Effect of salt stress on contents of soluble sugar (**a**) and Proline (**b**) in shoot of Dongdao-4 and Jigeng-88 seedlings grown at normal and salt stress conditions. Expression levels of putative proline synthase genes ($Os05g0455500$ and $Os01g0848200$; **c** and **d**), transporter genes ($Os03g0644400$ and $Os07g0100800$; **e** and **f**) in the two rice genotypes were analysed. Total RNA was extracted from rice seedlings grown under control and salt stress conditions for one week. Transcript levels were measured by real-time PCR. Actin was used as an internal control. *Error bars* are calculated based on three biological replicates. Data are means ±SE ($n \geq 4$). Means with different letters are significantly different ($P < 0.05$) within the same treatments. *Asterisks* (*) indicate significant differences between control and salt stress of the same genotype which were determined by Student's t-test (** $0.001 < P < 0.01$, *** $P < 0.001$)

both genotypes, with the magnitude of increase in Dongdao-4 greater than in Jigeng-88 (Fig. 1e). Expression levels of *Os07g0100800* in Dongdao-4 were significantly lower than those in Jigeng-88 under control conditions (Fig. 1f). Exposure to salt stress led to a significant increase in expression level of *Os07g0100800* in Dongdao-4 plants (Fig. 1f), while expression level of *Os07g0100800* in Jigeng-88 plants was constant upon exposure to salt stress, leading to no differences in expression levels of *Os07g0100800* in the two genotypes under conditions of salt stress (Fig. 1f).

Effect of mannitol on Dongdao-4 and Jigeng-88

The greater accumulation of compatible solutes (Pro and soluble sugars) in Dongdao-4 seedlings prompted us to test whether the genotype has greater capacity for osmoregulation, thus conferring it more tolerant to osmotic stress. Therefore, three-week-old seedling of both Dongdao-4 and Jigeng-88 were transferred to solution containing 120 mM mannitol, which approximately equal to the osmolality of solution containing 60 mM NaCl. Upon exposure of the two genotypes to 120 mM mannitol for 5 d, shoot and root biomass and survival rate of both Dongdao-4 and Jigeng-88 were reduced, with the magnitude of reduction in Dongdao-4 plants significantly less than that in Jigeng-88 plants (Fig. 2 and Table 1), suggesting that Dongdao-4 plants are more tolerant to osmotic stress than Jigeng-88.

The two genotypes exhibited a comparable shoot soluble sugars and Pro contents in the absence of mannitol (Table 1). There was a significant increases in shoot soluble sugars and Pro contents in both genotypes upon exposure to the medium supplemented with 120 mM mannitol, whereas the magnitude of treatment-induced increases in shoot soluble sugars and Pro contents in Dongdao-4 plants was greater than that in Jigeng-88 plants, leading to a significantly higher shoot soluble sugars and proline contents in Dongdao-4 than that in Jigeng-88 seedlings under osmotic stress conditions (Table 1).

Dongdao-4 plants exhibited greater tolerance to oxidative stress

Plants suffering from abiotic stress often exhibit symptoms of oxidative stress as evidenced by enhanced accumulation of reactive oxygen species (ROS) and malondialdehyde (MDA). Therefore, histochemical analysis was performed to detect ROS accumulation by NBT and DAB staining. As shown in Fig. 3, little ROS accumulation was observed in Dongdao-4 and Jigeng-88 seedlings under control conditions. After treatment with salt stress for one week, an evident ROS accumulation was detected in both Dongdao-4 and Jigeng-88 seedlings, with the levels of ROS in Dongdao-4 lower than Jigeng-88 (Fig. 3a, b). No significant differences in MDA contents in Dongdao-4 and Jigeng-88 seedlings were found when grown in control medium (Fig. 3c). Significant increases in MDA contents in Jigeng-88 seedlings were observed after exposure to salt stress, while Dongdao-4 plants maintain a relatively constant MDA contents when challenged by the identical salt stress, leading to a significantly higher MDA contents in Jigeng-88 than in Dogndao4 plants under salt conditions (Fig. 3c). These results indicate that Dongdao-4 plants are equipped with greater tolerance to the oxidative stress associated with salt stress.

The less accumulation of ROS and MDA in Dongdao-4 seedlings under salt stress prompted us to check the activities of the major antioxidant enzymes. Under control conditions, activities of SOD and CAT were comparable in shoots of Dongdao-4 and Jigeng-88 plants, while activities of POD were higher in shoots of Jigeng-88 than those in Dongdao-4 (Fig. 3d–f). There were marked increases in activities of these enzymes for rice seedlings upon exposure to salt stress, and the treatment-induced increases in activities of POD and SOD had no significant differences in shoots of Dongdao-4 and Jigeng-88 plants (Fig. 3e, f). In contrast, the magnitude of the treatment-induced increases in activities of CAT was greater in Dongdao-4 than in Jigeng-88 plants, such that a significantly higher content of CAT in Dongdao-4 than in Jigeng-88 was observed under salt stress (Fig. 3d). These results suggest that higher activities of CAT may contribute to greater

Fig. 2 Effects of osmotic stress on growth performance of Dongdao-4 and Jigeng-88 plants after transferred to solution containing 40 mM mannitol for 1 day, and 80 mM mannitol for 1 day, and then exposed to 120 mM mannitol for 5 days, finally recovered for another 3 and 15 days. Bars, 10 cm

Table 1 Survival rate, Dry shoot biomass, Dry root biomass, Shoot soluble sugar, Shoot Pro of Dongdao-4 and Jigeng-88 plants before, after exposure to medium supplemented with mannitol and recovery from the mannitol treatment. Three-week-old seedlings were transferred to solution containing 40 mM mannitol for 1 day, and 80 mM mannitol for 1 day, and then exposed to 120 mM mannitol for 5 days, finally recovered for 15 days. Data are means ±SE ($n \geq 4$). Means with different letters are significantly different ($P < 0.05$) within the same treatments. Asterisks (*) indicate significant differences between control and treatments of the same genotype which were determined by Student's t-test (*$0.01 < P < 0.05$, **$0.001 < P < 0.01$, ***$P < 0.001$)

	Before treatment		120 mM Mannitol		Recovery	
	Dongdao-4	Jigeng-88	Dongdao-4	Jigeng-88	Dongdao-4	Jigeng-88
Survival rate (%)	100 ± 0.00	100 ± 0.00	100 ± 0.00***a	72.92 ± 3.45b		
Dry shoot biomass (mg plant^{-1})	113.12 ± 4.47	100.68 ± 3.49	132.18 ± 7.68 **a	83.53 ± 2.99 *b	526.08 ± 18.83 ***a	175.03 ± 13.82 *b
Dry root biomass (mg plant^{-1})	12.81 ± 0.12	12.5 ± 0.62	19.95 ± 0.88**a	12.73 ± 0.35b	66.23 ± 3.85***a	16.48 ± 0.66b
Shoot soluble sugar (mg g^{-1} DW)	1.79 ± 0.11	1.94 ± 0.43	16.61 ± 0.73***a	13.52 ± 0.41***b		
Shoot proline (mg g^{-1} DW)	0.67 ± 0.06	0.75 ± 0.06	1.32 ± 0.14**a	0.64 ± 0.02b		

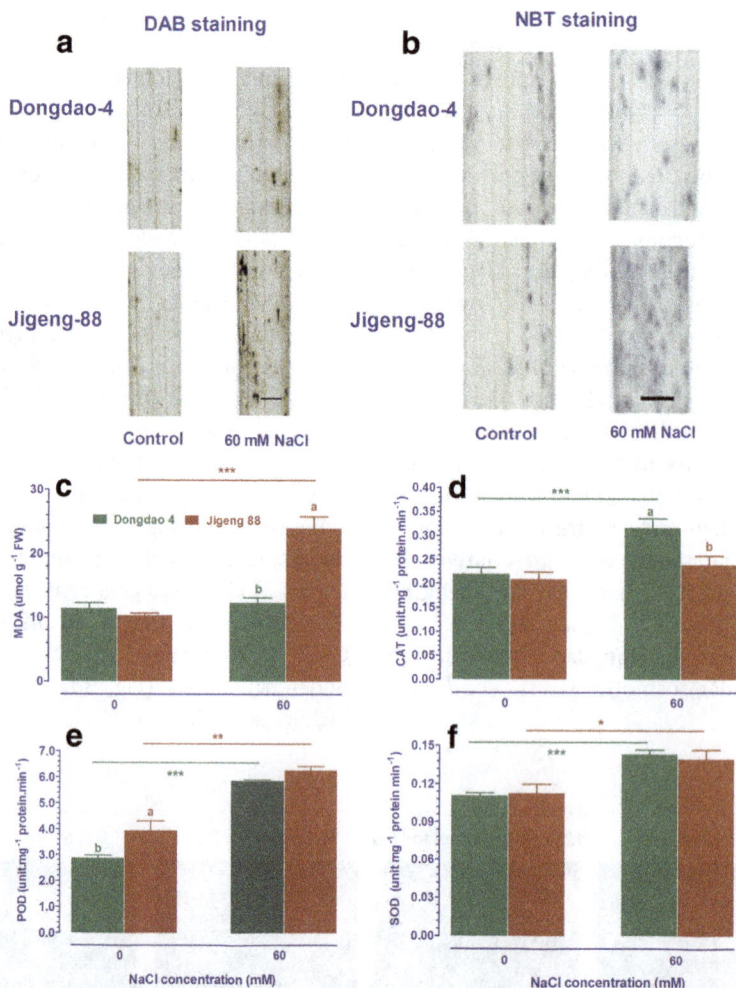

Fig. 3 Effect of salt stress on accumulation of reactive oxygen species (ROS) and contents of oxidants and antioxidant enzymes in shoot of Dongdao-4 and Jigeng-88 seedlings. ROS accumulation in Dongdao-4 and Jigeng-88 was detected at 5 day with 3,3'-diaminobenzidine (DAB) (**a**) and nitroblue tetrazolium (NBT) staining (**b**). Bar, 0.1 cm. **c** malondialdehyde (MDA), **d** catalase (CAT), **e** peroxidase (POD), **f** superoxide dismutase (SOD). Data are means ±SE ($n \geq 4$). Means with different letters are significantly different ($P < 0.05$) within the same treatments. *Asterisks* (*) indicate significant differences between control and salt stress of the same genotype which were determined by Student's t-test (*$0.01 < P < 0.05$, **$0.001 < P < 0.01$, *** $P < 0.001$)

tolerance of Dongdao-4 seedlings to salt stress by counteracting oxidative stress evoked by salt stress.

RNA-seq analysis

To gain a better understanding of the mechanism underlying salt tolerance of Dongdao-4 seedlings at global transcriptional level, the transcriptome of Dongdao-4 and Jigeng-88 seedlings grown in control and NaCl-supplemented media were investigated by RNA-Seq. In total, 916.86 milloin reads were generated. After trimming adapters and filtering out low quality reads, more than 875.10 million clean reads were retained for futher analysis. Among all the reads, more than 89% had Phred-like quality scores at the Q30 level (an error probability of 0.1%) (Table 2). In addition, principle component analysis (PCA) showed that the two replicates were highly comparable (Additional file 1: Figure S4). These results showed that the quality of throughput and sequencing is high enough for further analysis.

Validation of differentially regulated genes (DEGs) by real-time PCR analysis

To validate the data from RNA-sequencing, 15 DEGs were randomly selected for real-time PCR analysis in both genotypes in response to salt stress. The primers of selected genes are listed in Additional file 2: Table S1. A high degree of concordance was observed between the results generated by the two methods (Pearson correlation coefficients R^2 = 0.6843; Additional file 1: Figure S5).

Identification of DEGs of Dongdao-4 and Jigeng-88 plants under salt stress

Based on the criteria of a greater than 2-fold change and significance at P < 0.05 in t-tests, differentially regulated genes (DEGs) were identified in shoots of Dongdao-4 and Jigeng-88 under salt stress compared with control condidtions. There were more DEGs in salt-sensitive Jigeng-88 than in salt-tolerant Dongdao-4 under conditions of salt stress. More specifically, a total of 456 DEGs were detected in shoots of Dongdao-4 plants, including 217 up-regulated genes and 239 down-regulated genes (Fig. 4a). A total of

740 DEGs were found in shoots of Jigeng-88, including 399 up-regulated genes and 341 down-regulated genes (Fig. 4a). Venn diagram indicates that 110 and 292 genes were specifically up-regulated in shoots of Dongdao-4 and Jigeng-88 seedlings, respectively (Fig. 4b, c). A total of 107 genes were up-regulated in shoots of both Dongdao-4 and Jigeng-88 seedlings, with 3 genes having higher expresssion level in Dongdao-4 than that in Jigeng-88 (Additional file 2: Table S4). Of the down-regulated genes, 140 and 242 genes were specifically inhibited in shoots of Dongdao-4 and Jigeng-88 seedlings, respectively (Additional file 2: Tables S5 and S6). There were 99 genes that commonly suppressed in shoots of both genotypes, with 9 genes having higher expresssion level in Dongdao-4 than that in Jigeng-88 (Additional file 2: Table S7).

Gene ontology (GO) categorization of DEGs

To assign functional information to the DEGs that are diferently regulated between Dongdao-4 and Jigeng-88 in saline stress, Gene Ontology (GO) analysis was carried out. For the 110 specifically up-regulated in Dongdao-4 seedlings, only one GO term (response to stress) was significantly enriched (Fig. 5a). For the 140 specifically down-regulated genes, no GO term was significantly enriched (Fig. 5b). In Jigeng-88, the 292 specifically up-regulated and 242 specifically down-regulated genes were assigned to 13 and 11 terms, respectively (Fig. 5a and b). The most enriched GO terms of specifically up-regulated genes were diterpene phytoalexin metabolic process, while the most significantly overrepresented terms of specifically down-regulated genes was pentose-phosphate shunt (Fig. 5). For the commonly up-regulated and down-regulated genes, there are only 3 and 9 genes with higher expression level in Dongdao-4 than Jigeng-88, respectively (Additional file 2: Tables S4 and S7). These genes were also used to perform GO analysis and no GO terms were significantly enriched owning to the small percentage overall of the dataset.

Discussion

Dongdao-4 is an elite rice genotype that is capable of growing in saline-alkaline soils in the northeast of China.

Table 2 Summary of sequencing results and their matches in the *Oryza sativa Japonica. cv. Nipponbare* genome

Sample name	Raw reads	Clean reads	Q30%	Total mapped reads
Dongdao-4-control-Shoot-1	55,250,786	54,137,962	92.20	50,926,791
Dongdao-4-control-Shoot-2	58,323,950	57,208,312	92.66	53,987,360
Jigeng-88-control-Shoot-1	58,448,584	56,784,834	89.69	53,001,156
Jigeng-88-control-Shoot-2	54,904,288	53,432,852	90.41	49,824,285
Dongdao-4-salt-Shoot-1	51,848,598	48,385,532	89.52	50,869,413
Dongdao-4-salt-Shoot-2	57,200,640	54,379,656	90.49	44,764,519
Jigeng-88-salt-Shoot-1	62,512,034	57,515,542	90.94	54,147,061
Jigeng-88-salt-Shoot-2	58,009,966	54,798,810	89.09	51,173,210

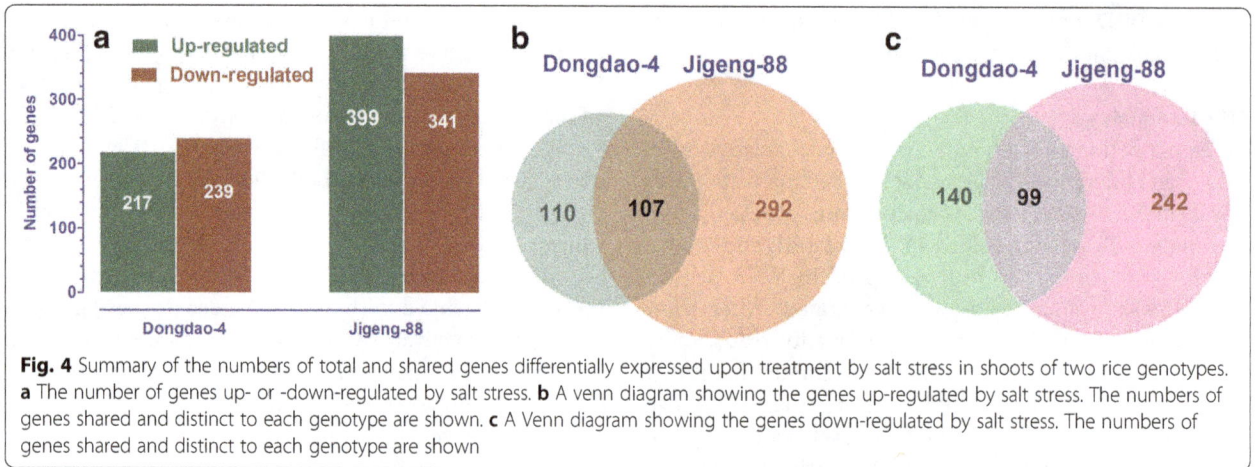

Fig. 4 Summary of the numbers of total and shared genes differentially expressed upon treatment by salt stress in shoots of two rice genotypes. **a** The number of genes up- or -down-regulated by salt stress. **b** A venn diagram showing the genes up-regulated by salt stress. The numbers of genes shared and distinct to each genotype are shown. **c** A Venn diagram showing the genes down-regulated by salt stress. The numbers of genes shared and distinct to each genotype are shown

Our previous studies revealed that Dongdao-4 plants are equipped with an efficient system for acquisition iron under saline-alkaline conditions, thus conferring tolerance to saline-alkaline stress [19]. In the present study, we evaluated the effect of moderate salt stress on Dongdao-4 plants and compared with that of a saline-alkaline-sensitive genotype, Jigeng-88. Our results demonstrated that, similar to saline-alkaline stress, Dongdao-4 plants were also more tolerant to salt stress than Jigeng-88 plants, as evidenced by less reduction in plant height and shoot biomass (Additional file 1: Figure S1).

The greater performance of Dongdao-4 plants may be accounted for by their higher foliar chlorophyll contents and photosynthetic rates than those in Jigeng-88 plants under conditions of salt stress (Additional file 1: Figure S2). Moreover, there were no difference in K^+ concentrations, Na^+ concentrations and Na^+/K^+ ratio in shoots between Dongdao-4 and Jigeng-88 plants under salt stress (Additional file 1: Figure S3), suggesting that acquisition of Na^+ and K^+ may not account for the greater tolerance of Dongdao-4 than Jigeng-88 to salt stress. In addition to Na^+ toxicity, plants suffering from salt stress also

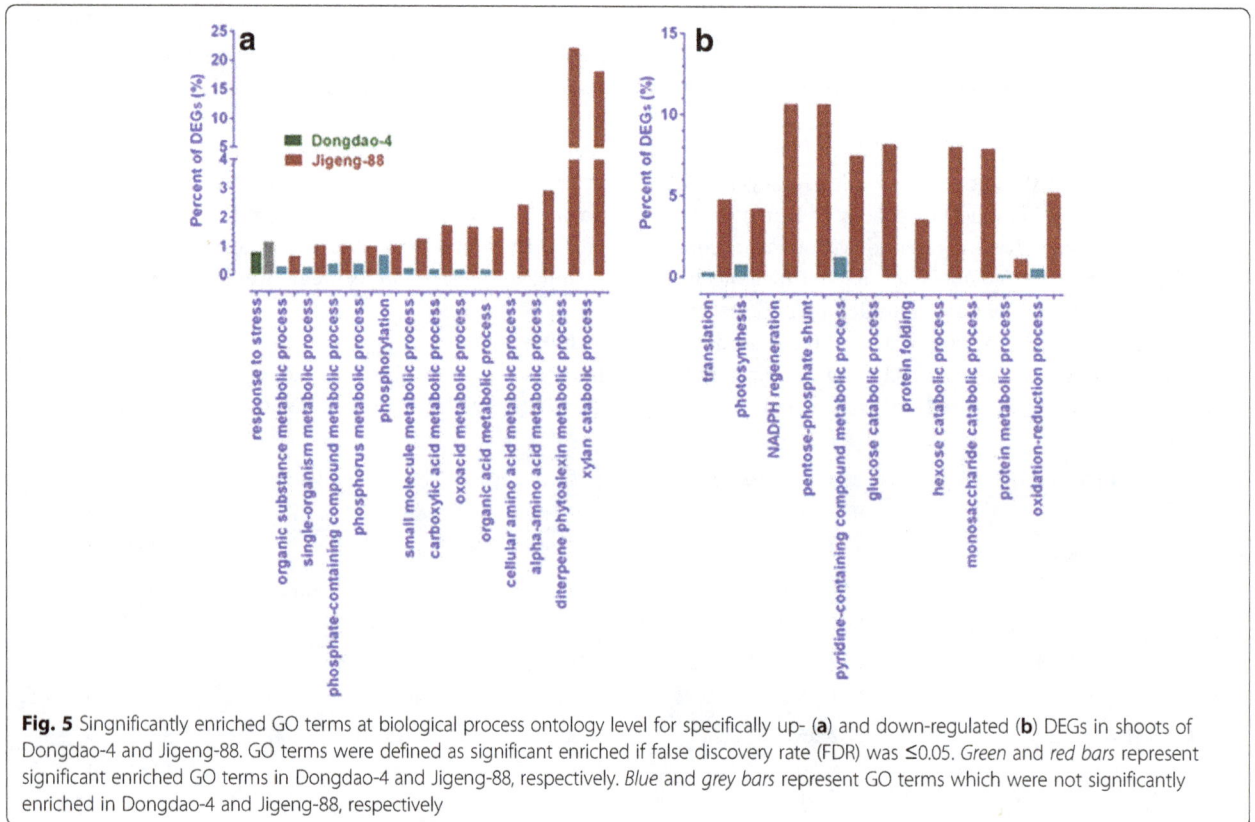

Fig. 5 Singnificantly enriched GO terms at biological process ontology level for specifically up- (**a**) and down-regulated (**b**) DEGs in shoots of Dongdao-4 and Jigeng-88. GO terms were defined as significant enriched if false discovery rate (FDR) was ≤0.05. *Green* and *red bars* represent significant enriched GO terms in Dongdao-4 and Jigeng-88, respectively. *Blue* and *grey bars* represent GO terms which were not significantly enriched in Dongdao-4 and Jigeng-88, respectively

need to deal with osmotic stress. One important finding in the present study is that Dongdao-4 seedlings were capable of accumulating greater amount of soluble sugars, Pro than Jigeng-88 when exposed to moderate salt stress (Figs. 1 and 3). The greater accumulation of soluble sugars and Pro would confer Dongdao-4 plants more effective osmoregulation, thus we further compared transcript profiles of Dongdao-4 with Jigeng-88 in response to salt stress. Our results revealed that Dongdao-4 showed some specific biological processes may underlie the mechanism by which Dongdao-4 has greater tolerance than Jigeng-88. These findings highlight that effective osmoregulation by accumulation of soluble sugars and Pro to maintain water potential and up-regulation of ROS detoxifying systems underlie the greater tolerance of Dongdao-4 to salt stress.

Plants suffering from salt stress have to cope with high concentrations of Na^+ in the growth medium. Foliar accumulation of Na^+ in the cytosol may be toxic by inhibiting photosynthesis and other metabolic processes. K^+ is the most abundant inorganic cation in plants, and plays a diverse role in many physiological processes. Therefore, maintaining an optimal K^+/Na^+ ratio is an important indicator of plant tolerance to saline stress. However, there are reports demonstrating independence of salt tolerance and foliar Na^+ concentration among wheat genotypes [6, 20]. In addition, Pires et al. (2015) observed a high variability of physiological response to salt stress between rice genotypes with similar foliar Na^+ concentration, suggested that other mechanisms may be responsible for salt tolerance [21]. Almeida et al. (2014) also reported that the maintenance of growth in the presence of 100 mM NaCl can be independent of the exclusion or accumulation of Na^+ among tomato varieties [22]. In the present study, a comparable Na^+ concentration and Na^+/K^+ ratio was observed in Dongdao-4 and Jigeng-88 when challenged by salt stress (Additional file 1: Figure S3c), suggesting that tolerance of Dongdao-4 plants to salt stress is not achieved by minimizing Na accumulation.

Roy et al. (2014) reported that Na^+ exclusion may be more effective under conditions of higher salinity, whereas 'osmotic tolerance' may be more important in moderately saline conditions [23]. In addition, the introgression of *TmHKT1;5-A* that encodes a Na^+ transporter from *Triticum monococcum* L. into the durum wheat resulted in a significant improvement in grain yield when growing in saline soils [24, 25]. However, the yield of durum wheat with *TmHKT1;5-A* was similar to that cultivar without the introgressed gene, under low and moderate saline conditions [24, 25], suggesting that osmotic stress has a greater effect on yield of these plants growing in low to moderate saline soils than Na^+ toxic effect [23]. Therefore, our results

highlight, that a higher osmoregulation capacity may confer greater tolerance of Dongdao-4 to moderate, neutral salt stress (i.e., 60 mM NaCl).

Accumulation of compatible solutes (e.g. Pro, raffinose, and glycine betaine) is a common phenomenon for plants to adapt to salt stress by maintaining turgor [2]. In addition, the compatible solutes can also be used to stabilize proteins and cellular structures, and to counteract oxidative stress associated with abiotic stress. In the present study, we found that Dongdao-4 plants accumulated greater amount of soluble sugars and Pro than Jigeng-88 when exposed to salt stress (Fig. 5). We further demonstrated that the expression levels of genes encoding putative Pro synthase and transporters in Dongdao-4 were higher than those in Jigeng-88 under salt stress (Fig. 5), suggesting that the higher expression levels of these genes may underpin the greater accumulation of Pro in Dongdao-4 plants, thus contributing to their greater tolerance to salt stress. The higher Pro and soluble sugar concentration in Dongdao-4 may confer Dongdao-4 plants more effectively osmo regulation, thus accounting for the greater tolerance to osmotic stress than Jigeng-88 (Table 1).

Reactive oxygen species (ROS), as byproducts of photosynthesis, respiration, photorespiration have to maintain in an optimal level to protect plants from oxidative stress and to function as signaling molecules mediating many physiological processes in plants [26, 27]. To avoid excessive accumulation of ROS under salt stress, ROS scavenging mechanisms including enzymatic and non-enzymatic (antioxidants) are often activated in plants. Recent studies reported that an increase in activity of antioxidative enzymes enhanced salt tolerance in rice [11, 28, 29], wheat [30, 31], cotton [32], barley [33], alfalfa [34]. Overexpressing genes encoding antioxidant enzymes enhanced salt tolerance in transgenic plants [12, 13, 35]. In present study, we found that salt stress-induced increases in activities of CAT were greater in Dongdao-4 than in Jigeng-88 plants (Fig. 6), which may allow Dongdao-4 plant to more efficiently regulate ROS, thus conferring it more tolerance to salt stress. No significant differences in the expression level of genes encoding CAT between Dongdao-4 and Jigeng-88 seedlings were found, suggesting that epigenetic and post-transcriptional regulation may underlies the activity of CAT.

To further explore the mechanism underlying the greater tolerance of Dongdao-4 at transcriptome level, we treated Dongdao-4 and Jigeng-88 seedlings with 60 mM NaCl for one week, and analyzed their transcriptome. A total of 110 genes were specifically up-regulated in Dongdao-4 (Additional file 2: Tables S2 and S10). Sun et al. (2013) reported that overexpresion of *GsSRK* in *Arabidopsis* enhanced salt tolerance and higher yields under salt stress [36]. In this study, one gene encoding

G-type lectin S-receptor-like serine/threonine-protein kinase (*GsSRK*) was up-regulated (Additional file 2: Table S2). As shown in Additional file 1: Figure S6a, the magnitude of the increase was greater in Jigeng-88 than in Dongdao-4, leading to a signifcantly higher expression levels in Dongdao-4. In addition, nine genes encoding lectin receptor-like kinases that play crucial roles in stress perception [37], were found to be up-regulated. Thaumatin-like proteins (TLPs) is involved in abiotic stresses including salinity and drought [38–42]. Here, one gene encoding thaumatin-like protein was specifically up-regulated by 2.6 fold in Dongdao-4 (Additional file 2: Table S2). Furthermore, WRKY transcription factor 6 and 46 were specifically up-regulated in Dongdao-4 respectively (Additional file 2: Table S2). Validation of *OsWRKY46* by Real-time PCR suggested that the expression level of *OsWRKY46* was up-regulated in both genotypes by salt stress, the increase was much greater in Dongdao-4 than in Jigeng-88 (Additional file 1: Figure S6b). *OsWRKY46* is invovled in regulation of a set of genes associated with cellular osmoprotection and oxidative detoxification under drought and salt stress [43]. To determine if a common transcription factors underlie the specifically up-regulation of genes in Dongdao-4, promoter of these genes were analysed by PlantCARE. As shown in the Additional file 2: Table S10, many motifs have been found in the promoters of these genes by promoter element analysis, such as ABRE (cis-acting element involved in the abscisic acid responsiveness), W-box (WRKY binding site), CGTCA-motif (cis-acting regulatory element involved in the MeJA-responsiveness), MBS (MYB binding site involved in drought-inducibility). However, there are no motif shared by all these genes. Therefore, we speculated that these specifically up-regulated genes may be regulated by multiple transcriptional factors rather than a common transcriptional factor.

A total of 107 genes were up-regulated in both Dongdao-4 and Jigeng-88 (Fig. 4). There are 3 genes with the magnitude of induction being higher in Dongdao-4 than in Jigeng-88, including *Os08g0518900*, *Os12g0564100* and *Os03g0847800*. *Os08g0518900* encoding xylanase inhibitor protein 1 is involved in defence responses that mainly due to biotic stress. Dimkpa et al. (2009) reported that defence response to biotic stress can alleviate abiotic stress conditions, including drought and salinity [44]. These cross-tolerance between abiotic and biotic stress may induce a positive effect in enhancing abiotic tolerance in plants [45–47]. Therefore, the higher transcription levels of genes involved in defence response may contribute to greater tolerance of Dongdao-4 seedlings. MYB transcription factors were reported to be involved in various abiotic stress in plants [48]. *Os12g0564100*, encoding a MYB transcription factor, may also underlie the mechanism of greater tolerance to salt stress in Dongdao-4.

Among the commonly suppressed genes in both genotypes, nine genes were inhibited greater in Dongdao-4 than in Jigeng-88 by salt treatment. These genes encoding myo-inositol oxygenase, mannose-1-phosphate guanylyltransferase and protein involving in photosynthesis, are reported to be involved in plant adaption to abiotic stresses. Myo-inositol oxygenase which catalyzes the oxidation of free myo-inositol to D-glucuronate, are related to osmotic balance and possibly to the transport of Na^+ from root to shoot in the common ice plant *Mesembryanthemum crystallinum* [49]. Cotsaftis et al. (2011) also found that transcript levels of genes encoding myo-inositol oxygenase was significantly down-regulated in salt-tolerant lines FL478 and Pokkli [50]. In addition, Das-Chatterjee et al. (2006) also reported that increased production of myo-inositol through overexpression of the L-*myo*-inositol 1-phosphate synthase gene in transgenic rice provided the plants with increased salinity tolerance [51]. In our results, one gene encoding myo-inositol oxygenase was significantly down-regulated in both genotypes, while the magnitude of reduction in Dongdao-4 was significantly higher than Jigeng-88 (Additional file 1: Figure S6c, Additional file 2: Table S7). This results indicated that decrease of the transcript levels of genes encoding enzymes that degrade or synthesize compatible solutes will be a mechanism to reduce the effects of the imposed stress, thus conferring greater tolerance of Dongdao-4 plants to salt stress.

Conclusions

In conclusion, we report here that Dongdao-4 was more tolerant to salt stress than Jigeng-88 in spite of an similar Na^+ concentration in shoot of both genotypes under salt stress. The greater salt-tolerant in Dongdao-4 may be caused by their greater capapcity to synthesize and accumulation of soluble sugars and Pro and higher activities of CAT under salt stress. Our findings demonstrate that the greater tolerance of Dongdao-4 to osmotic stress and efficient ROS detoxifying system underpin the higher tolerance to moderate salt stress than that of Jigene-88. The information about RNA-seq dataset also provides us potent and valuable guidance in underlying the mechanism with which Dongdao-4 has greater salt-tolerance than Jigeng-88.

Methods

Plant materials, growth conditions, and stress treatments

Two rice genotypes, *Oryza sativa* L. ssp. Japonica (cv. Dongdao-4 and Jigeng-88), were used in this study. Seeds were germinated in tap water at 37 °C for 2 days and transferred on moist tissue paper for 2 days at 30 °C in the dark. Thereafter the germinated seedlings were transferred to nutrient solution containing (mM): 1.425 NH_4NO_3, 0.42 NaH_2PO_4, 0.510 K_2SO_4, 0.998 $CaCl_2$, 1.643

MgSO$_4$, 0.168 Na$_2$SiO$_3$, 0.100 Fe-EDTA, 0.019 H$_3$BO$_3$, 0.009 MnCl$_2$, 0.155 CuSO$_4$, 0.152 ZnSO$_4$, and 0.075 Na$_2$MoO$_4$. This hydroponic experiment was carried out in a growth chamber with 30 °C/22 °C (day/night) with 14-h photoperiod, and the relative humidity was controlled at approximately 70%.

For analysis of tolerance to salt stress, Dongdao-4 and Jigeng-88 plants were grown in the culture solution for 3 weeks. Then, half of plants were transferred to culture solution supplemented with 20 mM NaCl for one day, and 40 mM NaCl for one day, and then exposed to 60 mM NaCl for one week. The remaining half of plants was kept in the medium without NaCl as controls. Medium pH was adjusted to 5.8 and the solution was renewed every 3 d for both control and NaCl-supplemented media. To evaluate the effects of osmotic stress on the two rice genotypes, Dongdao-4 and Jigeng-88 seedlings were cultured in the control hydroponic solution for three weeks, and then transferred to the hydroponic solution containing 40 mM mannitol for 1 day, and 80 mM mannitol for 1 day, and then exposed to 120 mM mannitol for 5 days. After treatment, height, shoot biomass, and root biomass were measured. Shoots and roots of the rice seedlings were harvested and oven-dried at 75 °C for 2 days until their weight reached constant for determination of dry weight.

Determination of Na$^+$ and K$^+$ concentration

Shoots and roots of rice seedlings were harvested and dried as described in the above section. They were digested with the 6 mL of nitric acid and 2 mL of perhydrol with microwave system (MARS, CEM) after grinding into fine powder. Total ion contents were determined using inductively coupled plasma mass spectrometry (ICAP6300; Thermo Scientific, Waltham, MA).

Measurements of total root length

To analyze total root length, roots were scanned with an Epson digital scanner (Expression 10000XL, Epson Inc.) and analyzed with the WinRHIZO/WinFOLIA software (Regent Instruments Inc.).

Measurements of chlorophyll (CHL) concentration

Newly formed leaves were harvested, weighed, and extracted with aqueous ethanol (95% v/v) in both Dongdao-4 and Jigeng-88 plants, in order to determine CHL concentration. Absorbance (A) readings of the supernatant was recorded at wavelengths of 663 and 645 nm. Total CHL concentration was calculated as 8.02A663 + 20.21A645, and was expressed as mg chlorophyll g^{-1} fresh weight.

Measurements of photosynthetic characteristics

Photosynthetic rates of rice seedlings were measured between 8:30–11:30 with a LI-6400 XT portable photosynthesis system equipped with a LED leaf cuvette (Li-Cor, 146 Lincoln, NE, USA). Artificial illumination was applied to the leaves in the chamber from a red-blue 6400-02B LED light source attached to the sensor head with continuous light (1000 µmol m^{-2} s^{-1} photosynthetic photon flux density) and ambient CO$_2$ concentration of approximately 500 µmol CO$_2$ per mol. At least 15 individual Dongdao-4 and Jigeng-88 plants in each stress treatment were selected for measuring photosynthetic rates.

Determination of Pro and soluble sugars

Approx. 100 mg dried shoot-materials were extracted in 5 mL 80% ethanol at 80 °C for 1 h, then cooled at room temperature. Approx. 50 mg active carbon were added to the extract for 30 min and then centrifuged at 6000 g for 10 min. The supernatants were filtered and used to determine concentrations of Pro and soluble sugars. Pro contents in rice leaves were determined by the method described previously [48, 52]. In brief, 2 mL filtrate was incubated with 2 mL ninhydrin reagent (2.5% (w/v) ninhydrin, 60% (v/v) glacial acetic acid, 40% 6 M phosphoric acid) and 2 mL of glacial acetic acid at 100 °C for 1 h, and the reaction were terminated in an ice bath. 4 mL toluene was added into the mixtures, followed by vibrating and incubation at room temperature. The absorbance was measured at wavelength of 520 nm using a spectrophotometer (SmartSpecTM Plus, BioRad).

Total soluble sugar content was measured following the methods used previously [48, 52]. Briefly, 5 mL anthrone reagent was added to 1 mL extract incubating at 95 °C for 15 min, and then cooled at room temperature. The absorbance was measured at wavelength of 625 nm using a spectrophotometer (SmartSpecTM Plus, BioRad).

Visualization of reactive oxygen species (ROS)

The formation of hydrogen peroxide and superoxide anion radicals was detected by 3,3'-diaminobenzidine (DAB) staining and nitroblue tetrazolium (NBT) staining, respectively, as described previously by Wohlgemuth et al. (2002) [53]. In brief, rice leaves from triplicate biological replicates of the samples were first cut into sections (c. 1 cm in length) and then immersed in 40 mL staining solution (0.1% (w/v) DAB, pH 6.5, or 0.1% (w/v) NBT, 10 mM sodium azide, 50 mM potassium phosphate, pH 6.4) in a desiccator. Infiltration was carried out by building up a vacuum (~100–150 mbar) and until the leaves were completely infiltrated. The incubation was conducted in a growth chamber in the dark overnight.

Determination of malondialdehyde (MDA)

Malondialdehyde (MDA) content in rice leaves was determined following the protocols described by Song et al. (2011) [52]. Briefly, rice leaves were weighed and homogenized in 5 mL of 10% TCA solution, and then centrifuged at 10,000 g for 10 min. Thereafter 2 mL supernatant was added in 2 mL of 10% trichloroacetic acid containing 0.6% thiobarbituric acid. The mixture was then incubated in water at 95 °C for 30 min and the reaction was stopped in an ice bath. The absorbance of the solution was measured at 450, 532, and 600 nm, respectively.

Determination of peroxidase, superoxide dismutase, and catalase activity

Approx. 0.5 g rice leaves were ground thoroughly with a cold mortar and pestle in 50 mM potassium phosphate buffer (pH 7.8) containing 1% polyvinylpyrrolidone. The homogenate was centrifuged at 15,000 g for 20 min at 4 °C. The supernatant was crude enzyme extraction. The activities of peroxidase (POD; EC 1.11.1.7), superoxide dismutase (SOD; EC 1.15.1.1), and catalase (CAT; EC 1.11.1.6) were measured using the protocols described by Yang et al. (2012) [48].

RNA isolation and real-time RT-PCR

Total RNA of shoots and roots was isolated using RNAiso reagent (Takara) and reverse-transcribed into first-strand cDNA with a PrimeScript® RT Reagent kit (Takara). Real-time PCR was performed in an optical 96-well plate with an Applied Biosystems Stepone™ Real-Time PCR system. Each reaction contained 7.5 µL of 2× SYBR Green Master Mix reagent, 0.5 µL of cDNA samples, and 0.6 µL of 10 µM gene-specific primers in a final volume of 15 µL. The thermal cycle used was as follows: 95 °C for 10 min, and 40 cycles of 95 °C for 30 s, 60 °C for 30 s, and 72 °C for 30s. All the primers used for quantitative RT-PCR are listed in Additional file 2: Table S1. The relative expression level was analyzed by the comparative Ct method.

Total RNA extraction, cDNA library construct ion and Illumina deep sequencing

Total RNA was extracted as described above. The preparation of whole transcriptome libraries and deep sequencing were performed by the Annoroad Gene Technology Corporation (Beijing, PR China). Whole transcriptome libraries were constructed using TruSeq Stranded Total RNA with Ribo-Zero Gold (Illumina, San Diego, CA, USA) according to the manufacturer's instructions. Libraries were controlled for quality and quantified using the BioAnalyzer 2100 system and qPCR (Kapa Biosystems, Woburn, MA, USA). The resulting libraries were sequenced initially on a HiSeq 2000 instrument that generated paired-end reads of 100 nucleotides.

Sequencing data analysis

The original image data were transferred into sequence data by base calling, which are defined as raw reads and saved as FastQ files. Prior to data analysis, the raw reads were filtered to obtain the clean reads by removing adaptors, tags of reads and low-quality reads. Then, the remaining high-quality reads were submitted for mapping analysis against the reference genome sequences (ftp:// ftp.ensemblgenomes.org/pub   /plants/release-30,   last accessed November 30, 2015) using Tophat [54]. Differentially expressed genes (DEGs) were identified using the empirical criterion of a greater than 2-fold change and a significant q value (false discovery rate-adjusted P value) of <0.05 based on two independent biological replicates. GO enrichment was performed using Cytoscape software (http://www.cytoscape.org, version 2.5.2) with Bingo plugin (http://www.psb.ugent.be/cbd/papers/BiNGO/, version 2.3). Hypergeometric test with Benjamini & Hochberg false discovery rate (FDR) were performed using the default parameters to adjust the P-value.

Statistical analysis

All data were analyzed by the analysis of variance using the SAS statistical software. Significant differences were evaluated using student's t-test.

Additional files

Additional file 1: Figure S1. Effects of salt stress on Dongdao-4 and Jigeng-88 seedlings. (a) Seedling growth performance, (b) Plant height, (c) Dry shoot biomass. Three-week-old rice seedlings grown in normal culture solution were transferred to culture solution supplemented with 20 mM NaCl for one day, and 40 mM NaCl for one day, and then exposed to 60 mM NaCl for one week. Bars, 10 cm. Data are means ±SE ($n \geq 4$). Means with different letters are significantly different ($P < 0.05$) within the same treatments. Asterisks (*) indicate significant differences between control and salt stress of the same genotype which were determined by Student's t-test (* $0.01 < P < 0.05$, ** $0.001 < P < 0.01$, *** $P < 0.001$). **Figure S2.** Foliar chlorophyll concentration (a) and Photosynthesis rates (b) of Dongdao-4 and Jigeng-88 plants grown at normal and salt stress conditions. Data are means ±SE ($n \geq 4$). Treatments and statistical analysis were as described in Additional file 1: Figure S1. **Figure S3.** Effects of salt stress on K^+ concentration (a), Na^+ concentration (b), Na^+/K^+ ratio (c) of shoot in Dongdao-4 and Jigeng-88 seedlings grown at normal and salt stress conditions. Data are means ±SE ($n \geq 4$). Treatments and statistical analysis were as described in Additional file 1: Figure S1. **Figure S4.** Principal component analysis (PCA) of the RNA sequencing data. The samples (two biological replicates) of each treatment were projected in four principal component; duplicates were projected together, which suggested that the duplicates were more similar. **Figure S5.** Verification of RNA-Seq results by real-time PCR. Correlation between data obtained from RNA-Seq and RT-PCR data. Data are mean ± SE of three replicates. **Figure S6.** Expression levels of GsSRK (a), WRKY46 (b) and Os06g0561000 (myo-inositol oxygenase) (c) in the two rice plants were analysed. Treatments and statistical analysis were as described in Additional file 1: Figure S1.

Additional file 2: Table S1. Primers used in real-time PCR to verify the expression pattern of differentially expressed genes. **Table S2.** Differentially expressed genes up-regulated only in Dongdao-4 under salt stress. **Table S3.** Differentially expressed genes up-regulated only in Jigeng-88 under salt stress. **Table S4.** Differentially expressed genes up-regulated in both Dongdao-4 and Jigeng-88 by salt stress. **Table S5.** Differentially expressed genes down-regulated only in Dongdao-4 under salt stress. **Table S6.** Differentially expressed genes down-regulated only in Jigeng-88 under salt stress. **Table S7.** Differentially expressed genes down-regulated in both Dongdao-4 and Jigeng-88 by salt stress. **Table S8.** List of all the differentially expressed genes in Dongdao-4 by salt stress. **Table S9.** List of all the differentially expressed genes in Jigeng-88 by salt stress. **Table S10.** Promoter analysis of specifically up-regulated genes in Dongdao-4 by PLantCARE (http://bioinformatics.psb.ugent.be/webtools/plantcare/html/).

Acknowledgements
This study was supported by the National Natural Science Foundation of China (31301832) and Chinese Academy of Sciences (XDA08010401).

Funding
The National Natural Science Foundation of China and Chinese Academy of Sciences.

Authors' contributions
QL, AY and WHZ designed the experiments; QL and AY conducted the experiments; QL and AY analyzed the data; QL and AY WHZ wrote the paper. All authors read and approved the final manuscript.

Competing interests
The authors declare that they have no competing interests.

Author details
[1]State Key Laboratory of Vegetation and Environmental Change, Institute of Botany, the Chinese Academy of Sciences, Beijing 100093, People's Republic of China. [2]University of Chinese Academy of Sciences, Beijing 100049, People's Republic of China. [3]Research Network of Global Change Biology, Beijing Institutes of Life Science, Chinese Academy of Sciences, Beijing 100093, China.

References
1. Rengasamy P. Soil processes affecting crop production in salt-affected soils. Funct Plant Biol. 2010;37:613–20.
2. Munns R, Tester M. Mechanisms of salinity tolerance. Annu Rev Plant Biol. 2008;59:651–81.
3. Kawasaki S, Borchert C, Deyholos M, Wang H, Brazille S, Kawai K, Galbraith D, Bohnert HJ. Gene expression profiles during the initial phase of salt stress in rice. Plant Cell. 2001;13:889–905.
4. Zeng L, Shannon MC, Grieve CM. Evaluation of salt tolerance in rice genotypes by multiple agronomic parameters. Euphytica. 2002;127:235–45.
5. Zhu JK. Abiotic stress signaling and responses in plants. Cell. 2016;167(2): 313–24.
6. Genc Y, McDonald GK, Tester M. Reassessment of tissue Na$^+$ concentration as a criterion for salinity tolerance in bread wheat. Plant Cell Environ. 2007; 30(11):1486–98.
7. Ashraf M, Foolad MR. Roles of glycine betaine and proline in improving plant abiotic stress resistance. Environ Exp Bot. 2007;59(2):206–16.
8. Apse MP, Blumwald E. Engineering salt tolerance in plants. Curr Opin Biotechn. 2002;13:146–50.
9. Nanjoa T, Kobayashia M, Yoshibab Y, Kakubaric Y, Yamaguchi-Shinozakid K, Shinozaki K. Antisense suppression of proline degradation improves tolerance to freezing and salinity in *Arabidopsis thaliana*. FEBS Lett. 1999;461:205–10.
10. Apel K, Hirt H. Reactive oxygen species: metabolism, oxidative stress, and signal transduction. Annu Rev Plant Biol. 2004;55:373–99.
11. Mishra P, Bhoomika K, Dubey RS. Differential responses of antioxidative defense system to prolonged salinity stress in salt-tolerant and salt-sensitive Indica rice (*Oryza sativa* L.) seedlings. Protoplasma. 2013;250(1):3–19.
12. Nagamiya K, Motohashi T, Nakao K, Prodhan SH, Hattori E, Hirose S, Ozawa K, Ohkawa Y, Takabe T, Takabe T, et al. Enhancement of salt tolerance in transgenic rice expressing an *Escherichia coli* catalase gene, katE. Plant Biotechnol Rep. 2007;1(1):49–55.
13. Shafi A, Chauhan R, Gill T, Swarnkar MK, Sreenivasulu Y, Kumar S, Kumar N, Shankar R, Ahuja PS, Singh AK. Expression of SOD and APX genes positively regulates secondary cell wall biosynthesis and promotes plant growth and yield in *Arabidopsis* under salt stress. Plant Mol Biol. 2015;87(6):615–31.
14. Nieves-Cordones M, Martinez V, Benito B, Rubio F. Comparison between *Arabidopsis* and rice for main pathways of K$^+$ and Na$^+$ uptake by roots. Front Plant Science. 2016;7:992.
15. Deinlein U, Stephan AB, Horie T, Luo W, Xu G, Schroeder JI. Plant salt-tolerance mechanisms. Trends Plant Sci. 2014;19(6):371–9.
16. Horie T, Hauser F, Schroeder JI. HKT transporter-mediated salinity resistance mechanisms in *Arabidopsis* and monocot crop plants. Trends Plant Sci. 2009; 14(12):660–8.
17. Rabbani MA, Maruyama K, Abe H, Khan MA, Katsura K, Ito Y, Yoshiwara K, Seki M, Shinozaki K, Yamaguchi-Shinozaki K. Monitoring expression profiles of rice genes under cold, drought, and high-salinity stresses and abscisic acid application using cDNA microarray and RNA gel-blot analyses. Plant Physiol. 2003;133(4):1755–67.
18. Ren ZH, Gao JP, Li LG, Cai XL, Huang W, Chao DY, Zhu MZ, Wang ZY, Luan S, Lin HX. A rice quantitative trait locus for salt tolerance encodes a sodium transporter. Nat Genet. 2005;37(10):1141–6.
19. Li Q, Yang A, Zhang WH: Efficient acquisition of iron confers greater tolerance to saline-alkaline stress in rice (Oryza sativa L.). J Exp Bot 2016.
20. Zhu M, Shabala S, Shabala L, Fan Y, Zhou MX. Evaluating predictive values of various physiological indices for salinity stress tolerance in wheat. J Agron Crop Sci. 2016;202(2):115–24.
21. Pires IS, Negrao S, Oliveira MM, Purugganan MD. Comprehensive phenotypic analysis of rice (*Oryza sativa*) response to salinity stress. Physiol Plantarum. 2015;155(1):43–54.
22. Almeida P, Feron R, Boer G-Jd, Boer AHd: Role of Na$^+$, K$^+$, Cl$^-$, proline and sucrose concentrations in determining salinity tolerance and their correlation with the expression of multiple genes in tomato. AoB Plants 2014, 6:plu039.
23. Roy SJ, Negrao S, Tester M. Salt resistant crop plants. Curr Opin Biotech. 2014;26:115–24.
24. Munns R, James RA, Xu B, Athman A, Conn SJ, Jordans C, Byrt CS, Hare RA, Tyerman SD, Tester M, et al. Wheat grain yield on saline soils is improved by an ancestral Na$^+$ transporter gene. Nature Biotechnol. 2012;30(4):360–4.
25. James RA, Blake C, Zwart AB, Hare RA, Rathjen AJ, Munns R. Impact of ancestral wheat sodium exclusion genes *Nax1* and *Nax2* on grain yield of durum wheat on saline soils. Functl Plant Biol. 2013;39(7):609–18.
26. Inupakutika MA, Sengupta S, Devireddy AR, Azad RK, Mittler R. The evolution of reactive oxygen species metabolism. J Exp Bot. 2016;67(21):5933–43.
27. Schieber M, Chandel NS. ROS function in redox signaling and oxidative stress. Curr Biol. 2014;24(10):R453–62.
28. Moradi F, Ismail AM. Responses of photosynthesis, chlorophyll fluorescence and ROS-scavenging systems to salt stress during seedling and reproductive stages in rice. Ann Bot. 2007;99(6):1161–73.
29. Demiral T, Türkan İ. Comparative lipid peroxidation, antioxidant defense systems and proline content in roots of two rice cultivars differing in salt tolerance. Environ Exp Bot. 2005;53(3):247–57.
30. Khaliq A, Zia-ul-Haq M, Ali F, Aslam F, Matloob A, Navab A, Hussain S. Salinity tolerance in wheat cultivars is related to enhanced activities of enzymatic antioxidants and reduced lipid peroxidation. CLEAN - Soil, Air, Water. 2015;43(8):1248–58.
31. Mandhania S, Madan S, Sawheny V. Antioxidant defense mechanism under salt stress in wheat seedlings. Biol Plantarum. 2006;50(2):227–31.
32. Meloni DA, Oliva MA, Martinez CA, Cambraia J. Photosynthesis and activity of superoxide dismutase, peroxidase and glutathione reductase in cotton under salt stress. Environ Exp Bot. 2003;49:69–76.

33. Gao R, Duan K, Guo G, Du Z, Chen Z, Li L, He T, Lu R, Huang J. Comparative transcriptional profiling of two contrasting barley genotypes under salinity stress during the seedling stage. Int J Genomics. 2013;2013:972852.

34. Wang WB, Kim YH, Lee HS, Kim KY, Deng XP, Kwak SS. Analysis of antioxidant enzyme activity during germination of alfalfa under salt and drought stresses. Plant Physiol Bioch. 2009;47(7):570–7.

35. Sultana S, Khew CY, Morshed MM, Namasivayam P, Napis S, Ho CL. Overexpression of monodehydroascorbate reductase from a mangrove plant (AeMDHAR) confers salt tolerance on rice. J Plant Physiol. 2012;169(3):311–8.

36. Sun XL, Yu QY, Tang LL, Ji W, Bai X, Cai H, Liu XF, Ding XD, Zhu YM. GsSRK, a G-type lectin S-receptor-like serine/threonine protein kinase, is a positive regulator of plant tolerance to salt stress. J Plant Physiol. 2013;170(5):505–15.

37. Vaid N, Macovei A, Tuteja N. Knights in action: lectin receptor-like kinases in plant development and stress responses. Mol Plant. 2013;6(5):1405–18.

38. Rajam MV, Chandola N, Goud PS, Singh D, Kashyap V, Choudhary ML, Sihachakr D. Thaumatin gene confers resistance to fungal pathogens as well as tolerance to abiotic stresses in transgenic tobacco plants. Biol Plantarum. 2007;51(1):135–41.

39. Munis MF, Tu L, Deng F, Tan J, Xu L, Xu S, Long L, Zhang X. A thaumatin-like protein gene involved in cotton fiber secondary cell wall development enhances resistance against Verticillium dahliae and other stresses in transgenic tobacco. Biochem Biophy Res Comm. 2010;393(1):38–44.

40. Rodrigo I, Vera P, Frank R, Conejero V. Identification of the viroid-induced tomato pathogenesis-related (PR) protein P23 as the thaumatin-like tomato protein NP24 associated with osmotic stress. Plant Mol Biol. 1991;16:931–4.

41. Mahmood T, Jan A, Komatsu S. Proteomic analysis of bacterial blight defence signalling pathway using transgenic rice overexpressing thaumatin-like protein. Biol Plantarum. 2009;53(2):285–93.

42. Tachi H, Fukuda-Yamada K, Kojima T, Shiraiwa M, Takahara H. Molecular characterization of a novel soybean gene encoding a neutral PR-5 protein induced by high-salt stress. Plant Physiol Bioch. 2009;47(1):73–9.

43. Ding ZJ, Yan JY, Xu XY, Yu DQ, Li GX, Zhang SQ, Zheng SJ. Transcription factor WRKY46 regulates osmotic stress responses and stomatal movement independently in Arabidopsis. Plant J. 2014;79(1):13–27.

44. Dimkpa C, Weinand T, Asch F. Plant-rhizobacteria interactions alleviate abiotic stress conditions. Plant Cell Environ. 2009;32(12):1682–94.

45. Rejeb IB, Pastor V, Mauch-Mani B. Plant responses to simultaneous biotic and abiotic stress: molecular mechanisms. Plants. 2014;3(4):458–75.

46. Fujita M, Fujita Y, Noutoshi Y, Takahashi F, Narusaka Y, Yamaguchi-Shinozaki K, Shinozaki K. Crosstalk between abiotic and biotic stress responses: a current view from the points of convergence in the stress signaling networks. Curr Opin Plant Biol. 2006;9(4):436–42.

47. Atkinson NJ, Urwin PE. The interaction of plant biotic and abiotic stresses: from genes to the field. J Exp Bot. 2012;63(10):3523–44.

48. Yang A, Dai X, Zhang WH. A R2R3-type MYB gene, OsMYB2, is involved in salt, cold, and dehydration tolerance in rice. J Exp Bot. 2012;63(7):2541–56.

49. Chauhan S, Forsthoefel N, Ran Y, Quigley F, Nelson DE, Bohnert HJ. Na$^+$/myo-inositol symporters and Na$^+$/H$^+$-antiport in Mesembryanthemum crystallinum. Plant J. 2000;24(4):511–22.

50. Cotsaftis O, Plett D, Johnson AA, Walia H, Wilson C, Ismail AM, Close TJ, Tester M, Baumann U. Root-specific transcript profiling of contrasting rice genotypes in response to salinity stress. Mol Plant. 2011;4(1):25–41.

51. Das-Chatterjee A, Goswami L, Maitra S, Dastidar KG, Ray S, Majumder AL. Introgression of a novel salt-tolerant L-myo-inositol 1-phosphate synthase from Porteresia coarctata (Roxb.) Tateoka (PcINO1) confers salt tolerance to evolutionary diverse organisms. FEBS Lett. 2006;580(16):3980–8.

52. Song SY, Chen Y, Chen J, Dai XY, Zhang WH. Physiological mechanisms underlying OsNAC5-dependent tolerance of rice plants to abiotic stress. Planta. 2011;234(2):331–45.

53. Wohlgemuth H, Mittelstrass K, Kschieschan S, Bender J, Weigel H-J, Overmyer K, Kangasjärvi J, Sandermann H, Langebartels C. Activation of an oxidative burst is a general feature of sensitive plants exposed to the air pollutant ozone. Plant Cell Environ. 2002;25:717–26.

54. Trapnell C, Pachter L, Salzberg SL. TopHat: discovering splice junctions with RNA-Seq. Bioinformatics. 2009;25(9):1105–11.

6

Prediction and analysis of three gene families related to leaf rust (*Puccinia triticina*) resistance in wheat (*Triticum aestivum* L.)

Fred Y Peng[1] and Rong-Cai Yang[1,2*]

Abstract

Background: The resistance to leaf rust (*Lr*) caused by *Puccinia triticina* in wheat (*Triticum aestivum* L.) has been well studied over the past decades with over 70 *Lr* genes being mapped on different chromosomes and numerous QTLs (quantitative trait loci) being detected or mapped using DNA markers. Such resistance is often divided into race-specific and race-nonspecific resistance. The race-nonspecific resistance can be further divided into resistance to most or all races of the same pathogen and resistance to multiple pathogens. At the molecular level, these three types of resistance may cover across the whole spectrum of pathogen specificities that are controlled by genes encoding different protein families in wheat. The objective of this study is to predict and analyze genes in three such families: NBS-LRR (nucleotide-binding sites and leucine-rich repeats or NLR), START (Steroidogenic Acute Regulatory protein [STaR] related lipid-transfer) and ABC (ATP-Binding Cassette) transporter. The focus of the analysis is on the patterns of relationships between these protein-coding genes within the gene families and QTLs detected for leaf rust resistance.

Results: We predicted 526 *ABC*, 1117 *NLR* and 144 *START* genes in the hexaploid wheat genome through a domain analysis of wheat proteome. Of the 1809 SNPs from leaf rust resistance QTLs in seedling and adult stages of wheat, 126 SNPs were found within coding regions of these genes or their neighborhood (5 Kb upstream from transcription start site [TSS] or downstream from transcription termination site [TTS] of the genes). Forty-three of these SNPs for adult resistance and 18 SNPs for seedling resistance reside within coding or neighboring regions of the *ABC* genes whereas 14 SNPs for adult resistance and 29 SNPs for seedling resistance reside within coding or neighboring regions of the *NLR* gene. Moreover, we found 17 nonsynonymous SNPs for adult resistance and five SNPs for seedling resistance in the *ABC* genes, and five nonsynonymous SNPs for adult resistance and six SNPs for seedling resistance in the *NLR* genes. Most of these coding SNPs were predicted to alter encoded amino acids and such information may serve as a starting point towards more thorough molecular and functional characterization of the designated *Lr* genes. Using the primer sequences of 99 known non-SNP markers from leaf rust resistance QTLs, we found candidate genes closely linked to these markers, including *Lr34* with distances to its two gene-specific markers being 1212 bases (to *cssfr1*) and 2189 bases (to *cssfr2*).

(Continued on next page)

* Correspondence: rong-cai.yang@ualberta.ca
[1]Department of Agricultural, Food and Nutritional Science, University of Alberta, 410 Agriculture/Forestry Centre, Edmonton, AB T6G 2P5, Canada
[2]Feed Crops Section, Alberta Agriculture and Forestry, 7000 - 113 Street, Edmonton, AB T6H 5T6, Canada

(Continued from previous page)

Conclusion: This study represents a comprehensive analysis of *ABC*, *NLR* and *START* genes in the hexaploid wheat genome and their physical relationships with QTLs for leaf rust resistance at seedling and adult stages. Our analysis suggests that the *ABC* (and *START*) genes are more likely to be co-located with QTLs for race-nonspecific, adult resistance whereas the *NLR* genes are more likely to be co-located with QTLs for race-specific resistance that would be often expressed at the seedling stage. Though our analysis was hampered by inaccurate or unknown physical positions of numerous QTLs due to the incomplete assembly of the complex hexaploid wheat genome that is currently available, the observed associations between (i) QTLs for race-specific resistance and *NLR* genes and (ii) QTLs for nonspecific resistance and *ABC* genes will help discover SNP variants for leaf rust resistance at seedling and adult stages. The genes containing nonsynonymous SNPs are promising candidates that can be investigated in future studies as potential new sources of leaf rust resistance in wheat breeding.

Keywords: ABC transporter, NLR (NBS-LRR), START, Rust resistance genes, Molecular markers, Single nucleotide polymorphism (SNP), Bread wheat, *Triticum aestivum*, Genome analysis

Background

Leaf rust is a fungal disease caused by *Puccinia triticina* (= *P. recondita* Roberge ex Desmaz. f. sp. *tritici.*) which has been a serious threat to the world production of bread wheat (*Triticum aestivum* L.) and other cereals over the past decades [1–3]. The leaf rust pathogen is a biotrophic parasite with many physiological races that often are highly specific to wheat cultivars with compatible resistance genes. To date, at least 75 leaf rust (*Lr*) resistance genes have been identified, and the majority of them confer the race-specific resistance in the seedling stage [4]. However, a few race-nonspecific *Lr* genes including *Lr34* and *Lr67* have been also found particularly at adult stage, conferring resistance to multiple pathogen species [5–9].

According to Krattinger et al. [10], the *Lr* genes in wheat and other cereals may be divided into three groups based on their specificity and durability. The first group contains genes that confer race-specific resistance against one but not other races of the same pathogen species. As mentioned above, the majority of the *Lr* genes are within this group. These *Lr* genes often encode intracellular immune receptor proteins with nucleotide-binding sites and leucine-rich repeats (NLR, also known as NBS-LRR or *R* genes). Proteins encoded by *R* genes directly or indirectly perceive pathogen-derived virulence effectors that are secreted into the cytoplasm of host cells in order to suppress basal immunity. The second group contains genes that confer race-nonspecific resistance against multiple fungal pathogens simultaneously. The well-studied example in this group is *Lr34*. This locus was first reported by Dyck [11] and it was mapped on wheat chromosome 7D. It has been subsequently known to confer resistance against multiple diseases including leaf rust with the resistance gene named as *Lr34*, stem rust (caused by *P. graminis*) with the resistance gene named as *Sr57*, stripe (yellow) rust (caused by *P. striiformis*) with the resistance gene

named as *Yr18*, powdery mildew (caused by *Blumeria graminis*) with the resistance gene named as *Pm38* and barley yellow dwarf virus with the resistance gene named as *Bdv*1 [11–14]. Thus, *Lr34* has other designations including *Lr34/Yr18*, *Lr34/Yr18/Sr57/Pm38* and *Lr34/Yr18/Sr57/Pm38/Bdv*1 in the literature. *Lr34* has been molecularly characterized and it encodes a putative ABC transporter containing transmembrane (TM) and nucleotide binding (NB) domains [6, 15]. The third group, like the second group, confers race-nonspecific resistance, but unlike the second group, such resistance is against all races within the same pathogen species. A known example in this group is a gene with resistance to stripe (yellow) rust in wheat (*Yr36*), and the resistance genes in the group are called START genes because they code for a START (steroidogenic acute regulatory [StAR] protein-related lipid transfer domain) protein [16]. Additional examples of *START* genes in other cereal species include the recessive rice blast resistance gene *pi21* encoding a small proline-rich protein [17], and the recessive barley powdery mildew resistance gene *mlo* coding for a membrane-anchored protein [18].

A common feature among the NLR, ABC and START proteins described above is the presence of distinct domains within each family. In the NLR group, a large number of potential R-genes or resistance gene analog (RGAs) encode R-proteins or effector-recognition receptors known as intracellular immune receptors and most belong to nucleotide-binding site-LRR (NBS-LRR or NLR) class [19] including seven domains or motifs: Toll/Interleukin-1 receptor (TIR-NBS-LRR or TNL), coiled-coil (CC- NBS-LRR or CNL), leucine zipper (LZ), NBS, LRR, TM and serinethreonine kinase (STK) [20]. In the ABC group, the only well-characterized gene (*Lr34*) encodes a full-size ATP-binding cassette (ABC) transporter and this protein is a member of the ABCG subfamily which is also known as the PDR (pleiotropic drug resistance) subfamily [6, 10, 12] and this protein impedes the invasion and spread of compatible pathogens in wheat and other cereals [21]. The functions of *Lr34* are constitutive rather than induced because the gene functions irrespective of whether or not pathogens are present. There is no known *Lr* gene that is yet available in the START group. *Yr36* is the only rust resistance gene in the group that is currently known to confer resistance to a broad spectrum of stripe rust races in wheat and this adult resistance is highly expressed at high temperatures (25-35 °C) [16]. In general, the *START* genes are not well studied but they are known to encode proteins for many functions in plants. The first function of the START proteins is, of course, the resistance to plant pathogens (e.g., the three START genes described above: *Yr36*, *pi21* [17] and *mlo* [18]). The second function is the response to abiotic stresses, e.g., increased expression of transmembrane START (TM-START) genes in chickpea in response to drought, salt, wound and heat stresses [22]. The third function is the ability to modulate transcription factor activity in Arabidopsis [23].

RNA-Seq is a more accurate method of quantifying gene expression levels than previous expression assay techniques such as microarray [24]. As RNA-Seq works without need for a genome sequence, it enables joint assays of host and pathogen transcriptomes, thereby gaining insights into how pathogens regulate the expression of their genes for disease progression and how they influence the host plant's circuitry during a defense response [25–27]. A recent study [27] reported a detailed RNA-Seq time-course for a susceptible wheat cultivar (Vuka) and a resistant line (Avocet-*Yr5*) inoculated with the wheat yellow rust pathogen *Puccinia striiformis* f. sp. *tritici* (*Pst*) at different days post-inoculation (dpi). These authors were able to identify clusters of differentially expressed genes in wheat plants and *Pst*. For example, they identified a total of seven clusters of genes with similar expression profiles that were enriched in GO (gene ontology) term annotations and KEGG (Kyoto Encyclopedia of Genes and Genomes) pathway memberships for the wheat host. In particular, their Cluster III of host genes had a peak expression at 11 dpi and those genes for membrane transport and ABC transporters and chitinases in this cluster were significantly enriched. Thus, it would be desired to identify the expression profiles of the host genes that belong to the *ABC*, *NLR* and *START* gene families.

Recent QTL mapping studies have reported several hundred QTLs for rust resistance in wheat populations [28–31]. Additionally, novel QTLs associated with wheat rust resistance have been reported in several genome-wide association studies (GWAS) [32–34], using the 9 K or 90 K single nucleotide polymorphism (SNP) chips [35, 36]. Most of the QTLs for leaf rust resistance at the seedling and adult stages are now stored in the T3 database [37]. However, little is known about the

physical relationships between these QTLs and the *ABC*, *NLR* and *START* genes. For example, it may be expected [10] that the *ABC* (and *START*) genes are more likely to share genomic regions with QTLs for race-nonspecific adult resistance whereas the *NLR* genes are more likely to share genomic regions with QTLs for race-specific resistance that is readily expressed in the seedlings. The objective of this study is to conduct bioinformatic prediction and gene annotation within *ABC*, *NLR* and *START* gene families in wheat that share genomic regions with QTLs for leaf rust resistance at the seedling and adult stages as obtained from the T3 database [37]. Specifically, we first predicted the putative *ABC*, *NLR* and *START* genes across the wheat genome. Then we attempted to establish physical relationships between these putative genes and the designated *Lr* genes from the sequences of SNP markers flanking the leaf rust QTLs. Together, this work provides an important framework for future studies to discover molecular functions of existing and new rust resistance genes in the wheat gene pool for their deployment in the development of wheat cultivars with improved rust resistance.

Results
Putative ABC, NLR and START proteins in wheat
In total, we predicted 526 *ABC* genes in the wheat genome. These ABC proteins were classified into eight subfamilies with the subfamily G further divided into Gwbc and Gpdr (Table 1). Our results show that the G (Gwbc and Gpdr), C, B and I subfamilies in wheat are the four largest in the ABC family, accounting for 30.4%, 23.0%, 18.8% and 16.2% of total 526 *ABC* genes, respectively. Notably, 78 ABC Gpdr transporters were predicted in the wheat genome, slightly more than a previous estimate of 60 full-size ABCG transporters estimated from the full-size ABCG gene numbers in Arabidopsis and rice [15, 38]. In comparison, we also performed a parallel analysis in Arabidopsis to validate the prediction pipeline. Our result showed that, of 131 ABC proteins found in Arabidopsis G (33.5%), B (20.8%), I (16.1%), C (12.8%) were the four largest subfamilies, which is consistent with the Arabidopsis result on the ABC proteins recently reported by Andolfo et al. [39]. These authors also reported 146 ABC proteins in rice [39], including 7 (4.8%), 29 (19.9%), 18 (12.3%), 3 (2.1%), 3 (2.1%), 7 (4.8%), 41 (28.1%), 22 (15.1%), and 16 (11.0%) in A, B, C, D, E, F, Gwbc, Gpdr and I families, respectively.

For the NLR family, we predicted 1117 NLR proteins in wheat (Table 1). This number is slightly higher than those in previous studies of the genome-wide analysis of NLR-coding genes in wheat [40, 41] but slightly lower than the 1185 NLR proteins predicted using the new gene models in the wheat genome assembly recently generated by

Table 1 Numbers of ABC, NLR and START family proteins and their subfamily classes over three wheat genomes (A, B and D) and chi-squared tests for even distributions across the genomes

Protein family	Class	Wheat genome			Total (%)	χ^2	P-value
		A	B	D			
ABC	A	6	6	7	19 (3.6)	0.105	0.949
	B	31	36	32	99 (18.8)	0.424	0.809
	C	32	49	40	121 (23.0)	3.587	0.166
	D	5	4	4	13 (2.5)	0.154	0.926
	E	3	4	4	11 (2.1)	0.182	0.913
	F	6	6	6	18 (3.4)	0.000	1.000
	Gpdr	24	28	26	82 (15.6)	0.308	0.857
	Gwbc	28	30	24	78 (14.8)	0.683	0.711
	I	30	26	29	85 (16.2)	0.306	0.858
	Total	165	189	172	526	1.738	0.419
NLR	CNL	320	346	313	979 (87.6)	1.853	0.396
	TNL	3	3	3	9 (0.8)	0.000	1.000
	N/A	47	46	36	129 (11.6)	1.721	0.423
	Total	370	395	352	1117	2.505	0.286
START	HD-ST	3	4	4	47 (32.6)	0.182	0.913
	HD-ST	2	2	3	54 (37.5)	0.286	0.867
	MINIM	16	16	15	11 (7.6)	0.043	0.979
	START	15	20	19	7 (4.9)	0.778	0.678
	OTHER	9	9	7	25 (17.4)	0.320	0.852
	Total	45	51	48	144	0.375	0.829

Abbreviations: *ABC* ATP-binding cassette; *NLR* nucleotide-binding site /leucine-rich repeat (NBS-LRR); *START* steroidogenic acute regulatory (StAR) protein-related lipid transfer domain

Earlham Institute [42]. In a concurrent analysis, we predicted 249 NLR proteins in Arabidopsis and the majority of them were TNL (Tir-NBS-LRR), accounting for nearly 69% (171), compared to less than 21% (52) of CNL (Coiled-coil-NBS-LRR) proteins. This high TNL/CNL ratio was consistent with previous studies of the NLR proteins in dicot species [40, 43, 44]. In contrast, nearly 88% (979) of the wheat NLR proteins were classified to the CNL class. However, nine (0.8%) wheat TNL proteins were detected using NLR-Parser, contrary to the commonly accepted belief that TNL genes might have been lost completely in monocots [19, 40, 45]. These nine proteins were further analyzed with PfamScan [46] and the analysis indicated that none of them contains a predicted TIR domain (PF01582), The analysis also showed that eight of them contain a truncated NB-ARC domain (PF00931) whose length ranges from 63 to 271 amino acids (the full length of the NB-ARC domain is 288 in Pfam) and the remaining one (Traes_3B_706A56165) was annotated by PfamScan as a Ceramidase family protein. Thus, this limited number of TNL proteins predicted in

wheat is likely due to NLR-Parser annotation error arising from the presence of truncated NB-ARC domains. In addition, 10.4% in Arabidopsis and 11.6% in wheat of the NLR proteins could not be classified to either CNL or TNL, which were designated as 'N/A' by NLR-Parser.

The *START* family genes were predicted exclusively based on the presence of START (IPR002913 and IPR005031) and/or START-like domain (IPR023393) following domain analysis of the wheat proteome using InterProscan [47]. From this analysis, we detected 139 and 144 *START* genes in Arabidopsis and wheat, respectively. Among the five classes, the START class was the largest in both Arabidopsis and wheat, with 75 (54%) in the former and 54 (37.5%) in the latter. HD-START was the second largest class in Arabidopsis with 21 members (15.1%), whereas MINIMAL-START was the second largest class in wheat with 47 members (32.6%). In both species, HD-START-MEKHLA was the smallest, each with seven members accounting for 5.0% in Arabidopsis and 4.9% in wheat. In an early study, only 35 Arabidopsis *START* genes were reported [48], but these authors used only five START-domain proteins (Arabidopsis GL2 and ATML1, rice ROC1 and AAP54082, and human PCTP) as queries for BLAST searches. Recently, Satheeth and colleagues [22] reported 36 *START* genes in chickpea (*Cicer arietinum* L.), using the START domain (~200 amino acids) encoded by the Arabidopsis gene *AT5G54170* (in our Arabidopsis *START* gene list) as query sequence in the TBLASTN search against the genome contigs database of chickpea. To the best of our knowledge, no genome-scale analysis of START proteins has been reported in wheat.

Our prediction was primarily based on domain analysis, with their domains summarized in Additional file 1. However, our classification of the putative ABC proteins in wheat into different subfamilies was based on protein sequence homology to the ABCs in Arabidopsis and rice (*Oryza sativa*). Therefore, we analyzed sequence similarity of ABC proteins and their two main types of domains (transmembrane domains or TMDs, and cytosolic nucleotide binding domains or NBDs) between wheat, Arabidopsis and rice (Fig. 1). The protein sequence identity was higher between wheat and rice than between wheat and Arabidopsis. On average, these entire ABC proteins shared 62.2% sequence identity (S.D. ± 14.9%) between wheat and Arabidopsis, and 71.3% (S.D. ± 19.1%) between wheat and rice (Fig. 1a). As expected, the ABC domains showed greater homology in both wheat-Arabidopsis and wheat-rice comparisons, especially for the NBDs including ABC transporter-like domain (IPR003439), AAA+ ATPase domain (IPR003593) and P-loop NTPase fold (IPR027417). For instance, the ABC transporter type 1 transmembrane domain (IPR011527) had an average sequence identity of 65.1%

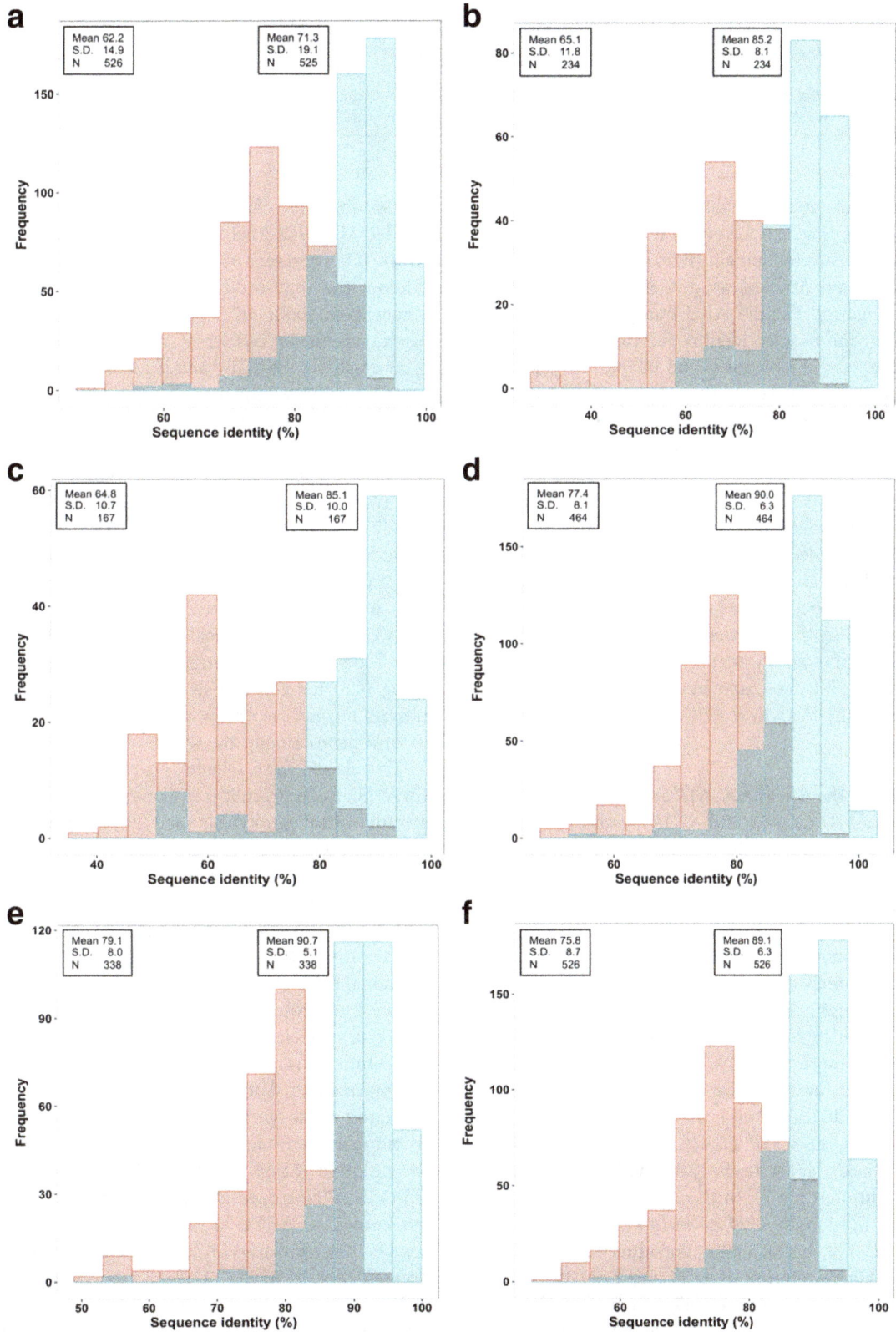

Fig. 1 (See legend on next page.)

between wheat and Arabidopsis, and 85.2% between wheat and rice (Fig. 1b). Likewise, the average identity of the ABC-2 transporter domain (IPR013525) was 64.8% between wheat and Arabidopsis, and 85.1% between wheat and rice (Fig. 1c). For the ABC transporter-like domain (IPR003439) the average identity increased to 77.4% between wheat and Arabidopsis, and 90.0% between wheat and rice (Fig. 1d). The homology of the AAA+ ATPase domain (IPR003593) was even higher between these species, with the average identity reaching 79.1% between wheat and Arabidopsis and 90.7% between wheat and rice (Fig. 1e). Lastly, the P-loop NTPase fold (IPR027417) showed slightly lower homology, with an average ideniity of 75.8% between wheat and Arabidopsis and 89.1% between wheat and rice (Fig. 1f). Although the sequence identity between wheat and rice ABC proteins is mostly higher than between wheat and Arabidopsis, the top hit of some wheat ABC proteins was from Arabidopsis. Thus, we used both Arabidopsis and rice ABCs to classfiy the wheat ABC proteins.

Annotation of the wheat *ABC, NLR* and *START* genes

As almost all *ABC, NLR* and *START* genes have no informative annotation (only ambiguous annotation such as uncharacterized protein or hypothetical protein) in the Ensembl Plants database [49], we enhanced their annotation using sequence analysis with BLAST [50]. We first performed BLAST searches against the known plant disease resistance genes in GenBank and PRGdb [51, 52]. Though 145 ABC proteins showed various degrees of sequence similarity to the known LR34 protein, only two had more than 97% identity (Additional file 2), suggesting that the respective *Lr34* and *Lr34-B* genes are *Traes_7DS_25B02BA46* and *Traes_4AL_603B6DC64* in Ensembl. *Traes_7DS_25B02BA46* is partial, encoding only 799 amino acids (aa) instead of 1401 aa for the full-length LR34 (Additional file 2). In the Ensembl release 33, we found the full-length *Lr34* gene (TRIAE_CS42_7DS_T GACv1_621754_AA2025300) encoding a protein of 1401 aa. Our annotation of *Lr34* and *Lr34-B* was confirmed by multiple sequence alignment (MSA) of the LR34 and LR34-B proteins in Ensembl as well as the known resistant (accession: ACN41354) and susceptible (ADK62371) LR34 proteins in GenBank (Fig. 2). Interestingly, this MSA indicated that the putative LR34 proteins

in Ensembl (both Traes_7DS_25B02BA46 in release 31 and TRIAE_CS42_7DS_TGACv1_621754_AA2025300 in release 33) represent the resistant version of LR34. In addition, the putative LR34-B (Traes_4AL_603B6DC64) protein has about 97% sequence identity with LR34 (Additional file 2), consistent with a previous study [15]. In contrast, in the NLR and START families, no genes showed over 90% sequence identity with known rust related proteins, and thus unlikely to represent the same genes. For example, Traes_1DS_5FF8D9E2D in the NLR family exhibited an identity of ~85% with the known stripe rust resistance protein YR10 in GenBank (accession: AAG42168). In the START family, only the sequence of a kinase-START domain protein from *T. dicoccoides* was found in GenBank (ACF33195), and the highest identity was ~63% at the protein sequence level with Traes_7-DL_A330C0F90 (Additional file 2). The relative low sequence identity may be attributed to (i) the use of Chinese Spring (CS) for genome sequencing and different wheat cultivars for gene isolation and (ii) increased allelic diversity of *R* genes among the wheat accessions. For example, *Lr1* was cloned from Glenlea [7] and *Lr10* from Thatcher*Lr10* [8]. In contrast, it is reasonable to expect that when the same wheat genotype is used for both genome sequencing and gene cloning perfect sequence matching may be found. The only example we have found so far for this case is *Lr34* whose sequencing and cloning was based on CS [6] and its protein sequence showed 100% identity with putative LR34 we annotated in this study (Fig. 2; Additional file 2).

Since numerous Arabidopsis genes have been characterized, especially for the ABC family, of which the major functions were largely known [21, 38, 53–55] (summarized in Additional file 3: Table S1), we then annotated these genes based on their best hit in the Arabidopsis genome (Additional file 2). For example, the putative LR34 in Ensmebl was annotated as ATPDR5 (Pleiotropic Drug Resistance 5, AT2G37280, also known as ABCG33 or ABC transporter G family member 33). Similarly, LR34-B (Traes_4AL_603B6DC64) was annotated as ATPDR9 (Pleiotropic Drug Resistance 9, AT3G53480, also known as ABCG37 or ABC transporter G family member 37). Thus, this annotation can be used to help select candidate genes for molecular characterization in wheat by leveraging the knowledge in the model plant.

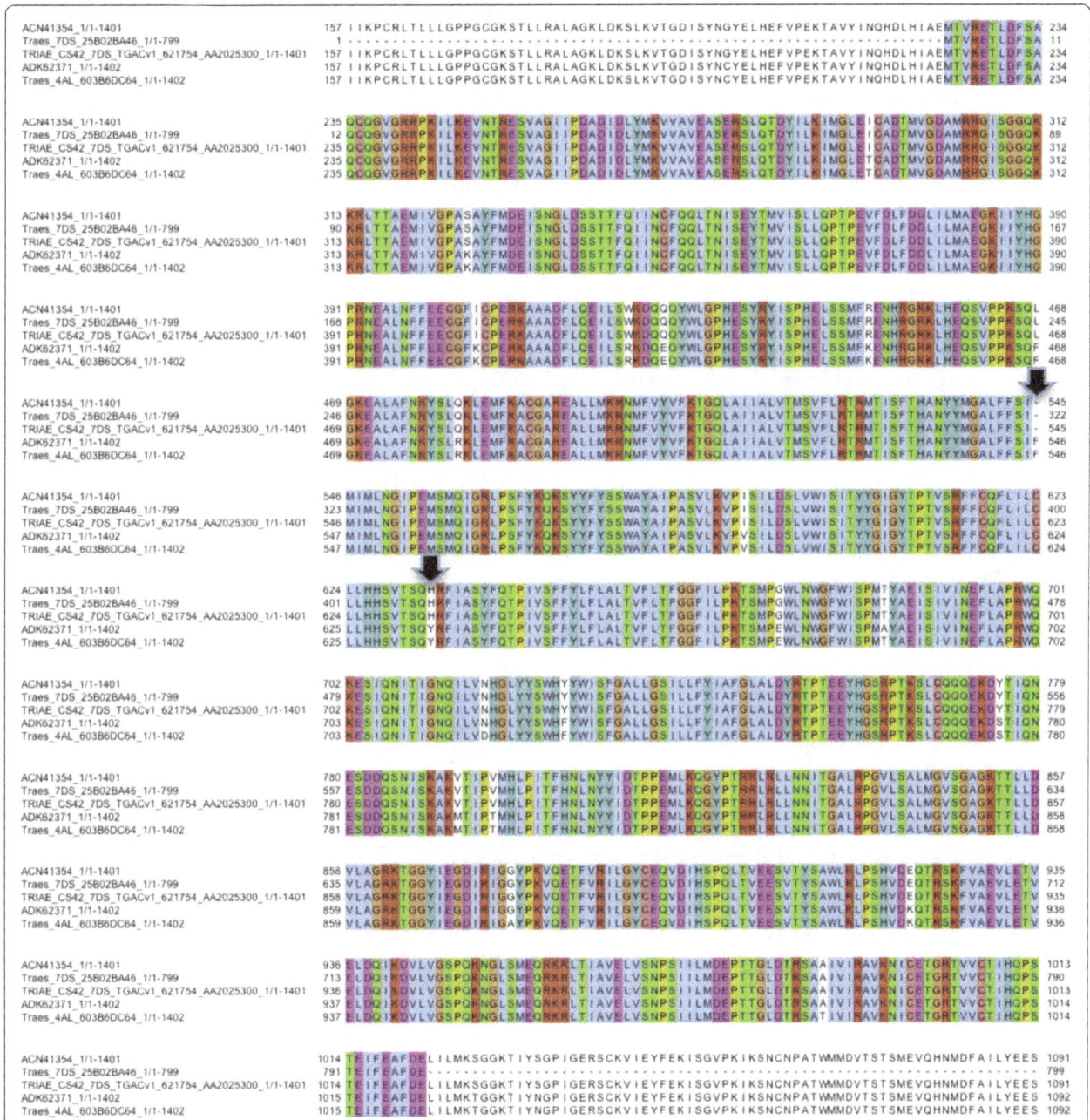

Fig. 2 Protein sequence alignment of the resistant and susceptible Lr34-D in Genbank and putative Lr34-D and Lr34-B in Ensembl. ACN41354: resistance version of GenBank LR34-D; ADK62371: susceptibility version of GenBank LR34-D; Traes_7DS_25B02BA46: LR34-D in Ensembl release 31 (partial); TRIAE_CS42_7DS_TGACv1_621754_AA2025300: LR34-D in the Ensembl release 33 (full length). Traes_4AL_603B6DC64: LR34-B in Ensembl release 31. The protein length is shown after the slash ('/') of each identifier. *Arrows* indicate the deletion of the phenylalanine residue (Phe/F) on position 546 and tyrosine (Y) substitution to histidine (His/H) on position 634 in LR34res-D. The beginning and ending regions of the sequence alignment were removed for clarity

Distribution of *ABC*, *NLR* and *START* genes in the three genomes of wheat and their physical proximity to SNPs of QTLs for leaf rust resistance

Of the *ABC* genes we predicted in this study, 165, 189 and 172 are from the A, B, and D genomes, respectively (Table 1). For *NLR* genes, 370 (CNL: 320, TNL: 3, N/A: 47), 395 (CNL: 346, TNL: 3, N/A: 46) and 352 (CNL: 313, TNL: 3, N/A: 36) are from A, B and D genomes respectively. For the *START* family: 45 (START: 15, HD-START: 3, HD-START-MEKHLA: 2, MINIMAL-START: 16, OTHER: 9), 51 (START: 20, HD-START: 4, HD-START-MEKHLA: 2, MINIMAL-START: 16, OTHER: 9), and 48 (START: 19, HD-START: 4, HD-START-MEKHLA: 3, MINIMAL-START: 15, OTHER: 7) are from A, B and D genomes respectively. The *ABC*, *NLR* and *START* genes were evenly distributed over the three wheat genomes (A, B and D), judging from insignificant chi-squared (χ^2) tests for individual subfamily classes.

From the two data sets each with 2500 leaf rust resistance QTLs being downloaded from the T3 database [37], we retained 1090 and 1542 significant ($P \leq 0.01$) QTLs for seedling and adult resistance, respectively. In these two data sets, a total of 1809 unique SNP markers were used (Additional file 4), and 451 of them are present in the QTLs of both seedlings and adult plants (Additional file 3: Fig. S1A). Because a total of 570 SNPs for seedling and adult resistance (representing 391 unique markers) have unknown (UNK) chromosomes (Additional file 3: Fig. S1B and C), we mapped their flanking sequences to the wheat genome to determine if these SNPs are located within the ABC, NLR and START genes analyzed in this study or their neighboring regions (5 Kb upstream of 5′ transcription start site [TSS] or downstream of 3′ transcription termination site [TTS] to the predicted genes).

The predicted *ABC*, *NLR* and *START* genes and the SNPs for seedling and adult resistance that reside within coding or neighboring (5 Kb upstream or downstream) regions of the predicted genes were mapped onto individual chromosomes in each of the three wheat genomes (A, B and D) (Fig. 3). The designated *Lr* genes previously mapped on each chromosome are also shown in the outer layer in Fig. 3 to indicate their physical affinity with the candidate genes and SNPs for leaf rust resistance. Unfortunately such affinity is very coarse due to the lack of mapping positions of the *Lr* genes on individual chromosomes. Some of the candidate genes with leaf rust SNPs only have positions on a scaffold (a set of ordered sequence contigs with gaps between them), and thus could not be depicted in Fig. 3 (Additional file 5). For this reason, we divided these SNPs into two categories: the genes with chromosome positions (WCP) and the genes with scaffold positions (WSP). The numbers of SNPs for seedling and adult resistance residing in coding regions of *ABC*, *NLR* and *START* genes or their neighborhood (within or 5 Kb upstream of 5′ end or downstream of 3′ end) showed different patterns (Table 2). For example, we found 18 SNPs in genic or neighboring regions of the *ABC* genes in seedlings and 39 in adult plants. Conversely, 29 and 14 SNPs for seedling and adult resistance were found in genic or neighboring regions of the *NLR* genes, respectively. In contrast, the START genes or their neighborhood contain roughly identical numbers of QTL SNPs for both seedlings and adult resistance.

Excluding the SNPs in non-coding and neighboring regions of *ABC* and *NLR* genes as shown in Table 2, we obtained a subset of SNPs that reside within coding regions and cause amino acid changes in these gene families (Table 3). These coding SNPs are the markers of QTLs identified for leaf rust resistance in wheat seedlings and adult plants as described above. The numbers of nonsynonymous SNPs in the coding regions of *ABC* and *NLR* genes exhibited different patterns in seedlings and adult plants. In the *ABC* genes, we found 17 missense SNPs for adult resistance and only five SNPs for seedling resistance. In contrast, in the *NLR* genes, we only predicted five nonsynonymous SNPs for adult resistance and six SNPs for seedling resistance (Table 3; Additional file 6). These candidate genes with nonsynonymous SNPs from the QTL data can be used in further studies to investigate their functional roles in *Lr* genes in wheat.

ABC, NLR and START genes linked to non-SNP molecular markers of rust resistance

The primer sequences of 99 non-SNP molecular markers linked to designated rust genes were used to identify corresponding candidate *ABC*, *NLR* and *START* genes in wheat. In total, we found 75 combinations of rust and Ensembl genes located near these markers, with the median physical distance of about 1.7 Mb (Additional file 7). Of these, eight genes are located less than 100 Kb away from the rust-associated markers, including two, five and one gene(s) in the *ABC*, *NLR* and *START* family, respectively (Table 4). For example, we found the *gwm350* marker is only 293 bp to *Traes_4AL_7AFBA6334*. This gene might therefore be *SrND643*, a temporarily designated stem rust gene conferring resistance to several Ug99 races of *Pgt* [56]. In addition, *Lr34* (*Yr18/Bdv1/Pm38/Ltn1*) on 7DS was associated with *Traes_7DS_25B02BA46*, with a distance of 1212 bp to *cssfr1* and 2189 bp to *cssfr2*, two known markers for *Lr34* [13, 57]. Along with Fig. 2 (above-mentioned MSA), this result further supports the annotation of *Traes_7DS_25B02BA46* as *Lr34*, demonstrating the usefulness of bioinformatics analyses in the discovery of rust resistance genes for functional characterization.

Gene expression

The expression differences of the putative *ABC*, *NLR* and *START* genes at different time points of infection (days post inoculation or dpi) in resistant and susceptible wheat plants are presented in Additional file 8. Our analysis revealed that the expression levels of the *NLR* genes for both resistant and susceptible plants were close to zero across all time points with the resistant plants being expressed only slightly more that the susceptible plants at comparable dpi (Fig. 4). In contrast, we observed higher levels of gene expression in the resistant wheat plants than in susceptible for ABC and START genes at early stages of infections (≤ 5 dpi) but

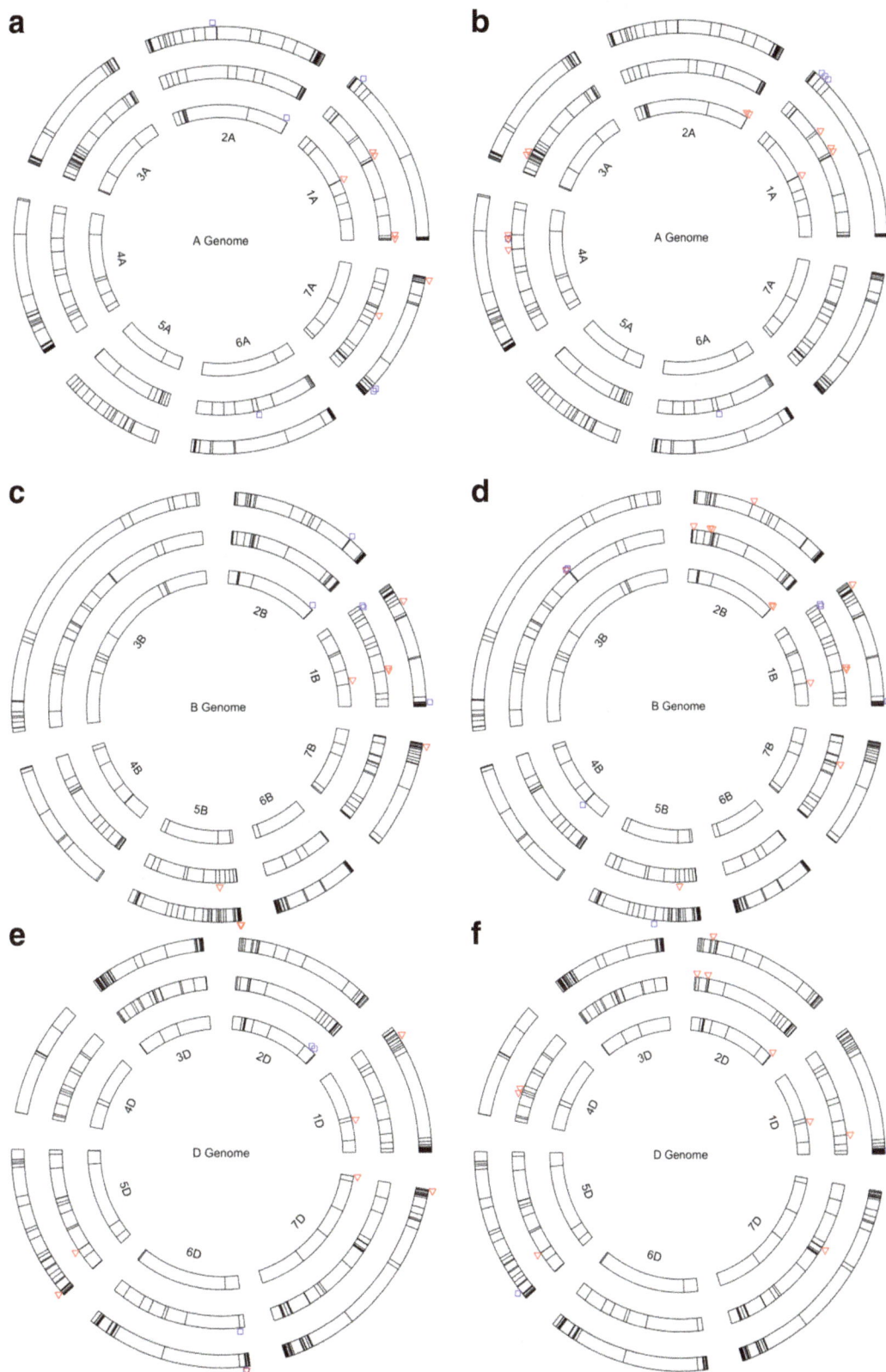

Fig. 3 (See legend on next page.)

susceptible plants had the similar amount of expression at the later stage (11 dpi). This is consistent with the finding in Fig. 3 of Dobon et al. [27] in which seven clusters of host genes were identified and one cluster (Cluster III) including the genes encoding ABC transporters had a peak expression at 11 dpi.

However, this result needs to be interpreted with care. Resistant and susceptible plants may have very different genetic backgrounds (i.e. with other rust resistance genes than *Yr5* across the wheat genome) because they are different cultivars but not isogenic lines. Similar gene expression analyses were performed in the wheat cultivars Thatcher (susceptible) and its near-isogenic line (NIL) carrying a leaf rust resistance gene *Lr9* (Thatcher*Lr9*) or *Lr10* (Thatcher*Lr10*) (resistant) [58, 59], but both using an EST (expressed sequence tag) based approach. These analyses showed some genes were expressed at higher levels in the resistant cultivar, including an *ABC* transporter gene, while several *NLR* genes were expressed in both the susceptible and resistant samples infected with *Ptr*. Additionally, gene expression changes induced by the rust resistance gene *Lr34/Yr18* were reported in the wheat NILs using Affymetrix wheat GeneChip microarray [60], which identified both upregulated and downregulated genes associated with resistance.

Discussion

The present study employed a domain analysis of wheat proteome to predict a total of 1787 genes, with 526 coding for ABC proteins, 1117 for NLR proteins and 144 for

START proteins over the three wheat genomes (A, B and D) (Table 1, Fig. 3). We were able to identify a number of SNPs that reside within protein-coding regions and cause amino acid changes of the predicted *ABC*, *NLR* and *START* genes or their neighborhood (5 Kb upstream from 5′ end or downstream from 3′ end). A total of 59 such SNPs were identified as the markers of QTLs for leaf rust resistance at the seedling stage and another 67 SNPs as the markers of QTLs for leaf rust resistance at the adult stage (Table 2; Fig. 3). The number of adult-resistance SNPs residing in the *ABC* genes or their neighborhood at the adult stage was more than doubled the number of seedling-resistance SNPs whereas the reverse pattern was observed for the SNP variants associated with the *NLR* genes (Table 2). This seems consistent with the expectation [10] that the QTLs for race-nonspecific, adult rust resistance are more likely to share genomic regions with the candidate genes in the *ABC* (and *START*) family whereas those for race-specific rust resistance in the seedlings are more likely to share genomic regions with the candidate genes in the *NLR* family.

This study represents a major step towards classification of the candidate genes for race-specific and race-nonspecific resistance in the three gene families and their chromosome-wise physical affinity with SNPs (and thus QTLs) for leaf rust resistance in wheat. Such information will be valuable for further characterizing QTLs at molecular level and their functional relationships with the *ABC*, *NLR* and *START* genes. We were already able to document our predicted amino acid changes

Table 2 Numbers of SNPs that reside within the *ABC*, *NLR* and *START* genes or their neighborhoods[a]

	Seedling			Adult		
	WCP[b]	WSP	Total	WCP	WSP	Total
ABC	14 (9)[c]	4 (3)	18 (12)	31 (19)	12 (8)	43 (27)
NLR	21 (15)	8 (6)	29 (21)	11 (8)	3 (2)	14 (10)
START	9 (7)	3 (3)	12 (10)	9 (7)	1 (1)	10 (8)

These SNPs are the markers of QTLs identified for leaf rust resistance in wheat seedlings and adult plants
[a]Neighborhood defined as 5 Kb upstream (from 5′ transcription start site or TSS) or downstream (from 3′ transcription termination site or TTS) of a gene, see Additional file 5 for details
[b]WCP, With known Chromosome Positions; WSP, With only Scaffold Positions
[c]The number of involved genes is indicated in parentheses

Table 3 Numbers of SNPs that reside within coding regions and cause amino acid changes in the ABC, NLR and START genes. These SNPs are the markers of QTLs identified for leaf rust resistance in wheat seedlings and adult plants[a]

	Seedling			Adult		
	WCP[b]	WSP	Total	WCP	WSP	Total
ABC	3 (3)[c]	2 (2)	5 (5)	10 (9)	7 (6)	17 (15)
NLR	4 (4)	2 (2)	6 (6)	3 (3)	2 (2)	5 (5)
START	1 (1)	0 (0)	1 (1)	2 (2)	1 (1)	3 (3)

[a]Seven synonymous SNPs (highlighted in Additional file 6) and their genes not included
[b]WCP, With known Chromosome Positions; WSP, With Scaffold Positions
[c]The number of involved genes is indicated in parentheses

Table 4 Candidate genes less than 100 Kb to rust resistance markers for designated rust genes in the ABC, NLR and START families

Designated rust gene	Chromosome	Marker	Ensembl gene[a]	Family[b]	Distance[c]
SrND643	4AL	gwm350	4AL_7AFBA6334	NLR (CNL)	293
Lr37/Sr38/Yr17	2AS	Ventriup-LN2	2AS_71B82606A	START (START)	503
Sr26	6AL	BE518379	6AL_91A76CBCC	ABC (I)	561
Lr34/Yr18/Bdv1/Pm38/Ltn1	7DS	cssfr1	7DS_25B02BA46[d]	ABC (Gpdr)	1212
Lr34/Yr18/Bdv1/Pm38/Ltn1	7DS	cssfr2	7DS_25B02BA46[d]	ABC (Gpdr)	2189
Sr33	1DS	Xcfd15	1DS_9EABB4B7A	NLR (CNL)	9523
Lr46/Yr29/Pm39/Ltn	1BL	Xwmc719	1BL_33D479058	NLR (CNL)	28,712
Sr13	6AL	CD926040	6AL_22F03E292	NLR (CNL)	29,257
SrND643	4AL	wmc219	4AL_7E627E983	NLR (CNL)	99,528

[a]The 'Traes_' prefix in each gene identifier removed
[b]Abbreviations: ABC, ATP-binding cassette; NLR, nucleotide-binding site /leucine-rich repeat (NB-LRR or NBS-LRR); START, steroidogenic acute regulatory (StAR) protein-related lipid transfer domain. The subgroup names were added in the parentheses. CNL: coiled-coil NBS-LRR; Gpdr: pleiotrpoic drug resitance class in G subfamily
[c]Physical distance in the number of bases between the gene start (5' transcription start site or TSS) and marker end (marker located upstream of a gene) or the gene end (3' transcription termination site or TTS) and marker start (marker located downstream of a gene)
[d]Traes_7DS_25B02BA46 represents the cloned Lr34 gene (cf. Figure 2)

due to alternation of single SNPs (point mutation) in the coding regions of the *ABC*, *NLR* and *START* genes (Fig. 2; Table 3). Our analysis was motivated by recent reports that changes in two amino acids caused the difference in proteins encoded by the resistance and susceptibility alleles of *Lr*34 [6, 15] and *Lr*67 [61]. Nevertheless, our prediction based on point mutation must be treated as a preliminary result as structural variations such as copy number variation (CNV) of resistance genes may possess a more dramatic impact on disease and other biotic stress [62]. Further complication arises from temporal variation among expressions of the putative *ABC*, *NLR* and *START* genes in compatible (susceptible) and incompatible (resistant) wheat-rust interactions as evident in our RNA-Seq analysis (Fig. 4). Thus, the predictions based on point mutation such as ours can only serve as a starting point towards more thorough molecular and functional characterization of resistance genes for their eventual cloning.

Our prediction indicates that the number of race-specific *NLR* genes (*R* genes) is almost doubled the

number of race-nonspecific genes (*ABC* and *START* genes). This seems consistent with the phenomenon that the majority of *Lr* genes identified so far in wheat are effective against single races of lest rust only. As Krattinger et al. [10] pointed out, this phenomenon is also true for resistances to other diseases in wheat and other cereals. The problem with the use of such race-specific *Lr* genes deployed in the wheat cultivars is that they have quickly become ineffective when new, more virulent races appear in leaf rust pathogen [63]. For this reason, wheat breeders and pathologists have focused on discovery, characterization and use of race-nonspecific genes for durable resistance. However, decades of genetic and breeding research have only been able to identify a limited number of genes in wheat with durable and broad-spectrum resistance to rusts including *Lr34/Yr18/Sr57/Pm38*, *Lr46/Yr29/Pm39*, *Lr67/Yr46/Sr55/Pm46*, *Lr68*, *Lr75*, and *Yr36* [5, 6, 15, 16, 61, 64, 65]. Thus far, *Lr34/Yr18/Sr57/Pm38* (or simply *Lr34* hereafter) is the most well-characterized multi-pathogen

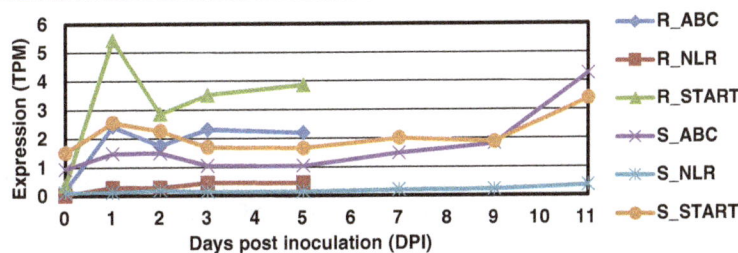

Fig. 4 RNA-seq expression profiles of *ABC*, *NLR* and *START* genes in resistant (R) wheat line (Avocet-Yr5) and susceptible (S) wheat cultivar (Vuka) inoculated with *yellow* rust pathogen *Puccinia striiformis* f. sp. *tritici* (Pst) during 5 and 11 days post inoculation, respectively. Abbreviations: DPI, days post inoculation; TPM, number of transcripts per million reads; R_ABC, *ABC* gene expression in resistant wheat; R_NLR, *NLR* gene expression in resistant wheat; R_START, *START* gene expression in resistant wheat; S_ABC, *ABC* gene expression in susceptible wheat; S_NLR, *NLR* gene expression in susceptible wheat; S_START, *START* gene expression in susceptible wheat

resistance gene located on wheat chromosome 7D [11]. Unfortunately we did not find any known SNPs for the leaf rust resistance in the T3 data sets [34] that would reside within coding regions of the *Lr34* gene or its neighborhood within 5 Kb upstream and downstream. Certainly, the wheat 9 K and 90 K SNP chips are hardly sufficient to cover large, complex wheat genomes, particularly the D genome (cf. Fig. 3). With ongoing international efforts on manifold sequencing of wheat genomes, exact physical positions and functions of those SNP variants within the *ABC* genes (Table 2) will be further clarified and characterized.

It may be questioned why *Lr67/Yr46/Sr55/Pm46* (or simply *Lr67*) was not included in this study as it is another well-characterized race-nonspecific gene at the molecular level just like *Lr34*. The reason is that it does not encode an ABC transporter, but rather it encodes a hexose transporter and confers a similar, but somewhat reduced partial resistance to all three wheat rust pathogens and powdery mildew [62]. Like in *Lr34*, only two amino acid substitutions (Arg144Gly and Leu387Val) distinguish the resistance (LR67res) from susceptibility (LR67sus) protein [62]. The hexose transport activity of LR67 has been demonstrated in a yeast (*Saccharomyces cerevisiae*) mutant lacking the glucose uptake ability [62, 66]. A preliminary analysis of five protein sequences encoded by *Lr67* [62] including three homoeologs (in A, B, D chromosomes) and the resistance/susceptibility versions from GenBank (accessions: ALL26327, ALL26328, ALL26329, ALL26330 and ALL26331) showed that none of them exhibited a high level of homology to the three family proteins (ABC, NLR and START) analyzed in this study (E-value $>1 \times 10^{-5}$ and sequence identity <36%). Since LR67 belongs to the sugar transporter protein (STP) family, a genome-wide search of the STP proteins using the major facilitator, sugar transport domain (InterPro accession IPR005828, a domain present in all the above five LR67 proteins), allowed for discovering 261 STP proteins (E-value $<1 \times 10^{-5}$) in wheat. This STP data set along with the ABC and other possible transporter families can be further examined in future studies for their physical relationships with QTLs for partial, adult resistance to multiple pathogens in wheat and other cereals.

This study has focused on annotating and characterizing the physical relationships between *ABC*, *NLR* and *START* genes and QTLs for resistance to leaf rust. Similar efforts can be made for stem and stripe (yellow) rusts. At the time of completion of this work as of February 2017, there is only a limited amount of QTL data available for stem and stripe rust resistance from the T3 database [37] that can be used for such annotation and characterization; in particular, imbalanced availability of QTLs for seedling and adult resistance makes a valid comparative assessment more difficult. As more QTL data become available for

stem and stripe rusts, it is conceivable that a similar pattern of the physical relationships between the gene families and QTLs will be identified, just like what we observed in leaf rust.

Our sequence analyses were greatly affected by the fragmentary status of the draft wheat genome assembly [3]. In the Ensembl release 31 of the wheat genome used in this study, the 21 wheat chromosomes were represented by a total of 317,977 assembled scaffold sequences. The average scaffold length was about 20,389 bp (S.D. ± 222 7384 bp), with the median of 3800 bp and maximum of 774,434,471 bp. Because of similar concerns about assembly quality, we chose not to use Ensembl release 33 (the latest release in October 2016), which incorporated short RNA-seq reads and full-length cDNA sequences [42] but seems more fragmented than release 31 according to our assessment. For example, the wheat genome assembly in release 33 consists of 735,943 supercontig sequences (compared with 317,977 in release 31), with an average supercontig length of 18,245 bp (S.D. ± 41,822 bp), median of 2431 bp and maximum of 823,974 bp. The fragmentary assembly of the wheat genome might also have affected our BLAST searches using the primer sequences against the wheat genomic sequences, and this may be the primary reason of why only seven *ABC*, *NLR* and *START* genes were identified within 100Kb to the markers of QTLs for rust resistance (Table 4, Additional File 7). Other studies based on similar fragmentary assemblies were not even able to identify candidate genes but rather some genomic regions for disease resistance and other agronomic traits. Two such studies are: (i) an in silico mapping of DArT marker sequences for identification of genomic regions harboring resistance to the three rusts [67] and (ii) flanking sequences of the SNPs of the wheat 90 K chip for identification of candidate genes and regions for preharvest sprouting (PHS) resistance in wheat [68], but with the candidate genes being inferred from Brachypodium (*Brachypodium distachyon*) and rice genomes via comparative mapping.

In addition to the *ABC*, *NLR* and *START* genes analyzed in this study, genes in other families may also be involved in rust resistance. We have already mentioned genes within the STP transporter family like *Lr67*. Additional genes include those encoding unusual kinases, receptor-like kinases (RLKs) and receptor-like proteins (RLPs) [69], and genes involving cell wall biosynthesis and metabolism, such as cysteine proteinase, phenylalanine ammonia-lyase, plasma membrane ATPase and chalcone synthase [58]. When a pathogen attacks its host, the first line of host defense is through secretion of cell surface pattern-recognition receptors (PRRs) and such defense-induced proteins may be recognized and overcome by specific pathogen effectors. The majority of characterized PRRs are either RLKs or RLPs [70–72]. A recently developed tool called RGAugury [73] focused on classification of these proteins and NLR protein family. However, our study had a different focus on the three major protein families related to leaf rust resistance in wheat and these families contain different domains suitable for large-scale analyses.

Certainly with ongoing international efforts for more in-depth wheat genome sequencing, the candidate genes we identified for each of the three gene families eventually will be physically mapped at accurate locations on individual chromosomes. In other words, Fig. 3 will be significantly clarified and updated. Such updating will enable wheat breeders and pathologists to use the well-characterized *ABC*, *NLR* and *START* genes for molecular characterization of the designated *Lr* genes in breeding for rust resistances. While *ABC* and *START* genes may be preferably used as reasoned above, their resistance is only partial and insufficient under high disease pressure when used alone. Strong, durable resistance has been achieved by combining these race-nonspecific genes with race-specific *NLR* genes such as 'Sr2-complex' [74]. However, little is known about the other combining genes and their interactions in the complex. Thus, it would be desirable that future studies will investigate how race-nonspecific genes in the *ABC* or *START* gene family interact with weak and strong *R* genes in the *NLR* gene family at the molecular level using the candidate genes we identified and characterized for the three gene families.

Conclusions

In this study we predicted the putative *ABC*, *NLR* and *START* genes in the hexaploid wheat genome, and performed an integrated analysis with the available genetic resources related to rust resistance, including mapped *Lr* genes, SNPs and other molecular markers. Our analysis suggests that the *ABC* (and *START*) genes are more likely to share genomic regions with QTLs for race-nonspecific, adult resistance whereas the *NLR* genes are more likely to share genomic regions with QTLs for race-specific resistance that would be often expressed at the seedling stage. Candidate genes such as those containing one or multiple missense coding SNPs can be tested in future studies. With continuing efforts for producing an improved assembly and annotation of the wheat genome, bioinformatics analyses such as ours can help identify novel genes for rust resistance breeding in wheat.

Methods

Prediction of ABC, NLR and START genes in wheat genome

The wheat proteome (release 31) was downloaded from the FTP archive site (ftp://ftp.ensemblgenomes.org/pub/

release-31/plants/fasta/triticum_aestivum/pep/Triticum_aestivum.IWGSC1+popseq.31.pep.all.fa) of Ensembl Plants [49]. This database contains a total of 100,344 predicted protein sequences. Using these protein sequences, we predicted the domains within each protein using the standalone software InterProScan (version 5.21–60.0, 64-bit) [47], installed on a CentOS (release 7.2) Linux system with 8-core CPU (1600 MHz) and 32 Gb RAM. The InterProscan E-value cutoff was set to 1×10^{-5} [47]. From the InterProscan output file, the presence of PF00005 (ATP-binding domain of ABC transporters) from the protein family database Pfam [46] within a protein sequence was used to select candidate ABC proteins. These candidates were then classified into eight families (A, B, C, D, E, F, G and I) based upon their best BLASTP hit against the known ABC proteins in Arabidopsis and *Oryza sativa* (rice), which were collected from the literature [54, 75]. The G family was further divided into Gwbc (White Brown Box) and Gpdr (Pleiotropic Drug Resistance). For the START family, we used the presence of the START domain (InterPro: IPR002913 and IPR005031) or START-like domain (IPR023393) as the signature, and grouped the putative START proteins into five categories: MINIMAL-START, START, HD-START, HD-START-MEKHLA and OTHER. This classification scheme is similar to those reported in previous studies [22, 48]. A MINIMAL-START protein only contains a single START or START-like domain, whereas a START protein contains multiple domains with at least one START or START-like domain. Moreover, a multidomain START protein was designated as HD-START if it contains a homeodomain-like domain (IPR009057) and as HD-START-MEKHLA if it also contains a MEKHLA domain (IPR013978). The NBS-LRR resistance proteins were predicted using the NLR-Parser tool [76] in a two-step process. First, the entire wheat proteome was scanned with the MEME-formatted motifs (meme.xml) using MAST (Motif Alignment and Scanning Tool) in the MEME (version 4.9.1) suite [77]. Second, the MAST output (mast.xml) was parsed and the NBSLRR genes were classified with NLR-Parser into three classes: CNL (COIL-NBS-LRR), TNL (TIR-NBS-LRR) and N/A.

In silico mapping of the QTL SNP markers to ABC, NLR and START genes on the wheat chromosomes

Two data sets, each with 2500 quantitative trait loci (QTLs) for leaf rust resistance at seedling or adult stage, were extracted from the "GWAS Results" in the Triticeae Toolbox (T3) database (https://triticaetoolbox.org/wheat/qtl/qtl_report.php) [37]. By selecting 'Biotic stress' in the "Category" column and by clicking appropriate leaf rust traits at adult or seedling stage in the "Traits" column, the T3 would display the results from the GWAS analysis for associations between markers (including Infinium 9 K, Infinium 90 K, and GBS restriction sites) and traits for individual trials (individual locations or inoculum types) within the T3 database. The GWAS analysis was carried out using rrBLUP GWAS package [78] for individual trials or the combined analysis across all trials with the genotype-by-environment interaction effect being adjusted by including those principle components that accounted for more than 5% of the environment-relationship matrix variance as fixed effects in the mixed-model analysis. For a better control of false positive rates in these preliminary QTL data sets, we applied a significance level of $P \leq 0.01$ which allowed for retaining 1090 QTLs in seedlings and 1542 QTLs in adult plants for subsequent analysis (Additional file 4).

The sequences of the SNP markers from the GWAS results described above were obtained in T3 and CerealsDB [79]. To predict their functional consequences, these SNPs were first converted into a VCF (Variant Call Format) and annotated with VEP [80]. Then, to better estimate the genomic position of each SNP and its proximity to the predicted genes in the *ABC*, *NLR* and *START* gene families, we mapped the sequence (removing the '/', B allele and the surrounding square brackets) flanking each SNP by alignment to the wheat genome assembly (Triticum_aestivum.IWGSC1 + popseq.31.dna_sm.toplevel.fa) using BLASTN (version 2.5.0+) in the BLAST+ package [50]. This assembly was generated by Ensembl Plants from the whole genome shotgun sequencing data [2], a chromosome-based draft sequence [3] and the POPSEQ (population sequencing) data [81]. The chromosomal positions of these genes were based on the Ensembl annotation file (Triticum_aestivum.IWGSC1 + popseq.31.gff3) in GFF3 (General Feature Format) format. Both files were downloaded from the Ensembl FTP archive site (ftp://ftp.ensemblgenomes.org/pub/release-31/plants/). As each flanking sequence often matches several genomic regions, we used E-value, alignment identity and query coverage as criteria to retain the top three candidates that usually represent one hit in each of the three wheat genomes.

For the SNPs located in the genomic regions covering the *ABC*, *NLR* and *START* genes, those in the coding regions were recognized through aligning their flanking sequences with their coding sequence (CDS) downloaded from Ensembl. We then compared the two alleles of each SNP marker with the allele at the same position of the Ensembl CDS and determined if the Ensembl allele (as reference) was the same as the A or B allele at the SNP site, or in some cases differed from both alleles. To predict the amino acid change, we replaced the Ensembl CDS with the A or B allele, translated the newly generated CDS into protein sequence using the standard codon table. Therefore, if the Ensembl reference allele matches the A or B allele at the SNP site,

two protein sequences were generated and otherwise three protein sequences were translated for each SNP marker. To verify this process, we utilized the known SNP (and deletion) in the *Lr34* CDS as well as the resistant and susceptible LR34 proteins [6, 15]. We then found the amino acid difference at the corresponding position of each protein translated from each CDS containing different SNP alleles.

Prediction of candidate genes based on sequences of non-SNP markers in wheat

We collected a set of 99 non-SNP markers that showed significant (a LOD score of ≥ 3.0) associations with wheat rust resistance from Liu et al. [57] and other sources. The primer sequences of these markers were found in GrainGenes [82], MASWheat (http://maswheat.ucdavis.edu/protocols/stemrust/) and the literature (Additional file 7). Similar to the above analysis of SNP flanking sequences, we searched the primer sequences against the wheat genome sequence assembly (Triticum_aestivum.IWGSC1 + popseq.31.dna_sm.toplevel.fa) using BLASTN [50] to identify genomic regions covered by these markers. Each primer query was generated by concatenating the forward primer and reverse primer with five 'N' letters inserted between them as gaps. Because of the short length of each primer pair (usually less than 40 bp), blastn-short, a BLASTN program optimized for sequences shorter than 50 bases [50], was used with settings of a word size of seven and a relaxed E-value of 100. To compare chromosomal positions of markers and candidate genes, we also used the Ensembl annotation file (Triticum_aestivum.IWGSC1 + popseq.31.gff3).

Transcription analysis

Dobon et al. [27] recently published a RNA-seq time-course data set for a resistant line (Avocet-*Yr5*) and a susceptible wheat cultivar (Vuka) inoculated with yellow rust pathogen *Puccinia striiformis* f. sp. tritici (*Pst*). The rust inoculum (isolate 87/66 of *Pst*) is avirulent to the resistant line (Avocet-*Yr5*), but virulent to the susceptible cultivar (Vuka). We analyzed the expression values (Transcripts Per Million reads or TPM) of 123,532 wheat transcripts (based on the annotation of Ensembl Plants [49]) taken from Supplemental Tables S20 and S21 of Dobon et al. [27]. We identified putative *ABC*, *NLR* and *START* genes from those host transcripts and analyzed their expression profiles at different days post inoculation (dpi) in the resistant and susceptible wheat plants. The original data on the expression profiles contain five time points (0, 1, 2, 3, and 5 dpi) for Avocet-*Yr5*, and eight (0, 1, 2, 3, 5, 7, 9, and 11 dpi) for Vuka. We then extracted the TPM values for the putative *ABC*, *NLR* and *START* genes, eliminating those with zero values on all dpi. For each of the putative genes in the three gene families, we calculated a fold change as the sum of the five expression values in the resistant host divided by the sum in the susceptible host. In addition, we identified *ABC*, *NLR* and *START* genes expressed exclusively in the resistant host or the susceptible host.

Additional files

Additional file 1: Summary of the family classification of the ABC, NLR and START proteins in wheat and their predicted InterPro domains as well as the description for each InterPro identifier.

Additional file 2: Annotation of the putative *ABC*, *NLR* and *START* genes in wheat using the sequences of plant resistance genes in GenBank and PRDdb.

Additional file 3: Table S1. Summary of the ABC transporter family, domain structures and major functions in plants. **Table S2.** The VEP annotation of the 1809 unique SNPs associated with leaf rust resistance in seedling and adult wheat plants from the T3 database. **Figure S1.** The Venn diagram (A) of the SNP markers of leaf rust resistance QTLs in seedlings and adult plants in the T3 database and their distribution on different wheat chromosomes (B: seedlings; C: adult plants).

Additional file 4: The leaf rust resistance QTLs ($P \leq 0.01$) from the T3 database of the wheat seedlings and adult plants and the flanking sequences of their SNP markers from CerealsDB and T3 database.

Additional file 5: The SNP markers located within (marker located between the gene start and the gene end on the same chromosome; highlighted) ABC, NLR and START genes and 5 Kb upstream (from transcription start site or TSS) or downstream (from transcription termination site or TTS) of them.

Additional file 6: Predicted effects of rust SNPs on the amino acid changes in the proteins encoded by the putative ABC, NLR and START genes. Genes with multiple SNPs were highlighted.

Additional file 7: The molecular markers and their primer sequences found in GrainGenes and candidate genes near these makers.

Additional file 8: RNA-seq expression values (in unit of TPM or transcripts per million) of *ABC*, *NLR* and *START* genes in resistant and susceptible wheat hosts at five time points (0 dpi, 1 dpi, 2 dpi, 3 dpi and 5 dpi or days post inoculation), fold changes of genes expressed in both resistant and susceptible wheat, and genes expressed only in the resistant or susceptible wheat.

Abbreviations
ABC: ATP-binding cassette; CDS: coding sequence; DPI: days post inoculation; NLR: nucleotide-binding site /leucine-rich repeat (NBS-LRR); Pdr: pleiotropic drug resistance; QTL: quantitative trait locus; SNP: single nucleotide polymorphism; START: steroidogenic acute regulatory (StAR) protein-related lipid transfer domain; STP: sugar transporter; UTR: untranslated region; Wbc: white brown complex

Acknowledgements
We wish to thank three reviewers for critical and constructive comments on an earlier draft, and Dr. Zhiqiu Hu for helpful discussion and for his help with the preparation of Fig. 3.

Funding
This research is funded by the *Growing Forward* 2 Research Opportunities and Innovation Internal Initiatives of Alberta Agriculture and Forestry to R-C Yang.

Authors' contributions
Conceived and designed the analyses: RCY and YFP; Performed the analyses: FYP and RCY; Writing of the manuscript: YFP and RCY; Both authors read and approved the final version of the manuscript.

Competing interests
The authors declare that they have no competing interests.

References
1. Choulet F, Alberti A, Theil S, Glover N, Barbe V, Daron J, et al. Structural and functional partitioning of bread wheat chromosome 3B. Science. 2014;345(6194):1249721.
2. Brenchley R, Spannagl M, Pfeifer M, Barker GLA, D'more R, Allen AM, et al. Analysis of the bread wheat genome using whole-genome shotgun sequencing. Nature. 2012;491(7426):705–10.
3. The International Wheat Genome Sequencing Consortium. A chromosome-based draft sequence of the hexaploid bread wheat (*Triticum aestivum*) genome. Science. 2014;345(6194):1251788.
4. McIntosh R, Dubcovsky J, Rogers W, Morris C, Appels R, Xia X, AZUL B: Catalogue of gene symbols for wheat: 2013–2014. 2013.
5. Kolmer JA, Singh RP, Garvin DF, Viccars L, William HM, Huerta-Espino J, et al. Analysis of the Lr34/Yr18 rust resistance region in wheat germplasm. Crop Sci. 2008;48(5):1841–52.
6. Krattinger SG, Lagudah ES, Spielmeyer W, Singh RP, Huerta-Espino J, McFadden H, et al. A putative ABC transporter confers durable resistance to multiple fungal pathogens in wheat. Science. 2009;323(5919):1360–3.
7. Cloutier S, McCallum BD, Loutre C, Banks TW, Wicker T, Feuillet C, et al. Leaf rust resistance gene Lr1, isolated from bread wheat (*Triticum aestivum* L.) is a member of the large psr567 gene family. Plant Mol Biol. 2007;65(1–2):93–106.
8. Feuillet C, Travella S, Stein N, Albar L, Nublat A, Keller B. Map-based isolation of the leaf rust disease resistance gene Lr10 from the hexaploid wheat (*Triticum aestivum* L.) genome. Proc Natl Acad Sci U S A. 2003;100(25):15253–8.
9. Huang L, Brooks SA, Li WL, Fellers JP, Trick HN, Gill BS. Map-based cloning of leaf rust resistance gene Lr21 from the large and polyploid genome of bread wheat. Genetics. 2003;164(2):655–64.
10. Krattinger SG, Sucher J, Selter LL, Chauhan H, Zhou B, Tang MZ, et al. The wheat durable, multipathogen resistance gene Lr34 confers partial blast resistance in rice. Plant Biotechnol J. 2016;14(5):1261–8.
11. Dyck PL. The association of a gene for leaf rust resistance with the chromosome 7D suppressor of stem rust resistance in common wheat. Genome. 1987;29(3):467–9.
12. Risk JM, Selter LL, Chauhan H, Krattinger SG, Kumlehn J, Hensel G, et al. The wheat Lr34 gene provides resistance against multiple fungal pathogens in barley. Plant Biotechnol J. 2013;11(7):847–54.
13. Lagudah ES, Krattinger SG, Herrera-Foessel S, Singh RP, Huerta-Espino J, Spielmeyer W, et al. Gene-specific markers for the wheat gene Lr34/Yr18/Pm38 which confers resistance to multiple fungal pathogens. Theor Appl Genet. 2009;119(5):889–98.
14. Spielmeyer W, McIntosh RA, Kolmer J, Lagudah ES. Powdery mildew resistance and Lr34/Yr18 genes for durable resistance to leaf and stripe rust cosegregate at a locus on the short arm of chromosome 7D of wheat. Theor Appl Genet. 2005;111(4):731–5.
15. Krattinger SG, Lagudah ES, Wicker T, Risk JM, Ashton AR, Selter LL, et al. Lr34 multi-pathogen resistance ABC transporter: molecular analysis of homoeologous and orthologous genes in hexaploid wheat and other grass species. Plant J. 2011;65(3):392–403.
16. Fu DL, Uauy C, Distelfeld A, Blechl A, Epstein L, Chen XM, et al. A kinase-START Gene confers temperature-dependent resistance to wheat stripe rust. Science. 2009;323(5919):1357–60.
17. Fukuoka S, Saka N, Koga H, Ono K, Shimizu T, Ebana K, et al. Loss of function of a proline-containing protein confers durable disease resistance in Rice. Science. 2009;325(5943):998–1001.
18. Buschges R, Hollricher K, Panstruga R, Simons G, Wolter M. Frijters A, vanDaelen R, vanderLee T, Diergaarde P, Groenendijk J *et al*: The barley mlo gene: A novel control element of plant pathogen resistance. Cell. 1997; 88(5):695–705.
19. Kim J, Lim CJ, Lee BW, Choi JP, Oh SK, Ahmad R, et al. A genome-wide comparison of NB-LRR type of resistance gene analogs (RGA) in the plant kingdom. Mol Cells. 2012;33(4):385–92.
20. van Ooijen G, van den Burg HA, Cornelissen BJC, Takken FLW. Structure and function of resistance proteins in solanaceous plants. Annu Rev Phytopathol. 2007;45:43–72.
21. Kang J, Park J, Choi H, Burla B, Kretzschmar T, Lee Y, et al. Plant ABC transporters. The Arabidopsis Book / American Society of Plant Biologists. 2011;9:e0153.
22. Satheesh V, Chidambaranathan P, Jagannadham PT, Kumar V, Jain PK,

Chinnusamy V, Bhat SR, Srinivasan R: Transmembrane START domain proteins: in silico identification, characterization and expression analysis under stress conditions in chickpea (*Cicer arietinum* L.). *Plant Signal Behav* 2016, 11(2):e992698.

23. Schrick K, Bruno M, Khosla A, Cox PN, Marlatt SA, Roque RA, et al. Shared functions of plant and mammalian StAR-related lipid transfer (START) domains in modulating transcription factor activity. BMC Biol. 2014;12:70.

24. Wang Z, Gerstein M, Snyder M. RNA-Seq: a revolutionary tool for transcriptomics. Nat Rev Genet. 2009;10(1):57–63.

25. Xu GR, Strong MJ, Lacey MR, Baribault C, Flemington EK, Taylor CM: RNA CoMPASS: A Dual Approach for Pathogen and Host Transcriptome Analysis of RNA-Seq Datasets. PLoS One 2014, 9(2).

26. Westermann AJ, Gorski SA, Vogel J. Dual RNA-seq of pathogen and host. Nat Rev Microbiol. 2012;10(9):618–30.

27. Dobon A, Bunting DCE, Cabrera-Quio LE, Uauy C, Saunders DGO. The host-pathogen interaction between wheat and yellow rust induces temporally coordinated waves of gene expression. BMC Genomics. 2016;17:380.

28. Soriano JM, Royo C. Dissecting the genetic architecture of leaf rust resistance in wheat by QTL meta-analysis. Phytopathology. 2015;105(12):1585–93.

29. Bajgain P, Rouse MN, Bhavani S, Anderson JA. QTL mapping of adult plant resistance to Ug99 stem rust in the spring wheat population RB07/MN06113-8. Mol Breed. 2015;35(8):170.

30. Rosewarne GM, Herrera-Foessel SA, Singh RP, Huerta-Espino J, Lan CX, He ZH. Quantitative trait loci of stripe rust resistance in wheat. Theor Appl Genet. 2013;126(10):2427–49.

31. Njau PN, Bhavani S, Huerta-Espino J, Keller B, Singh RP. Identification of QTL associated with durable adult plant resistance to stem rust race Ug99 in wheat cultivar 'Pavon 76'. Euphytica. 2013;190(1):33–44.

32. Zegeye H, Rasheed A, Makdis F, Badebo A, Ogbonnaya FC. Genome-wide association mapping for seedling and adult plant resistance to stripe rust in synthetic Hexaploid wheat. PLoS One. 2014;9(8):e105593.

33. Gao LL, Turner MK, Chao SM, Kolmer J, Anderson JA. Genome wide association study of seedling and adult plant leaf rust resistance in elite spring wheat breeding lines. PLoS One. 2016;11(2):e0148671.

34. Maccaferri M, Zhang JL, Bulli P, Abate Z, Chao SM, Cantu D, Bossolini E, Chen XM, Pumphrey M, Dubcovsky J: A Genome-Wide Association Study of Resistance to Stripe Rust (Puccinia striiformis f. sp tritici) in a Worldwide Collection of Hexaploid Spring Wheat (*Triticum aestivum* L.). *G3-Genes Genom Genet* 2015, 5(3):449–465.

35. Cavanagh CR, Chao S, Wang S, Huang BE, Stephen S, Kiani S, et al. Genome-wide comparative diversity uncovers multiple targets of selection for improvement in hexaploid wheat landraces and cultivars. Proc Natl Acad Sci U S A. 2013;110(20):8057–62.

36. Wang SC, Wong DB, Forrest K, Allen A, Chao SM, Huang BE, et al. Characterization of polyploid wheat genomic diversity using a high-density 90 000 single nucleotide polymorphism array. Plant Biotechnol J. 2014;12(6):787–96.

37. Blake VC, Birkett C, Matthews DE, Hane DL, Bradbury P, Jannink JL, et al., 2. Plant Genome-Us. 2016;9(2).

38. Crouzet J, Trombik T, Fraysse AS, Boutry M. Organization and function of the plant pleiotropic drug resistance ABC transporter family. FEBS Lett. 2006; 580(4):1123–30.

39. Andolfo G, Ruocco M, Di Donato A, Frusciante L, Lorito M, Scala F, et al. Genetic variability and evolutionary diversification of membrane ABC transporters in plants. BMC Plant Biol. 2015;15

40. Sekhwal MK, Li P, Lam I, Wang X, Cloutier S, You FM. Disease resistance Gene analogs (RGAs) in plants. Int J Mol Sci. 2015;16(8):19248–90.

41. Bouktila D, Khalfallah Y, Habachi-Houimli Y, Mezghani-Khemakhem M, Makni M, Makni H. Large-scale analysis of NBS domain-encoding resistance gene analogs in Triticeae. Genet Mol Biol. 2014;37(3):598–610.

42. Clavijo BJ, Venturini L, Schudoma C, Accinelli GG, Kaithakottil G, Wright J, et al. An improved assembly and annotation of the allohexaploid wheat genome identifies complete families of agronomic genes and provides genomic evidence for chromosomal translocations. Genome Res. 2017;27:885–96.

43. Yu J, Tehrim S, Zhang F, Tong C, Huang J, Cheng X, et al. Genome-wide comparative analysis of NBS-encoding genes between brassica species and *Arabidopsis thaliana*. BMC Genomics. 2014;15:3.

44. Meyers BC, Kozik A, Griego A, Kuang HH, Michelmore RW. Genome-wide analysis of NBS-LRR-encoding genes in Arabidopsis. Plant Cell. 2003;15(4):809–34.

45. Tarr DEK, Alexander HM. TIR-NBS-LRR genes are rare in monocots: evidence from diverse monocot orders. BMC Res Notes. 2009;2:197.

46. Finn RD, Coggill P, Eberhardt RY, Eddy SR, Mistry J, Mitchell AL, et al. The Pfam protein families database: towards a more sustainable future. Nucleic Acids Res. 2016;44(D1):D279–85.

47. Jones P, Binns D, Chang HY, Fraser M, Li WZ, McAnulla C, et al. InterProScan 5: genome-scale protein function classification. Bioinformatics. 2014;30(9):1236–40.

48. Schrick K, Nguyen D, Karlowski WM, Mayer KFX. START lipid/sterol-binding domains are amplified in plants and are predominantly associated with homeodomain transcription factors. Genome Biol. 2004;5(6):R41.

49. Cunningham F, Amode MR, Barrell D, Beal K, Billis K, Brent S, et al. Ensembl 2015. Nucleic Acids Res. 2015;43(D1):D662–9.

50. Camacho C, Coulouris G, Avagyan V, Ma N, Papadopoulos J, Bealer K, et al. BLAST plus : architecture and applications. BMC Bioinformatics. 2009; 10:421.

51. Clark K, Karsch-Mizrachi I, Lipman DJ, Ostell J, Sayers EW. GenBank. Nucleic Acids Res. 2016;44(D1):D67–72.

52. Sanseverino W, Hermoso A, D'Alessandro R, Vlasova A, Andolfo G, Frusciante L, et al. PRGdb 2.0: towards a community-based database model for the analysis of R-genes in plants. Nucleic Acids Res. 2013;41(Database issue):D1167–71.

53. Rea PA. Plant ATP-binding cassette transporters. Annu Rev Plant Biol. 2007;58:347–75.

54. Verrier PJ, Bird D, Buria B, Dassa E, Forestier C, Geisler M, et al. Plant ABC proteins - a unified nomenclature and updated inventory. Trends Plant Sci. 2008;13(4):151–9.

55. Kretzschmar T, Burla B, Lee Y, Martinoia E, Nagy R. Functions of ABC transporters in plants. Essays Biochem. 2011;50:145–60.

56. Basnet BR, Singh S, Lopez-Vera EE, Huerta-Espino J, Bhavani S, Jin Y, et al. Molecular mapping and validation of SrND643: a new wheat Gene for resistance to the stem rust pathogen Ug99 race group. Phytopathology. 2015;105(4):470–6.

57. Liu SY, Rudd JC, Bai GH, Haley SD, Ibrahim AMH, Xue QW, et al. Molecular markers linked to important genes in hard winter wheat. Crop Sci. 2014; 54(4):1304–21.

58. Manickavelu A, Kawaura K, Oishi K, Shin IT, Kohara Y, Yahiaoui N, et al. Comparative gene expression analysis of susceptible and resistant near-isogenic lines in common wheat infected by Puccinia triticina. DNA Res. 2010;17(4):211–22.

59. Dmochowska-Boguta M, Alaba S, Yanushevska Y, Piechota U, Lasota E, Nadolska-Orczyk A, et al. Pathogen-regulated genes in wheat isogenic lines differing in resistance to brown rust Puccinia triticina. BMC Genomics. 2015;16:742.

60. Hulbert SH, Bai J, Fellers JP, Pacheco MG, Bowden RL. Gene expression patterns in near isogenic lines for wheat rust resistance gene Lr34/Yr18. Phytopathology. 2007;97(9):1083–93.

61. Moore JW, Herrera-Foessel S, Lan C, Schnippenkoetter W, Ayliffe M, Huerta-Espino J, et al. A recently evolved hexose transporter variant confers resistance to multiple pathogens in wheat. Nat Genet. 2015;47(12):1494–8.

62. Zmienko A, Samelak A, Kozlowski P, Figlerowicz M. Copy number polymorphism in plant genomes. Theor Appl Genet. 2014;127(1):1–18.

63. Moore G. Cereal genome evolution - pastoral pursuits with Lego genomes. Curr Opin Genet Dev. 1995;5(6):717–24.

64. Herrera-Foessel SA, Singh RP, Huerta-Espino J, Rosewarne GM, Periyannan SK, Viccars L, et al. Lr68: a new gene conferring slow rusting resistance to leaf rust in wheat. Theor Appl Genet. 2012;124(8):1475–86.

65. Singla J, Luthi L, Wicker T, Bansal U, Krattinger SG, Keller B. Characterization of Lr75: a partial, broad-spectrum leaf rust resistance gene in wheat. Theor Appl Genet. 2017;130(1):1–12.

66. White FF, Frommer W. Deciphering durable resistance one R gene at a time. Nat Genet. 2015;47(12):1376–7.

67. Li HH, Singh S, Bhavani S, Singh RP, Sehgal D, Basnet BR, et al. Identification of genomic associations for adult plant resistance in the background of popular south Asian wheat cultivar, PBW343. Front Plant Sci. 2016;7:1674.

68. Cabral AL, Jordan MC, McCartney CA, You FM, Humphreys DG, MacLachlan R, et al. Identification of candidate genes, regions and markers for pre-harvest sprouting resistance in wheat (*Triticum aestivum* L.). BMC Plant Biol. 2014;14:340.

69. Krattinger SG, Keller B. Molecular genetics and evolution of disease resistance in cereals. New Phytol. 2016;212(2):320–32.
70. Zipfel C. Plant pattern-recognition receptors. Trends Immunol. 2014;35(7):345–51.
71. Monaghan J, Zipfel C. Plant pattern recognition receptor complexes at the plasma membrane. Curr Opin Plant Biol. 2012;15(4):349–57.
72. Bohm H, Albert I, Fan L, Reinhard A, Nurnberger T. Immune receptor complexes at the plant cell surface. Curr Opin Plant Biol. 2014;20:47–54.
73. Li PC, Quan XD, Jia GF, Xiao J, Cloutier S, You FM. RGAugury: a pipeline for genome-wide prediction of resistance gene analogs (RGAs) in plants. BMC Genomics. 2016;17(1):852.
74. Singh RP, Hodson DP, Jin Y, Lagudah ES, Ayliffe MA, Bhavani S, et al. Emergence and spread of new races of wheat stem rust fungus: continued threat to food security and prospects of genetic control. Phytopathology. 2015;105(7):872–84.
75. Andolfo G, Ruocco M, Di Donato A, Frusciante L, Lorito M, Scala F, et al. Genetic variability and evolutionary diversification of membrane ABC transporters in plants. BMC Plant Biol. 2015;15:51.
76. Steuernagel B, Jupe F, Witek K, Jones JDG, Wulff BBH. NLR-parser: rapid annotation of plant NLR complements. Bioinformatics. 2015;31(10):1665–7.
77. Bailey TL, Boden M, Buske FA, Frith M, Grant CE, Clementi L, et al. MEME SUITE: tools for motif discovery and searching. Nucleic Acids Res. 2009;37:W202–8.
78. Endelman JB. Ridge regression and other kernels for genomic selection with R package rrBLUP. Plant Genome-Us. 2011;4(3):250–5.
79. Wilkinson PA, Winfield MO, Barker GLA, Tyrrell S, Bian X, Allen AM, et al. CerealsDB 3.0: expansion of resources and data integration. BMC Bioinformatics. 2016;17:256.
80. McLaren W, Gil L, Hunt SE, Riat HS, Ritchie GRS, Thormann A, et al. The Ensembl variant effect predictor. Genome Biol. 2016;17(1):122.
81. Chapman JA, Mascher M, Buluc A, Barry K, Georganas E, Session A, et al. A whole-genome shotgun approach for assembling and anchoring the hexaploid bread wheat genome. Genome Biol. 2015;16:26.
82. Carollo V, Matthews DE, Lazo GR, Blake TK, Hummel DD, Lui N, et al. GrainGenes 2.0. An improved resource for the small-grains community. Plant Physiol. 2005;139(2):643–51.

7

Diversity, distribution of *Puroindoline* genes and their effect on kernel hardness in a diverse panel of Chinese wheat germplasm

Xiaoling Ma[1,2], Muhammad Sajjad[1,3], Jing Wang[1,4], Wenlong Yang[1], Jiazhu Sun[1], Xin Li[1], Aimin Zhang[1*] and Dongcheng Liu[1*]

Abstract

Background: Kernel hardness, which has great influence on the end-use properties of common wheat, is mainly controlled by *Puroindoline* genes, *Pina* and *Pinb*. Using EcoTILLING platform, we herein investigated the allelic variations of *Pina* and *Pinb* genes and their association with the Single Kernel Characterization System (SKCS) hardness index in a diverse panel of wheat germplasm.

Results: The kernel hardness varied from 1.4 to 102.7, displaying a wide range of hardness index. In total, six *Pina* and nine *Pinb* alleles resulting in 15 genotypes were detected in 1787 accessions. The most common alleles are the wild type *Pina-D1a* (90.4%) and *Pina-D1b* (7.4%) for *Pina*, and *Pinb-D1b* (43.6%), *Pinb-D1a* (41.1%) and *Pinb-D1p* (12. 8%) for *Pinb*. All the genotypes have hard type kernel hardness of SKCS index (>60.0), except the wild types of *Pina* and *Pinb* combination (*Pina-D1a/Pinb-D1a*). The most frequent genotypes in Chinese and foreign cultivars was *Pina-D1a/Pinb-D1b* (46.3 and 39.0%, respectively) and in Chinese landraces was *Pina-D1a/Pinb-D1a* (54.2%). The frequencies of hard type accessions are increasing from 35.5% in the region IV, to 40.6 and 61.4% in the regions III and II, and then to 77.0% in the region I, while those of soft type are accordingly decreasing along with the increase of latitude. Varieties released after 2000 in Beijing, Hebei, Shandong and Henan have higher average kernel hardness index than that released before 2000.

Conclusion: The kernel hardness in a diverse panel of Chinese wheat germplasm revealed an increasing of kernel hardness generally along with the latitude across China. The wild type *Pina-D1a* and *Pinb-D1a*, and one *Pinb* mutant (*Pinb-D1b*) are the most common alleles of six *Pina* and nine *Pinb* alleles, and a new double null genotype (*Pina-D1x/Pinb-D1ah*) possessed relatively high SKCS hardness index. More hard type varieties were released in recent years with different prevalence of *Pin-D1* combinations in different regions. This work would benefit the understanding of the selection and molecular processes of kernel hardness across China and different breeding stages, and provide useful information for the improvement of wheat quality in China.

Keywords: Common wheat, Kernel hardness, *Puroindoline* genes, EcoTILLING, Allelic variants

Background

Wheat (*Triticum aestivum* L.) is one of the most widely grown food crop all over the world with a wide array of food products for human consumption. The kernel hardness is a major determinant of end-use food properties of wheat grain. Kernel hardness refers to the texture of the grain (caryopsis), that is, whether the endosperm is physically hard or soft. This difference in grain texture is due to a 13–15 kDa marker protein, friabilin, which is highly present on the surface of water-washed starch of soft wheat, lower on hard wheat starch and absent on durum wheat starch [1]. The N-terminal sequence analysis of friabilin revealed a mixture of two or more distinct polypeptides [2, 3], which were found to be identical to the two lipid binding proteins, Puroindoline a (PINA) and b (PINB) [4]. The transcripts of *PINA* and *PINB* are

* Correspondence: amzhang@genetics.ac.cn; dcliu@genetics.ac.cn
[1]State Key Laboratory of Plant Cell and Chromosome Engineering, Institute of Genetics and Developmental Biology, Chinese Academy of Sciences, 1 West Beichen Road, Chaoyang District, Beijing 100101, China
Full list of author information is available at the end of the article

controlled by two linked genes, *Pina* and *Pinb*, respectively, located on short arm of chromosome 5D [5]. The presence of wild type *Pina-D1a* and *Pinb-D1a* genes are both necessary to the grain softening of allohexaploid nature of wheat (AABBDD, 2n = 6× =42). However, homologous of the *Pina* and *Pinb* genes are absent on the wheat 5A and 5B chromosomes, and thus durum wheats (*Triticum turgidum* L.) (AABB, 2n = 4× = 28) lacking *Pina* and *Pinb* genes have hard textured grains [6].

Soft grain wheat varieties have wild type (WT) alleles of both *Pina* and *Pinb* genes and any mutation in WT alleles at one or both *Pin* genes gives rise to hard grain texture leading to changed food technological properties [7]. The variation in degree of grain texture hardness is due to a spectrum of alleles and their combinations at *Pina* and *Pinb*, and the number of alleles at *Pina* and *Pinb* detected so far has increased to 23 and 33, respectively [7–10]. Since the basic mechanism by which *Puroindolines* induce soften endosperm is not well known, genotype-phenotype associations are useful to estimate the kernel hardness effect of a *Pin* allele. The absence or altered primary structure of one of PINA and PINB will result in a hard grain texture, and among commercial wheat cultivars the most prevalent hard genotypes are the absence of PINA or the altered primary structure of PINB with null alleles of *Pina* and *Pina-D1a/Pinb-D1b*, respectively [11–13]. However, cultivars with null alleles of *Pina* have been proposed harder than those with *Pina-D1a/Pinb-D1b* [13–19]. Moreover, the genotype with *Pina-D1b* (the null allele of *Pina*) may have poor milling quality and relatively inferior processing quality for steamed bread, pan bread, and Chinese noodles [16]. This null allele has been molecular characterized with a large deletion of 15,380 bp through a primer walking strategy and a diagnostic sequence tagged site (STS) marker has also been developed spanning the deletion fragment [20]. Moreover, on chromosomes 7A, 7B and 7D in bread wheat, homologous *Pinb* genes have been found with more than 70.0% similarity, though their function is elusive [21–23]. Although most of the known hardness alleles confer large and somewhat similar changes in endosperm texture relative to soft wheat, the discovery of new alleles could broaden the genetic background for kernel hardness and provide industry with grains more suitable for a variety of end-uses. Furthermore, different combinations of *Pina* and *Pinb* alleles in common wheat determine the grain textural classes with diverse end-use characteristics [24]. Thus, knowledge on the *Puroindoline* allelic composition in a diverse panel of germplasm is prerequisite for the parental selection for developing varieties with desired kernel hardness.

The research herein presents the analysis of *Puroindoline* allelic variations, genotypes and their association with kernel hardness in a diverse panel of wheat accessions comprising 1539 Chinese cultivars, 107 Chinese landraces and 141 foreign accessions. A subset of 623 accessions was evaluated for two consecutive years to assess environmental effect on SKCS index of kernel hardness. Accessions collected in Beijing, Hebei, Shandong and Henan were analyzed in detail to reveal the hardness trends along with breeding stages and the prevalence of *Pina* and *Pinb* alleles. The knowledge generated in this study about the allelic effect on and dynamics of kernel hardness, and the allele distribution across China would enhance breeder's choice for suitable parent selection to develop cultivars of desired kernel hardness, and improve the understanding on the patterns of kernel hardness across different wheat zones in China. The new double null mutants would provide an alternative option for kernel hardness improvement.

Results
High phenotypic variation in kernel hardness
To assess the kernel hardness in Chinese wheat germplasm, 1646 accessions were collected from nine wheat cultivation regions in China and grown in Beijing in 2009–2010 cropping season, along with 141 foreign accessions from USA, Australia, Europe and Japan (Table 1). Of the Chinese accessions, 1582 accessions were from cultivation region I (Northern winter wheat region), II (Yellow and Huai River Valley winter wheat region), III (Middle and Low Yangtze Valley winter wheat region) and IV (Southwestern winter wheat region), and the remaining 64 accessions were from other spring wheat cultivation regions since most spring wheat from these regions hardly survive through the cold winter season in Beijing. The kernel hardness of harvested samples in 2009–2010 growing season measured with

Table 1 Number and types of accessions for kernel hardness measurement and *Puroindoline-D1* detection

	Region[a]	Cultivars	Landraces	Total
China	I	244	4	248
	II	1065	25	1090
	III	81	25	106
	IV	95	43	138
	V	2	1	3
	VI	9	3	12
	VIII	32	4	36
	IX	1	2	3
	X	10	0	10
Foreign				141
Total		1539	107	1787

[a]Definition of wheat cultivation regions: I (Northern winter wheat region), II (Yellow and Huai River Valley winter wheat region), III (Middle and Low Yangtze Valley winter wheat region), IV (Southwestern winter wheat region), V (Southern winter wheat region), VI (Northeastern spring wheat region), VIII (Northwestern spring wheat region), IX (Qing-Tibetan Plateau spring-winter wheat region) and X (Xinjiang winter-spring wheat region)

SKCS showed that 1787 accessions contained a wide range of SKCS index varying from 1.4 to 102.7 (Table 2). The major class was of hard type including 1075 accessions (60.2%), which were from most wheat cultivation regions of China, except Qing-Tibetan Plateau spring-winter wheat region (cultivation region IX) (Table 3). The SKCS index of hard type ranged from 60.1 to 102.7 with mean value of 72.4, and about 93.0% accessions have a hardness index lower than 85.0. The soft type class consisted of 523 accessions (29.2%) with SKCS index ranging from 1.4 to 40.0 and mean value of 25.1, while the medium type class with medium kernel hardness included only 189 accessions (10.6%), ranging from 40.0 to 60.0 with mean value of 54.1 (Table 2).

To determine the environment effect on kernel hardness, 623 accessions were grown on the same station during 2009–2010 and 2010–2011 growing seasons using the same agronomic practices and phenotyped for SKCS. The ranges of SKCS index for the year 2009–2010 were of 1.4 to 97.8 with an average of 49.2 and these for 2010–2011 were 1.1 to 98.0 with an average of 48.0. For the environment effect, the analysis of variance (ANOVA) showed non-significant with P value of 0.39 (Additional file 1: Table S1) between 2 years' data, and thus we considered the SKCS data of 2009–2010 valid for association analysis between SKCS index of kernel hardness and genotypes of *Pina* and *Pinb*.

Classified regional distribution of hard and soft kernel accessions

Based on the kernel hardness, accessions from nine wheat growing regions in China were analyzed, region I (Northern winter wheat region) had the highest number of hard textured type wheat accessions, 77.0%, and the regions II (Yellow and Huai River Valley winter wheat region), VIII (Northwestern spring wheat region), X (Xinjiang winter-spring wheat region) and foreign accessions were also dominantly represented by hard type wheat, around 60.0% (Table 3). Conversely, the regions III (Middle and Lower Yangtze Valley winter wheat region), IV (Southwestern winter wheat region), V (Southern winter wheat region) and VI (Northeastern spring wheat region) were represented by soft type wheat with about 50.0% prevalence (Table 3). All the regions and foreign accessions were represented by all three kernel

hardness types, except regions V and VI that did not contain medium type accessions and IX did not have hard and medium types. This might be due to only a few available accessions collected from these regions (Table 3).

In China, the winter wheat growing regions IV (Southwestern winter wheat region), III (Middle and Low Yangtze Valley winter wheat region), II (Yellow and Huai River Valley winter wheat region) and I (Northern winter wheat region) extends from South to North along with the increase of latitude. Regarding to hardness type in different wheat regions, we found that the frequencies of hard type accessions are increasing from 35.5% in the region IV, to 40.6 and 61.4% in the regions III and II, and then to 77.0% in the region I, while those of soft type are accordingly decreasing along with the increase of latitude (Table 3). This phenomenon was also observed in landraces (Additional file 1: Table S2). More soft accessions present in the regions III and IV than that in the region II, and no soft accessions were detected in the region I (Additional file 1: Table S2).

Allelic variants of *Pina* and *Pinb* genes

The nested PCR and modified EcoTILLING analysis was performed with allele specific primers to detect allelic variants of *Pina* and *Pinb* genes (Fig. 1; Additional file 1: Table S3). Of the analyzed 1787 accessions, six *Pina* allelic variants including the wild type (*Pina-D1a*) and five mutants were characterized (Table 4). Except *Pina-D1l* has a cytosine deletion as comparing to *Pina-D1a* (Additional file 4: Figure S3A), the remaining mutants belong to different types of *Pina* null mutant (Table 4). The wild type *Pina-D1a* was the most common allele observed in 1616 accessions (90.4%) followed by *Pina-D1b* in 133 accessions (7.4%), while the remaining variants, *Pina-D1l*, *Pina-D1r*, *Pina-D1s* and *Pina-D1x* were rare alleles, present in a few samples (Table 4). Nine allelic variants were identified at *Pinb*, of which *Pinb-D1b*, *Pinb-D1c*, *Pinb-D1d*, *Pinb-D1p*, *Pinb-D1q* and *Pinb-D1u* have SNP mutations, and *Pinb-D1ah* and *Pinb-D1ai* belong to different types of *Pinb* null mutant (Table 4; Additional file 4: Figure S3B). Only three variants *Pinb-D1b*, *Pinb-D1a* and *Pinb-D1p* (43.6, 41.1 and 12.8%, respectively) were common in these allelic variants (Table 4). For the null alleles of *Pina* and *Pinb*, the *Pina-D1x* as a novel null variant was different from previously reported alleles with various fragment deletions (*Pina-D1b*, *Pina-D1r* and *Pina-D1s*) and detected in seven Chinese cultivars and three foreign accessions. There are two novel *Pinb-null* alleles, one is *Pinb* absent but *Pina* present, so this *Pinb-null* allele was designated as *Pinb-D1ai*, and detected only in two accessions; the other *Pinb-null* allele was named as *Pinb-D1ah* and observed in the ten accessions containing the *Pina-D1x* allele, in which the

Table 2 Distribution of kernel hardness in and percentage of soft, hard and medium wheats

Type	Number	Freq. (%)	SKCS index	
			Mean	Range
Soft	523	29.2	25.1	1.4–40.0
Hard	1075	60.2	72.4	60.1–102.7
Medium	189	10.6	54.1	40.0–60.0

Table 3 Region-wise percent distribution of soft, hard and medium wheats in two different origins and nine wheat cultivation regions in China[a]

	China (1646)									Foreign (141)	Freq. (%)
	I	II	III	IV	V	VI	VIII	IX	X		
Soft	10.5	27.6	51.9	55.8	66.7	66.7	30.6	100.0	30.0	25.5	29.2 (522)
Hard	77.0	61.4	40.6	35.5	33.3	33.3	63.9	0.0	60.0	63.1	60.2 (1075)
Medium	12.5	11.0	7.5	8.7	0.0	0.0	5.6	0.0	10.0	11.3	10.6 (190)

[a]The values in parentheses are the number of accessions and the wheat cultivation regions are defined in Table 1

simultaneous deletion of the large segment containing *Pina* and *Pinb* genes might have occurred.

Regarding the distribution of allelic variants across Chinese cultivars, landraces and foreign accessions, frequency of each allelic variant of *Pina* and *Pinb* genes was calculated. For *Pina* gene, the wild type allele *Pina-D1a* was predominant in Chinese cultivars, landraces and foreign accessions with frequencies of 91.9, 80.4 and 82.3% in the respective groups. Among the mutants, the most frequent allele was *Pina-D1b* in Chinese cultivars and foreign accessions with frequency of 7.3 and 11.3%, respectively, whereas *Pina-D1r* (7.5%) was the most in landraces, followed by *Pina-D1s* (5.6%) and *Pina-D1b* (3.7%). The variants *Pina-D1l* and *Pina-D1x* were missing in foreign and landrace accessions, respectively (Table 4). For *Pinb* gene, the highest frequency of the wild type allele *Pinb-D1a* was present in Chinese landraces (72.9%), followed by foreign accessions (48.2%) and Chinese cultivars (38.3%), which has an opposite pattern to that of *Pina-D1a*. Among the mutants, *Pinb-D1b* and *Pinb-D1p* had frequencies of 46.8 and 13.1% in Chinese cultivars, respectively, and other four variants were rarely present (totally less than 2.0%); In foreign accessions, *Pinb-D1b* and *Pinb-D1p* were also prevalent with frequencies of 39.0 and 6.4%, but other three variants (*Pinb-D1c*, *Pinb-D1d* and *Pinb-D1ah*) had relatively high frequencies as compared with those in Chinese accessions, whereas *Pinb-D1b* was present in only 3.7% Chinese landraces, and *Pinb-D1p* was the most present mutant (16.8%) in Chinese landraces (Table 4). Interestingly, the allele *Pinb-D1c* was missing in Chinese

cultivars and landraces but was found in foreign accessions, while the allele *Pinb-D1u* was completely absent in Chinese cultivars and foreign accessions and was only detected in Chinese landraces, Zhushimai, Zhugoumai and Dajinbaihuazi (Table 4). Those three landraces are found in Yunnan and Sichuan provinces, belonging to Southwestern winter wheat region (IV), and they could be exploited to incorporate this novel allele into Chinese wheat cultivars. Moreover, the variants *Pina-D1l*, *Pinb-D1q* and *Pinb-D1ai* were missing in foreign accessions, and the double mutant *Pina-D1x/Pinb-D1ah* was not detected in Chinese landraces. The different allelic frequencies in three type accessions suggested that the *Pina* and *Pinb* were artificially selected by breeders.

To assess the diversity at *Pina* and *Pinb* genes, Nei's diversity index (*H*) was calculated in the total collection of 1787 accessions. The *H* value at *Pina* and *Pinb* were 0.18 and 0.62, respectively. At *Pina*, the highest diversity (0.34) was observed in 107 Chinese landraces, followed by 0.31 in the group of 141 foreign accessions, and the least (0.15) was observed in 1539 Chinese cultivars. At *Pinb*, Chinese cultivars and foreign accessions had similar high diversity (0.62), and Chinese landraces possessed a lower diversity (0.43), which was different from that of *Pina*. These data demonstrated that more allelic variants of *Pinb* with higher frequencies were retained in breeding process than those of *Pina*.

Association analysis between genotype and kernel hardness

To investigate the effect of *Pina* and *Pinb* on kernel hardness, genotypes were assayed on six *Pina* and nine

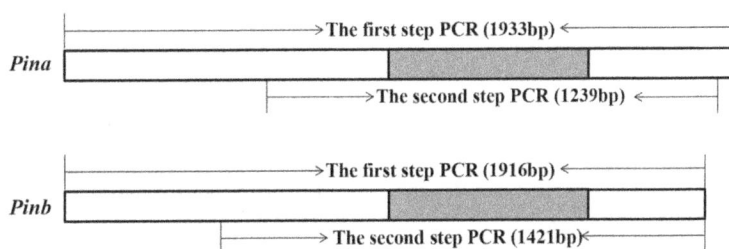

Fig. 1 Diagrams of PCR amplification for *Pina* and *Pinb*. The gray boxes show coding regions for *Pina* and *Pinb*, respectively. The first step PCR reaction of *Pina* and *Pinb* (Pina-OutF/R, Pinb-OutF/R) used for amplifying *Pina* (1933 bp) and *Pinb* (1916 bp) genomic sequence. The second step PCR reaction of *Pina* and *Pinb* (Pina-InF/R, Pinb-InF/R) used for amplifying *Pina* (1239 bp) and *Pinb* (1421 bp) genomic sequence

Table 4 Distribution of *Pina* and *Pinb* alleles in total 1787 accessions

	Pina[a]						Pinb[a]								
	D1a	D1b	D1l	D1r	D1s	D1x	D1a	D1b	D1c	D1d	D1p	D1q	D1u	D1ah	D1ai
Cultivars	1414	113	1	2	2	7	589	721	0	13	201	7	0	7	1
Landraces	86	4	3	8	6	0	78	4	0	1	18	2	3	0	1
Foreign	116	16	0	1	5	3	68	55	2	4	9	0	0	3	0
Total	1616	133	4	11	13	10	735	780	2	18	228	9	3	10	2

[a]The values are the number of accessions for each allele

Pinb variants. Of the 15 *Pina* and *Pinb* combinations (Table 5), eight were the *Pina* wild type (*Pina-D1a*) with eight *Pinb* variants, occupying a 90.4% accessions, three were *Pina-D1b* and three *Pinb* variants with a total frequency of 7.4%, and the remaining four combinations were formed individually by four *Pina* variants with corresponding *Pina* variants. At *Pinb* genotypes, five were formed by the wild type (*Pinb-D1a*) with five *Pina* variants, and two were *Pinb-D1b* and two *Pina* variants, in which *Pinb-D1a* and *Pinb-D1b* genotypes occurred in 41.1 and 43.6 accessions, respectively (Table 5). The most abundant combination was *Pina-D1a/Pinb-D1b* (43.1%), followed by *Pina-D1a/Pinb-D1a* (32.8%), *Pina-D1a/Pinb-D1p* (12.5%) and *Pina-D1b/Pinb-D1a* (6.7%) with corresponding average kernel hardness of 67.7, 31.0, 68.0 and 79.3. The other 11 combinations were rare (≤1.0%) with a total frequency of 4.8%, however these combinations had higher average kernel hardness (≥60.0), of which that of *Pina-D1x/Pinb-D1ah* was 91.1, at least 10 hardness index higher than all the remaining genotypes (Table 5).

Regarding to the distribution of 15 combinations of *Pina/Pinb* alleles in three groups, 13, 12 and 9 were observed in Chinese cultivars, Chinese landraces and foreign accessions, respectively (Table 5). In Chinese cultivars, *Pina-D1a/Pinb-D1b* (46.3%), *Pina-D1a/Pinb-D1a* (31.4%), *Pina-D1a/Pinb-D1p* (12.9%) and *Pina-D1b/Pinb-D1a* (6.6%) were predominant, and those predominant combinations also had high frequencies in foreign accessions, while *Pina-D1r/Pinb-D1a* (7.5%) and *Pina-D1s/Pinb-D1a* (5.6%) were predominant in Chinese landraces, along with *Pina-D1a/Pinb-D1a* (54.2%) and *Pina-D1a/Pinb-D1p* (15.9%) (Table 5). In addition, some rare combinations were observed in some specific samples, e.g. *Pina-D1a/Pinb-D1u* (2.8%) in Chinese landraces, *Pina-D1a/Pinb-D1c* (1.4%) in foreign accessions, and *Pina-D1b/Pinb-D1b* (0.6%) in Chinese cultivars. All in together, the Nei's diversity, allelic and genotype frequency results show the promising use of Chinese landraces and foreign accessions to broaden the genetic base of modern Chinese cultivars, and the frequency difference of combinations in three groups suggested that the *Pina* and *Pinb* genes might have been broadly selected during the breeding process (Table 5).

Table 5 Association of *Puroindoline-D1* alleles with SKCS index and their distributions in Chinese cultivars, Landraces and foreign accessions

Genotype	SKCS index		Cultivars	Landraces	Foreign
	Mean	Range			
Pina-D1a/Pinb-D1a	31.0	1.4–92.5	483	58	46
Pina-D1a/Pinb-D1b	67.7	11.0–102.7	712	4	55
Pina-D1a/Pinb-D1c	60.0	45.1–74.9	0	0	2
Pina-D1a/Pinb-D1d	63.6	48.5–73.6	13	1	4
Pina-D1a/Pinb-D1p	68.0	16.1–88.6	198	17	9
Pina-D1a/Pinb-D1q	66.2	57.0–78.3	7	2	0
Pina-D1a/Pinb-D1u	65.9	58.4–78.4	0	3	0
Pina-D1a/Pinb-D1ai	78.6	65.6–91.6	1	1	0
Pina-D1b/Pinb-D1a	79.3	19.7–95.6	101	3	16
Pina-D1b/Pinb-D1b	68.9	19.9–75.5	9	0	0
Pina-D1b/Pinb-D1p	71.1	66.3–82.4	3	1	0
Pina-D1l/Pinb-D1a	74.3	68.8–84.2	1	3	0
Pina-D1r/Pinb-D1a	76.7	59.9–94.1	2	8	1
Pina-D1s/Pinb-D1a	67.1	53.4–86.2	2	6	5
Pina-D1x/Pinb-D1ah	91.1	77.1–97.8	7	0	3
Total			1539	107	141

Selection of kernel hardness and allele preference during breeding process

To investigate the hardness trend and allelic variant preference during Chinese wheat breeding process, 1108 cultivars bred in Beijing, Hebei, Shandong and Henan with clear released time were selected (Table 6). In each province, based on the released year, cultivars were separated into two groups, before year 2000 and after 2000 (including 2000), for the kernel hardness was considered as one of important criterions for new hard wheat varieties released in north China in year 2000 and later. The average kernel hardness was calculated in each group, along with numbers and hardness of three kernel categories (hard, medium and soft) (Table 6). Varieties released after 2000 in Beijing, Hebei, Shandong and Henan had an average kernel hardness index of 69.2, 68.3, 63.7 and 61.1, respectively, significantly higher than that released before 2000 in all four provinces, which

Table 6 Frequencies of hard, medium and soft wheats bred in different periods in Beijing, Hebei, Shandong and Henan, and their prevalent genotypes in hard wheat

Location	Period[a] (year 2000)	Number	SKCS index		Freq. (%)			Aver. Hardness Freq. (%)			Genotype freq. in hard wheat		
			Mean	Range	Hard	Medium	Soft	Hard	Medium	Soft	Pina-D1a/ Pinb-D1b	Pina-D1a/ Pinb-D1p	Pina-D1b/ Pinb-D1a
Beijing	after	83	69.2	10.0–85.5	90.4	4.8	4.8	72.5	57.1	20.7	37.3	44.0	10.7
	before	106	58.8	12.1–92.5	65.1	16.0	18.9	69.7	52.6	26.7	59.4	20.3	14.5
Hebei	after	281	68.3	13.0–94.2	86.5	8.2	5.3	72.0	56.1	26.6	64.2	16.9	9.9
	before	128	48.4	11.0–90.4	42.2	18.8	39.1	68.4	53.0	24.5	61.1	29.6	1.9
Shandong	after	108	63.7	1.4–95.6	73.2	6.5	20.4	76.5	55.0	20.5	55.7	13.9	20.3
	before	156	50.4	11.1–92.7	48.7	10.3	41.0	71.6	52.9	24.7	75.0	17.1	1.3
Henan	after	107	61.1	6.7–95.0	67.3	13.1	19.6	72.8	53.2	26.3	75.0	8.3	12.5
	before	140	42.2	1.5–83.4	30.7	13.6	55.7	71.1	52.5	23.7	67.4	11.6	16.3

[a]after year 2000 means accessions or varieties bred and released in 2000 or later, and before year 2000 means released in 1999 or before

demonstrated that the kernel hardness has been greatly improved in the new century. The increase of hardness index after 2000 might be due to the higher frequency of hard type varieties, e.g. 90.4 Vs 65.1% in Beijing, 86.5 Vs 42.2% in Hebei,73.2 Vs 48.7% in Shandong, and 67.3 Vs 30.7% in Henan. Interestingly, the frequencies of hard type varieties released after 2000 were increasing along with the latitude, resulting in an increase of the average hardness index from Henan (generally lower latitude) to Beijing (higher latitude). Moreover, slight differences of the average hardness index were observed in hard type varieties in different groups and provinces, and this could be attributed mainly to the frequencies of combinations of *Pina* and *Pinb* alleles since they preserved different effects on kernel hardness (Table 5). In the hard type varieties, *Pina-D1a/Pinb-D1b*, *Pina-D1a/Pinb-D1p* and *Pina-D1b/Pinb-D1a* were three dominant genotypes, occupying 90.0% or more varieties in each groups. However, these genotypes were differentially selected by breeders in different stages and provinces. For example, *Pina-D1a/Pinb-D1p* was a dominant genotype with a percentage of 44.0% in the group of varieties released after 2000 in Beijing, while the preference of *Pina-D1a/ Pinb-D1b* was observed before 2000; the frequency of *Pina-D1b/Pinb-D1a* has been greatly improved in the group after 2000 in Shandong; *Pina-D1b/Pinb-D1a* was seldom detected in varieties from Hebei and released before 2000 in Shandong. These data demonstrated that combinations of *Pina* and *Pinb* genes have been widely selected in different breeding stages in different regions.

The novel alleles at *Pina* and *Pinb*
Several new types of alleles of *Pina* and *Pinb* were also detected in this study. According to the catalogue of gene

symbols [6, 9], a synonymous allele 'C 321 T' of *Pina-D1a* was designated as *Pina-D1y* (Additional file 2: Figure S3), and its kernel hardness index was 20.5 ± 12.7, categorized to the soft wheat class, which indicated that the synonymous allele of *Pina-D1a* had non-significant effect on kernel hardness. The null alleles at both *Pina* and *Pinb* loci (*Pina-D1x/Pinb-D1ah*) were observed in ten accessions (0.6%), of which three were detected in foreign accessions and seven were from Chinese cultivars, including five from Yellow and Huai River Valley winter wheat region (II), and one each from region III and VI (Table 5; Additional file 2: Figure S1; Additional file 3: Figure S2). The SKCS index of these double null mutants ranged from 77.1 to 97.8 with mean of 91.1, which had highest average kernel hardness among 15 combinations of *Pina* and *Pinb* genes and were comparable with that of durum wheat (88.1 ± 15.4) (Additional file 1: Table S4; Table 5).

In effort to further illustrate the deletion size and position in these double null mutants, nine primer pairs (Additional file 1: Table S3) specific to chromosome 5D were designed surrounding the *Pina* and *Pinb* genes with the availability of the BAC sequence (CT009735) in NCBI. Based on the primer walking strategy, amplicons with expected fragment size could be observed with primer sets Pina-1, Pina-7, Pina-8 and Pina-9, whereas no targeted fragment was amplified with primer sets through Pina-2 to Pina-6 in four Chinese cultivars Yunfengzao 21, 06–01216, Kelao 4 and Shan 150 and one foreign accession NIL-Novos 67 (Additional file 1: Table S4; Additional file 2: Figure S1). These absent amplicons revealed a 25-kb deletion at least from −435 to +24,592 bp (reference to ATG of *Pina*) on these five accessions compared with the BAC sequence of Chinese

Spring (CT009735) containing the *Pina* and *Pinb* genes. However, no expected fragments were detected with all the primer sets through Pina-1 to Pina-9 in three Chinese accessions Hedong TX-008, Xinong 8925–13, and 91G149/Chang 128,865, and one foreign accession Vendvr, which provided a large fragment deletion in these accessions (at least 90-kb, from −21,803 to +68,481 bp referring to ATG of *Pina* in Chinese Spring BAC sequence CT009735 in NCBI). Moreover, expected fragments of amplicons were only gained with primer sets Pina-7, Pina-8 and Pina-9, but not through Pina-1 to Pina-6 in foreign accession Victory, demonstrating an at least 63-kb deletion from −21,803 to 41,844 bp of Chinese Spring BAC sequence (CT009735) on chromosome 5DS (Additional file 1: Table S4). All these data with primer walking strategy suggested that the *Pina* and *Pinb* gene regions were completely deleted in these double null mutants, which confirmed the EcoTILLING data and the durum-like kernel hardness of these accessions, though further analysis is needed to elucidate the exact deletion sizes. However, these double null mutants with extremely high kernel hardness provided an elite germplasm resource for kernel hardness improvement in wheat breeding though molecular marker assisted selection.

Double null mutants have durum-like starch granules and protein matrix

SEM has been considered as a straightforward method for determining the adhesion between protein matrix and starch granules [25, 26]. In this study, SEM was performed with one wild type accession (Chinese Spring), two double null accessions (Xinong-8925-13 and Yunfengzao 21) and one durum wheat accession (Neixiang 4184). The SEM images revealed large, medium and small round to spherical starch granules in all four accessions (Fig. 2a-d), and the basic differences were of the adhesion between starch granules and protein matrix among the wild type (*Pina-D1a/Pinb-D1a*) (Fig. 2a) and the double null mutants (*Pina-D1x/Pinb-D1ah*) (Fig. 2c, d). Starch granules and protein matrix are clearly separated without significant adhesions in-between in the wild type soft textured Chinese Spring, while starch granules were strongly adhered to protein matrix in durum wheat Neixiang 4184 (Fig. 2a, b). More medium and small starch granules were observed in durum wheat, which was prone to be adhered to protein matrix, resulting in higher kernel hardness. SEM images of the double null mutants Xinong 8925-13 (*Pina-D1x/Pinb-D1ah*) and Yunfengzao 21 (*Pina-D1x/Pinb-D1ah*) revealed that both double null mutants had strong

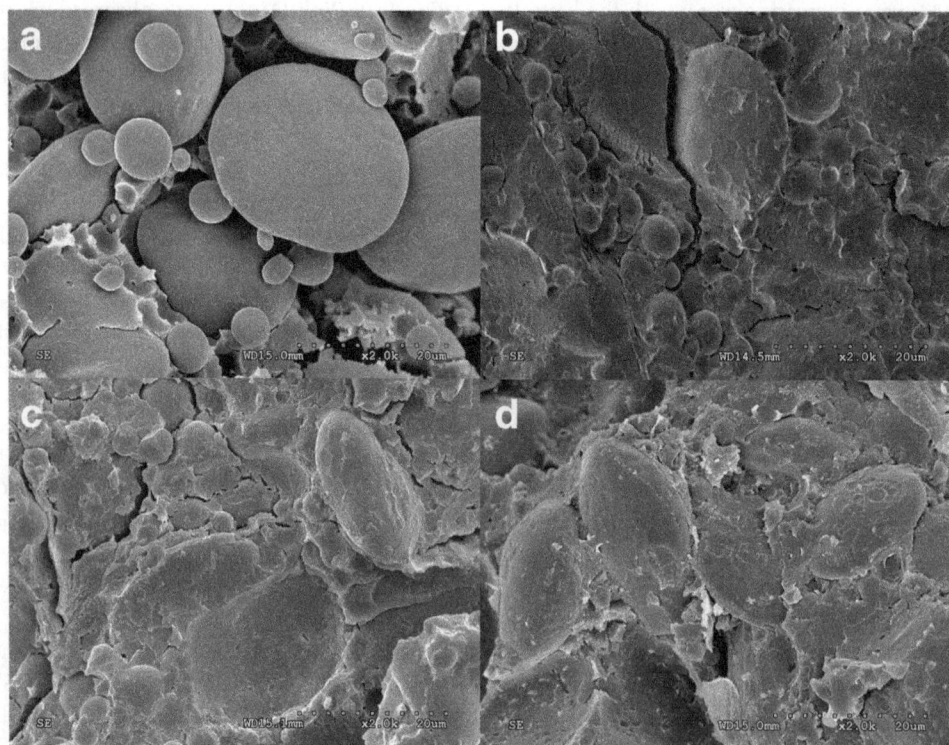

Fig. 2 Scanning electron micrographs of the central endosperm region of double null mutants of *Pina* and *Pinb* genes. **a** Freeze-fractured grain of wild type accession (Chinese Spring, soft type control). **b** Durum wheat accession (Neixiang 4184; *Pin null*: both genes deleted, very hard control). **c** Xinong 8925–13 (*Pina-D1x/Pinb-D1ah*). **d** Yunfengzao 21 (*Pina-D1x/Pinb-D1ah*)

adhesion between protein matrix and starch granules, narrow space between starch granules, and more medium and small starch granules, which were similar to that of durum wheat but completely different from that of Chinese Spring. This phenomena and extreme high kernel hardness confirmed the deletion of both *Pina* and *Pinb* genes in Xinong 8925-13 and Yunfengzao 21 for durum wheat lacking both *Pina* and *Pinb* genes used to have high kernel hardness. Moreover, the protein matrix appeared differently between hexaploid and durum wheat, which has more flake-shape fragments on the surface of starch granules in Chinese Spring and both mutants (Fig. 2a-d). The SKCS and SEM analysis conclusively demonstrated that high kernel hardness was attributed to durum-like grain structure of double null mutants, and both *Pina* and *Pinb* genes might play a vital role in kernel hardness development.

Discussion

Due to their important roles in determining kernel hardness, the allelic variation of *Pina* and *Pinb* genes have been widely investigated using wheat germplasms from different countries and regions [5, 9–11, 13, 24, 27]. The methods used for characterizing *Pina* and *Pinb* alleles were improved from SDS-PAGE analysis, cloning and sequencing of the alleles in a few lines to EcoTILLING [5]. The optimized EcoTILLING approach was exploited to investigate *Pina* and *Pinb* alleles in the 225 micro-core collection (MCC) accessions of Chinese wheat germplasm [10]. Compared to previous techniques, Eco-TILLING approach is highly efficient on time and cost for largely reducing the repetitive work of sequencing the most frequent *Pina-D1a*, *Pinb-D1a* and *Pinb-D1b* alleles. On the other hand, the detection of novel alleles was directly based on the nucleotide sequences, which improved the discovery of novel variants with electrophoretic mobility similar to previously known alleles in SDS-PAGE analysis. The differentiation of *Pinb-D1x* allele (14.5 kDa) from *Pinb-D1ab* (14.4 kDa) was due to the distinct advantage of EcoTILLING approach [10] over SDS-PAGE analysis that might had regarded the two alleles as identical. So the high throughput EcoTIL-LING approach was exploited in our research with minor modifications. On the one hand, gene cloning method was applied to get the plasmid DNA containing wild type genotype (*Pina-D1a/Pinb-D1a*) which minimized the use of control DNA templates and made the adjustment of PCR products concentration more effective than the wild type genomic DNA. On the other hand, nested PCR was used to improve the sensitivity and specificity.

The wide range of kernel hardness index varying from 1.4 to 102.7 was observed in overall collection of 1787 accessions. The high phenotypic variation in kernel hardness index is a reflection of the diverse nature of the scanned germplasm consisting of Chinese cultivars and landraces from 9 geographical regions and 141 foreign accessions from USA, Australia, Europe, and Japan. The distribution of hard and soft type wheats across nine geographical regions broadly divided China into two categories as North and South China. The first category included regions I, II, VIII, and X, where hard type accessions had higher frequency than soft type accession, and the second category consisted of regions III, IV, VI and IX, having more soft type accessions. This pattern of distribution of hard and soft type wheat accessions reflected the breeding preference. The hard type accessions are suitable for wheat noodles and steamed bread, which are preferential daily staple food for people living in North China. Conversely, people in South China prefer rice as staple food and use wheat as secondary food usually for making bakery products like biscuits and cookies etc. The similar distribution of hard and soft type accessions across North and South China was reported previously [10].

In total, six *Pina* and nine *Pinb* variants were detected in 1787 accessions, *Pina-D1a* and *Pinb-D1b* were the most abundant genotypes at *Pina* and *Pinb*, respectively (Table4). Corresponding to this, of 15 *Pina* and *Pinb* combinations, the most extensive distributed combinations were *Pina-D1a/Pinb-D1b* (43.1%), followed by *Pina-D1a/Pinb-D1a* (32.9%), *Pina-D1a/Pinb-D1p* (12.5%) and *Pina-D1b/Pinb-D1a* (6.7%) with corresponding average kernel hardness index of 67.7, 31.0, 68.0 and 79.3 (Table 5), which suggest that the order of kernel hardness index between these four combinations from high to low were *Pina-D1b/Pinb-D1a*, *Pina-D1a/Pinb-D1p*, *Pina-D1a/Pinb-D1b* and *Pina-D1a/Pinb-D1a* [24].

Regarding to the effects of these allelic variations on kernel hardness, the average kernel hardness index of each combinations are 60.0 or more, except *Pina-D1a/Pinb-D1a*, which supporting that wild type genotype (*Pina-D1a/Pinb-D1a*) was necessary for showing the soft kernel texture [9, 24]. In addition, the combination *Pina-D1x/Pinb-D1ah* had the highest average kernel hardness than all the remaining genotypes, indicating the null mutations of *Pina* or *Pinb* containing higher kernel hardness than others' mutations [28].

The wild type *Pina-D1a* was detected as the most frequent allele of *Pina* with frequencies of 91.9, 80.4 and 82.3% in Chinese cultivars, landraces and foreign accessions, respectively (Table 5). Chinese cultivars are prone to have this wild type allele, which could be concluded from previous reports. In Chinese micro-core collection mainly consisting of landraces, *Pina-D1a* had a relatively low frequency (83.6%) [10], while this allele was found to be completely dominant (95.0%) in wheat cultivars released recently from the Yellow and Huai valley of China

[11]. The phenomena suggested that this wild type *Pina-D1a* might preserve beneficial agronomic effects selected by breeders. The wild type of *Pinb* (*Pinb-D1a*) and its two mutants (*Pinb-D1b* and *Pinb-D1p*) were frequently detected in Chinese wheat germplasm [10, 11, 15, 19], and the frequency of mutated *Pinb* alleles was much higher than that of mutated *Pina* alleles, supporting the notion that hard wheats in China were mostly due to *Pinb* mutations rather than the ones arising from *Pina* [10, 15, 19]. Moreover, comparing to Chinese landraces and foreign accessions [10, 27], recently released varieties in this work (46.9%) and the Yellow and Huai valley (50.9%) [11] had a relatively high frequency in *Pinb-D1b*, which played a vital role on the improvement of kernel hardness in Chinese wheat breeding. These results are different from the gene pool investigations in North America, Europe, India, CIMMYT and Australia, where both *Pina-D1b* and *Pinb-D1b* alleles are the main sources of kernel hardness [13, 24, 29, 30].

Interestingly, the allele *Pinb-D1c* was completely missing in Chinese cultivars and landraces and was detected only in two foreign accessions. The reports of *Pinb-D1c* in genotypes from Ukraine, Portugal, Finland and European countries [24, 27, 29, 31] supported our finding. The allele *Pinb-D1p* was reported as restricted to Chinese wheat germplasm [10, 11, 15, 17, 19, 27, 28], but in this work we found nine foreign accessions with *Pinb-D1p* (Table 4). Similarly, though the allele *Pina-D1r* and *Pina-D1s* were found restricted to Chinese accessions in international collections of 803 landraces [24], 493 wheat cultivars [27], and 267 wheat cultivars and advanced lines [11], *Pina-D1r* and *Pina-D1s* were characterized in one and five foreign accessions, respectively. These data provided a broad representativeness of wheat accessions collected in this work. The stratified distribution of *Pina* and *Pinb* variants in different geographical regions might have been influenced by repeated use of the core germplasm in breeding and consumer's preference in a particular region [31].

The simultaneous deletions at both *Pina* and *Pinb* loci are rarely reported so far [27, 28]. In 2005, Ikeda et al. reported the first double null mutation with the presence of D genome and absolute absence of *Pina* and *Pinb* proteins [28]. Later, three Chinese landraces and one Netherlands cultivar were detected absent of both *Pina-D1* and *Pinb-D1* genes, and this mutation was further deduced by primer walking strategy as an approximate 33-kb fragment deletion containing *Pina* and *Pinb* coding regions when compared with the BAC sequence of Chinese Spring on chromosome 5DS [27]. In this work, seven Chinese cultivars and three foreign accessions, designated as *Pina-D1x/Pinb-D1ah*, were observed lacking both *Pina-D1* and *Pinb-D1* genes, and these accessions were divided into three groups based on the presence or absence of amplicons surrounding

the *Ha* locus on chromosome 5DS, which resulted in three deletion sizes of at least 25-kb, 63-kb and 90-kb, respectively. The accessions with *Pina-D1x/Pinb-D1ah* have the highest SKCS index among all 15 genotypes, and these grain textures similar with durum wheat were further observed high degree of adhesion between starch granules and protein matrix through SEM. However, these double null mutants are different from previously characterized double null mutants for the deletion size and kernel hardness [27]. Therefore, these double null mutants could be incorporated into quality improvement in bread wheat though further analysis is needed to clarify their exact deletion size.

Conclusion
The study herein verified many previous results regarding higher allelic diversity at *Pinb* locus; predominant prevalence of *Pina-D1a* and *Pinb-D1b* alleles; and association of various mutations at *Pina* and *Pinb* loci with SKCS index. The regions I, II, VIII, X foreign accessions were dominantly represented with hard type wheat while the soft type accessions were more frequent in regions III, IV, VI and IX. The Nei's diversity, allelic and genotype frequency results together show the promising use of Chinese landraces and foreign accessions to broaden the genetic base of modern Chinese cultivars. The wild type *Pina-D1a* and *Pinb-D1a*, and one *Pinb* mutant (*Pinb-D1b*) are the most common alleles of six *Pina* and nine *Pinb* alleles, and a new double null genotype (*Pina-D1x/Pinb-D1ah*) possessed relatively high SKCS hardness index. More hard type varieties were released in recent years with different prevalence of genotypes in different regions. This work would benefit the understanding of the selection and molecular processes of kernel hardness across China and different breeding stages, and provide useful information for the improvement of wheat quality in China.

Methods
Wheat germplasm
A total of 1787 accessions were used for this study, including 1539 Chinese cultivars, 107 landraces and 141 accessions originated from USA, Australia, Europe and Japan and grouped together as foreign accessions (Table 1). This panel was obtained from the Institute of Genetics and Developmental Biology, Chinese Academy of Sciences (IGDB, CAS), and the information about the origin and nature of each accession was collected from Chinese Crop Germplasm Resources Information System (http://www.cgris.net/cgris_english.html). All the accessions were grown at the experimental station of the Institute of Genetics and Developmental Biology, Chinese Academy of Sciences, Beijing, in the 2009–2010 cropping season according to local crop management

practices. Most accessions are from the winter wheat regions, but a few accessions are from the spring or winter-spring wheat regions, and to secure the survival of all the accessions, the seedlings were covered with a plastic film during the winter season. After harvesting, 300 physiologically mature seeds for each accession were subjected to measuring the kernel hardness using the Perten's Single Kernel Characterization System (SKCS) 4100. In the successive 2010–2011 growing season, a subset of 623 accessions was grown and harvested for the kernel hardness measurement to assess the effect of years on kernel hardness. Each accession was planted on a 2-m row with inter row spacing of 0.25 cm and plant-to-plant distance of 5 cm within a row.

DNA isolation

Genomic DNA was extracted from single seedlings of each accessions growing in the greenhouse using CTAB procedure [32]. The isolated DNA was measured with the Nanodrop spectrophotometer (Thermo Fisher Scientific Inc., Wilmington, DE) and diluted to 100 ng/μl for further PCR analysis.

The modified EcoTILLING

High throughput EcoTILLING analysis was applied to identify *Pina* and *Pinb* alleles in the selected germplasm. The method was followed with Li et al. (2013) with minor modifications based on gene cloning to minimize the use of control DNA templates. Firstly, three dominant allelic variants of *Pina* (*Pina-D1a* from Chinese Spring) and *Pinb* (*Pinb-D1a* from Chinese Spring and *Pinb-D1b* from Neimai 11) were cloned into the pGEM-T Easy vector (Transgen, Beijing, China) as the control samples, respectively. The nested PCR was used to detect *Pina* and *Pinb* allelic variants (Fig. 1; Additional file 1: Table S3). The full length of *Pina* and *Pinb* were amplified individually from genomic DNA of tested samples with primers (Pina-Out-F/R, Pinb-Out-F/R) in the first step PCR reaction. Then, the PCR products and the plasmids (containing *Pina-D1a*, *Pinb-D1a* and *Pinb-D1b*, respectively) were diluted for 100-fold as templates for the second step PCR reaction individually. To improve the amplification efficiencies, the fluorescently labeled primers (LI-COR Biosciences, Lincoln, USA) were mixed with unlabeled primers (Pina-In-F/R, Pinb-In-F/R) in 1:1 ratio for the second step PCR reaction.

Screening the single nucleotide polymorphism of *Pina* and *Pinb*

For identifying the null alleles of *Pina* and *Pinb*, all of the products of nested PCR were detected with 1% agarose gels electrophoresis. Specific primers were further applied to identify the known null alleles, *Pina-D1b* [20], *Pina-D1r* and *Pina-D1s* [12] (Additional file 1: Table S3). For those samples whose nested PCR products of *Pina* and *Pinb* contained the target bands, different strategies were attempted to identify *Pina* and *Pinb* alleles [10]. The nested PCR products of four tested samples and one control were mixed at the same volumes, and the mixture were denatured and re-annealed to allow the formation of hetero-duplex between the wild type and mutant DNA molecules in a thermocycler as follows: 99 °C for 10 min, 70 cycles of 70 °C for 20 s decreasing 0.3 °C per cycle [33]. The resulted hetero-duplex mixtures were digested with CEL1 enzyme [34], and the cleavage reaction was stopped by adding 5 μl stop solution (2.5 μl 0.25 M EDTA and 2.5 μl formamide loading dye) [10]. To visualize the polymorphisms between the tested samples and the control, 1 μl of CEL1 enzyme digestion product was loaded into 6.5% polyacrylamide gels, and separated on the LI-COR 4300 DNA analyzer (LI-COR Biosciences, Lincoln, USA) at 1500 V/40 watts/45 °C for 4 h. The new allelic variants of *Pina* and *Pinb* were further confirmed by sequencing. The genetic diversity at each locus was calculated using Nei's index [35] with formula $H = 1 - \sum Pi^2$, where H and Pi denote the genetic variation index and the frequency of the number of alleles at the locus, respectively.

For *Pina* and *Pinb* double null mutants, the BAC sequences (CT009735) flanking the *Pina* and *Pinb* genes were download from NCBI to design genome-specific primers surrounding these two genes. Four pairs of specific primers which located on different position of *Pina* and *Pinb* coding sequences were explored to investigate the deletion of *Pina* and *Pinb* in the wheat genomes (Additional file 1: Table S3), which could largely detect the *Pina* and *Pinb* genes in spite of chromosomal rearrangement. Moreover, Nine pairs of primer sets spanning an approximately 90-kb region were designed between −21,803 bp (reference to the ATG of the *Pina* gene) and +68,481 to clarify the molecular mechanism of the *Pina* and *Pinb* double null mutants (Additional file 1: Table S3). The size and position of deletions in these double null mutants were deduced based on the PCR amplification.

Kernel hardness measurement

The harvested seeds of 1787 accessions were cleaned and kept at dry indoor ventilation for 3 days to bring the moisture content to 11–13%. For each sample, kernel hardness index was measured with 300 seeds through the Perten's Single Kernel Characterization System (SKCS) 4100 according to the manufacturer's procedure (Perten Instruments North America Inc., Springfield, IL, USA). Chinese Spring was included as the reference for soft wheat with SKCS index of 25.0 ± 17.0. Regarding to kernel hardness based classification of the germplasm [24, 36], the categories normally include <40.0 (soft), 40.0–60.0 (medium), and >60.0 (hard) though different classification systems have been adopted in different countries [10, 24].

According to these categories, 1787 accessions with different SKCS index were classified into soft (<40.0), medium (40.0–60.0) and hard wheat (>60.0).

Assessment of grain texture by scanning electron microscopy (SEM)

The SEM images of physiologically mature grains from wheat accessions with double null mutants at *Pina* and *Pinb* loci were compared with those of wild type alleles at *Pina* and *Pinb* loci (*Pina-D1a/Pinb-D1a*) and of durum wheat lacking *Pina* and *Pinb* loci. Two grains from each accession were transversely sliced and placed onto glass microscope slides. The slides were fixed with double sided tape, and coated with gold in Dynavac CS300 coating unit [37]. Photographic images were captured at 2000-fold magnification using a ZEISS supra 10 vp field emission scanning electron microscope (Carl Zeiss Microscopy, NY, USA) at 10 kV.

Additional files

Additional file 1: Table S1. Year effect on SKCS value in 623 accessions. **Table S2.** Number of hard, soft and medium wheat in wheat cultivation regions when regarding to accession type. **Table S3.** Sequence, product size and annealing temperature of PCR primers used for *Pina* and *Pinb* amplification. **Table S4.** SKCS hardness index of *Pina-D1x/Pinb-D1ah* genotype.

Additional file 2: Figure S1. Identification of *Pina* and *Pinb* deletions in some accessions by a set of selected markers. (A) Primers AGPS-1 was used to check all of DNA quality. (B) Primers Pina-part amplified part of *Pina* coding sequence. (C) Primers Pina-cds amplified the *Pina* coding sequence. (D) Primers Pinb-part amplified part of *Pinb* coding sequence. (E) Primers Pinb-cds amplified the *Pinb* coding sequence. (F) Primers Pina-4 was used to check the deletion of *Pina* and *Pinb* downstream sequence. 1–7 show accessions Chinese Spring, NIL-Novos 67, Yunfengzao 21, Shan 150, 91G 149/Chang 128,865, Hedong TX-008, Xinong 8925–13 respectively.

Additional file 3: Figure S2. Wheat seed protein analysis. (A) Western blot analysis of PINB protein. (B) SDS-PAGE gel of total proteins from wheat mature seeds. Lanes were loaded with 20 μg protein, 1–7 show accessions Chinese Spring, NIL-Novos 67, Yunfengzao 21, Shan 150, 91G 149/Chang 128,865, Hedong TX-008, Xinong 8925–13, respectively.

Additional file 4: Figure S3. Sequence alignments of *Pina* (A) and *Pinb* (B) alleles. (A) *Pina-D1a* (DQ363911), *Pina-D1l* ([6, 15]), and *Pina-D1y*. (B) *Pinb-D1u* (EF620911), *Pinb-D1a* (DQ363913), *Pinb-D1b* (DQ363914), *Pinb-D1c* (KC585019), *Pinb-D1d* (KR259645), *Pinb-D1p* (AY581889), and *Pinb-D1q* (EF620909).

Abbreviations

ANOVA: Analysis of variance; EcoTILLING: Ecotype targeting induced local lesions in genomes; *H*: Nei's index of allelic diversity index; NCBI: National Center for Biotechnology Information; nt: Nucleotide; PINA: Puroindoline a; PINB: Puroindoline b; SEM: Scanning electron microscopy; SKCS: Single kernel characterization system; STS: Sequence tagged site; WT: Wild type

Acknowledgements
We thank two anonymous reviewers for their critical reviews and suggestions.

Funding
This work was supported by Ministry of Agriculture of China for transgenic research (2016ZX08009-003) and the National Science Foundation of China (31371610).

Authors' contributions
XM carried out most experiments and analyzed the data. XM, MS and DL wrote the manuscript. JW and XM analyzed the data. WY, XL and JS grew samples and measured the kernel hardness. AZ and DL conceptualized the experiments and revised the manuscript. All authors read and approved the final manuscript.

Competing interests
The authors declare that they have no competing interests.

Author details
[1]State Key Laboratory of Plant Cell and Chromosome Engineering, Institute of Genetics and Developmental Biology, Chinese Academy of Sciences, 1 West Beichen Road, Chaoyang District, Beijing 100101, China. [2]University of Chinese Academy of Sciences, Beijing 100049, China. [3]Department of Environmental Sciences, COMSATS Institute of Information Technology, Vehari 61100, Pakistan. [4]The Institute of Forestry and Pomology, Beijing Academy of Agriculture and Forestry Sciences, Beijing 100093, China.

References
1. Pauly A, Pareyt B, Fierens E, Delcour JA. Wheat (Triticum Aestivum L. and T. Turgidum L. Ssp. Durum) kernel hardness: II. Implications for end-product quality and role of Puroindolines therein. Compr Rev Food Sci F. 2013;12(4):427–38.
2. Jolly CJ, Rahman S, Kortt AA, Higgins TJV. Characterization of the wheat Mr 15000 "grain-softness protein" and analysis of the relationship between its accumulation in the whole seed and grain softness. Theor Appl Genet. 1993;86(5):589–97.
3. Morris CF, Greenblatt GA, Bettge AD, Malkawi HI. Isolation and characterization of multiple forms of Friabilin. J Cereal Sci. 1994;20(2):167–74.
4. Gautier MF, Aleman ME, Guirao A, Marion D, Joudrier P. Triticum Aestivum puroindolines, two basic cystine-rich seed proteins: cDNA sequence analysis and developmental gene expression. Plant Mol Biol. 1994;25(1):43–57.
5. Bhave M, Morris CF. Molecular genetics of puroindolines and related genes: allelic diversity in wheat and other grasses. Plant Mol Biol. 2008;66(3):205–19.
6. Morris CF, Bhave M. Reconciliation of D-genome puroindoline allele designations with current DNA sequence data. J Cereal Sci. 2008;48(2):277–87.
7. Bhave M, Morris CF. Molecular genetics of puroindolines and related genes: regulation of expression, membrane binding properties and applications. Plant Mol Biol. 2008;66(3):221–31.
8. Ali I, Sardar Z, Rasheed A, Mahmood T. Molecular characterization of the puroindoline-a and b alleles in synthetic hexaploid wheats and in silico functional and structural insights into Pina-D1. J Theor Biol. 2015;376:1–7.
9. Kumar R, Arora S, Singh K, Garg M. Puroindoline allelic diversity in Indian wheat germplasm and identification of new allelic variants. Breed Sci. 2015; 65(4):319–26.
10. Wang J, Sun JZ, Liu DC, Yang WL, Wang DW, Tong YP, Zhang AM. Analysis of Pina and Pinb alleles in the micro-core collections of Chinese wheat germplasm by Ecotilling and identification of a novel Pinb allele. J Cereal Sci. 2008;48(3):836–42.
11. Chen F, Zhang FY, Xia XC, Dong ZD, Cui DQ. Distribution of puroindoline alleles in bread wheat cultivars of the yellow and Huai valley of China and discovery of a novel puroindoline a allele without PINA protein. Mol Breeding. 2011;29(2):371–8.
12. Ikeda TM, Cong H, Suzuki T, Takata K. Identification of new Pina null mutations among Asian common wheat cultivars. J Cereal Sci. 2010;51(3):235–7.
13. Morris CF, Lillemo M, Simeone MC, Giroux MJ, Babb SL, Kidwell KK. Prevalence of puroindoline grain hardness genotypes among historically significant north American spring and winter wheats. Crop Sci. 2001;41(1):218–28.
14. Cane K, Spackman M, Eagles HA. Puroindoline genes and their effects on grain quality traits in southern Australian wheat cultivars. Aust J Agric Res. 2004;55(1):89–95.
15. Chen F, He ZH, Xia XC, Xia LQ, Zhang XY, Lillemo M, Morris CF. Molecular and biochemical characterization of puroindoline a and b alleles in Chinese landraces and historical cultivars. Theor Appl Genet. 2006;112(3):400–9.
16. Chen F, Yu YX, Xia XC, He ZH. Prevalence of a novel puroindoline b allele in Yunnan endemic wheats (Triticum Aestivum Ssp. Yunnanense king). Euphytica. 2007;156(1):39–46.

17. Chang C, Zhang HP, Xu J, Li WH, Liu GT, You MS, Li BY. Identification of allelic variations of puroindoline genes controlling grain hardness in wheat using a modified denaturing PAGE. Euphytica. 2006;152(2):225–34.

18. Martin JM, Frohberg RC, Morris CF, Talbert LE, Giroux MJ. Milling and bread baking traits associated with puroindoline sequence type in hard red spring wheat. Crop Sci. 2001;41(1):228–34.

19. Xia LQ, Chen F, He ZH, Chen XM, Morris CF. Occurrence of puroindoline alleles in Chinese winter wheats. Cereal Chem. 2005;82(1):38–43.

20. Chen F, Zhang FY, Morris C, He ZH, Xia XC, Cui DQ. Molecular characterization of the Puroindoline a-D1b allele and development of an STS marker in wheat (Triticum Aestivum L.). J Cereal Sci. 2010;52(1):80–2.

21. Wilkinson M, Wan YF, Tosi P, Leverington M, Snape J, Mitchell RAC, Shewry PR. Identification and genetic mapping of variant forms of puroindoline b expressed in developing wheat grain. J Cereal Sci. 2008;48(3):722–8.

22. Chen F, Zhang FY, Cheng XY, Morris C, Xu HX, Dong ZD, Zhan KH, Cui DQ. Association of Puroindoline b-B2 variants with grain traits, yield components and flag leaf size in bread wheat (Triticum Aestivum L.) varieties of the yellow and Huai valleys of China. J Cereal Sci. 2010;52(2): 247–53.

23. Chen F, Xu HX, Zhang FY, Xia XC, He ZH, Wang DW, Dong ZD, Zhan KH, Cheng XY, Cui DQ. Physical mapping of puroindoline b-2 genes and molecular characterization of a novel variant in durum wheat (Triticum Turgidum L.). Mol Breeding. 2010;28(2):153–61.

24. Qamar ZU, Bansal UK, Dong CM, Alfred RL, Bhave M, Bariana HS. Detection of puroindoline (Pina-D1 and Pinb-D1) allelic variation in wheat landraces. J Cereal Sci. 2014;60(3):610–6.

25. Chen F, He ZH, Xia XC, Lillemo M, Morris C. A new puroindoline b mutation present in Chinese winter wheat cultivar Jingdong 11. J Cereal Sci. 2005; 42(2):267–9.

26. Morris C, Beecher B. The distal portion of the short arm of wheat (Triticum Aestivum L.) chromosome 5D controls endosperm vitreosity and grain hardness. Theor and Appl Genet. 2012;125(2):247–54.

27. Chen F, Li HH, Cui DQ. Discovery, distribution and diversity of Puroindoline-D1 genes in bread wheat from five countries (Triticum Aestivum L.). BMC Plant Biol. 2013;13(1):125.

28. Ikeda TM, Ohnishi N, Nagamine T, Oda S, Hisatomi T, Yano H. Identification of new puroindoline genotypes and their relationship to flour texture among wheat cultivars. J Cereal Sci. 2005;41(1):1–6.

29. Lillemo M, Chen F, Xia XC, William M, Pena RJ, Trethowan R, He ZH. Puroindoline grain hardness alleles in CIMMYT bread wheat germplasm. J Cereal Sci. 2006;44(1):86–92.

30. Ram S, Boyko E, Giroux MJ, Gill BS. Null mutation in puroindoline a is prevalent in Indian wheats: Puroindoline genes are located in the distal part of 5DS. J Plant Biochem Biot. 2002;11(2):79–83.

31. Ma DY, Zhang Y, Xia XC, Morris CF, He ZH. Milling and Chinese raw white noodle qualities of common wheat near-isogenic lines differing in puroindoline b alleles. J Cereal Sci. 2009;50(1):126–30.

32. Dellaporta SL, Wood J, Hicks JB. A plant DNA minipreparation: version II. Plant Mol Biol Rep. 1983;1(4):19–21.

33. Li AX, Yang WL, Lou XY, Liu DC, Sun JZ, Guo XL, Wang J, Li YW, Zhan KH, Ling HQ, Zhang AM. Novel natural allelic variations at the Rht-1 loci in wheat. J Integr Plant Biol. 2013;55(11):1026–37.

34. Yang B, Wen X, Kodali NS, Oleykowski CA, Miller CG, Kulinski J, Besack D, Yeung JA, Kowalski D, Yeung AT. Purification, cloning, and characterization of the CEL I nuclease. Biochemistry. 2000;39(13):3533–41.

35. Nei M. Analysis of gene diversity in subdivided populations. Proc Nat Acad Sci USA. 1973;70(12):3321–3.

36. Gaines CS, Finney PF, Fleege LM, Andrews LC. Predicting a hardness measurement using the single-kernel characterization system. Cereal Chem. 1996;73(2):278–83.

37. Brennan CS, Harris N, Smith D, Shewry PR. Structural differences in the mature endosperms of good and poor malting barley cultivars. J Cereal Sci. 1996;24(2):171–7.

8

Catalytic and functional aspects of different isozymes of glycolate oxidase in rice

Zhisheng Zhang[1], Xiangyang Li[2], Lili Cui[2], Shuan Meng[1], Nenghui Ye[1] and Xinxiang Peng[2*]

Abstract

Background: Glycolate oxidase (GLO) is a key enzyme for photorespiration in plants. There are four GLO genes encoding and forming different isozymes in rice, but their functional differences are not well understood. In this study, enzymatic and physiological characteristics of the GLO isozymes were comparatively analyzed.

Results: When expressed heterologously in yeast, GLO1, GLO4 and GLO1 + 4 showed the highest activities and lowest K_m for glycolate as substrate, whereas GLO3 displayed high activities and affinities for both glycolate and L-lactate, and GLO5 was catalytically inactive with all substrates tested. To further reveal the physiological role of each GLO isozyme in plants, various GLO genetically modified rice lines were generated and functionally analyzed. GLO activity was significantly increased both in GLO1 and GLO4 overexpression lines. Nevertheless, when either GLO1 or GLO4 was knocked out, the activity was suppressed much more significantly in GLO1 knockout lines than in GLO4 knockout lines, and both knockout mutants exhibited obvious dwarfism phenotypes. Among GLO3 and GLO5 overexpression lines and RNAi lines, only GLO3 overexpression lines showed significantly increased L-lactate-oxidizing activity but no other noticeable phenotype changes.

Conclusions: These results indicate that rice GLO isozymes have distinct enzymatic characteristics, and they may have different physiological functions in rice.

Keywords: Glycolate oxidase isozymes, Enzymatic characteristics, Physiological functions, Rice (*Oryza sativa*)

Background

Glycolate oxidase (GLO, EC 1.1.3.15) is an important peroxisomal FMN-dependent oxidase involved in photorespiration. Plant photorespiration begins with the oxygenating reaction of ribulose 1, 5-bisphosphate carboxylase-oxygenase (Rubisco) in chloroplasts. This process produces a toxic intermediate metabolite phosphoglycolate (2-PG), which is further converted to glycolate by 2-PG phosphatase (PGP). Glycolate is transferred to peroxisomes and oxidized into glyoxylate by GLO with equimolar amount of hydrogen peroxide (H_2O_2) released [1–3]. In addition to its metabolic function in photorespiration, GLO has been reported to play roles in plant photosynthetic regulation and stress resistance. Suppression of GLO leads to glyoxylate accumulation and inhibits photosynthesis, while overexpressing GLO confers improved photosynthesis under high light and high temperature in rice [4, 5]. GLO has been significantly induced in cowpea, tobacco and pea under drought stress [6–8], while in rice and barley, GLO was induced notably by pathogen infection [9–11]. Furthermore, because of the high metabolite flux of photorespiration, about 70% of the total H_2O_2 in C_3 plants comes from the oxidation of glycolate as catalyzed by GLO, and this value could be even higher under some stress conditions such as drought and high temperature [3, 12–14]. Therefore, GLO may also play an important role in plant H_2O_2-related pathways.

The sequencing of *Arabidopsis thaliana*, *Nicotiana benthamiana* and rice revealed that GLO are encoded by a gene family in these plant species [15, 16]. Moreover, GLO isozymes have been observed in several plant species such as *Arabidopsis thaliana*, maize, and spinach [16–19]. The expressions of isozymes are usually tissue-specific, which may satisfy metabolic behavior of the cells in which each isozyme is expressed [20, 21]. For example, the *Arabidopsis* 1-Amino-cyclopropane-1-carboxylate synthase (ACS)

* Correspondence: xpeng@scau.edu.cn
[2]State Key Laboratory for Conservation and Utilization of Subtropical Agro-bioresources, College of Life Sciences, South China Agricultural University, Guangzhou, Guangdong 510642, China
Full list of author information is available at the end of the article

isozymes are biochemically distinct, have tissue-specific expression, and function in different cellular environments for C_2H_4 synthesis [21]. The GLO isozymes have been reported to show tissue-specific expression in maize and *Arabidopsis* (e.g., there are two different GLO isozymes exist in the bundle sheath and mesophyll tissues of maize leaves) [17, 22], while the enzymatic and physiological characteristics of their isozymes have not been comparatively studied. It is not well understood why there are different tissue-specific GLO isozymes in these plant species. In addition, GLO isozymes are related with resistances to various stresses, wherein each GLO isozyme may perform different functions. Rojas et al. (2012) found that each *Arabidopsis* GLO isozyme could play different roles in the H_2O_2 signal transduction to induce defense responses during the nonhost resistance of *Arabidopsis thaliana* [16]. *Arabidopsis* GOX1 and GOX2 have been reported to perform different functions in the oxidative stress-related cell death [23]. Accordingly, the potential tissue- or environment-specific expression and enzymatic diversity of rice GLO isozymes would be relevant to the distinct physiological function of each GLO isozyme during various biological processes.

A total of four *GLO* genes have been identified in rice genome (i. e., Os03g0786100, Os04g0623500, Os07g0152900 and Os07g0616500, encoding *OsGLO1*, *OsGLO3*, *OsGLO4* and *OsGLO5*), each of which has a peroxisomal targeting signal, PTS1 [24]. In this study, we comparatively investigated the enzymatic characteristics of each GLO isozyme, and furthermore, different genetically modified rice lines of these *GLO* genes were generated and analyzed for functionality. Our results demonstrate that rice GLO isozymes have distinct enzymatic characteristics, and their physiological functions are nonredundant in rice.

Results

1. Transcriptional expression patterns of *GLO* genes and their responses to stresses

The rice genome contains four *GLO* genes located on three different chromosomes [15], both the mRNA sequences and polypeptides of these four *GLO* genes are highly similar (Additional file 1). Our previous transcriptional analyses have shown that *GLO1* and *GLO4* were predominantly expressed in leaves, while *GLO3* and *GLO5* were mainly expressed in roots [15, 24]. In this study, we further noticed that *GLO1* and *GLO4* were abundantly expressed in leaves and leaf sheaths, and moderately expressed in stems and husks. *GLO3* was primarily expressed in stems and leaf sheaths, and *GLO5* was only slightly expressed in leaves (Fig. 1a). Furthermore, transcription profiles of *GLO* genes were analyzed for leaves at different growth stages. The expression of *GLO1* increased about 65-75% in the booting and

heading stages but not in other developmental periods, while *GLO4* only showed a 35-40% increase in the booting and heading stages (Fig. 1b). The expression levels of *GLO3* and *GLO5* were very low in rice leaves throughout all developmental periods (Fig. 1b).

Because GLO is reported to be involved in stress resistance in plants [6–11], we investigated responses of different *GLO* gene members to various stresses here (PEG 6000, 10%; NaCl, 100 mM; ABA, 10 μM; H_2O_2, 5 mM; AlCl$_3$, 2 mM; CK, Control group). The results showed that *GLO1* transcripts increased by 60% under PEG treatment, but decreased about 35-40% under other treatments such as NaCl and ABA treatments (Fig. 1c). *GLO4* expression showed a 35% increase under NaCl treatment, but was suppressed by 25-40% under ABA, H_2O_2 and AlCl$_3$ treatments (Fig. 1c). Elevated expression level of *GLO3* was observed under all treatments, the expression of *GLO5* was also increased under these treatments, except for the AlCl$_3$ treatment (Fig. 1c). Meanwhile, GLO activities were correspondingly assayed for the above samples. Inconsistent with the changes in *GLO* gene expression, the GLO activity only increased 5-8% under the PEG, NaCl and H_2O_2 treatments (Fig. 1d).

2. Enzymatic differences of GLO isozymes

Enzymatic characteristics of the GLO isozymes in plants are seldom investigated. Previous research found that there are three types of GLO isozymes present in rice leaves, including two that are homo-oligomers composed of either GLO1 or GLO4 subunits, and the others are hetero-oligomers composed of interacted GLO1 and GLO4 subunits [24]. Here the kinetic properties of rice GLO isozymes were comparatively analyzed. A 6 × His-tag was fused to the N-terminus of each GLO and expressed in the yeast *Saccharomyces cerevisiae*, since it has been proved that the N-terminal His-tag rarely influences the properties of the fused protein [25]. Yeast cell lysates were prepared using acid-washed glass beads, and the western blot analyses showed that all GLO isozymes could be heterologously expressed in *S.cerevisiae* (Fig. 2a). Substrate screens using crude enzyme revealed that GLO1, GLO4 and GLO1 + 4 displayed the highest activity with glycolate as substrate, and showed appreciably high activity on glycerate and less activity on L-lactate and glyoxylate (Fig. 2b). In contrast, GLO3 showed the highest activity on L-lactate, and then on glycolate and glycerate, respectively (Fig. 2b). GLO5 was completely inactive to all substrates tested. Each GLO isozyme was further purified from the yeast crude extraction by immobilized metal-affinity chromatography. SDS-PAGE analysis, which guarantees purity for each isozyme, showed an identical subunit molecular weight of about 40 kDa for various isozymes (Fig. 2c). A preliminary analysis showed that the optimum pH of 7.8 was

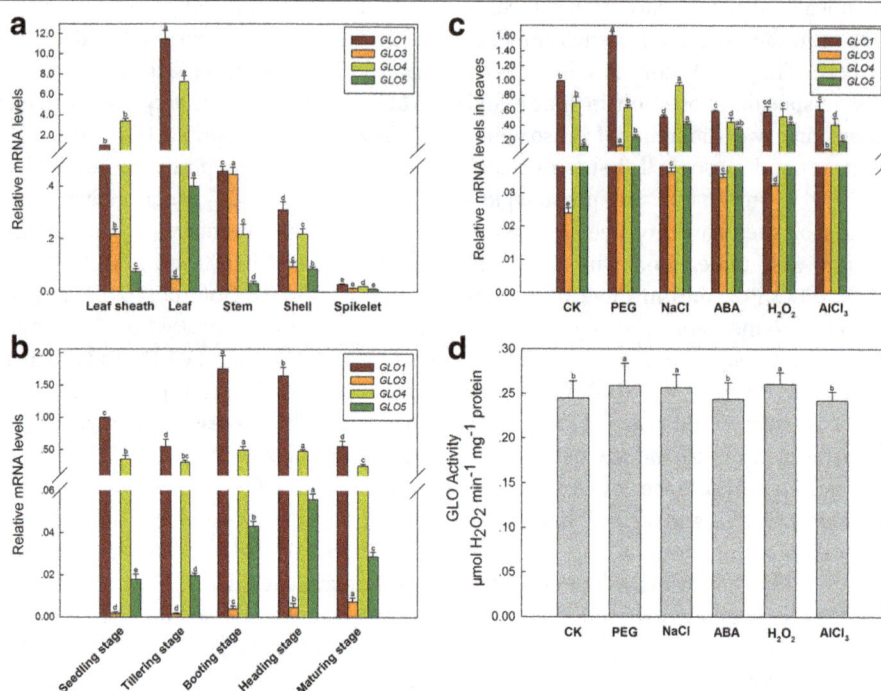

Fig. 1 Expression patterns of *GLO* genes and their response to stresses in rice. **a** Relative mRNA levels in different leaf tissues were graphed based on the *GLO1* mRNA level in leaf sheath as 1. **b** Relative mRNA levels at different developmental stages were graphed based on the *GLO1* mRNA level in seedling stage as 1. **c** mRNA transcriptional levels of *GLO* genes in response to various stress treatments for 2 days (PEG 6000, 10%; NaCl, 100 mM; ABA, 10 μM; H_2O_2, 5 mM; $AlCl_3$, 2 mM; CK, Control group). Relative mRNA levels were graphed based on the *GLO1* mRNA level in CK as 1. **d** GLO activities in response to various stress treatments (the same samples as in Fig.1c). Values are means ± SD ($n = 3$). Means denoted by the same letter did not significantly differ at $P < 0.05$ according to Duncan's multiple range tests

identical for all GLO isozymes (Additional file 2), and the optimum temperature for GLO1, GLO4 and GLO1 + 4 was 42 °C, while the optimum temperature of GLO3 was 47 °C (Additional file 2).

The kinetic studies were performed at 30 °C and pH 7.8. GLO1, GLO4 and GLO1 + 4 have the highest affinity for glycolate, with $K_{m(glycolate)}$ values of 0.499 mM, 0.613 mM and 0.423 mM, respectively (Table 1), and the $V_{max(glycolate)}$ values of GLO1 + 4 and GLO1 were higher than that of GLO4 (Additional file 3). The affinities of GLO1, GLO4 and GLO1 + 4 for glyoxylate, L-lactate and glycerate were much lower than for glycolate (Table 1). GLO3 also exhibited highest affinity to glycolate, meanwhile, it showed a high affinity to L-lactate, with K_m values of 0.470 mM and 1.104 mM, respectively (Table 1). Oxalate is prevalent in the plant kingdom [26], and is a competitive inhibitor of GLO [27]. Here K_i values for oxalate were detected to range from 4.572 to 6.337 mM when using glycolate as substrate (Table 2). Oxalate could more strongly inhibit glyoxylate-oxidizing activity of each GLO isozyme, with K_i values between 1.887 and 3.018 mM (Table 2).

3. Functional analysis of GLO isozymes

As described above, rice GLO isozymes have distinct enzymatic properties. It is more interesting to know

whether these isozymes may play distinct physiological roles in plants. As such, each of the 4 isozymes was overexpressed in rice to determine their contribution to the glycolate metabolism. As shown in Additional file 4, each *GLO* gene was up-regulated as expected at the mRNA level in the corresponding transgenic line. Overexpression of *GLO1* and *GLO4* increased GLO activity by 110% and 65% in rice leaves, respectively. However, overexpression of *GLO3* increased GLO activity by only about 12% (Fig. 3a) but increased the L-lactate-oxidizing activity by more than 140% (Fig. 3a). Overexpression of *GLO5* had no effect on GLO activity, consistent with the result of enzymatic assay. In addition, both *GLO3* and *GLO5* were silenced by RNAi (Additional file 4), *GLO1* and *GLO4* were knocked out using pYLCRISPR/Cas9P$_{ubi}$ system (Additional file 4). In leaves of *GLO1* and *GLO4* knockout lines, GLO activity was decreased by about 65% and 20%, respectively, while suppression of *GLO3* and *GLO5* had no effect (Fig. 3a). The GLO isozymes zymogram analysis of different *GLO* genetically modified rice lines verified that GLO1 and GLO4 were completely knocked out in the corresponding transgenic lines (Fig. 3b), further supporting our previous results [24]. In contrast, suppression of *GLO3* and *GLO5* did not alter GLO isozyme patterns (Fig. 3c), implying that in leaves of wild rice plants

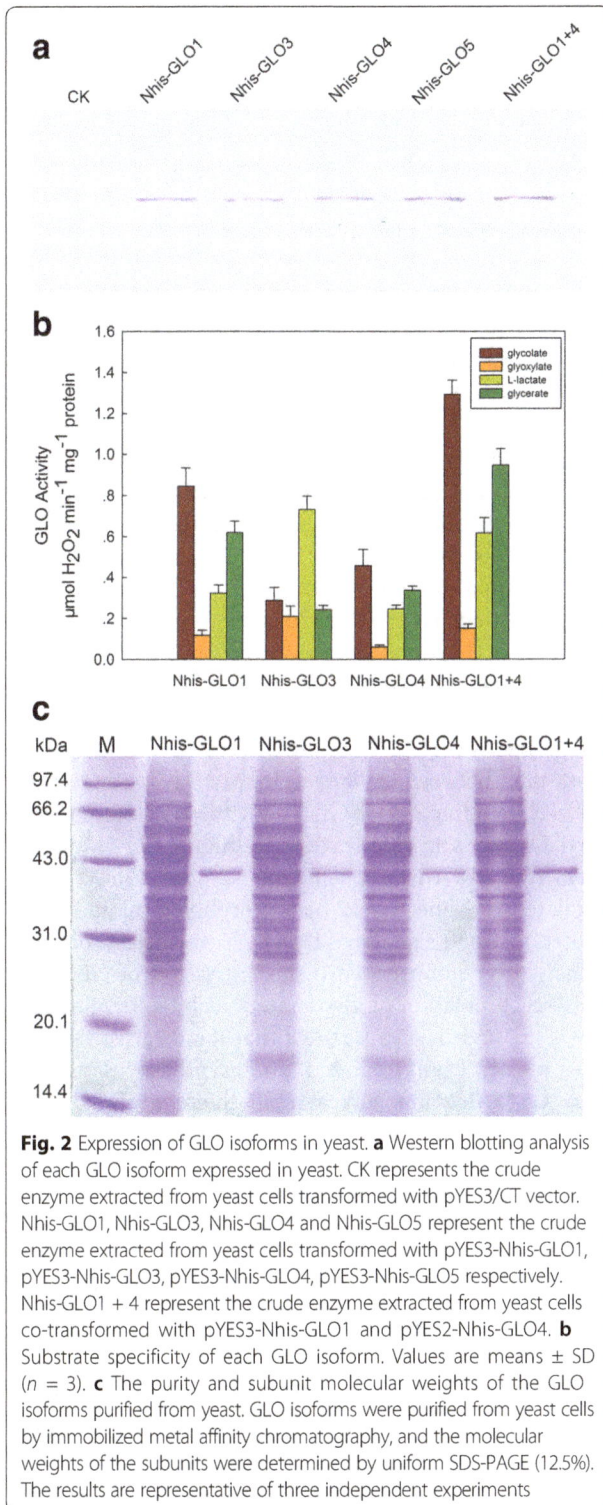

Fig. 2 Expression of GLO isoforms in yeast. **a** Western blotting analysis of each GLO isoform expressed in yeast. CK represents the crude enzyme extracted from yeast cells transformed with pYES3/CT vector. Nhis-GLO1, Nhis-GLO3, Nhis-GLO4 and Nhis-GLO5 represent the crude enzyme extracted from yeast cells transformed with pYES3-Nhis-GLO1, pYES3-Nhis-GLO3, pYES3-Nhis-GLO4, pYES3-Nhis-GLO5 respectively. Nhis-GLO1 + 4 represent the crude enzyme extracted from yeast cells co-transformed with pYES3-Nhis-GLO1 and pYES2-Nhis-GLO4. **b** Substrate specificity of each GLO isoform. Values are means ± SD (n = 3). **c** The purity and subunit molecular weights of the GLO isoforms purified from yeast. GLO isoforms were purified from yeast cells by immobilized metal affinity chromatography, and the molecular weights of the subunits were determined by uniform SDS-PAGE (12.5%). The results are representative of three independent experiments

observations [5], the rice lines with GLO activities suppressed displayed dwarfism phenotype (data not shown), and reduced H_2O_2 content was also detected in these rice lines (Additional file 5). However, it was noticed that GLO3 had high activity on L-lactate, meaning that it might participate in the L-lactate metabolism in rice, particularly in the roots (Fig. 2b; Table 1). So we tested if GLO3 contributed to L-lactate tolerance as recently reported by Engqvist et al. (2015). As shown in Additional file 6, the phenotype of *GLO3* overexpression and RNAi lines were not different from that of WT plants under lactate treatment, though lactate-oxidizing activity was markedly increased in both leaves and roots of the overexpression lines (Fig. 3a; Additional file 7).

Discussion

Members of a gene family can have different tissue-, development- or environment-specific expression patterns. Using the *GLO* gene family that is present in the rice genome [15, 24], we show that the four rice *GLO* genes are differently expressed in various tissues and developmental stages of the rice plant (Fig. 1a and b), as well as in response to stresses (Fig. 1c). Our findings suggest that the physiological roles of these *GLO* genes are not redundant, however, the exact biological significance of the rice *GLO* gene family needs further investigation.

Generally speaking, gene family members that have different expression patterns would encode various isozymes with diverse enzymatic characteristics and physiological roles [21, 28]. In plants, GLO was first purified from spinach [29, 30], the primary structure of spinach GLO (SpGLO) was identified by peptide sequencing, which contains only one uniform peptide [31], and its K_m for glycolate was detected to be 0.38 mM [32]. GLO was also isolated from *Parthenium hysterophorus* and *Pisum sativum*, which both consisted of two different peptides. The $K_{m(glycolate)}$ of the *Parthenium hysterophorus* GLO is 0.2 mM and that of the *Pisum sativum* GLO is 0.3 mM [33]. However, the enzymatic characteristics of each GLO isozyme in plants are seldom comparatively analyzed. *Arabidopsis* contains five GLO members, i. e. GOX1, GOX2, GOX3, HAOX1 and HAOX2 [4, 34], only GOX1, GOX2 and GOX3 have been heterologously expressed and purified in *E.coli*, respectively. Enzymatic assays of these purified enzymes showed that they all had high affinity for glycolate but with distinct K_m values. In addition, GOX3 possesses high catalytic efficiency for L-lactate [22]. In this study, we comparatively investigated enzymatic characteristics of all rice GLO isozymes. GLO1 + 4 showed highest catalytic efficiency using glycolate as substrate, followed by GLO1 and GLO4, respectively (Fig. 2b; Table 1 and Additional file 3). Indeed, the transgenic rice plants that had up-regulated or knocked-out GLO1 or GLO4 exhibited much higher or lower GLO activity,

(WT), *GLO3* and *GLO5* may have not contributed to GLO activities. Consistent with changes in GLO activities as addressed above, phenotypes were not obviously altered in all the *GLO3* and *GLO5* up-regulated or down-regulated transgenic lines (Fig. 4). However, similar to our previous

Table 1 K_m of each GLO isozyme with various substrates

GLO isozymes	$K_{m(glycolate)}$ (mM)	$K_{m(glyoxylate)}$ (mM)	$K_{m(L\text{-}lactate)}$ (mM)	$K_{m(glycerate)}$ (mM)
GLO1	0.499 ± 0.041^b	6.505 ± 0.412^a	5.128 ± 0.315^a	4.976 ± 0.305^a
GLO3	0.470 ± 0.035^{bc}	1.799 ± 0.084^d	1.104 ± 0.056^c	2.762 ± 0.311^c
GLO4	0.613 ± 0.050^a	5.983 ± 0.333^b	4.720 ± 0.291^b	4.367 ± 0.395^b
GLO1 + 4	0.423 ± 0.038^c	4.883 ± 0.321^c	5.227 ± 0.448^a	4.511 ± 0.307^b

Values are means ± SD of three replicates. Means denoted by the same letter did not significantly differ at $P < 0.05$ according to Duncan's multiple range tests

respectively. Our previous studies also observed that the phenotypes were altered in the GLO1 and GLO4 up-regulated or down-regulated transgenic lines [4, 5]. In contrast, there were no changes in the GLO activity and phenotypes in GLO3 and GLO5 up-regulated or RNAi rice lines (Fig. 3a; Fig 4). In addition, the GLO isozymes zymogram analysis further supported that the GLO isozymes in rice leaves consisted of GLO1 and GLO4 subunits (Fig. 3b and c), and GLO1 was more abundant than GLO4 in rice leaves (Fig. 3b). In combination with our previous results [24], it can be concluded that GLO1, GLO4 and GLO1 + 4 are the GLO isozymes for photorespiration in rice. In addition, this study further revealed that GLO1 + 4 and GLO1 have higher catalytic efficiency on glycolate-oxidation than GLO4 (Fig. 2a and b; Table 1). Therefore it can be further implied that GLO1 could be the major contributor to GLO activity and consequently to photorespiration and the associated H_2O_2 production.

GLO3 is the rice homolog of *Arabidopsis* GOX3. It was recently reported that GOX3 in *Arabidopsis* functions as an L-lactate oxidase catalyzing the conversion of L-lactate to pyruvate, in order to maintain low levels of L-lactate in roots under normoxic conditions [22]. We observed that the rice GLO3 was also able to efficiently catalyze the oxidation of L-lactate to pyruvate (Fig. 2b; Table 1). Our previous results have demonstrated that *GLO3* is predominantly expressed in roots [15, 24], but unexpectedly no GLO activities could be detected in wild type rice roots (Additional file 7). We further noticed that the GLO3 overexpression rice plants conferred no improved resistance to L-lactate toxicity (Additional file 6). This was inconsistent with the *Arabidopsis* GOX3 overexpression plants which were shown to be more tolerant to L-lactate toxicity [22]. As a semi-aquatic plant,

rice produces low lactate as compared with wheat, potato and *Arabidopsis* [35–38], which might explain why GLO3 is not associated with lactate toxicity in rice as reported for *Arabidopsis* [22].

While the primary metabolic role of GLO is well known, the physiological function is still not well understood. Rojas et al. (2012) suggested that each GLO isozyme played different roles in the non-host disease resistance in *Arabidopsis* [16]. A more recent study demonstrated that the *Arabidopsis* GOX1 and GOX2 played distinct roles in the oxidative stress-related cell death [23]. GLO has also been reported to be involved in some other biological processes such as protein repair responses and salicylic acid signaling pathway [39, 40]. More interestingly, recent studies found that there was cross-talk between photorespiratory H_2O_2 and auxin [41, 42]. We found that GLO activity was closely related to H_2O_2 levels in rice leaves (Additional file 5), and, as such, the dwarfism phenotype of the GLO down-regulated rice lines could be a morphological aberration related to auxin signaling [43, 44]. Nevertheless, our previous work demonstrated that suppression of GLO may cause accumulation of glyoxylate that inhibits photosynthesis [5], so it is also possible that the dwarfism phenotype results from inhibited photosynthesis. Therefore, each GLO isozyme may exhibit different physiological functions in these various biological processes in rice, but more direct experimental evidence is needed to elucidate the potential mechanisms.

Conclusions

Our findings suggested that rice GLO isozymes have distinct enzymatic characteristics and different physiological functions. GLO1, GLO4 and GLO1 + 4 are the photorespiration GLO isozymes, moreover, GLO1 is the major contributor to GLO activity and the related photorespiratory H_2O_2 production. In addition, there may be interplay between the photorespiration glycolate-H_2O_2 metabolism and plant development. However, the certain functions of GLO3 and GLO5 remain to be fully elucidated.

Methods

Plant materials and growth conditions

The seeds of rice (*Oryza sativa*) cv. Zhonghua 11 (japonica cultivar-group) provided by the state key

Table 2 K_i of GLO with oxalate

GLO isozymes	$K_{i(oxalate)}$ (mM) Glycolate as substrate	$K_{i(oxalate)}$ (mM) Glyoxylate as substrate
GLO1	4.572 ± 0.930^c	1.887 ± 0.326^c
GLO3	5.604 ± 1.229^b	3.018 ± 1.359^a
GLO4	6.337 ± 1.736^a	2.272 ± 0.524^b
GLO1 + 4	5.491 ± 1.041^b	2.316 ± 0.608^b

Values are means ± SD of three replicates. Means denoted by the same letter did not significantly differ at $P < 0.05$ according to Duncan's multiple range tests

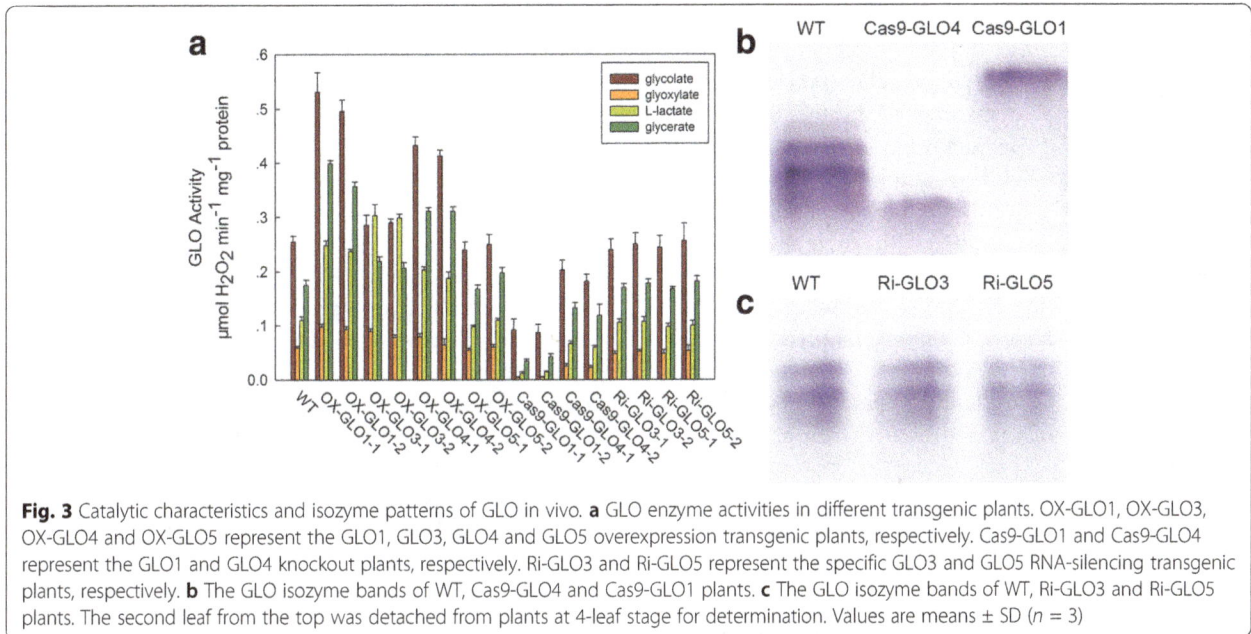

Fig. 3 Catalytic characteristics and isozyme patterns of GLO in vivo. **a** GLO enzyme activities in different transgenic plants. OX-GLO1, OX-GLO3, OX-GLO4 and OX-GLO5 represent the GLO1, GLO3, GLO4 and GLO5 overexpression transgenic plants, respectively. Cas9-GLO1 and Cas9-GLO4 represent the GLO1 and GLO4 knockout plants, respectively. Ri-GLO3 and Ri-GLO5 represent the specific GLO3 and GLO5 RNA-silencing transgenic plants, respectively. **b** The GLO isozyme bands of WT, Cas9-GLO4 and Cas9-GLO1 plants. **c** The GLO isozyme bands of WT, Ri-GLO3 and Ri-GLO5 plants. The second leaf from the top was detached from plants at 4-leaf stage for determination. Values are means ± SD ($n = 3$)

laboratory for conservation and utilization of subtropical agro-bioresources were used for the construction of transgenic lines. Rice seeds were germinated in the dark for 4-6 days at 25 °C, and then the seedlings were grown in Kimura B complete nutrient solution [45] in plant growth chambers with 14 h light (30 °C) /10 h dark (25 °C), 800 µmol m^{-2} s^{-1} average light intensity, and 60-70% relative humidity. After reaching the 4-leaf stage, seedlings were transplanted, either being continuously grown in Kimura B complete nutrient solution in the plant growth chambers, or grown in soil under natural condition. The seedlings grown in Kimura B complete nutrient solution in plant growth chambers were exposed to various stress treatments (PEG 6000, 10%; NaCl, 100 mM; ABA, 10 µM; H_2O_2, 5 mM; $AlCl_3$, 2 mM). The seedlings grown in soil under natural condition were used for GLO isozyme zymogram and growth phenotype analyses.

Plasmid construction

Total RNA was extracted from rice leaves using RNAprep Pure Kit (TIANGEN, China). The quality and quantity of the purified RNA were assessed with a NanoDrop-1000 (NanoDrop, USA). First-strand cDNA was synthesized using ReverTra Ace (Toyobo, Japan). Primers were designed to cover the complete open reading frame of each *GLO* gene (Additional file 8). To construct the vectors for protein expression in yeast, a 6 × His-tag was fused to the N-terminus of *GLO1*, *GLO3* and *GLO4* [23], and then these modified sequences (*NHisGLO1*, *NHisGLO3* and *NHisGLO4*) were cloned into pYES3 and pYES2 vectors. To generate *GLO*-overexpression transgenic lines, each *GLO* sequence was cloned into pYLox.5 vector. To generate *GLO*-silencing transgenic lines, primers were designed to amplify the interfering fragment to guarantee the specificity of the silencing (Additional file 8), each specific fragment was then cloned into the RNAi vector pYLRNAi.5. To generate CRISPR-Cas9 knockout lines, specific targeting sequences were synthesized and cloned into pYLCRISPR/Cas9P$_{ubi}$ vector (Additional file 4) [46]. (pYLox.5, pYLRNAi.5 and pYLCRISPR/Cas9 vectors were kindly provided by Dr. Yao-Guang Liu, College of Life Sciences, South China Agricultural University).

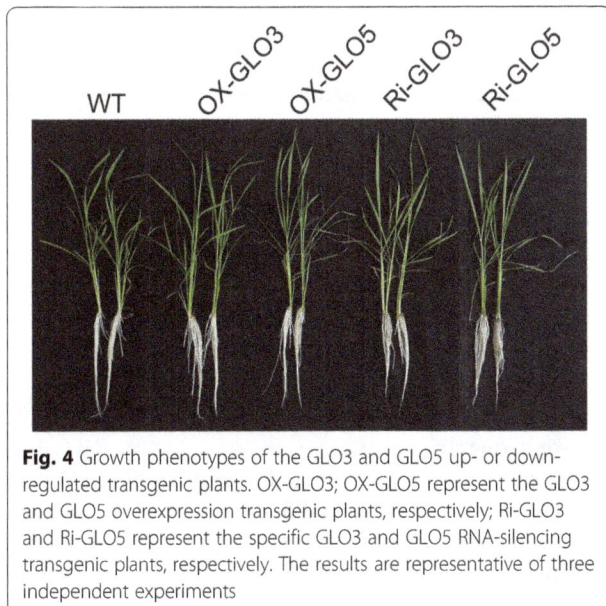

Fig. 4 Growth phenotypes of the GLO3 and GLO5 up- or downregulated transgenic plants. OX-GLO3; OX-GLO5 represent the GLO3 and GLO5 overexpression transgenic plants, respectively; Ri-GLO3 and Ri-GLO5 represent the specific GLO3 and GLO5 RNA-silencing transgenic plants, respectively. The results are representative of three independent experiments

Protein expression in *Saccharomyces cerevisiae*

The constructed GLO expression vectors were transformed into *Saccharomyces cerevisiae* INVSc1 (*his3Δ1/*

his3Δ1 leu2/leu2 trp1-289/trp1-289 ura3-52/ura3-52, Invitrogen) using the lithium acetate/carrier DNA method [47]. After that, yeast cells transformed with GLO constructs were maintained in the SC selective plates (SC-Dropout Medium without Trp/Ura) with 2% glucose, at 30 °C for 3-4 days to isolate the positive clones. The proteins were inductively expressed in *S.cerevisiae* as previously described [24, 48] with some modifications. Briefly, positive clones were transferred to 10 mL SC selective medium containing 2% glucose, and incubated at 30 °C for 16 h with shaking at 250 rpm. Appropriate volumes of the culture were transferred to 50 mL SC inductive culture medium (SC selective medium containing 2% galactose) to obtain an OD600 of 0.4, and then shaken at 30 °C with 250 rpm for 24 h. Cells were harvested by centrifuging the culture at 5000 rpm for 5 min at 4 °C, the cell pellets were collected and stored at −75 °C until ready to use.

Western blot analysis
Proteins were extracted from yeast cells using acid-washed glass beads as described previously [48]. Supernatant of the yeast cells extract was collected and used for SDS polyacrylamide gel (SDS-PAGE) electrophoresis. Protein samples separated on 10% SDS-PAGE were electroblotted onto a nitrocellulose membrane using wet transfer [49, 50]. GLO was detected using a GLO-antibody. The anti-GLO primary antibody was generated in rabbits by immunization with GLO1 proteins, and this antibody was able to recognize all the rice GLO polypeptides due to the high similarity between these polypeptides.

Purification of active GLO from yeast cells
Supernatant of the yeast cell lysate was filtered through a 0.45 μm membrane for subsequent purification. The Ni-IDA resin (Bio-Rad) was packed in a Bio-Scale MT5 column (10 × 64 mm) for bed volumes up to 2 mL, equilibrate the column with 5 column volumes (CV) of wash buffer (50 mM PBS, pH 8.0, 10 mM imidazole, 300 mM NaCl). The filtered supernatant were prepared by mixing them with an equal volume of binding buffer (100 mM PBS, pH 8.0, 20 mM imidazole, 600 mM NaCl), and then loading them onto the Ni-IDA resin column. The column was washed with 10 CV of wash buffer at a flow rate of 1.0 mL min^{-1}. The bound proteins were eluted with 5 CV of 50 mM PBS (pH 8.0) containing 150 mM imidazole and 300 mM NaCl. The eluted fractions were desalted by ultrafiltration and checked by 10% SDS-PAGE. All the purified enzymes were stored in 50 mM PBS (pH 8.0) containing 0.1 mM FMN and 10% glycerol at −75 °C for subsequent assay.

GLO isozyme zymogram analysis
To identify GLO isozymes a Caps-ammonium Clear Native-PAGE system (CN-PAGE) with a running pH of 10.2 was used [51]. The separation gel concentration is 6%, the electrophoresis was performed at 100 V for 9 h at 4 °C. Activity staining was carried out by incubating the gel at 30 °C for 20 min in a staining solution which contained 0.016% (W/V) NBT, 0.003% (W/V) PMS,0.1 mM FMN,10 mM glycolate (pH 6-8),and 50 mM PBS (pH 8.0).

GLO activity assay
GLO catalytic activity was measured in an enzyme-coupled assay according to Hall et al. (1985) unless specific variations were needed in different kinetic and characterization investigations [52]. The typical reaction mixture containing 50 mM PBS (pH 7.8), 5 units of horseradish peroxidase, 1 mM 4-amino-antipyrine, 2 mM phenol, 0.1 mM FMN and 10 mM substrate. The reaction was started by adding the enzyme, distilled water was substituted for substrate as the blank. Formation of H_2O_2 produced in the reaction which reflects the catalytic activity of GLO was monitored spectrophotometrically at 520 nm under 30 °C.

The effects of varying pH were assayed with two buffers: pH of 50 mM PBS from 6 to 8, and 50 mM Tris-HCl from 8 to 9. As for the effects of temperature, the reaction mixtures were pre-incubated at various temperatures (22-60 °C) for 5 min, and then the GLO activities were measured at the same temperature.

For kinetic parameter determination, oxidations of the substrate were performed over a range of concentration (0.1-0.8 mM for glycolate, 1.0-8.0 mM for glyoxylate, 0.5-6.0 mM for L-lactate, 0.5-6.0 mM for glycerate), the K_m was calculated from double-reciprocal plots according to the method of Lineweaver and Burk, moreover, slopes of the reciprocal plots were then plotted against the concentration of oxalate (range 1.0-8.0 mM), to evaluate K_i data [53].

Generation of GLO transgenic line
The constructed vectors were introduced into rice by *Agobacterium*-mediated infection (strain EHA105) [54]. The seeds from the positive T0 lines were germinated in complete Kimura B nutrient solution and then transplanted to soil. The T2 or T3 plants were used for the determination of GLO activity and the Real-time quantitative PCR (qRT-PCR) assay. Besides, the T3 plants of GLO3-transgenic lines were used for lactate toxicity test (L-lactate, 2.0 mM).

Real-time quantitative PCR
The specific primer pairs were designed for the qRT-PCR of each *GLO* gene (Additional file 8). Total RNA was purified from rice using TRIzol® reagent (Life Technologies, Carlsbad, USA), and further treated with DNase I (RNase free, Toyobo, Osaka, Japan). The quality of the isolated

RNA was assessed with a NanoDrop-1000 (Thermo Fisher Scientific, Bremen, Germany). One microgram of RNA was used as a template for first-strand cDNA synthesis using ReverTra Ace (Toyobo, Osaka, Japan). The qRT-PCR reaction mixture consisted of 0.2 μM (each) primer, 10 μL of 2 × SYBR Green PCR Master Mix (Toyobo, Osaka, Japan), and 2 μL of appropriate diluted cDNA. The analysis was conducted using a DNA Engine Opticon 2 Real-Time PCR Detection system and Opticon Monitor software (Bio-Rad, Hercules, CA). The data were normalized to the amplification of the *OsActin1* gene (Os03g0718100).

3, 3′-diaminobenzidine staining

The leaf H_2O_2 abundance was estimated by the 3, 3′-diaminobenzidine (DAB) uptake method [55]. The youngest fully expanded leaves of the five-leaf stage rice were detached (10 cm), and the cut end was dipped into 4 mL of DAB solution (1 mg mL^{-1}, pH 3.8) for 2 h in light at 30 °C. The experiment was terminated by boiling the leaves in ethanol for 30 min.

Quantification of proteins

The protein content was determined according to Bradford (1976) with bovine serum albumin as a standard [56], and the experiments were repeated three times with at least three replicates for each sample.

Additional files

Additional file 1: Similarities of rice *GLO* gene members at the level of mRNA and protein.

Additional file 2: (a) Effect of varying pH on activities of GLO isozymes. Each buffer (50 mM) was made of respective PBS (pH 6.0-8.0) and Tris-HCl (pH 8.0-9.0). (The highest activity of each GLO isozyme at pH 7.8 was set as 1). (b) Effect of temperature on activities of GLO isozymes. Enzymes in 50 mM PBS buffer (pH 7.8) were pre-incubated at various temperatures (22-60 °C) for 5 min, and then activities were measured at the same temperature (For GLO1 and GLO1 + 4, the highest activity at 45 °C was set as 1; For GLO3, the highest activity at 47 °C was set as 1; For GLO4, the highest activity at 42 °C was set as 1). Values are means ± SD (n = 3).

Additional file 3: The V_{max} of purified GLO isozymes with various substrates.

Additional file 4: (a) Relative mRNA levels were graphed based on the *GLO1* mRNA level in WT leaves as 1. (b) Relative mRNA levels were graphed based on the *GLO4* mRNA level in WT leaves as 1. (c) Mutation of GLO1 and GLO4 knockout lines generated by pYLCRISPR/Cas9P$_{ubi}$ system. (d) Relative mRNA levels were graphed based on the *GLO3* mRNA level in WT leaves as 1. (e) Relative mRNA levels were graphed based on the *GLO5* mRNA level in WT leaves as 1. Values are means ± SD (n = 3).

Additional file 5: H_2O_2-3, 3′-diaminobenzidine (DAB) staining in rice leaves. Cas9-GLO1 and Cas9-GLO4 represent the GLO1 and GLO4 knockout plants, respectively. The result is representative of three independent experiments.

Additional file 6: L-lactate toxicity test in GLO3 transgenic plants. Different transgenic rice lines (4-leaf stage) were grown in Kimura B complete nutrient solution containing 2.0 mM L-lactate for one week. OX-GLO3 and Ri-GLO3 represent the GLO3 overexpression transgenic plants and the specific GLO3 RNA-silencing transgenic plants, respectively. The results are representative of three independent experiments.

Additional file 7: GLO activity in root of GLO3 over expression plants.

Additional file 8: The primers used for the plasmid construction and real-time quantitative PCR.

Abbreviations

2-PG: Phosphoglycolate; CN-PAGE: Clear-native polyacrylamide gel electrophoresis; DAB: 3, 3′-diaminobenzidine; FMN: Flavin mononucleotide; GLO: Glycolate oxidase; PGP: 2-PG phosphatase; qRT-PCR: Real-time quantitative PCR

Acknowledgements
None.

Funding
This work was supported in part by the National Natural Science Foundation of China (Grant Nos. 31,600,193 and 31,470,343) and the Hunan Provincial Natural Science Foundation of China (Grant No. 2017JJ3109). The funders had no role in study design, data collection and analysis, decision to publish, or preparation of the manuscript.

Authors' contributions
PXX conceived the study, edited the manuscript and supervised the work. ZZS carried out most analyses, contributed to the design of the study and drafted the manuscript. LXY prepared protein samples for analysis, prepared the rice seeds, grew rice plants and performed stress treatment. CLL performed the qRT-PCR analysis. MS and YNH drafted and revised the manuscript. All authors read and approved the final manuscript.

Competing interests
The authors declare that they have no competing interests.

Author details
[1]Southern Regional Collaborative Innovation Center for Grain and Oil Crops in China, Hunan Agricultural University, Changsha, Hunan 410128, China. [2]State Key Laboratory for Conservation and Utilization of Subtropical Agro-bioresources, College of Life Sciences, South China Agricultural University, Guangzhou, Guangdong 510642, China.

References

1. Richardson KE, Tolbert NE. Oxidation of glyoxylic acid to oxalic acid by glycolic acid oxidase. J Biol Chem. 1961;236:1280–4.
2. Macheroux P, Massey V, Thiele DJ, Volokita M. Expression of spinach glycolate oxidase in *Saccharomyces cerevisiae*: purification and characterization. Biochemistry. 1991;30:4612–9.
3. Foyer CH, Bloom AJ, Queval G, Noctor G. Photorespiratory metabolism. Genes, mutants, Energetics, and Redox signaling. Annu Rev Plant Biol. 2009;60:455–84.
4. Cui L, Lu Y, Li Y, Yang C, Peng X. Overexpression of Glycolate Oxidase confers improved photosynthesis under high light and high temperature in Rice. Front Plant Sci. 2016;7:1165.
5. Lu Y, Li Y, Yang Q, Zhang Z, Chen Y, Zhang S, Peng X. Suppression of glycolate oxidase causes glyoxylate accumulation that inhibits photosynthesis through deactivating Rubisco in rice. Physiol Plant. 2014;150:463–76.
6. Mittler R, Zilinskas BA. Regulation of pea cytosolic ascorbate peroxidase and other antioxidant enzymes during the progression of drought stress and following recovery from drought. Plant J. 1994;5:397–405.
7. Rizhsky L, Liang H, Mittler R. The combined effect of drought stress and

heat shock on gene expression in tobacco. Plant Physiol. 2002;130:1143–51.

8. Mukherjee SP, Choudhuri MA. Implications of water stress-induced changes in the levels of endogenous ascorbic acid and hydrogen peroxide in *Vigna* seedlings. Physiol Plant. 1983;58:166–70.

9. Bohman S, Wang M, Dixelius C. *Arabidopsis thaliana*-derived resistance against *Leptosphaeria maculans* in a *Brassica napus* genomic background. Theor Appl Genet. 2002;105:498–504.

10. Taler D, Galperin M, Benjamin I, Cohen Y, Kenigsbuch D. Plant eR genes that encode photorespiratory enzymes confer resistance against disease. Plant Cell. 2004;16:172–84.

11. Schafer P, Huckelhoven R, Kogel KH. The white barley mutant albostrians shows a supersusceptible but symptomless interaction phenotype with the hemibiotrophic fungus *Bipolaris sorokiniana*. Mol Plant-Microbe Interact. 2004;17:366–73.

12. Peterhansel C, Horst I, Niessen M, Blume C, Kebeish R, Kürkcüoglu S, Kreuzaler F. Photorespiration. Arabidopsis Book. 2010;8:e130.

13. Noctor G. Drought and oxidative load in the leaves of C3 plants: a predominant role for photorespiration? Ann Bot. 2002;89:841–50.

14. Corpas FJ, Barroso JB, Del Río LA. Peroxisomes as a source of reactive oxygen species and nitric oxide signal molecules in plant cells. Trends Plant Sci. 2001;6:145–50.

15. Xu H, Zhang J, Zeng J, Jiang L, Liu E. Inducible antisense suppression of glycolate oxidase reveals its strong regulation over photosynthesis in rice. J Exp Bot. 2009; 60:1799–809.

16. Rojas CM, Senthil-Kumar M, Wang K, Ryu CM, Kaundal A, Mysore KS. Glycolate Oxidase modulates reactive oxygen species-mediated signal transduction during nonhost resistance in *Nicotiana benthamiana* and *Arabidopsis*. Plant Cell. 2012;24:336–52.

17. Popov VN, Dmitrieva EA, Eprintsev AT, Igamberdiev AU. Glycolate oxidase isoforms are distributed between the bundle sheath and mesophyll tissues of maize leaves. J Plant Physiol. 2003;160:851–7.

18. Havir EA. Evidence for the presence in tobacco leaves of multiple enzymes for the oxidation of glycolate and glyoxylate. Plant Physiol. 1983;71:874–8.

19. Lindqvist Y. Refined structure of spinach glycolate oxidase at 2 Å resolution. J Mol Biol. 1989;209:151–66.

20. Kamada T, Nito K, Hayashi H, Mano S, Hayashi M, Nishimura M. Functional differentiation of Peroxisomes revealed by expression profiles of Peroxisomal genes in *Arabidopsis thaliana*. Plant Cell Physiol. 2003;44:1275–89.

21. Yamagami T, Tsuchisaka A, Yamada K, Haddon WF, Harden LA, Theologis A. Biochemical diversity among the 1-amino-cyclopropane-1-carboxylate synthase isozymes encoded by the *Arabidopsis* gene family. J Biol Chem. 2004;278:49102–12.

22. Engqvist MK, Schmitz J, Gertzmann A, Florian A, Jaspert N, Arif M, Balazadeh S, Mueller-Roeber B, Fernie AR, Maurino VG. GLYCOLATE OXIDASE3, a Glycolate OXIDASE homolog of yeast L-lactate Cytochrome c Oxidoreductase, supports L-lactate oxidation in roots of *Arabidopsis*. Plant Physiol. 2015;169:1042–61.

23. Kerchev PI, Waszczak C, Lewandowska A, Willems P, Shapiguzov A, Li Z, Alseekh S, Mühlenbock P, Hoebrichts F, Huang JJ, Van Der Kelen K, Kangasjärvi J, Fernie AR, De Smet R, Van de Peer Y, Messens J, Van Breusegem F. Lack of GLYCOLATE OXIDASE 1, but not GLYCOLATE OXIDASE 2, attenuates the photorespiratory phenotype of CATALASE2-deficient *Arabidopsis*. Plant Physiol. 2016;171:1704–19.

24. Zhang Z, Lu Y, Zhai L, Deng R, Jiang J, Li Y, He Z, Peng X. Glycolate oxidase isozymes are coordinately controlled by GLO1 and GLO4 in rice. PLoS One. 2012;7:e39658.

25. Gräslund S, Nordlund P, Weigelt J, Bray J, Gileadi O, Knapp S. Protein production and purification. Nat Methods. 2008;5:135–46.

26. Franceschi VR, Nakata PA. Calcium oxalate in plants: formation and function. Annu Rev Plant Biol. 2005;56:41–71.

27. Nishimura M, Akhmedov YD, Strzalka K, Akazawa T. Purification and characterization of glycolate oxidase from pumpkin cotyledons. Arch Biochem Biophys. 1983;222:397–402.

28. Odanaka S, Bennett AB, Kanayama Y. Distinct physiological roles of fructokinase isozymes revealed by gene-specific suppression of Frk1 and Frk2 expression in tomato. Plant Physiol. 2002;129:1119–26.

29. Frigerio NA, Harbury HA. Preparation and some properties of crystalline glycolic acid oxidase of spinach. J Biol Chem. 1958;231:135–57.

30. Zelitch I, Ochoa S. Oxidation and reduction of glycolic and glyoxylic acids in plants I. Glycolic acid oxidase. J Biol Chem. 1953;201:707–18.

31. Cederlund E, Lindqvist Y, Söderlund G, Brändén CI, Jörnvall H. Primary structure of glycolate oxidase from spinach. Eur J Biochem. 1988;173:523–7.

32. Macheroux P, Mulrooney SB, Williams CJ, Massey V. Direct expression of active spinach glycolate oxidase in *Escherichia coli*. Biochim Biophys Acta. 1992;1132:11–6.

33. Devi MT, Rajagopalan AV, Raghavendara AS. Purification and properties of glycolate oxidase from plants with different photosynthetic pathways: distinctness of C4 enzyme from that of a C3 species and a C3-C4 intermediate. Photosynth Res. 1996;47:231–8.

34. Reumann S. Specification of the Peroxisome targeting signals type 1 and type 2 of plant Peroxisomes by bioinformatics analyses. Plant Physiol. 2004;135:783–800.

35. Sweetlove LJ, Dunford R, Ratcliffe RG, Kruger NJ. Lactate metabolism in potato tubers deficient in lactate dehydrogenase activity. Plant Cell Environ. 2000;23:873–81.

36. Mustroph A, Jr BG, Kaiser KA, Larive CK, Bailey-Serres J. Characterization of distinct root and shoot responses to low-oxygen stress in *Arabidopsis* with a focus on primary C- and N-metabolism. Plant Cell Environ. 2014;37:2366–80.

37. Menegus F, Cattaruzza L, Mattana M, Beffagna N, Ragg E. Response to anoxia in rice and wheat seedlings changes in the pH of intracellular compartments, glucose-6-phosphate level, and metabolic rate. Plant Physiol. 1991;95:760–7.

38. Menegus F, Cattaruzza L, Chersi A, Fronza G. Differences in the anaerobic lactate-succinate production and in the changes of cell sap pH for plants with high and low resistance to anoxia. Plant Physiol. 1989;90:29–32.

39. Sewelam N, Jaspert N, Van Der Kelen K, Tognetti VB, Schmitz J, Frerigmann H, Stahl E, Zeier J, Van Breusegem F, Maurino VG. Spatial H_2O_2 signaling specificity: H_2O_2 from chloroplasts and peroxisomes modulates the plant transcriptome differentially. Mol Plant. 2014;7:1191–210.

40. Zhang Z, Xu Y, Xie Z, Li X, He Z, Peng X. Association-dissociation of Glycolate Oxidase with Catalase in Rice: a potential switch to modulate intracellular H_2O_2 levels. Mol Plant. 2016;9:737–48.

41. Kerchev P, Muhlenbock P, Denecker J, Morreel K, Hoeberichts FA, Kelen KVD, Vandorpe M, Long N, Audenaert D, Breusegem FV. Activation of auxin signaling counteracts photorespiratory H_2O_2 dependent cell death. Plant Cell Environ. 2013;38:253–65.

42. Gao X, Hong M, Hu Y, Li J, Lu Y. Mutation of *Arabidopsis* CATALASE2 results in hyponastic leaves by changes of auxin levels. Plant Cell Environ. 2014;37:175–88.

43. Zhang S, Li C, Jia C, Zhang Y, Zhang S, Xia Y, Sun D, Ying S. Altered architecture and enhanced drought tolerance in rice via the down-regulation of indole-3-acetic acid by TLD1/OsGH3.13 activation. Plant Physiol. 2009;151:1889–901.

44. Zhao S, Xiang J, Xue H. Studies on the Rice LEAF INCLINATION1 (LC1), an IAA-amido Synthetase, reveal the effects of Auxin in leaf INCLINATION control. Mol Plant. 2013;6:174–87.

45. Yoshida S, Forno DA, Cock JH, Gomez KA. Laboratory manual for physiological studies of rice. Manila: International Rice Research Institute; 1976.

46. Ma X, Zhang Q, Zhu Q, Wei L, Yan C, Rong Q, Wang B, Yang Z, Li H, Lin Y. A robust CRISPR/Cas9 system for convenient, high-efficiency multiplex genome editing in monocot and dicot plants. Mol Plant. 2015;8:1274–84.

47. Schiestl RH, Gietz RD. High efficiency transformation of intact yeast cells using single stranded nucleic acids as a carrier. Curr Genet. 1989;16:339–46.

48. Zhang Z, Mao X, Ou J, Ye N, Zhang J, Peng X. Distinct photorespiratory reactions are preferentially catalyzed by glutamate:glyoxylate and serine: glyoxylate aminotransferases in rice. J Photochem Photobiol B. 2015;142:110–7.

49. Towbin H, Staehelin T, Gordon J. Electrophoretic transfer of proteins from polyacrylamide gels to nitrocellulose sheets: procedure and some applications. Proc Natl Acad Sci U S A. 1979;76:4350–4.

50. Laemmli UK. Cleavage of structural proteins during the assembly of the head of bacteriophage T4. Nature. 1970;227:680–5.

51. McLellan T. Electrophoresis buffers for polyacrylamide gels at various pH. Anal Biochem. 1982;126:94–9.

52. Hall NP, Reggiani R, Lea PJ. Molecular weights of glycollate oxidase from C3 and C4 plants determined during early stages of purification. Phytochemistry. 1985;24:1645–8.

53. Lineweaver H, Burk D. The determination of enzyme dissociation constants. J Am Chem Soc. 1934;56:658–66.

54. Hiei Y, Ohta S, Komari T, Kumashiro T. Efficient transformation of rice (*Oryza sativa* L.) mediated by Agrobacterium and sequence analysis of the boundaries of the T-DNA. Plant J. 1994;6:271–82.

55. Thordal Christensen H, Zhang Z, Wei Y, Collinge DB. Subcellular localization of H_2O_2 in plants. H_2O_2 accumulation in papillae and hypersensitive response during the barley-powdery mildew interaction. Plant J. 1997;11:1187–94.

56. Bradford MM. A rapid and sensitive method for the quantitation of microgram quantities of protein utilizing the principle of protein-dye binding. Anal Biochem. 1976;72:248–54.

9

Seed maturation associated transcriptional programs and regulatory networks underlying genotypic difference in seed dormancy and size/weight in wheat (*Triticum aestivum* L.)

Yuji Yamasaki[1], Feng Gao[1], Mark C. Jordan[2] and Belay T. Ayele[1*]

Abstract

Background: Maturation forms one of the critical seed developmental phases and it is characterized mainly by programmed cell death, dormancy and desiccation, however, the transcriptional programs and regulatory networks underlying acquisition of dormancy and deposition of storage reserves during the maturation phase of seed development are poorly understood in wheat. The present study performed comparative spatiotemporal transcriptomic analysis of seed maturation in two wheat genotypes with contrasting seed weight/size and dormancy phenotype.

Results: The embryo and endosperm tissues of maturing seeds appeared to exhibit genotype-specific temporal shifts in gene expression profile that might contribute to the seed phenotypic variations. Functional annotations of gene clusters suggest that the two tissues exhibit distinct but genotypically overlapping molecular functions. Motif enrichment predicts genotypically distinct abscisic acid (ABA) and gibberellin (GA) regulated transcriptional networks contribute to the contrasting seed weight/size and dormancy phenotypes between the two genotypes. While other ABA responsive element (ABRE) motifs are enriched in both genotypes, the prevalence of G-box-like motif specifically in tissues of the dormant genotype suggests distinct ABA mediated transcriptional mechanisms control the establishment of dormancy during seed maturation. In agreement with this, the bZIP transcription factors that co-express with ABRE enriched embryonic genes differ with genotype. The enrichment of SITEIIATCYTC motif specifically in embryo clusters of maturing seeds irrespective of genotype predicts a tissue specific role for the respective TCP transcription factors with no or minimal contribution to the variations in seed dormancy.

Conclusion: The results of this study advance our understanding of the seed maturation associated molecular mechanisms underlying variation in dormancy and weight/size in wheat seeds, which is a critical step towards the designing of molecular strategies for enhancing seed yield and quality.

Keywords: Embryo, Endosperm, Genotype, Seed maturation, Transcriptome, *Triticum aestivum*, Wheat

* Correspondence: belay.ayele@umanitoba.ca
[1]Department of Plant Science, University of Manitoba, 222 Agriculture Building, Winnipeg, MB R3T 2N2, Canada
Full list of author information is available at the end of the article

Background

Wheat (*Triticum aestivum* L.) is one of the most economically important cereal crops in the world, and its seeds serve as basic units of propagation, and source of food, feed and raw material for a wide-range of industrial products. Owing to all these agronomic and economic importance, understanding the genetic and molecular mechanisms regulating wheat seed developmental programs, and thereby yield and quality has been the subject of recent wheat genomic studies. In cereals such as wheat, seed development starts with fertilization events and terminates with the formation of mature quiescent seeds, and the entire developmental process is comprised of three phases [1, 2]. The first phase, which is also referred to as early phase, comprises double fertilization that forms the embryo and endosperm, syncytium formation and endosperm cellularization events. The second phase is characterized by differentiation, which includes events associated with the formation of different cell types, endoreduplication and deposition of storage reserves. Maturation forms the third phase of seed development and it is characterized mainly by programmed cell death, dormancy and desiccation. The degree of dormancy induced during seed maturation is tightly associated with the level of tolerance to preharvest sprouting, which is defined as the germination of seeds on the spike prior to harvest, that causes significant losses in seed yield and quality. Dormancy is regulated by genetic and environmental factors, and the plant hormones abscisic acid (ABA) and gibberellin (GA) are reported to have important roles in the induction, maintenance and release of dormancy. Another seed related trait of economic importance is seed size/weight that forms one of the major components of grain yield. It is determined by deposition of storage reserves that occurs following histodifferentiation through to physiological maturity [3]. Developing wheat grains appear to attain physiological maturity at about 42 DAA and harvest maturity by 1–2 weeks thereafter [4, 5].

Transcriptomic studies in different plant systems have led to the identification of transcriptional programs and regulatory networks underlying molecular functions associated with cellular activities. In this context, seed development in the model plant Arabidopsis has been shown to be characterized by distinct and overlapping functional identities and regulatory networks in different regions and sub-regions of seeds [6]. For example, genes upregulated specifically in the embryos and endosperms of mature seeds are enriched with ABA responsive element (ABRE) motif, which acts as a binding site for bZIP transcription factors (TFs), and molecular functions associated with ABA stimulus and oil biogenesis. Furthermore, seed specific TFs with unknown and known roles including LEAFY COTYLEDON and FUSCA3 have

been implicated in mediating the regulatory networks required for programming seed development [7]. Transcriptomic analysis of developing seed tissues in cereals such as maize, rice and barley have also led to the identification of genes involved in the programming of seed developmental and maturation processes, and elucidation of the underlying functional transitions [8–10]. Furthermore, different tissue types of developing seeds of cereals such as embryo and endosperm have been shown to exhibit distinct gene expression profiles and therefore molecular functions [9], suggesting the ditinct roles they play in regulating seed traits. For example, genes involved in the synthesis and signaling of ABA, which influences starch biosynthesis in the endosperm and desiccation tolerance in the embryo, differ with tissue types.

Similarly, large scale gene expression studies of developing wheat seeds have provided important insights into the developmental shifts in gene expression, distinct and overlapping transcriptional reprogramming between different tissues such as aleurone and endosperm, and cell-type and stage-dependent transcriptional dynamics and genome interplays [11–13]. However, these studies analyzed either the whole seed, or different tissue types only during the early and differentiation/grain filling phases of seed development. Previous studies also investigated the transcriptome of wheat seed, for example, with respect to dormancy [14], however, most of the dormancy studies have been focused on post-harvest of seeds. As a result, the spatiotemporal transcriptional programs and regulatory networks underlying the establishment and maintenance of dormancy during the maturation phase of wheat seed development are poorly understood. To this effect, this study performed comparative transcriptomic analysis of the endosperm and embryo of maturing seeds between two wheat genotypes, AC Domain and RL4452, characterized by contrasting degree of dormancy at four developmental stages (20–50 DAA) that are critical to the acquisition of dormancy and desiccation tolerance. Owing to their difference in seed dormancy, these genotypes have been used as parental lines to generate breeding populations and thereby map QTLs associated with tolerance to preharvest sprouting [15, 16]. Since the two genotypes also exhibit contrasting final seed size/weight and represent different genetic materials than those studied previously, our transcriptomic analysis was also aimed at elucidating molecular features that regulate deposition of storage reserves during the seed maturation phase.

Result and discussion

Comparative analysis of seed phenotypes during and after maturation

Seed maturation in both AC Domain and RL4452 genotypes was studied from 20 to 50 DAA (Fig. 1a). The

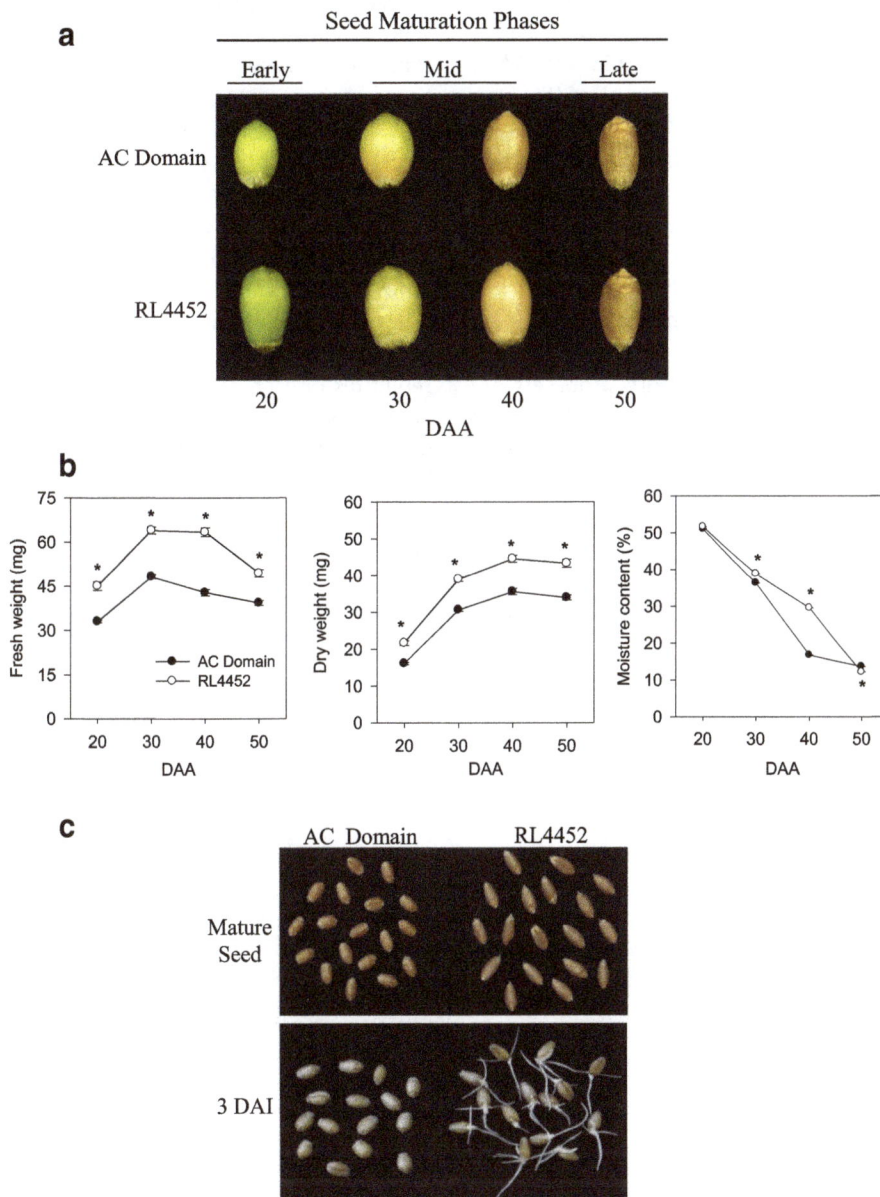

Fig. 1 Maturing seeds of wheat. Seeds of AC Domain and RL4452 genotypes from 20 to 50 DAA (**a**); changes in seed fresh weight, dry weight and moisture content during maturation - data are means ± SE (*n* = 20–23) and asterisks indicate statistically significant difference in seed fresh weight, dry weight and moisture content between the two genotypes (*P* < 0.05; t-Student test) (**b**); and difference in seed size and dormancy/germination phenotypes between mature seeds of the two genotypes (**c**)

seeds of both genotypes showed increases in fresh and dry weights from 20 to 30 DAA, and after 30 DAA the fresh weights starts to decline in both genotypes, while the dry weights increased slightly from 30 to 40 DAA and remained constant thereafter (Fig. 1b) as observed before [17]. Overall, the RL4452 seeds exhibited consistently significantly higher fresh and dry weights over the entire seed maturation period than that of AC Domain, suggesting more storage reserves deposition in RL4452 seeds and therefore production of larger seeds. The two genotypes exhibited similar rate of water loss from 20 to 30 DAA (Fig. 1b). However, AC Domain seeds desiccated at a higher rate from 30 to 40 DAA (2% moisture loss/day for AC Domain vs. ~1% moisture loss/day for RL4452) while the RL4452 seeds desiccated at a higher rate from 40 to 50 DAA (1.7% moisture loss/day for RL4452 vs. 0.3% moisture loss/day for AC Domain). The overall desiccation rate appeared to be higher in RL4452 seeds (1.32% moisture loss/day for RL4452 vs. 1.25% moisture loss/day for AC Domain), leading to seed moisture contents at 50 DAA of 13.6% and 12.2% in AC Domain and RL4452, respectively. Therefore, the time points studied here represent the stage from late reserve accumulation to desiccation, and we divided this period into early (20 DAA), mid (30–40 DAA) and late (50 DAA) phases of seed maturation (Fig. 1a). Freshly harvested mature seeds of RL4452 exhibited over 93% germination within 24 h imbibition, indicating that they are non-dormant, while only 15% of the corresponding AC Domain seeds germinated even after 5-day imbibition, indicating their dormant phenotype [18, 19] (Fig. 1c).

Genotype- and tissue-specific gene expression profiles during wheat seed maturation

It appeared from our analysis that more probesets are expressed in the embryo and/or endosperm tissues at least at one time point during seed maturation in RL4452 (32,721 probesets; 53.4% of the probesets on the Wheat GeneChip) than in AC Domain (31,828 probesets; 51.9% of the probesets on the Wheat GeneChip) (Fig. 2), suggesting the prevalence of more transcriptional activity in maturing seeds of RL4452, a genotype that produces larger seeds (on fresh and dry weight basis). The number of probesets expressed in the embryo of each genotype overall or at each stage of development was more than that observed in the endosperm (Fig. 2), suggesting that the two tissue types of maturing wheat seeds are characterized by differing gene transcriptional states. Scatter-plot expression analysis revealed high reproducibility between replicates of each sample ($r^2 > 0.95$), and a strong correlation was observed between the microarray and qPCR data for 10 randomly selected differentially expressed probesets (Additional file 1: Fig. S1).

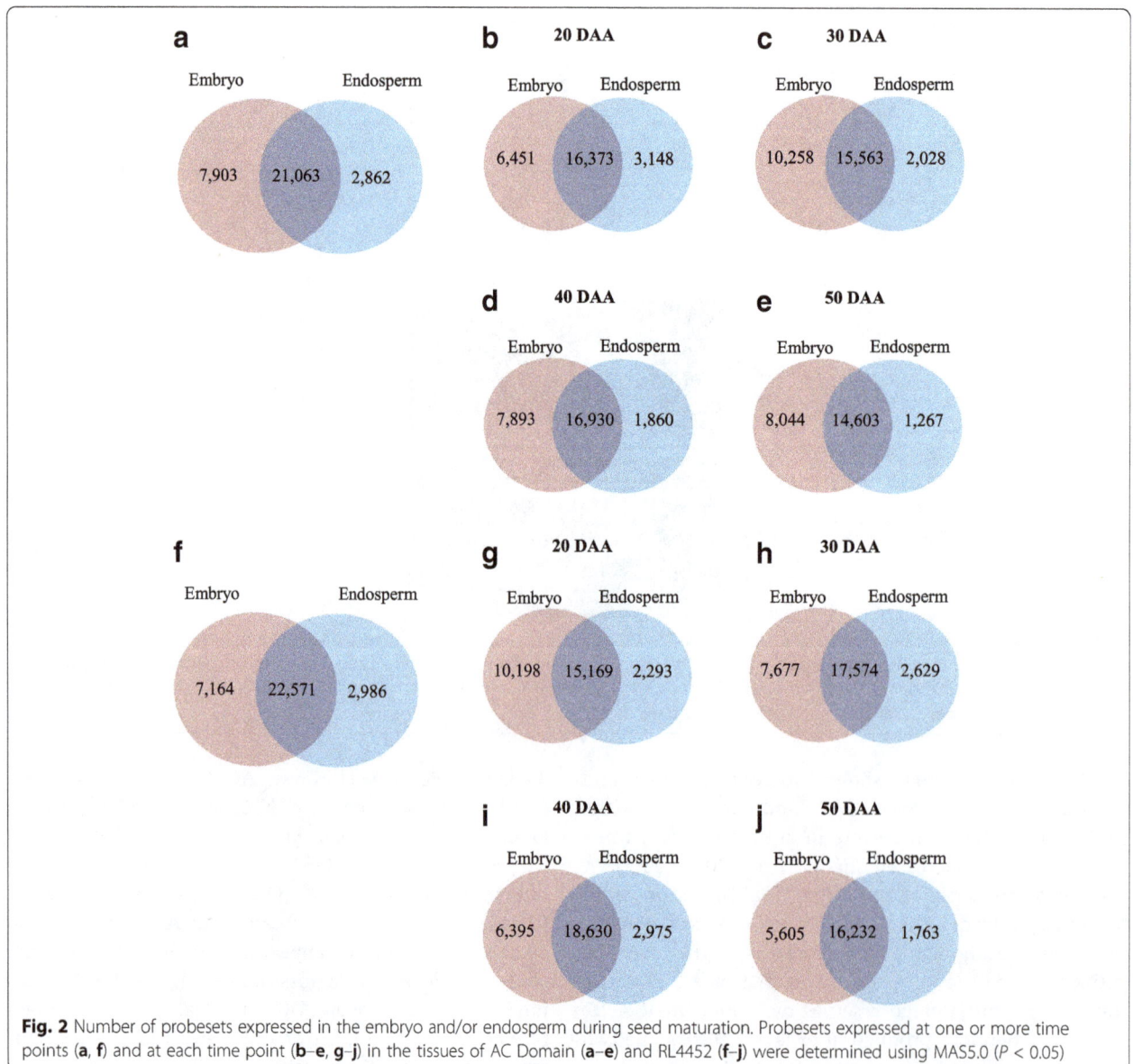

Fig. 2 Number of probesets expressed in the embryo and/or endosperm during seed maturation. Probesets expressed at one or more time points (a, f) and at each time point (b–e, g–j) in the tissues of AC Domain (a–e) and RL4452 (f–j) were determined using MAS5.0 (P < 0.05)

Temporal shifts in gene expression profiles during seed maturation varies with genotype and tissue

Principal component analysis (PCA) of the transcriptomic data revealed three different expression patterns for the endosperm of AC Domain; the 20 and 30 DAA endosperms are more closely related while each of those at 40 DAA and 50 DAA represent a separate and distinct group, indicating deviation in gene expression as maturation proceeds from the mid to late phase (Fig. 3a). In RL4452, the endosperm at the early to mid-phases (20 to 40 DAA) of seed maturation appeared to have a close relationship while those at the late phase (50 DAA) represent a distinct group, indicating that a major shift in gene expression occurs only as the seed desiccates (Fig. 3b). The two genotypes differ in seed size/weight, which is determined mainly by the deposition of storage reserves in the endosperm that has been shown to

express genes involved in carbohydrate metabolic processes [11]. It is therefore likely that the differential shifts in endospermic gene expression between the two genotypes might contribute to the difference in seed size/weight. Embryos of AC Domain at each time point of seed maturation appeared to represent a distinct group, implying a constant shift in gene expression profile during the seed maturation phase. In RL4452, however, embryos at the mid phase (30 and 40 DAA) appeared to be closely related while each of those at the early (20 DAA) and late (50 DAA) phases formed a distinct group, suggesting that major shifts in gene expression occur as seeds transitioned from early to mid and from mid to late phases of maturation. Given that the embryo can represents a major component of seed dormancy [20], the distinct gene expression shift observed in the developing AC Domain embryos might underlie

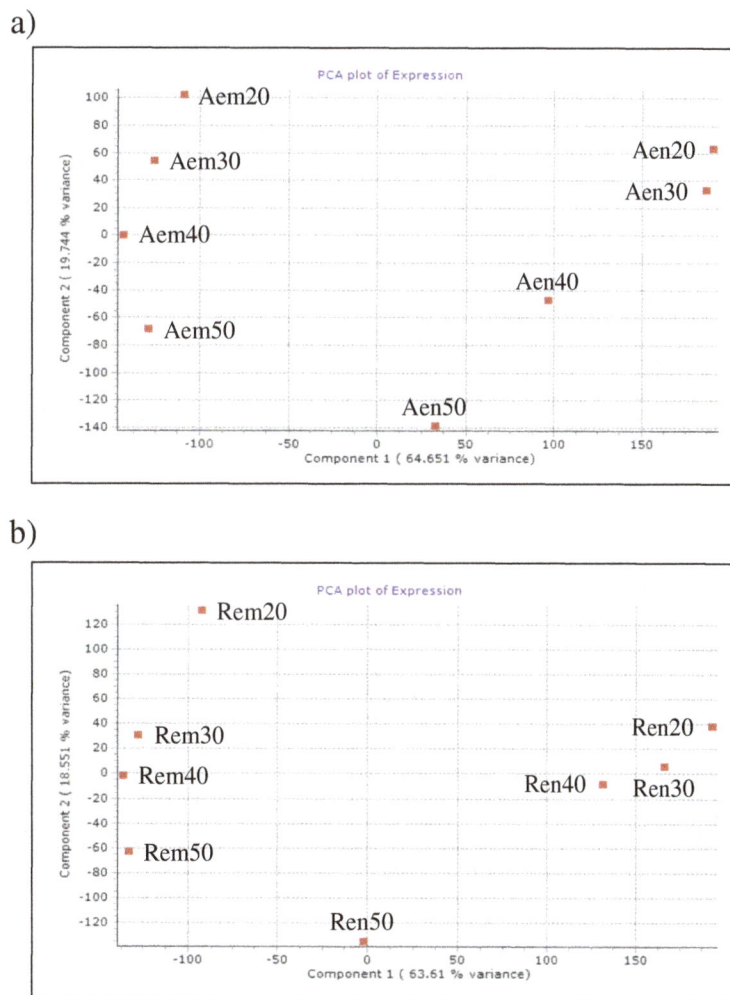

Fig. 3 Principal component analysis of seed maturation transcripts. Embryo and endosperm of maturing AC Domain (**a**) and RL4452 (**b**) seeds. Aem, AC Domain embryo; Aen, AC Domain endosperm; Rem, RL4452 embryo; Ren, RL4452 endosperm at 20, 30, 40 and 50 days after anthesis. Principal components (PC)1 representing tissue identity (64.65% in AC Domain and 63.61% in RL4452) and PC2 representing developmental stages (19.74% in AC Domain and 18.55% in RL4452) with a cumulative percentage > 80% in each genotype

the induction of dormancy exhibited by the mature seeds of this genotype [18, 19] (Fig. 1c). Our data overall suggest that the genotype specific temporal shift in gene expression profile thereby molecular function in each of the endosperm and embryo tissues of maturing wheat seeds represent the variations in seed size/weight and dormancy level.

Tissue types of maturing seeds are characterized by distinct but temporally and genotypically overlapping gene expression profiles

Our PCA analysis also showed larger difference in gene expression pattern between the endosperm and embryo tissues mainly at the early- and mid-phases of seed maturation irrespective of genotype, suggesting that the two tissues are characterized by distinct molecular functions. However, as the seed enters into the late phase of maturation the difference in gene expression profile between the two tissues became smaller in both genotypes but slightly more pronounced in RL4452, which exhibited a more rapid rate of desiccation during transition from mid to late phases, suggesting desiccation induced repression of gene transcription [21, 22]. Likewise, difference in gene expression profile between the endosperm and embryo tissues have been shown to shrink during the late phases of seed maturation in maize [8].

Hierarchical clustering of the embryo and endosperm tissues also revealed overlap between the embryo samples and 50 DAA endosperm in RL4452 while the clustering in AC Domain is based primarily on tissue identity (Additional file 2: Fig. S2).

Genotype and tissue specific co-expression clusters during wheat seed maturation

To better understand the commonalities and differences in transcriptional programs underlying variation in seed size/weight and dormancy between the two genotypes, we performed cluster analysis of all probesets expressed at least at one time point of seed maturation in each of the embryo (28,966 probesets in AC Domain; 29,735 probesets in RL4452) and endosperm (23,925 probesets in AC Domain; 25,557 probesets in RL4452) tissues (Figs. 4 and 5). To confirm tissue specificity of the probesets, further cluster analysis was performed by combining those expressed in both embryo and endosperm (31,828 probesets in AC Domain; 32,721 probesets in RL4452). Accordingly, we generated 16 clusters for each of the embryo, endosperm, and embryo and endosperm categories of AC Domain; and 14, 18 and 16 clusters for each of the embryo, endosperm, and embryo and endosperm categories of RL4452, respectively (Figs. 4 and 5, Additional file 7: Table S1); promoter motif enrichment

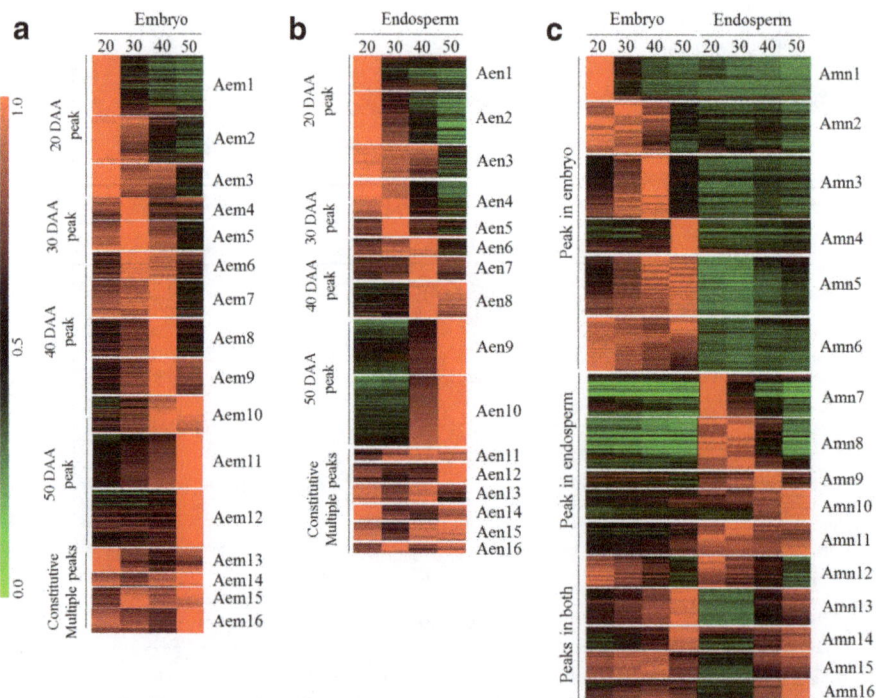

Fig. 4 Co-expression gene clusters of AC Domain (A). All probesets expressed in the embryo (**a**), endosperm (**b**) or both (**c**) at one or more time points were clustered based on the peak expression at a given time point(s) during seed maturation. The expression level of probesets in each of the embryo (em) (**a**), endosperm (en) (**b**) or both (mn) (**c**) tissues was determined relative to a time point with the highest RMA-normalized signal intensity, which was arbitrarily set to a value of 1 and represented by red color in the red-green scale shown on the left side of the heat map

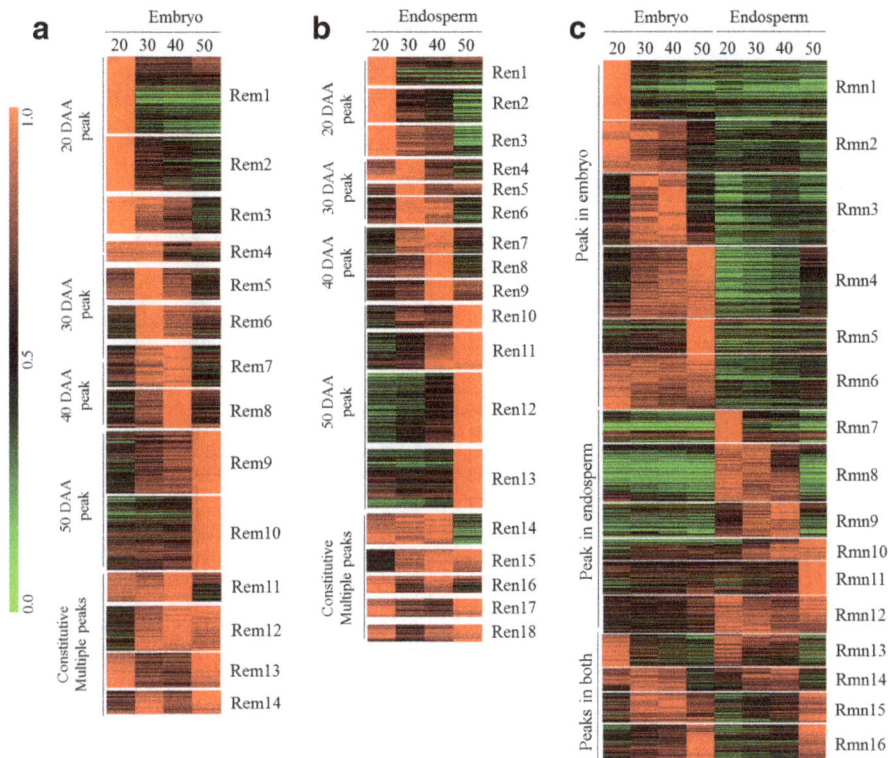

Fig. 5 Co-expression gene clusters of RL4452 (R). Clustering of probesets expressed in the embryo (em) (**a**), endosperm (en) (**b**) or both (mn) (**c**) at one or more time points. Figure descriptions are as shown in Fig. 4

was used as a criterion to determine the total number of clusters per category. The embryo or endosperm clusters in both genotypes exhibited peak expression at one or more time points during seed maturation (Figs. 4a,b and 5a,b) while gene clusters generated by combining the embryo and endosperm are roughly divided into three categories in both genotypes (Figs. 4c and 5c). The first category represents clusters with peak expression at least at one time point specifically in the embryo, the second category represents clusters with peak expression at least at one time point specifically in the endosperm, and the third category represents clusters with peak expression at least at one time point in both embryo and endosperm tissues.

Functional identity of genotypically and temporally overlapping embryo-specific clusters

To gain insights into distinct and overlapping tissue specific biological processes/molecular functions, gene ontology (GO) terms enriched in the embryo and endosperm gene clusters of both genotypes were identified (Figs. 6 and 7, Additional file 8: Table S2). Embryo clusters characterized by peak expression at the early phase with strong repression in the subsequent phases of seed maturation irrespective of genotype (Aem1 and Amn1; Rem1 and Rmn1) are enriched with chromatin/

chromosome organization (GO:0031497, GO:0006333, GO:0006325, GO:0051276; $P = 2.4e^{-31} \sim 3.10e^{-18}$), DNA-dependent DNA replication (GO:0006261; $P = 1.1e^{-6} \sim 1.2e^{-4}$) and protein-DNA complex assembly (GO:0065004; $P = 1.1e^{-32} \sim 8.9e^{-21}$) GO terms. Given that genes in these GO terms such as those encoding histone proteins (H2A/H2B/H3/H4) and DNA polymerases are involved in cellular proliferation [23], it is likely that their transcriptional activation at the early phase of maturation forms a mechanism underlying rapid growth of embryos and initiation of axial growth, which has been shown to occur during 15 to 30 DAA in wheat seeds [24] with no major role in the regulation of dormancy. Embryo clusters exhibiting peak expression during the mid-phase of seed maturation in both genotypes (Aem6–8 and Amn3; Rem7 and Rmn3) are enriched with ATPase activity (GO:0016887; $P = 3.7e^{-16} \sim 1.2e^{-4}$) and helicase activity (GO:0008026, GO:0004386, GO:0070035; $P = 5.3e^{-24} \sim 3.1e^{-7}$) GO terms that include genes encoding RNA helicase, SNF2 type helicase and ABC transporter proteins; and many of the RNA helicase probesets are annotated as DEAD-box type (Fig. 6, Additional file 8: Table S2). Given that the DEAD-box type RNA helicases are implicated in the regulation of growth and developmental events, and participate in the control of stress responsive genes [25, 26] and ABA biosynthesis [27], it is likely that the

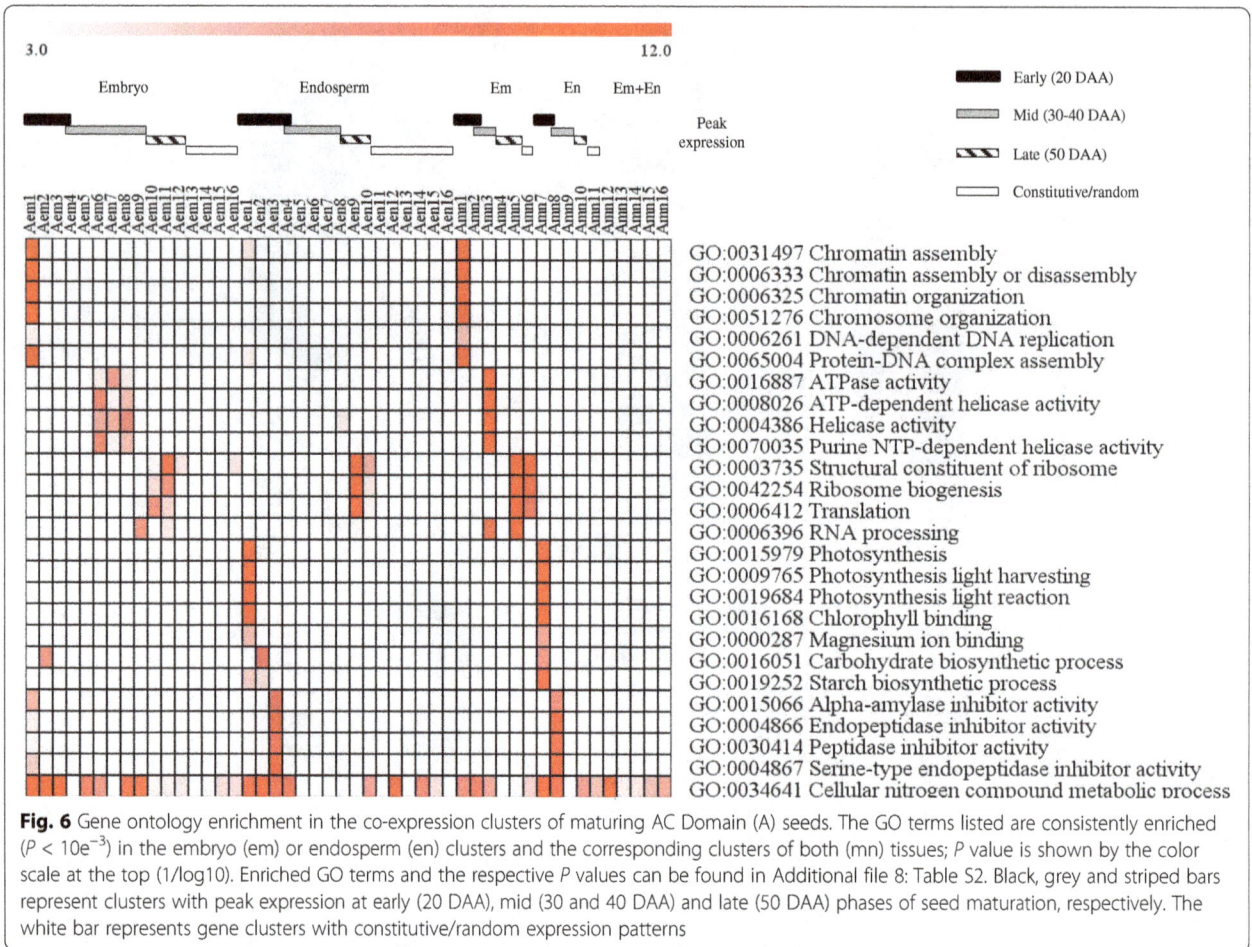

Fig. 6 Gene ontology enrichment in the co-expression clusters of maturing AC Domain (A) seeds. The GO terms listed are consistently enriched ($P < 10e^{-3}$) in the embryo (em) or endosperm (en) clusters and the corresponding clusters of both (mn) tissues; P value is shown by the color scale at the top (1/log10). Enriched GO terms and the respective P values can be found in Additional file 8: Table S2. Black, grey and striped bars represent clusters with peak expression at early (20 DAA), mid (30 and 40 DAA) and late (50 DAA) phases of seed maturation, respectively. The white bar represents gene clusters with constitutive/random expression patterns

genes in these clusters form an integral part of a mechanism regulating embryo growth and desiccation-induced stress tolerance but not the induction of dormancy in maturing seeds.

Ribosome biogenesis and translation GO terms (GO:0003735, GO:0042254, GO:0006412; $P = 2.2e^{-61} \sim 2.8e^{-4}$) are prevalent in embryo clusters of both genotypes exhibiting peak expression at mid to late phases of seed maturation (Aem10, Aem11, Amn5 and Amn6; Rem9, Rem12, Rmn4 and Rmn6). Furthermore, the RNA processing GO term (GO:0006396; $P = 2.8e^{-14} \sim 4.3e^{-4}$), which includes RNA splicing and exosome genes, is enriched in embryo clusters exhibiting peak expression around the same time (Aem9, Aem11, Amn3 and Amn5; Rem6, Rem9, Rem12, Rmn3 and Rmn4). These results suggest the ability of the embryo in the mid to late phases of seed maturation to employ post-transcriptional regulatory mechanisms for the synthesis of maturation/desiccation associated proteins from extant transcripts [28, 29]. Alternatively, the results might imply maturation-mediated storage of translational transcripts in the seeds to help with the synthesis of proteins required for germination upon imbibition of non-dormant seeds

[30]. In agreement with this hypothesis, the transcripts of ribosomal proteins are among the most abundant mRNAs stored in mature dry seeds of barley, and these genes exhibit transcriptional induction during imbibition of non-dormant seeds [31].

Functional identity of genotypically and temporally overlapping endosperm-specific clusters

The endosperm gene clusters of both genotypes with peak expression at the early phase of seed maturation (Aen1 and Amn7; Ren1 and Rmn7) are enriched with photosynthesis (GO:0015979, GO:0009765, GO:0019684; $P = 4.7e^{-23} \sim 9.3e^{-7}$), chlorophyll binding (GO:0016168; $P = 4.4e^{-20} \sim 1.9e^{-13}$) and magnesium ion binding (GO:0000287; $P = 1.30\times e^{-7} \sim 6.50\times e^{-6}$) GO categories. Given that immature seeds of wheat and other cereals have photosynthetically active pericarp [32], which is part of the endosperm in this study, it is likely that the genes in these clusters are exclusively expressed in the pericarp. However, the absence of differential expression pattern of these genes between the two genotypes might suggest their minimal role in inducing the variation in seed size/weight. Endospermic gene clusters of both

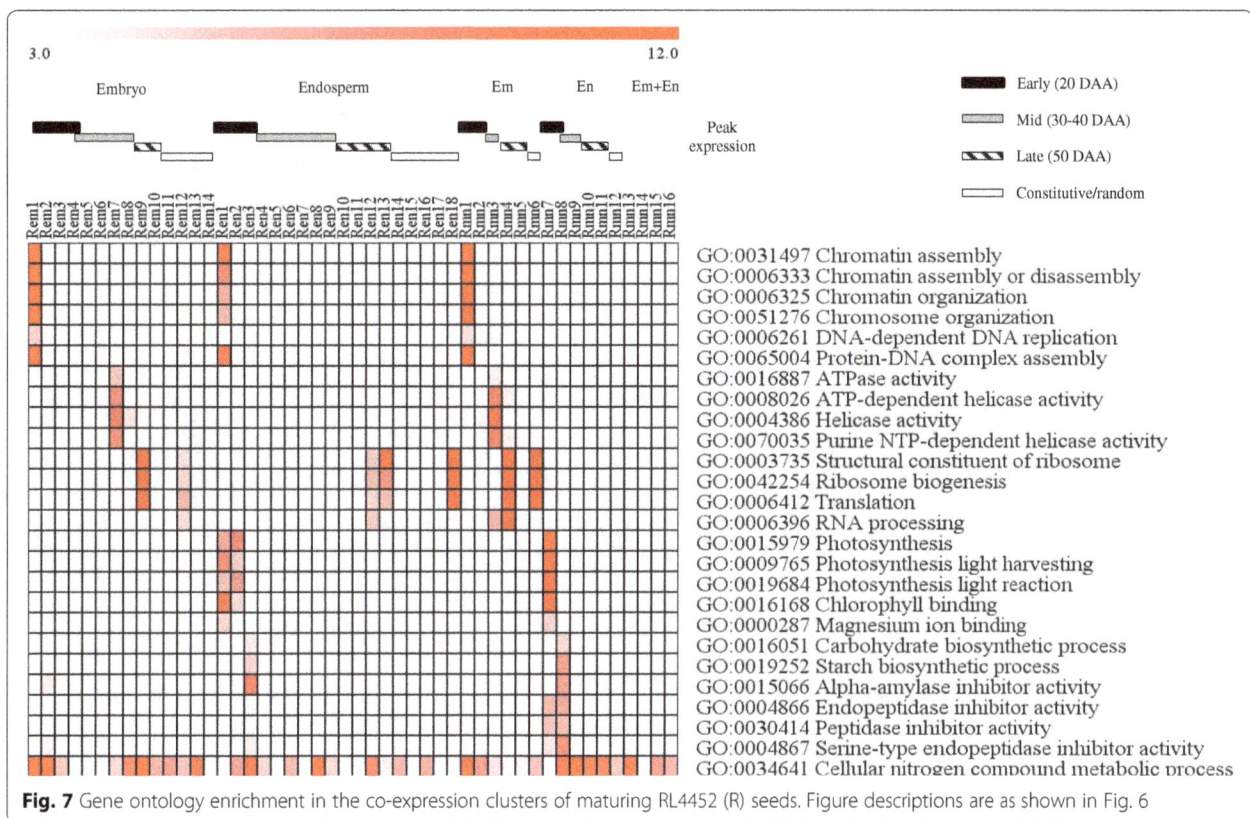

Fig. 7 Gene ontology enrichment in the co-expression clusters of maturing RL4452 (R) seeds. Figure descriptions are as shown in Fig. 6

genotypes whose expression is peaked at the early phase and remain enhanced during the mid-phase (Aen3 and Amn8; Ren3 and Rmn8) are enriched with alpha-amylase inhibitor activity (GO:0015066; $P = 7.3e^{-10} \sim 3.6e^{-8}$) and endopeptidase inhibitor activity (GO:0004866, GO:0030414, GO:0004867; $P = 1.7e^{-11} \sim 7.3e^{-4}$) GO terms, suggesting the significance of repressing starch degradation during the seed maturation process. Similar to that observed in the embryo, the cellular nitrogen compound metabolic process (GO:0034641; $P = 8.6e^{-27}$) appeared to be enriched in several endosperm clusters of both genotypes exhibiting peak expressions at different time points. It is likely that these nitrogen metabolism genes are involved in the supply of nitrogen to the maturing endosperm but are less likely to contribute to the variation in seed size/weight. Owing to the enrichment of the same GO term in germinating wheat and barley seeds, the nitrogen metabolism genes have also been suggested to play a role of providing nitrogen to the growing embryo [33, 34]. No GO term enrichment is evident in endosperm specific clusters, as confirmed by combining probesets expressed in both tissues, of both genotypes exhibiting enhanced/peak expression during mid to late phase of maturation (Amn9–11; Rmn9–12), and this might be attributed to cellular death and desiccation/quiescence, leading to a pause or a decrease in some cellular metabolic activities [35].

Gene regulatory networks during the maturation phase of wheat seed development

Genotypically distinct ABA regulated gene networks might contribute to the modulation of starch synthesis and thereby difference in seed weight and size

Endospermic gene clusters whose expression is peaked at early phase (20 DAA) in both genotypes, and maintained elevated level of expression in subsequent phases (30–40 DAA) only in RL4452 (Aen2 and Amn7, and Ren3 and Rmn8) are enriched with carbohydrate and starch biosynthetic processes (GO:0016051 and GO:0019252; $P = 2.5e^{-11} \sim 4.0e^{-5}$). Previous studies have shown that the expression pattern of starch biosynthesis genes is closely associated with that of starch accumulation in the endosperm of developing wheat seeds [17, 36]. Thus, our data might suggest the maintenance of elevated starch deposition in the maturing endosperm of RL4452 that produces larger seeds than that of AC Domain (Fig. 1). However, comparative analysis of the kinetics of starch accumulation between the two genotypes is required to clarify this. Given that ABA has been implicated in reducing starch synthesis ability of developing wheat seeds [17, 37] and this role of ABA has been suggested to be mediated by SNF1 kinase [9], the prevalence of ABRE motif specifically in the AC Domain clusters (Aen2 and Amn7) (Fig. 8) might suggest increased ABA sensitivity and therefore decreased accumulation of starch.

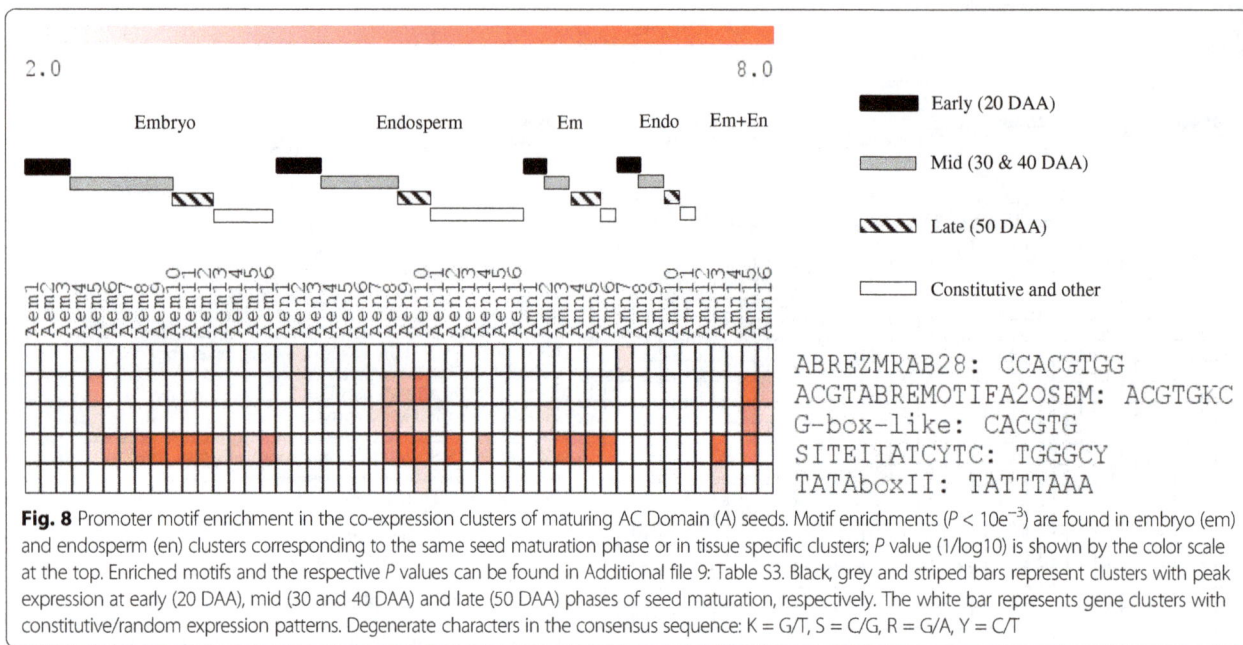

Fig. 8 Promoter motif enrichment in the co-expression clusters of maturing AC Domain (A) seeds. Motif enrichments ($P < 10e^{-3}$) are found in embryo (em) and endosperm (en) clusters corresponding to the same seed maturation phase or in tissue specific clusters; P value (1/log10) is shown by the color scale at the top. Enriched motifs and the respective P values can be found in Additional file 9: Table S3. Black, grey and striped bars represent clusters with peak expression at early (20 DAA), mid (30 and 40 DAA) and late (50 DAA) phases of seed maturation, respectively. The white bar represents gene clusters with constitutive/random expression patterns. Degenerate characters in the consensus sequence: K = G/T, S = C/G, R = G/A, Y = C/T

Genotypically, temporally and spatially distinct ABA regulated gene networks might contribute to the modulation of ABA response and thereby difference in seed dormancy

To predict gene regulatory networks underlying variation in seed size/weight and dormancy between the two genotypes, the 500 bp upstream region of rice homologs corresponding to the wheat probesets ($P < 10e^{-30}$) in each cluster was analyzed for motif enrichment using Osiris promoter database [38] (Figs. 8 and 9, Additional file 9: Table S3); motif enrichment was limited to rice due to insufficient database support for other cereal crops. Our data showed that several ABRE motifs containing the core ACGT sequence are enriched in several embryo and endosperm clusters of maturing seeds in both genotypes including ABREOSRAB21 (ACGTSSSC), ABREZMRAB28 (CCACGTGG), ACGTABREMOTI-FA2OSEM (ACGTGKC) and G-box-like (CACGTG), which act as binding sites for bZIP TFs such as ABI5 [39, 40]. In agreement with this result, genes highly expressed in mature seeds of Arabidopsis have been shown to be enriched with ABRE motifs [41] and enhanced expression of *ABI5* is prevalent during seed maturation [42]. Interestingly, the ACGTABREMOTIFA2OSEM (in both genotypes) and G-box-like (only in AC Domain) motifs are enriched in both embryo and endosperm clusters (Aem5, Aen8–10, Amn15 and Amn16; Rem12, Ren12 and Rmn15) while the prevalence of ABREZMRAB28 appeared to be endosperm specific in both genotypes (Aen2 and Amn7; Ren2 and Rmn7). Given that the seeds of AC Domain are dormant at maturity [19, 20] (Fig. 1c), our data might suggest the significance of ABA mediated transcriptional regulation specifically via G-box-like motif in

the control of dormancy induction and maintenance in maturing wheat seeds of AC Domain. Although ABA has also been implicated in the control of starch synthesis during endosperm maturation [9], genes in the endospermic clusters that are enriched with ACGTABREMOTIFA2O-SEM and ABREZMRAB28 motif appeared to show similar expression pattern between the two genotypes. It is therefore likely that variation in seed size/weight between the two genotypes is not influenced by ABA induced transcriptional programs mediated by the two motifs. The AC Domain gene clusters enriched with ACGTABREMOTI-FA2OSEM ($P = 10e^{-3} \sim 10e^{-9}$) and G-box-like ($P = 10e^{-3} \sim 10e^{-5}$) motifs exhibited peak expression at the mid and late phases in the embryo and endosperm, respectively (Fig. 4). Similar expression patterns were also observed for ACGTABREMOTIFA2OSEM enriched ($P = 10e^{-3} \sim 10e^{-4}$) embryo and endosperm clusters of RL4452, except that the embryo cluster (Rem12) exhibited enhanced expression through the late phase of seed maturation (Fig. 5). Given that some of these clusters are enriched with translation GO terms, our data might suggest that temporally distinct ABA regulated post-transcriptional mechanisms characterize tissue types. Furthermore, the difference between the two genotypes in the temporal window of the ABA mediated post-transcriptional regulation of embryonic genes might contribute to their variation in seed dormancy. However, the role of ABA in the control of the translation machinery during wheat seed maturation needs to be elucidated. The endospermic clusters characterized by overrepresentation of the ABREZMRAB28 motif ($P = 10e^{-3} \sim 10e^{-4}$) are enriched with photosynthesis related GO terms and exhibited peak expression specifically at the early phase of seed maturation in both genotypes (Figs. 4 and 5),

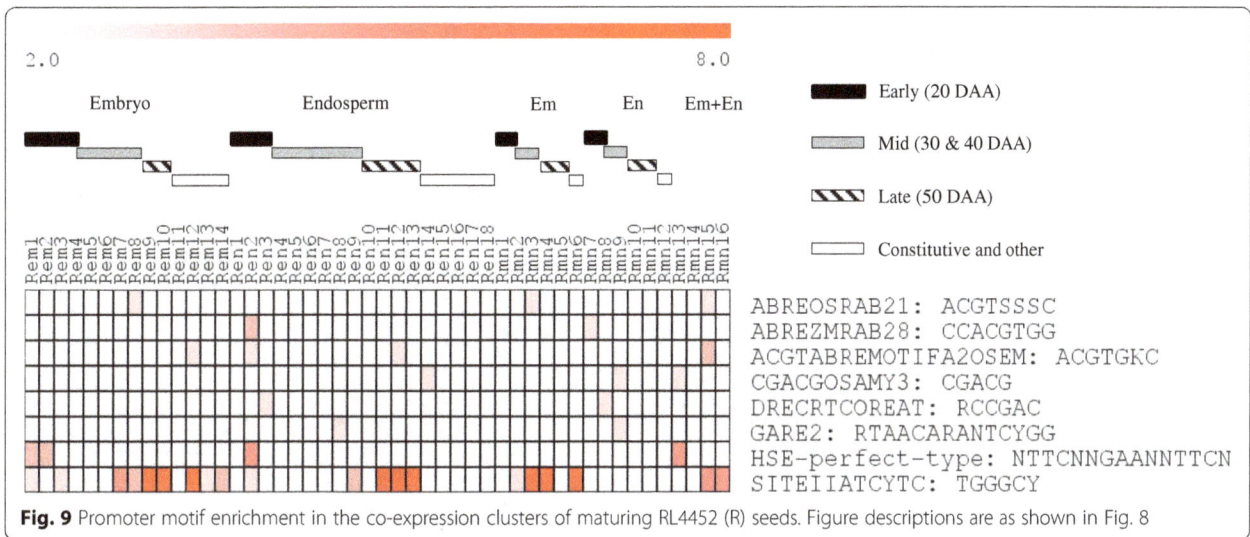

Fig. 9 Promoter motif enrichment in the co-expression clusters of maturing RL4452 (R) seeds. Figure descriptions are as shown in Fig. 8

implying the role of ABA in modulating photosynthetic activity in the pericarp [32], a constituent of the endosperm in this study. Although the pericarp is suggested to play a key role in photosynthesis and thereby providing storage reserves to the developing endosperm [43], it appears from their expression pattern that the pericarp localized photosynthetic genes do not contribute to the difference in seed size/weight between the two genotypes.

Genotypically distinct motifs suggest gene regulatory networks underlying variation in seed desiccation tolerance and dormancy

Our analysis also revealed genotype specific enrichment of promoter motifs in gene clusters. For example, the DRECRTCOREAT motif (RCCGAC), which acts as a binding site for DREB/CBF TFs [44], is enriched ($P = 10e^{-3}$) specifically in the endosperm clusters of RL4452 (Ren3 and Rmn8) (Figs. 8 and 9) that showed peak expression at the early phase and the expression remain elevated through the mid phase of seed maturation (Fig. 5). The HSE-perfect-type motif (NTTCNNGAANNTTCN), which acts as a binding site for members of heat shock factors (HSF) such as HsfA1a [45], is enriched ($P = 10e^{-4} \sim 10e^{-5}$) specifically in the embryo and endosperm clusters of RL4452 (Rem1, Rem2, Ren2 and Rmn13) that exhibit peak expression at the early phase of seed maturation. Given that the DREB and HSF TFs are co-induced by dehydration and desiccation [44, 46, 47] and the rate of seed desiccation is more pronounced in RL4452 than AC Domain (Fig. 1), the enrichment of DRECRTCOREAT and HSF-perfect-type motifs specifically in RL4452 might suggest the role of DREB/CBF and HSF in the acquisition of desiccation tolerance. The GA response element (GARE2, RTAACARANTCYGG) motif, which contains the core TAACAAA sequence and acts as a binding site for GAMYB in the promoters of GA regulated *α-amylase* genes [48, 49], is

overrepresented ($P = 10e^{-3}$) specifically in the endosperm clusters of RL4452 (Ren8 and Rmn9) that show peak expression at the mid phase, and this might form a part of the transcriptional mechanisms underlying the difference in seed response to GA, which has been implicated in playing a critical role in the regulation of seed dormancy [50].

bZIP TFs mediating ABA regulated embryonic but not endospermic gene networks of maturing seeds appear to vary with genotypic seed dormancy status

To identify TFs targeting the motifs identified, we referred to the PlantTFDB [51], which consists of amino acid sequences of 1940 wheat TFs belonging to 56 families. Blast searching of these TFs against PLEXdb revealed the presence of 285 TF probesets on Wheat GeneChip (E-value $<1e^{-100}$) encoding 49 TF families. It appeared from our data that 208 and 210 TF probesets are expressed in the embryo and endosperm tissues of AC Domain, respectively, at one or more time points during seed maturation; while 216 and 210 TF probesets are expressed in the embryo and endosperm tissues of RL4452, respectively. Of these probesets, 206 and 201 are found to be expressed in the embryo and endosperm of both genotypes, respectively (Additional file 10: Table S4). Our cluster analysis showed that the *ABI5* probeset (Ta.23671.1.S1_x_at) is co-expressed with probesets in ABRE enriched embryonic and endospermic clusters (Aen10, Rem12 and Ren12), which are also enriched with several GO terms (Fig. 10, Table 1), suggesting the significance of ABI5- and ABRE-mediated gene network in regulating maturation associated molecular functions. However, ABI5 is absent in the ABRE enriched embryonic cluster of AC Domain (Aem5) (Table 1). Given that seeds of AC Domain are dormant at maturity (Fig. 1c) and the induction of this adaptive trait during seed maturation is regulated by ABA [52], our data might imply

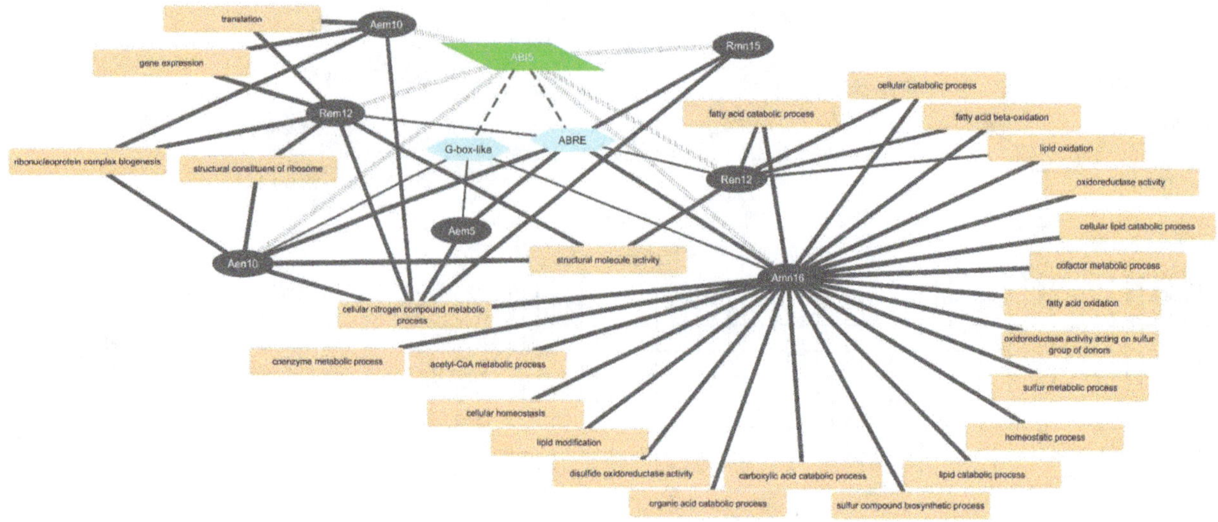

Fig. 10 Predicted ABI5-ABRE mediated gene regulatory network during seed maturation. The transcription factor ABI5 is represented by the green parallelogram, the ABRE motif by the light blue hexagon, the seed maturation gene clusters by grey oval, and GO terms by the orange rectangles. Solid lines connect GO terms and motifs enriched in the respective gene cluster ($P < 10e^{-3}$); the thickness of solid lines represent the P value in which thicker lines correspond to lower P values. Slash lines represent co-expression of ABI5 with the corresponding gene cluster. Dashed lines connect the ABI5 transcription factor with the predicted binding motifs

that ABA regulation of dormancy during seed maturation in this genotype is mediated by distinct transcriptional mechanisms. Consistent with this hypothesis, another bZIP TF, *G-box binding factor1* (*GBF1*), is co-expressed with probesets in the ABRE enriched embryonic clusters of AC Domain (Aem5). Alternatively, the absence of *ABI5* in the ABRE enriched Aem5 cluster might imply its genotype-specific post-transcriptional regulation, for example, through phosphorylation or protein degradation [53–55]. The *ABI5* probeset is also co-expressed with probesets in the embryonic clusters of AC Domain (Aem10) and RL4452 (Rem10), which exhibited peak expression at the late phase of seed maturation and are enriched with translation GO terms (Fig. 10). Although the ABRE motif is not overrepresented in these clusters, our result might imply the role of ABA in mediating post-transcriptional control of molecular functions during seed desiccation irrespective of genotype.

SITEIIATCYTC motif suggests genotypically overlapping role for TCP TFs in post-transcriptional regulation of embryo maturation

The SITEIIATCYTC motif (TGGGCY), which acts as a target for TEOSINTE BRANCHED1 (TB1), CYCLOIDEA (CYC) and PROLIFERATING CELL NUCLEAR ANTIGEN FACTOR1 (PCF) (TCP) TFs [56], is enriched ($P = 10e^{-3} \sim 10e^{-11}$) in several embryo and endosperm clusters of both genotypes that showed peak expression especially during the mid to late phases of seed maturation (Figs. 8 and 9). Interestingly, this motif is not enriched in Amn7–11 and Rmn7–12 clusters that consist of probesets

predominantly expressed in the endosperm with peak expressions occurring at different time points (Figs. 4 and 5), suggesting that genes with SITEIIATCYTC motif are embryo specific in maturing wheat seeds, however, further study is required to clarify this hypothesis. Using the amino acid sequences of 14 wheat TCPs available in the PlantTFDB, we were unable to identify a *TCP* probeset on Wheat GeneChip. However, gene clusters enriched with the SITEIIATCYTC motif are also found to be enriched with ribosome biogenesis and structure, gene expression and translation GO terms (Fig. 11), suggesting the role of TCP in the post-transcriptional regulation of embryonic molecular functions during the mid to late phases of seed maturation irrespective of the genotypic variation in dormancy and seed size/weight, likely to facilitate the synthesis of proteins required for acquisition of desiccation tolerance. Consistently, the SITEIIATCYTC motif is found in genes related to protein synthesis [56, 57]. Furthermore, the UP1 site (GGCCCAWW), which is almost identical with the SITEIIATCYTC motif [56, 58], has been shown to be overrepresented in clusters consisting of protein synthesis related genes upregulated in germinating seeds, and the SITEIIATCYTC targeting TCP14 regulates embryo growth potential during germination [59].

Genotypic distribution of gene expression reveals distinct and overlapping transcriptional programs and regulatory networks underlying variation in seed dormancy and size/weight

To gain further insights into distinct and overlapping transcriptional programs and regulatory networks underlying

Table 1 bZIP transcription factor probesets and the respective co-expressed gene clusters of maturing seeds in AC Domain and RL4452[a]

Probeset ID	AC Domain			RL4452			Arabidopsis		
	Aem	Aen	Amn	Rem	Ren	Rmn	AGI ID	E-Value	Gene name
Ta.808.1.S1_at	Aem16	Aen14	Amn4	Rem13	Ren9	Rmn1	AT5G06950.4	1E-104	AHBP-1B
Ta.161.1.S1_at	Aem1	Aen4	Amn12	Rem1	Ren2	Rmn13	AT2G35530.1	6E-72	bZIP transcription factor family protein
Ta.10054.1.S1_at	Aem10	*Aen9*	Amn10	*Rem12*	Ren15	Rmn10	AT2G36270.1	8E-55	ABI5 (ABA INSENSITIVE 5)
Ta.23671.1.S1_x_at	Aem10	*Aen10*	*Amn16*	*Rem12*	*Ren12*	*Rmn15*	AT2G36270.1	1E-44	ABI5 (ABA INSENSITIVE 5)
Ta.29331.1.S1_a_at	Aem1	Aen2	Amn12	Rem1	Ren2	Rmn8	AT3G56850.1	8E-29	AREB3 (ABA-RESPONSIVE ELEMENT BINDING PROTEIN 3)
Ta.29331.1.S1_at	Aem1	Aen2	Amn1	Rem1	Ren16	Rmn1	AT3G56850.1	8E-29	AREB3 (ABA-RESPONSIVE ELEMENT BINDING PROTEIN 3)
Ta.5410.1.S1_at	Aem15	*Aen8*	Amn10	*Rem12*	Ren15	Rmn10	AT5G06950.4	1E-135	AHBP-1B
Ta.1612.1.S1_at	Aem1	Aen14	Amn1	Rem7	Ren7	Rmn3	AT2G35530.1	2E-68	bZIP transcription factor family protein
Ta.23670.1.A1_at	NA[b]	NA	NA	NA	NA	NA	AT5G06950.4	1E-133	AHBP-1B
Ta.4604.1.S1_at	Aem1	Aen13	Amn1	Rem1	Ren6	Rmn1	AT1G06850.1	3E-47	AtbZIP52 (*Arabidopsis thaliana* basic leucine zipper 52)
Ta.6518.1.S1_at	Aem1	Aen14	Amn6	Rem1	Ren17	Rmn7	AT4G38900.3	2E-80	bZIP protein
TaAffx.24695.1.S1_at	Aem15	Aen2	Amn7	Rem6	Ren3	Rmn8	AT5G24800.1	5E-35	BZIP9 (BASIC LEUCINE ZIPPER 9)
TaAffx.88821.1.S1_at	Aem15	*Aen9*	Amn13	Rem13	Ren13	Rmn11	AT5G06950.4	1E-101	AHBP-1B
Ta.100.1.S1_at	Aem10	*Aen10*	Amn5	Rem9	*Ren12*	Rmn4	AT4G36730.1	1E-35	GBF1
Ta.13267.1.S1_at	NA	NA	NA	Rem10	NA	Rmn5	AT4G35040.1	1E-43	bZIP transcription factor family protein
Ta.13357.2.S1_at	Aem6	Aen1	Amn2	Rem5	Ren16	Rmn2	AT2G46270.1	6E-48	GBF3 (G-BOX BINDING FACTOR 3)
Ta.140.1.S1_at	*Aem5*	Aen11	*Amn16*	Rem5	Ren7	Rmn14	AT4G36730.1	3E-27	GBF1
Ta.19597.1.S1_at	Aem14	*Aen8*	*Amn16*	*Rem12*	Ren9	Rmn10	AT1G45249.1	1E-35	ABF2 (ABSCISIC ACID RESPONSIVE ELEMENTS-BINDING FACTOR 2)
Ta.19597.1.S1_x_at	Aem14	*Aen10*	Amn10	Rem14	Ren18	Rmn12	AT1G45249.1	1E-35	ABF2 (ABSCISIC ACID RESPONSIVE ELEMENTS-BINDING FACTOR 2)
Ta.23670.1.S1_at	Aem1	NA	Amn6	Rem1	NA	Rmn1	AT5G06950.4	1E-133	AHBP-1B
Ta.25454.1.S1_at	Aem1	NA	Amn1	Rem2	NA	Rmn2	AT2G42380.2	1E-49	bZIP transcription factor family protein
Ta.28910.1.S1_s_at	Aem3	Aen4	Amn2	Rem7	Ren6	Rmn3	AT2G35530.1	2E-64	bZIP transcription factor family protein
Ta.6443.2.S1_at	Aem8	Aen12	Amn3	Rem7	Ren13	*Rmn15*	AT2G40950.1	7E-62	BZIP17
Ta.6919.1.S1_at	Aem11	*Aen9*	Amn13	Rem9	Ren13	Rmn4	AT1G58110.2	6E-60	bZIP family transcription factor
Ta.893.1.S1_at	NA	Aen2	Amn8	NA	Ren3	Rmn8	AT5G28770.1	1E-30	BZO2H3

[a]The amino acid sequences of the wheat transcription factors available in the Plant Transcription Factor database (PlantTFDB 3.0) were searched against the wheat microarray platform in the Plant Expression Database (PLEXdb) to identify the respective probeset ($<e^{-100}$)
[b]NA not expressed in either or both tissues as determined by MAS5.0. Bold italicized clusters represent clusters enriched with ABRE motifs

seed maturation between the two genotypes, we determined the distribution of probesets in each of the AC Domain cluster across the embryo and endosperm clusters of RL4452 and vice versa (Additional files 3, 4, 5 and 6: Figs. S3–S6). Our result shows that over 50% of the probesets in AC Domain embryo clusters exhibiting peak expression at early (Aem1) and mid to late (Aem10 and Aem11) phases of seed maturation are overrepresented in the RL4452 embryo clusters with peak expression at early (Rem1) and mid to late (Rem9, Rem10 and Rem12) phases (Additional file 3: Figs. S3). Probesets overlapping between the Aem1 and Rem1 clusters are enriched with chromatin/chromosome ($P = 4.2e^{-28} \sim 2.9e^{-24}$) and protein-DNA

complex assembly ($P = 1.6e^{-29}$) GO terms while those overlapping between Aem10 and Aem11, and Rem9, Rem10 and Rem12 clusters are enriched with ribosome biogenesis and translation ($P = 2e^{-14} \sim 8.9e^{-14}$) GO terms (Additional file 11: Table S5), indicating that gene regulatory networks underlying embryo growth during the early phase and post-transcriptional regulation of molecular functions during the late phase of seed maturation are temporally conserved between the two genotypes despite their difference in seed dormancy and size/weight. Consistently, proteomic analysis of developing embryos of rice indicated that proteins involved in cell growth/division are highly expressed during the early stages of rice embryo

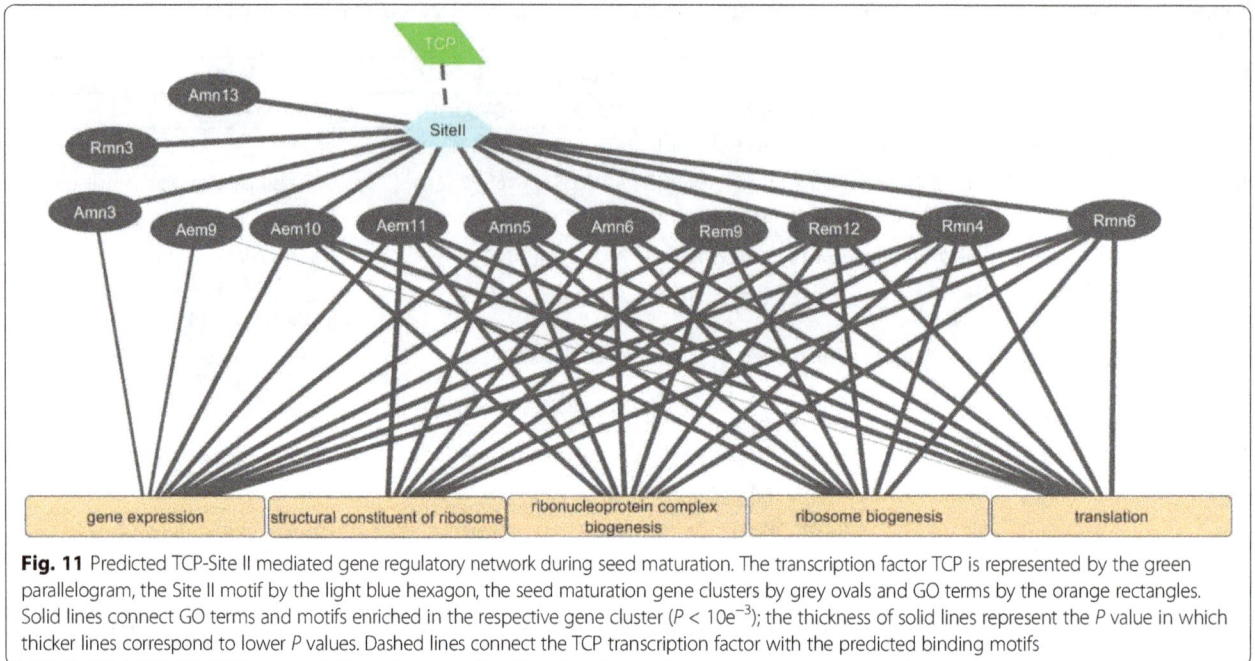

Fig. 11 Predicted TCP-Site II mediated gene regulatory network during seed maturation. The transcription factor TCP is represented by the green parallelogram, the Site II motif by the light blue hexagon, the seed maturation gene clusters by grey ovals and GO terms by the orange rectangles. Solid lines connect GO terms and motifs enriched in the respective gene cluster ($P < 10e^{-3}$); the thickness of solid lines represent the P value in which thicker lines correspond to lower P values. Dashed lines connect the TCP transcription factor with the predicted binding motifs

development [60]. Furthermore, enrichment of chromatin assembly GO term or upregulation of the related genes has been reported in germinating wheat and barley seeds that are characterized by growing/expanding embryos [34]. In contrast, embryonic clusters of AC Domain with peak expression at mid-phase (30 DAA) of seed maturation (Aem4 and Aem5) are distributed not only to the corresponding clusters of RL4452 (Rem4 and Rem5) that show similar expression pattern, but also to the adjacent clusters (Rem3 and Rem6); for example, over 21% of probesets in Aem5 cluster are overrepresented in Rem3 cluster that show peak expression mainly at the early phase, suggesting that temporally distinct transcriptional reprograming in maturing embryos characterizes genotypic differences in seed dormancy. Furthermore, although the Aem5 cluster is enriched with ACGTABREMOTIFA2O-SEM motif (Fig. 8), only ~1% of its probesets is found in the ACGTABREMOTIFA2OSEM enriched embryo cluster of RL4452 (Rem12) that exhibited peak expression during the mid to late (30 to 50 DAA) phases of seed maturation (Fig. 9). Likewise, over 50% of probesets in Rem12 cluster are overrepresented in the Aem9 and Aem10 clusters, which show peak expression during the mid to late (40 to 50 DAA) phases (Fig. 4), while only <1% Rem12 probesets are distributed to Aem5 cluster (Additional files 5 and 6: Figs. S5 and S6). Since ABA is a major player for the establishment of dormancy during seed maturation [52], these results might imply that temporally distinct, yet overlapping, ABA regulated embryonic gene networks in maturing seeds underlie the variation in the induction and maintenance of dormancy between the two genotypes.

Similar to that observed in the embryo, over 40% of the probesets in AC Domain endospermic clusters with peak expression in the early (Aen1 and Aen2) and late (Aen9 and Aen10) phases of seed maturation (Fig. 4) are overrepresented in the RL4452 clusters that exhibit peak expression at the early (Ren1, Ren2 and Ren3) and late (Ren12 and 13) phases, respectively (Additional file 4: Figs. S4). Probesets overlapping between the endospermic clusters of both genotypes exhibiting peak expression at the early phase are enriched with photosynthesis ($P = 1.1e^{-21}$) GO term (Additional file 12: Table S6), suggesting that pericarp derived photosynthate does not contribute to the difference in seed size/weight between the two genotypes. On the other hand, probesets overrepresented in the AC Domain endospermic clusters exhibiting peak expression at the mid phase (Aen4 and Aen5) are distributed not only to the corresponding RL4452 clusters (Ren4, Ren5 and Ren6) but also to clusters that show peak expression primarily at early phase of seed maturation (Ren2 and Ren3). Given that probesets overlapping between all these clusters are enriched with nutrient reservoir activity (GO:0045735; $P = 1.8e^{-15}$) GO term (Additional file 12: Table S6), our results suggest that transcriptional programs underlying reactions or pathways involved in the storage of nutritious substrates are shared by the two genotypes but operate in a temporally distinct manner that may affect seed size/weight. Unlike that observed in the embryo, over 30% of the probesets in each of the ABRE enriched endospermic clusters of AC Domain with peak expression at the late phase (Aen9 and Aen10) are shared with the ABRE enriched endospermic cluster of RL4452 (Ren12) exhibiting peak expression at the same phase of seed maturation. Since probesets overlapping between the Aen9, Aen10 and Ren12 clusters are enriched with gene expression and translation ($P = 3.1e^{-10} \sim 8.3e^{-8}$) GO terms (Additional file 12: Table S6), our data suggest that ABA regulated endospermic gene networks underlying post-transcriptional regulation of molecular functions during the late phase of seed maturation are conserved both genotypically and temporally, and therefore exert minimal effect on seed dormancy and size/weight.

Conclusions

The present study showed that temporal shifts in gene expression within the embryo and endosperm tissues vary with genotype, implying their role in governing phenotypic variations in seed size/weight and dormancy. However, tissue types appeared to be characterized by distinct but temporally and genotypically overlapping expression profiles and therefore molecular functions. It can be inferred from our data that genotypically distinct ABA and GA regulated gene networks modulate starch biosynthesis and acquisition of dormancy, leading to variations in seed weight/size and tolerance to preharvest sprouting, respectively. Our ABRE motif and bZIP TF data imply that genotypically distinct ABA regulated gene networks underlie the induction of dormancy during wheat seed maturation. Given that maturation associated biological processes significantly affect seed yield and quality, the findings of this study advance our knowledge of the transcriptional programs and regulatory networks regulating seed dormancy and seed size/weight during the maturation phase of seed development in wheat, a critical step to design molecular strategies for improving its yield and quality.

Methods

Plant material and growth conditions

Wheat plants of two genotypes with spring growth habit, AC Domain (a cultivar registered and widely cultivated in western Canada) and RL4452 (unregistered backcross derivative of 'Glenlea' and 'Kitt' cultivars [Glenlea*6/Kitt]) [61], were grown and managed as described previously[18] except that a growth room under a 22 °C/18 °C (day/night) with a 16/8 h photoperiod was used. Seeds of the two genotypes were obtained from Dr. Mark Jordan of Agriculture and Agri-Food Canada (AAFC)-Morden Research and Development Center (Morden, Manitoba, Canada). Seed developmental stages were determined using the first extrusion of the yellow anthers in the spikes as 0 DAA; maturing seeds were then harvested at 20, 30, 40 and 50 DAA. The endosperm (including the aleurone and pericarp) and the embryo (including the scutellum) were dissected from maturing seeds harvested from the middle region of each spike in

liquid nitrogen, and then stored at -80 °C until RNA extraction. Two to three independent biological replicates were collected from each tissue and genotype at each time point of seed maturation (~40 seeds per 2–3 spikes per 2–3 plants per replicate for 20, 30 and 40 DAA samples; ~100–120 seeds per 4–6 spikes per 4–6 plants per replicate for 50 DAA samples). Seed germination assays were performed as described previously [62].

Changes in seed weights and moisture content

Changes in fresh and dry weights of maturing seeds of the two genotypes were determined at each time point. Fresh weights were obtained from 20 to 23 seeds by weighing the individual maturing seed on a four decimal place analytical balance (Denver Instrument, Bohemia, NY, USA). To determine dry weights, the same seed samples were dried in an oven at 105 °C for 72 h and then reweighed. Moisture content was determined as a percentage of seed fresh weight. Test for statistically significant differences in seed fresh weight, dry weight and moisture content between the two genotypes was performed using t-Student test at a probability of $P < 0.05$.

RNA extraction

Total RNA samples of the embryonic tissues were extracted using the RNeasy Plant Mini Kit (Qiagen, Hilden, Germany). The integrity and purity of the resulting total RNA samples was verified by gel electrophoresis and spectrophotometrically before the samples are used for the microarray analysis. For the endosperm, total RNA samples were first extracted as described previously [17, 63]. The RNA samples were digested with DNase (Ambion, Austin, TX, USA) at 37 °C for 30 min to eliminate genomic DNA contamination. The purity and concentration, and integrity of the endospermic RNA samples was determined as described above. The RNA samples treated with DNase were then subjected to mRNA isolation, which was performed using PolyATract Kit (Promega, Madison, WI, USA) following the manufacturer's protocol.

DNA microarray analysis

The total RNA and mRNA samples of the embryo and the endosperm, respectively, were subjected to labeling and hybridization to the Affymetrix GeneChip Wheat Genome Array (Affymetrix, Santa Clara, CA, USA) exactly as described previously [64]. After hybridization, washing and staining of the arrays was performed in an Affymetrix Fluidics Station 450 following the manufacturer's protocol. Arrays were then scanned with an Affymetrix Scanner 3000. The microarray experiment for each tissue derived from each genotype at each stage of seed maturation involved at least two independent biological replicates.

Data analysis

Converting the data from all probe pairs into as single hybridization intensity and representation in CEL format, and adjustment of the total signal intensity per chip and determination of the number of probesets with present detection call was performed using the Affymetrix GeneChip Operating Software as described before [64]. Reproducibility of the data from the independent biological replicates was confirmed with scatter plot expression analysis. The MAS5 statistical algorithm of FlexArray was used to determine the number of probesets expressed in each tissue of both genotypes at least at one stage of seed maturation in all replications ($P < 0.05$). Following normalization of the raw data with the Robust Multi-array Average (RMA) procedure, the average RMA value of each probeset was used to perform Principal Component Analysis using FlexArray, and hierarchical clustering using MultiExperiment Viewer (MeV) software version 4.9 [65] with default setting. The FlexArray software version 1.6.3 [66] was used to calculate the expression level of each probeset at different time points of seed maturation relative to that exhibiting the highest RMA normalized signal intensity, which was arbitrarily set to a value of 1. The relative expression values were subjected to cluster analysis by K-means clustering and Pearson correlation coefficient using MeV software; MeV was also used to generate heat maps.

Gene ontology and motif enrichment analysis

GO enrichment for each gene cluster was performed with the AgriGO analysis toolkit (http://bioinfo.cau.edu.cn/agriGO/analysis.php) [67] using the default parameters of Fisher exact test (P < 0.05) and False Discovery Rate (FDR) correction by Yekutieli method. Candidate gene annotations for each probeset was obtained using HarvEST Wheat Chip version 1.59 (http://harvest.ucr.edu/) [68]. Promoter motif enrichment for each gene cluster was performed with Osiris [38] using the 500 bp upstream promoter region of the corresponding rice homologs in each cluster ($P < 10e^{-30}$). Visualization of the predictive gene regulatory networks was generated using Cytoscape (version 3.2.1) [69].

Identification of transcription factor probesets

To identify probesets representing TFs on Wheat GeneChip, amino acid sequences of wheat TFs available in the Plant TF Database version 3.0 (PlantTFDB; http://planttfdb.cbi.pku.edu.cn/index.php?sp=Tae) [51] were blast searched against the Plant Expression Database (PLEXdb; http://www.plexdb.org/index.php) [70] using E-value $< 1e^{-100}$.

Validation of microarray results by qPCR

Validation of the microarray data was performed with 10 randomly selected differentially expressed probesets using real-time qPCR and the same RNA samples used for the microarray analysis. Preparation of the cDNA samples and the qPCR assay was performed as described previously [71] using thermal cycling conditions described before [19]. Relative transcript level of the target genes was determined as described in Livak and Schmittgen [72] using β-actin as a reference gene. DNA sequences of the specific probeset IDs were obtained from HarvEST and used for designing the qPCR primers (Additional file 13: Table S7).

Additional files

Additional file 1: Fig. S1. Validation of the microarray data with qPCR.

Additional file 2: Fig. S2. Hierarchical clustering of the embryo and endosperm by expression patterns of probesets during seed maturation.

Additional file 3: Fig. S3. Distribution of probesets in each AC Domain embryonic cluster across the RL4452 clusters.

Additional file 4: Fig. S4. Distribution of probesets in each AC Domain endospermic cluster across the RL4452 clusters.

Additional file 5: Fig. S5. Distribution of probesets in each RL4452 embryonic cluster across the AC Domain clusters.

Additional file 6: Fig. S6. Distribution of probesets in each RL4452 endospermic cluster across the AC Domain clusters.

Additional file 7: Table S1. Robust multi-array average values and gene clusters for probesets expressed in the embryo and endosperm of maturing AC Domain and RL4452 seeds.

Additional file 8: Table S2. Gene ontology enrichment in the embryo and/or endosperm clusters of AC Domain and RL4452.

Additional file 9: Table S3. Promoter motif enrichment in the embryo and endosperm clusters of AC Domain and RL4452.

Additional file 10: Table S4. Identification and annotation of wheat transcription factor probesets.

Additional file 11: Table S5. Gene ontology enrichment of probesets overlapping between embryo clusters of AC Domain and RL4452.

Additional file 12: Table S6. Gene ontology enrichment of probesets overlapping between endosperm clusters of AC Domain and RL4452.

Additional file 13: Table S7. qPCR primers used for validation of the microarray data.

Abbreviations

ABA: Abscisic acid; ABRE: ABA responsive element; DAA: Days after anthesis; GA: Gibberellin; GO: Gene ontology; PlantTFDB: Plant Transcription Factor database; RMA: Robust Multi-array Average; TF: Transcription factor

Acknowledgments

The authors would like to thank Tran-Nguyen Nguyen and Zhen Yao for their technical assistance.

Funding

This work was supported by a grant from the Natural Sciences and Engineering Research Council of Canada, Western Grains Research Foundation & Agriculture and Agri-Food Canada through the National Wheat Improvement Program of Agri-Science Cluster, and Manitoba Wheat and Barley Growers Association & Agriculture Rural Development Initiatives Growing Forward 2 to BTA.

Authors' contributions
BTA conceived and designed the research, BTA, FG and MCJ performed the experiments, YY and BTA analyzed the data, BTA and YY wrote the manuscript, and all authors read and approved the final version of the manuscript.

Competing interests
The authors declare that they have no competing interests.

Author details
[1]Department of Plant Science, University of Manitoba, 222 Agriculture Building, Winnipeg, MB R3T 2N2, Canada. [2]Morden Research and Development Centre, Agriculture and Agri-Food Canada, Morden, MB R6M 1Y5, Canada.

References
1. Dominguez F, Cejudo FJ. Programmed cell death (PCD): an essential process of cereal seed development and germination. Front Plant Sci. 2014;5:366.
2. Sabelli PA, Larkins BA. The development of endosperm in grasses. Plant Physiol. 2009;149:14–26.
3. Bewley JD, Bradford K, Hilhorst H, Nonogaki H. Seeds: physiology of development, germination and dormancy. 3rd ed. New York: Springer-Verlag; 2013.
4. Evers T, Millar S. Cereal grain structure and development: some implications for quality. J Cereal Sci. 2002;36:261–84.
5. Wheat: The big picture. Available: http://www.cerealsdb.uk.net/cerealgenomics/WheatBP/Documents/DOC_WheatBP.php. Accessed 20 Mar 2016.
6. Belmonte MF, Kirkbride RC, Stone SL, et al. Comprehensive developmental profiles of gene activity in regions and subregions of the *Arabidopsis* seed. Proc Natl Acad Sci U S A. 2013;110:435–44.
7. Le BH, Cheng C, Bui AQ, et al. Global analysis of gene activity during *Arabidopsis* seed development and identification of seed-specific transcription factors. Proc Natl Acad Sci U S A. 2010;107:8063–70.
8. Chen J, Zeng B, Zhang M, Xie S, Wang G, Hauck A, Lai J. Dynamic transcriptome landscape of maize embryo and endosperm development. Plant Physiol. 2014;166:252–64.
9. Sreenivasulu N, Radchuk V, Strickert M, Miersch O, Weschke W, Wobus U. Gene expression patterns reveal tissue-specific signaling networks controlling programmed cell death and ABA- regulated maturation in developing barley seeds. Plant J. 2006;47:310–27.
10. Xue LJ, Zhang JJ, Xue HW. Genome-wide analysis of the complex transcriptional networks of rice developing seeds. PLoS One. 2012;7:e31081.
11. Gillies SA, Futardo A, Henry RJ. Gene expression in the developing aleurone and starchy endosperm of wheat. Plant Biotechnol J. 2012;10:668–79.
12. Pfeifer M, Kugler KG, Sandve SR, Zhan B, Rudi H, Hvidsten TR, Mayer KF, Olsen OA. Genome interplay in the grain transcriptome of hexaploid bread wheat. Science. 2014;345:1250091.
13. Wan Y, Poole RL, Huttly AK, et al. Transcriptome analysis of grain development in hexaploid wheat. BMC Genomics. 2008;9:121.
14. Gao F, Ayele BT. Functional genomics of seed dormancy and preharvest sprouting in wheat: advances and prospects. Front Plant Sci. 2014;5:458.
15. Cabral AL, Jordan MC, McCartney CA, You FM, Humphreys DG, MacLachlan R, Pozniak CJ. Identification of candidate genes, regions and markers for pre-harvest sprouting resistance in wheat (*Triticum aestivum* L). BMC Plant Biol. 2014;14:340.
16. Rasul G, Humphreys DG, Brûlé-Babel AL, McCartney CA, Knox RE, DePauw RM, Somers DJ. Mapping QTLs for pre-harvest sprouting traits in the spring wheat cross 'RL4452/AC domain'. Euphytica. 2009;168:363–78.
17. Mukherjee S, Liu A, Deol KK, Kulichikhin K, Stasolla C, Brule-Babel A, Ayele BT. Transcriptional coordination and abscisic acid mediated regulation of sucrose transport and sucrose-to-starch metabolism related genes during grain filling in wheat (*Triticum aestivum* L). Plant Sci. 2015;240:143–160.
18. Gao F, Rampitsch C, Chitnis VR, Humphreys GC, Jordan MC, Ayele BT. Integrated analysis of seed proteome and mRNA oxidation reveals distinct post-transcriptional features regulating dormancy in wheat (*Triticum aestivum* L). Plant Biotechnol J. 2013;11:921–932.
19. Son S, Chitnis VR, Liu A, Gao F, Nguyen TN, Ayele BT. Abscisic acid metabolic genes of wheat (*Triticum aestivum* L.): identification and insights into their functionality in seed dormancy and dehydration tolerance. Planta. 2016;244:429–447.
20. Barrero JM, Jacobsen JV, Talbot MJ, White RG, Swain SM, Garvin DF, Gubler F. Grain dormancy and light quality effects on germination in the model grass *Brachypodium distachyon*. New Phytol. 2012;193:376–86.
21. Meimoun P, Mordret E, Langlade NB, Balzergue S, Arribat C, Bailly S, El-Maarouf-Bouteau H. Is gene transcription involved in seed dry after-ripening. PLoS One. 2014;9:e86442.
22. Vertucci CW, Farrant JM. Acquisition and loss of desiccation tolerance. In: Kigel J, Galili G, editors. Seed development and germination. New York: Marcel Dekker; 1995. p. 237–71.
23. Raynaud C, Mallory AC, Latrasse D, Jegu T, Bruggeman Q, Delarue M, Bergounioux C, Benhamed M. Chromatin meets the cell cycle. J Exp Bot. 2014;65:2677–89.
24. Smart MG, O'Brien TP. The development of the wheat embryo in relation to the neighbouring tissues. Protoplasma. 1983;114:1–13.
25. Huang CK, Shen YL, Huang LF, Wu SJ, Yeh CH, Lu CA. The DEAD-box RNA helicase AtRH7/PRH75 participates in pre-rRNA processing, plant development and cold tolerance in Arabidopsis. Plant Cell Physiol. 2016;57:174–91.
26. Zhu M, Chen G, Dong T, Wang L, Zhang J, Zhao Z, Hu Z. *SlDEAD31*, a putative DEAD-box RNA helicase gene, regulates salt and drought tolerance and stress-related genes in tomato. PLoS One. 2015;10:e0133849.
27. Hsu YF, Chen YC, Hsiao YC, Wang BJ, Lin SY, Cheng WH, Jauh GY, Harada JJ, Wang CS. AtRH57, A DEAD-box RNA helicase, is involved in feedback inhibition of glucose-mediated abscisic acid accumulation during seedling development and additively affects pre-ribosomal RNA processing with high glucose. Plant J. 2014;77:119–35.
28. Dever TE. Gene-specific regulation by general translation factors. Cell. 2002; 108:545–56.
29. Mata J, Marguerat S, Bahler J. Post-transcriptional control of gene expression: a genome-wide perspective. Trends Biochem Sci. 2005;30:506–14.
30. Rajjou L, Gallardo K, Debeaujon I, Vandekerckhove J, Job C, Job D. The effect of alpha-amanitin on the Arabidopsis seed proteome highlights the distinct roles of stored and neosynthesized mRNAs during germination. Plant Physiol. 2004;134:1598–613.
31. An YQ, Lin L. Transcriptional regulatory programs underlying barley germination and regulatory functions of gibberellin and abscisic acid. BMC Plant Biol. 2011;11:105.
32. Radchuk V, Borisjuk L. Physical, metabolic and developmental functions of the seed coat. Front Plant Sci. 2014;5:510.
33. Sreenivasulu N, Usadel B, Winter A, et al. Barley grain maturation and germination: metabolic pathway and regulatory network commonalities and differences highlighted by new MapMan/PageMan profiling tools. Plant Physiol. 2008;146:1738–58.
34. Gao F, Jordan MC, Ayele BT. Transcriptional programs regulating seed dormancy and its release by after-ripening in common wheat (*Triticum aestivum* L). Plant Biotechnol J. 2012;10:465–476.
35. Alpert P, Oliver MJ. Drying without dying. In: Black M, Pritchard HM, editors. Desiccation and survival in plants: drying without dying. Wallingford: CABI Publishing; 2002. p. 3–43.
36. Kang G, Liu G, Peng X, Wei L, Wang C, Zhu Y, Ma Y, Jiang Y, Guo T. Increasing the starch content and grain weight of common wheat by overexpression of the cytosolic AGPase large subunit gene. Plant Physiol Biochem. 2013;73:93–8.
37. Ahmadi A, Baker DA. Effects of abscisic acid (ABA) on grain filling processes in wheat. Plant Growth Regul. 1999;28:187–97.
38. Morris RT, O'Connor TR, Wyrick JJ. Osiris: an integrated promoter database for *Oryza sativa* L. Bioinformatics. 2008;24:2915–7. Accessed 27 Oct 2015
39. Carles C, Bies-Etheve N, Aspart L, Leon-Kloosterziel KM, Koornneef M, Echeverria M, Delseny M. Regulation of *Arabidopsis thaliana* Em genes: role of ABI5. Plant J. 2002;30:373–83.
40. Yang X, Yang YN, Xue LJ, Zou MJ, Liu JY, Chen F, Xue HW. Rice ABI5-like1 regulates ABA and auxin responses by affecting the expression of ABRE-containing genes. Plant Physiol. 2011;156:1397–409.
41. Nakabayashi K, Okamoto M, Koshiba T, Kamiya Y, Nambara E. Genome-wide profiling of stored mRNA in *Arabidopsis thaliana* seed germination: epigenetic and genetic regulation of transcription in seed. Plant J. 2005;41:697–709.

42. Peng FY, Weselake RJ. Gene coexpression clusters and putative regulatory elements underlying seed storage reserve accumulation in Arabidopsis. BMC Genomics. 2011;12:286.

43. Kong LA, Xie Y, Sun MZ, Si JS, Hu L. Comparison of the photosynthetic characteristics in the pericarp and flag leaves during wheat (*Triticum aestivum* L.) caryopsis development. Photosynthetica. 2016;54:40–6.

44. Dubouzet JG, Sakuma Y, Ito Y, Kasuga M, Dubouzet EG, Miura S, Seki M, Shinozaki K, Yamaguchi-Shinozaki K. *OsDREB* genes in rice, *Oryza sativa* L., encode transcription activators that function in drought-, high-salt- and cold-responsive gene expression. Plant J. 2003;33:751–63.

45. Hubel A, Schoffl F. *Arabidopsis* heat shock factor: isolation and characterization of the gene and the recombinant protein. Plant Mol Biol. 1994;26:353–62.

46. Li KQ, Xu XY, Huang XS. Identification of differentially expressed genes related to dehydration resistance in a highly drought-tolerant pear, *Pyrus betulaefolia*, as through RNA-Seq. PLoS One. 2016;11:e0149352.

47. González-Morales SI, Chávez-Montes RA, Hayano-Kanashiro C, Alejo-Jacuinde G, Rico-Cambron TY, de Folter S, Herrera-Estrella L. Regulatory network analysis reveals novel regulators of seed desiccation tolerance in *Arabidopsis thaliana*. Proc Natl Acad Sci U S A. 2016;113:E5232–41.

48. Gubler F, Kalla R, Roberts JK, Jacobsen JV. Gibberellin-regulated expression of a *myb* gene in barley aleurone cells: evidence for Myb transactivation of a high-pI alpha-amylase gene promoter. Plant Cell. 1995;7:1879–91.

49. Huang N, Sutliff TD, Litts JC, Rodriguez RL. Classification and characterization of the rice alpha-amylase multigene family. Plant Mol Biol. 1990;14:655–68.

50. Finch-Savage WE, Leubner-Metzger G. Seed dormancy and the control of germination. New Phytol. 2006;171:501–23.

51. Jin J, Zhang H, Kong L, Gao G, Luo J. PlantTFDB 3.0: a portal for the functional and evolutionary study of plant transcription factors. Nucleic Acids Res. 2014;42:D1182–1187 (2014). Accessed 21 Nov 2015.

52. Holdsworth MJ, Bentsink L, Soppe WJ. Molecular networks regulating Arabidopsis seed maturation, after-ripening, dormancy and germination. New Phytol. 2008;179:33–54.

53. Johnson RR, Wagner RL, Verhey SD, Walker-Simmons MK. The abscisic acid-responsive kinase PKABA1 interacts with a seed-specific abscisic acid response element-binding factor, TaABF, and phosphorylates TaABF peptide sequences. Plant Physiol. 2002;130:837–46.

54. Lopez-Molina L, Mongrand S, Kinoshita N, Chua NH. AFP is a novel negative regulator of ABA signaling that promotes ABI5 protein degradation. Genes Dev. 2003;17:410–8.

55. Rodriguez MV, Mendiondo GM, Maskin L, Gudesblat GE, Iusem ND, Benech-Arnold RL. Expression of ABA signalling genes and ABI5 protein levels in imbibed *Sorghum bicolor* caryopses with contrasting dormancy and at different developmental stages. Ann Bot. 2009;104:975–85.

56. Tremousaygue D, Garnier L, Bardet C, Dabos P, Herve C, Lescure B. Internal telomeric repeats and 'TCP domain' protein-binding sites co-operate to regulate gene expression in *Arabidopsis thaliana* cycling cells. Plant J. 2003;33:957–66.

57. Kosugi S, Ohashi Y. PCF1 And PCF2 specifically bind to cis elements in the rice proliferating cell nuclear antigen gene. Plant Cell. 1997;9:1607–19.

58. Tatematsu K, Ward S, Leyser O, Kamiya Y, Nambara E. Identification of cis-elements that regulate gene expression during initiation of axillary bud outgrowth in Arabidopsis. Plant Physiol. 2005;138:757–66.

59. Tatematsu K, Nakabayashi K, Kamiya Y, Nambara E. Transcription factor AtTCP14 regulates embryonic growth potential during seed germination in *Arabidopsis thaliana*. Plant J. 2008;53:42–52.

60. Xu H, Zhang W, Gao Y, Zhao Y, Guo L, Wang J. Proteomic analysis of embryo development in rice (*Oryza sativa*). Planta. 2012;235:687–701.

61. McCartney CA, Somers DJ, Humphreys DG, Lukow O, Ames N, Noll J, Cloutier S, McCallum BD. Mapping quantitative trait loci controlling

agronomic traits in the spring wheat cross RL4452 x 'AC domain'. Genome. 2005;48:870–83.

62. Liu A, Gao F, Kanno Y, Jordan MC, Kamiya Y, Seo M, Ayele BT. Regulation of wheat seed dormancy by after-ripening is mediated by specific transcriptional switches that induce changes in seed hormone metabolism and signaling. PLoS One. 2013;8:e56570.

63. Li Z, Trick HN. Rapid method for high-quality RNA isolation from seed endosperm containing high levels of starch. BioTechniques. 2005;38:872–6.

64. Jordan MC, Somers DJ, Banks TW. Identifying regions of the wheat genome controlling seed development by mapping expression quantitative trait loci. Plant Biotechnol J. 200;75:442–453.

65. Saeed AI, Bhagabati NK, Braisted JC, Liang W, Sharov V, Howe EA, Li J, Thiagarajan M, White JA, Quackenbush J. TM4 Microarray software suite. Methods in Enzymol. 2006;411:134–93.

66. Blazejczyk M, Miron M, Nadon R. FlexArray: a statistical data analysis software for gene expression microarrays. Genome Quebec, Montreal, Canada. 2007;2007

67. Du Z, Zhou X, Ling Y, Zhang Z, Su Z. agriGO: a GO analysis toolkit for the agricultural community. Nucleic Acids Res. 2010;38:W64–70. Accessed 27 Oct 2015

68. HarvEST: WheatChip version 1.59. Available: http://harvest.ucr.edu/. Accessed 7 Sept 2014.

69. Shannon P, Markiel A, Ozier O, Baliga NS, Wang JT, Ramage D, Amin N, Schwikowski B, Ideker T. Cytoscape: a software environment for integrated models of biomolecular interaction networks. Genome Res. 2003;13:2498–504.

70. Plant Expression Database (PLEXdb) tools. Available: http://www.plexdb.org/. Accessed 21 Nov 2015 .

71. Nguyen TN, Son S, Jordan MC, Levin DB, Ayele BT. Lignin biosynthesis in wheat (*Triticum aestivum* L.): its response to waterlogging and association with hormonal levels. BMC Plant Biol. 2016;16:28.

72. Livak KJ, Schmittgen TD. Analysis of relative gene expression data using real-time quantitative PCR and the 2(−Delta Delta C(T)) method. Methods. 2001;25:402–8.

10

Characterization of four rice *UEV1* genes required for Lys63-linked polyubiquitination and distinct functions

Qian Wang[1†], Yuepeng Zang[1†], Xuan Zhou[1] and Wei Xiao[1,2*] ⓘ

Abstract

Background: The error-free branch of the DNA-damage tolerance (DDT) pathway is orchestrated by Lys63-linked polyubiquitination of proliferating cell nuclear antigen (PCNA), and this polyubiquitination is mediated by a Ubc13-Uev complex in yeast. We have previously cloned *OsUBC13* from rice, whose product functions as an E2 to promote Lys63-linked ubiquitin chain assembly in the presence of yeast or human Uev.

Results: Here we identify four highly conserved *UEV1* genes in rice whose products are able to form stable heterodimers with OsUbc13 and mediate Lys63-linked ubiquitin chain assembly. Expression of *OsUEV1*s is able to rescue the yeast *mms2* mutant from death caused by DNA-damaging agents. Interestingly, OsUev1A contains a unique C-terminal tail with a conserved prenylation site not found in the other three OsUev1s, and this post-translational modification appears to be required for its unique subcellular distribution and association with the membrane. The analysis of *OsUEV1* expression profiles obtained from the Genevestigator database indicates that these genes are differentially regulated.

Conclusions: We speculate that different OsUev1s play distinct roles by serving as a regulatory subunit of the Ubc13-Uev1 complex to respond to diverse cellular, developmental and environmental signals.

Keywords: Rice, Uev1, Ubc13, Lys63-linked polyubiquitination, DNA-damage response, Prenylation

Background

Ubiquitination is a critical post-translational protein modification process in eukaryotic cells, which involves a small protein modifier named ubiquitin (Ub). Although ubiquitination is well known to target proteins for degradation [1, 2], several non-proteolytic roles have also been found including manipulating protein interaction, activities and localization [3–5]. Different fates of the target protein after ubiquitination are often dictated by whether it is monoubiquitinated, or additional ubiquitins are attached to form a poly-Ub chain. In the latter case, the C-terminus of an incoming Ub can be linked to one of seven surface lysine residues (Lys6, Lys11, Lys27, Lys29, Lys33, Lys48 and Lys63) on the previous Ub [3, 6]. It was found that different poly-Ub chains have different topological and chemical properties; for example, while Lys11,

Lys29 and Lys48 linked chains lead to protein degradation [2, 3, 7], the Lys63-linked chain is generally involved in signal transduction [5].

Ubiquitination was initially implicated in DNA-damage response when Rad6, an E2 enzyme, was found to be required for post-replication repair (PRR) in budding yeast [8]. Rad6, along with its cognate E3 Rad18, monoubiquitinates proliferating cell nuclear antigen (PCNA) at the Lys164 residue in response to replication-blocking DNA damage; this monoubiquitination leads to translesion DNA synthesis (TLS). The monoubiquitinated PCNA can be further polyubiquitinated at the same residue by the E2-E3 complex Mms2-Ubc13-Rad5 [9, 10], which is required for error-free lesion bypass [11–14] via template switch [15, 16]. This process appears to be conserved in eukaryotic organisms from yeast to human, and is named DNA-damage tolerance (DDT) [17, 18].

Owing to their sessile nature, plants are continuously under different types of stresses, such as DNA damage by UV exposure. These stresses severely compromise plant survival, reduce crop yield and threaten food

---
* Correspondence: wei.xiao@usask.ca
†Equal contributors
[1]College of Life Sciences, Capital Normal University, Beijing 100048, China
[2]Department of Microbiology and Immunology, University of Saskatchewan, Saskatoon, SK S7N 5E5, Canada

security. Plants have established several strategies to cope with DNA-damage stresses, including various DNA repair pathways and tolerance of replication blocks by efficient TLS polymerases [19–24]. Meanwhile a few reports also indicate the conservation of error-free DDT in *Arabidopsis* [25–27]; however, little is known about the underlying mechanisms. We previously reported the cloning and characterization of rice *UBC13*, a putative error-free DDT gene, and showed that it is able to functionally complement the corresponding yeast *ubc13* mutant's defect in PRR, and its product mediates Lys63-linked polyubiquitination in vitro [28]. In both cases, rice Ubc13 has to rely on a heterologous Ubc-E2 variant (Uev). Indeed, Ubc13 and Uev proteins from yeast or mammalian cells form a stable heterodimer, which is absolutely required for Lys63-linked poly-Ub chain assembly [29–31], and this process appears to be highly conserved in eukaryotes [18]. In this study, four rice *UEV1* genes are identified and functionally characterized. Interestingly, one of the four rice *UEV1* products, Uev1A, is deemed to be post-translationally modified in its C-terminus, which makes it functionally different from other three Uev1s, suggesting that they are involved in multiple cellular processes, that they have distinct functions and that rice Uevs may serve as a regulatory subunit to modulate Ubc13 activities.

Results

The rice genome encodes four highly conserved *UEV1* genes

Our previous work has identified the *UBC13* gene in rice, which is predicted to produce a protein which is highly conserved with Ubc13s from some other species [28]. In general, Ubc13 works with Uev as a heterodimer to catalyze the assembling of Lys63-linked Ub chains, and OsUbc13 was proved to be able to interact with Uevs from yeast and human to achieve this goal. Therefore, it is reasonable to predict that the rice genome contains its own conserved *UEV* gene(s). In this study, the *Arabidopsis UEV1A* gene was used to BLAST the rice genome in the Rice Annotation Project Database (RAP-DB, http://rapdb.dna.affrc.go.jp/index.html). Four genes were retrieved and named *OsUEV1A* (Os03g0712300, GenBank accession number XM_015777395.1), *OsUEV1B* (Os12g0605400, XM_015764791.1), *OsUEV1C* (Os09g0297100, XM_015756494.1) and *OsUEV1D* (Os04g0684800, XM_015780422.1). The exon-intron organization and coding sequences of these rice loci were determined through sequence comparison with the PCR-amplified corresponding full-length cDNAs and available sequences from RAP-DB. Based on the cDNA PCR products detected, all the *OsUEV1* cDNA products were identical to corresponding annotations on RAP-DB.

Phylogenetic analysis was performed on the ORF sequences of *OsUEV1*s, *Arabidopsis thaliana* (*At*) *UEV1*s

[26] and *Brachypodium distachyon* (*Bd*) *UEV1*s [32] (Fig. 1a), which reveals that *OsUEV1A* is evolved from the same *UEV* ancestor as *AtUEV1A*, *AtUEV1B* and *BdUEV1A*, while the other three *OsUEV1*s are closely related to *AtUEV1C*, *AtUEV1D*, *BdUEV1B* and *BdUEV1C*, suggesting that they were duplicated and further evolved within each species. Of particular interest is that *OsUEV1*s are more closely related to their respective *BdUEV1* partners than *AtUEV1*s, consistent with a notion of parallel evolution within monocotyledon and dicotyledon plants.

Based on the ORF sequences of all four *UEV1* genes from rice, *OsUEV1B*, *OsUEV1C* and *OsUEV1D* are predicted to encode proteins with 146, 148 and 147 amino acids, respectively, whereas the predicted OsUev1A protein contains 161 amino acids with a C-terminal extension, which was also found in AtUev1A, AtUev1B and BdUev1A (Fig. 1b). It was noted that all plant Uev1s with the C-terminal extension contain a conserved CaaX motif predicted to be a target of prenylation, a protein lipid modification that facilitates the protein-protein or protein-membrane interaction by attaching the isoprenoid groups (a 15-carbon farnesyl or 20-carbon geranlygeranyl) to the Cys residue (blue asterisk) [33, 34]. In addition, several critical residues implicated in Uev activity are also conserved among all Uev1s, including hMms2-F13 (red asterisk) known to be required for the physical interaction with Ubc13 [29], and hMms2-S32 and I62 (green asterisks) required for non-covalent interaction with Ub and poly-Ub chain assembling [35, 36] (Fig. 1b).

OsUev1s physically interact with OsUbc13 to form a stable heterodimer

Lys63-linked polyubiquitination is thought to regulate target proteins in a non-proteolytic manner, and Ubc13 is the only known E2 dedicated to mediating Lys63-linked polyubiquitin chain assembly. However, the prerequisite of this activity is that Ubc13 must be associated with a Uev to form a stable heterodimer [30, 31]. To test whether the four predicted rice Uev1s function in a similar manner, we first assessed their ability to interact with OsUbc13 by a yeast two-hybrid assay. Indeed all four OsUev1s are able to interact with OsUbc13, as none of the negative controls are able to grow under same experimental conditions (Fig. 2a). However, the strength of association appears to be different among OsUev1s; in the high-stringent -Ade medium, OsUev1A grows better than other three OsUev1s (Fig. 2a).

To further confirm the direct interaction between OsUbc13 and OsUev1s, we performed an in vitro glutathione S-transferase (GST) pull-down assay with recombinant proteins purified from *Escherichia coli*. Indeed, all GST-tagged recombinant OsUev1s are able to pull-down His$_6$-tagged OsUbc13, while as a control, GST protein alone fails

a

```
                                          48 ┌─ AtUEV1C
                                    53 ┌────┤
                                       │    └─ AtUEV1D
                              78 ┌─────┤    ┌─ BdUEV1C
                           ┌────┤      └─76─┤
                           │    │           └─ OsUEV1C
                           │    └────────────── OsUEV1D
                           │          ┌──────── BdUEV1B
                           │     ┌─98─┤
                           │     │    └──────── OsUEV1B
                           │     │         ┌─── AtUEV1A
                           │  99 ┤     ┌99─┤
                           └─────┤     │   └─── AtUEV1B
                                 │ 99 ─┤
                                 └─────┤   ┌─── BdUEV1A
                                       └100┤
                                           └─── OsUEV1A
```

b

```
              10        20        30        40        50        60
AtUEV1A  --MSSEE--ARVVVPRNFRLLEELERGEKGIGDGTVSYGMDD-ADDIYMQSWTGTILGPP  55
AtUEV1B  --MGSEE--ERVVVPRNFRLLEELERGEKGIGDGTVSYGMDD-ADDIMQSWTGTILGPH  55
AtUEV1C  MTLGSG---SSVVVPRNFRLLEELERGEKGIGDGTVSYGMDD-GDDIYMRSWTGTIIGPH  56
AtUEV1D  MTLGSGG--SSVVVPRNFRLLEELERGEKGIGDGTVSYGMDD-GDDIYMRSWTGTIIGPH  57
BdUEV1A  --MGSEGSAG-VVVPRNFRLLEELERGEKGIGDGTVSYGMDD-ADDIYMRSWTGTIIGPP  56
BdUEV1B  --MASSGDAAGVVVPRNFRLLEELERGEKGIGDGTVSYGMDD-ADDIYMRSWTGTIIGPH  57
BdUEV1C  MTLGSSGAGSSVVVPRNFRLLEELERGEKGIGDGTVSYGMDD-ADDIYMRSWTGTIIGPH  59
OsUEV1A  --MGSEGSAGVVVVPRNFRLLEELERGEKGIGDGTVSYGMDD-ADDIYMRSWTGTIIGPP  57
OsUEV1B  --MASSGDPSAVVVPRNFRLLEELERGEKGIGDGSVSYGMDD-ADDIYMRSWTGTIIGPH  57
OsUEV1C  MTLGSSGAGSSVVVPRNFRLLEELERGEKGIGDGTVSYGMDD-ADDIYMRSWTGTIIGPH  59
OsUEV1D  MTIGGAS--SSVVVPRNFRLLEELERGEKGIGDGTVSYGMDDDGDDIFMRSWTGTIIGPL  58
              *                              *

              70        80        90        100       110       120
AtUEV1A  N-TAYEGKILQLKLFCGKEYPESPPTVRFQTRINMACVNPETGVVEPSLFPMITNVRREY  114
AtUEV1B  N-TAYEGKILQLKLFCGKDYPESPPTVRFQSRINMACVNPENGVVDPSHFPMISNVRREF  114
AtUEV1C  NVTVHEGRIYQLKLFCDKDYPEKPPTVRFHSRINMTCVNHDTGVVDSKFPGVLANWQRQY  116
AtUEV1D  NVTVHEGRIYQLKLFCDKDYPEKPPTVRFHSRVNMACVNHETGVVDPKKFPGLIANWQREY  117
BdUEV1A  N-TVHEGRIYQLKLFCDTDYPDKPPTVRFQARVNMTCVNQETGMVDPSRFPMIGNVRREH  115
BdUEV1B  N-TVHEGRIYQLKLFCDKDYPDRPPTVRFHSRINMTCVNPETGLVDQRKFPGLISNWRREY  116
BdUEV1C  N-TVHEGRIYQLKLFCDKDYPEKAPSVRFHSRINMTCVNHETGAVDPKKFPGVLANWQRDY  118
OsUEV1A  N-TVHEGRIYQLKLFCDTDYPDRPPTVRFQTRINMSCVNQETGMVEPSLFPMIGNVRREY  116
OsUEV1B  N-TVHEGRIYQLKLFCDKDYPDRPPTVRFHSRINMTCVNPENGLVDQRKFSLISNWRREY  116
OsUEV1C  N-SVHEGRIYQLKLFCDKDYPEKPPSVRFHSRINMSCVNHETGAVDSRKFPGVIANWQREY  118
OsUEV1D  N-SVHEGRIYQLKLFCDKDYPDKPPTVRFHSRINMFCVNPDTGLVESKRFHMIANWQREY  117
            *

              130       140       150       160
AtUEV1A  TMEDILVRLKKEMMTSHNRKLAQPPEGNEEARADPKG-PAKCCVM-  158
AtUEV1B  TMEDILLIQLKKEMMSSQNRKLAQPLEGNEEGRTDPKGLVVKCCVM-  159
AtUEV1C  TMEDILTQLKKEMAASHNRKLVQPPEG---------------TFF-  146
AtUEV1D  TMEDILVQLKKEMSTSHNRKLVQPPEG---------------TCF-  147
BdUEV1A  TMEDILTSLKKEMSAPQNRRLHQPHEGNDDQRVEQKGLAIRCVIM-  160
BdUEV1B  TMENILTQLKKEMAASHNRKLVQPPEG---------------TFYG  147
BdUEV1C  TMEHVISQLKRDMSAQQNRKLVQPPEG---------------TFF-  148
OsUEV1A  TMQDILIGLKKEMSAPQNRRLHQPHDGNEDQRVEQKGLSLRCVIM-  161
OsUEV1B  TMEAILTQLKKEMAASHNRKLVQPPEG---------------TFF-  146
OsUEV1C  TMETILTQLKKEMATPQNRKLVQPPEG---------------TFF-  148
OsUEV1D  TMENILTQLKKEMAAPHSRKLVQPPEG---------------TFF-  147
                                      *
```

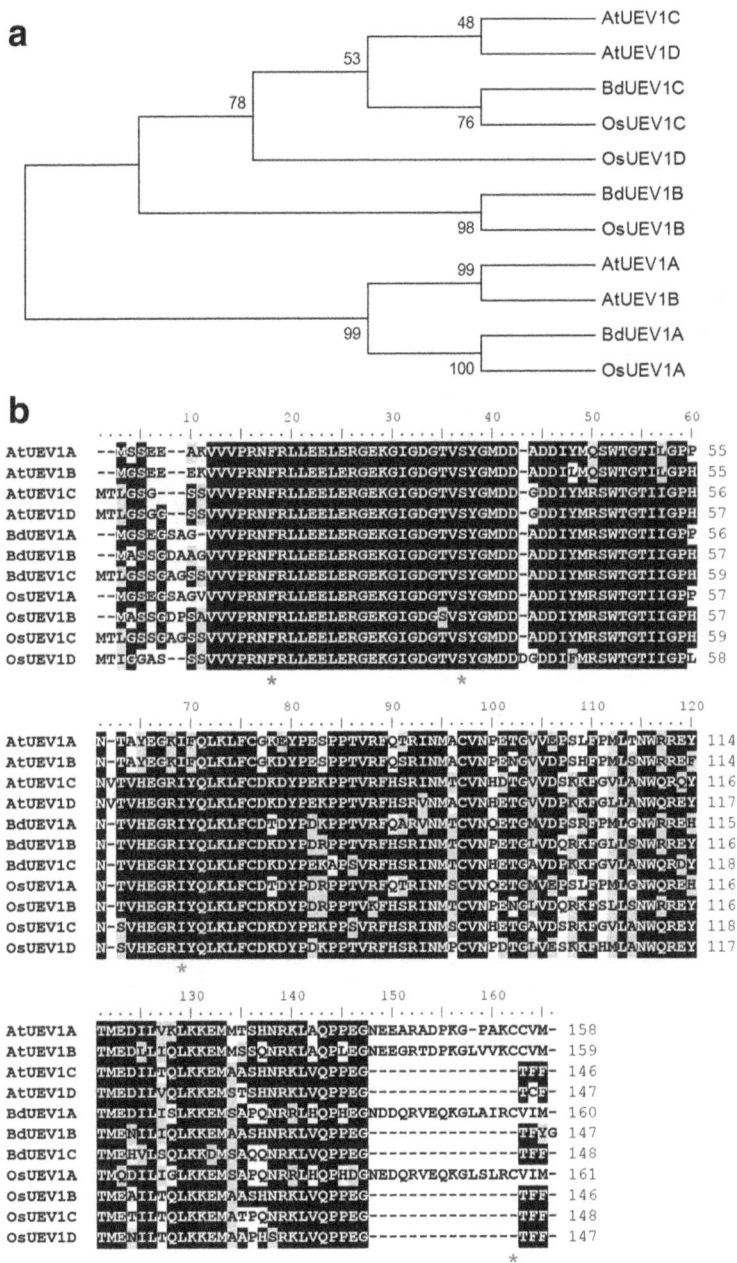

Fig. 1 *OsUEV1* phylogenetic and sequence analyses. **a** Phylogenetic analysis among OsUev1s, AtUev1s and BdUev1s. The tree is built by MEGA6 software. **b** Protein sequence alignment of OsUev1s, AtUev1s and BdUev1s. The sequences were processed using the BioEdit program v7.0.9. Identical residues are highlighted in black while conserved residues are in grey. Asterisks indicate known functional residues defined in yeast and human Uevs

to do so (Fig. 2b). From the above observations, we conclude that all four OsUev1s are able to interact with OsUbc13 directly and form stable heterodimers.

OsUev1s are required for Lys63-linked polyubiquitin chain assembly in vitro by OsUbc13

We previously reported that OsUbc13 is a functional E2 capable of assembling Lys63-linked Ub chains along with yeast Mms2 or human Uev1A in vitro [28]. In *Arabidopsis*, both Ubc13A and Ubc13B are also able to promote Lys63-linked poly-Ub chain formation in the presence of Uev1s [26]. To test whether these four OsUev1s are biochemically active, we performed an in vitro ubiquitination assay with recombinant OsUbc13 and OsUev1s. As shown in Fig. 3, OsUbc13 or any one of the OsUev1s alone cannot trigger the Ub chain formation in the presence of E1 and ATP in the reaction buffer, but OsUbc13 with any one of the OsUev1s can generate free poly-Ub chains. Furthermore, to determine which kind of Ub chain linkage the E2 complexes

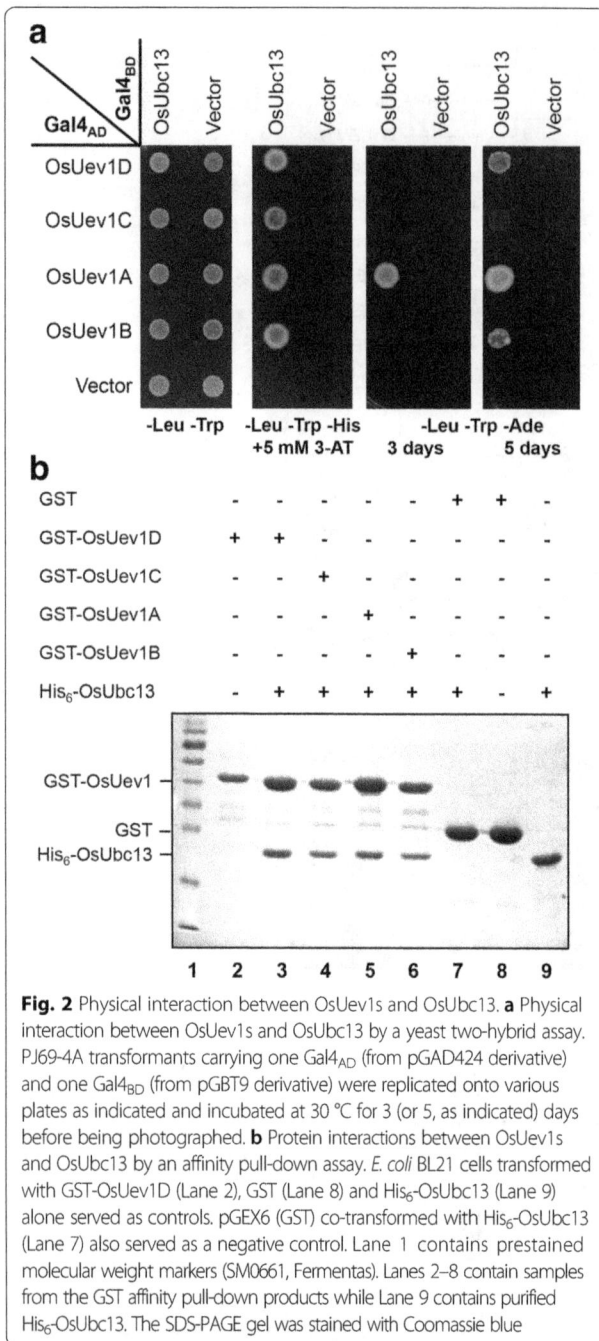

Fig. 2 Physical interaction between OsUev1s and OsUbc13. **a** Physical interaction between OsUev1s and OsUbc13 by a yeast two-hybrid assay. PJ69-4A transformants carrying one Gal4$_{AD}$ (from pGAD424 derivative) and one Gal4$_{BD}$ (from pGBT9 derivative) were replicated onto various plates as indicated and incubated at 30 °C for 3 (or 5, as indicated) days before being photographed. **b** Protein interactions between OsUev1s and OsUbc13 by an affinity pull-down assay. *E. coli* BL21 cells transformed with GST-OsUev1D (Lane 2), GST (Lane 8) and His$_6$-OsUbc13 (Lane 9) alone served as controls. pGEX6 (GST) co-transformed with His$_6$-OsUbc13 (Lane 7) also served as a negative control. Lane 1 contains prestained molecular weight markers (SM0661, Fermentas). Lanes 2–8 contain samples from the GST affinity pull-down products while Lane 9 contains purified His$_6$-OsUbc13. The SDS-PAGE gel was stained with Coomassie blue

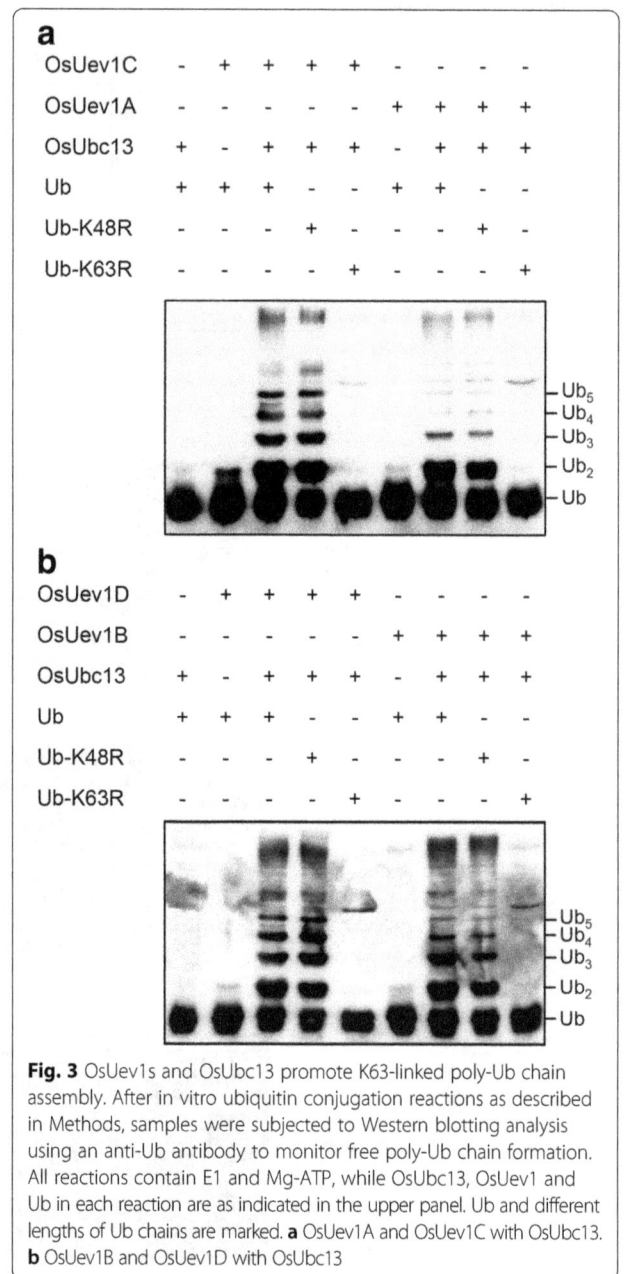

Fig. 3 OsUev1s and OsUbc13 promote K63-linked poly-Ub chain assembly. After in vitro ubiquitin conjugation reactions as described in Methods, samples were subjected to Western blotting analysis using an anti-Ub antibody to monitor free poly-Ub chain formation. All reactions contain E1 and Mg-ATP, while OsUbc13, OsUev1 and Ub in each reaction are as indicated in the upper panel. Ub and different lengths of Ub chains are marked. **a** OsUev1A and OsUev1C with OsUbc13. **b** OsUev1B and OsUev1D with OsUbc13

assemble, we utilized site-specific Ub-Lys mutations. As shown in Fig. 3, each OsUbc13-OsUev1 complex is still able to mediate poly-Ub chain assembly with Ub-K48R but not with Ub-K63R, confirming that the Ub chains are linked through Lys63. It is noted that among the four OsUev1s, OsUev1A primarily promotes the di-Ub chain formation and its ability to promote poly-Ub chains is relatively weaker than other three OsUev1s (Fig. 3a, b).

Functional complementation of yeast *mms2* by *OsUEV1* genes

To test whether *OsUEV1*s are functionally conserved between different species, we performed a DNA-damage sensitivity assay to determine whether *OsUEV1*s could functionally complement the error-free DDT defect in a yeast *mms2* null mutant. As shown in Fig. 4a, expression of any one of the *OsUEV1* genes is capable of rescuing the *mms2* null mutant from death caused by treatment with methyl methanesulfonate (MMS) to a level comparable with wild-type cells, whereas *mms2* null mutant cells carrying the empty vector are not rescued, indicating that OsUev1s are functionally conserved with yeast Mms2 and likely able to form a heterodimer with yeast Ubc13.

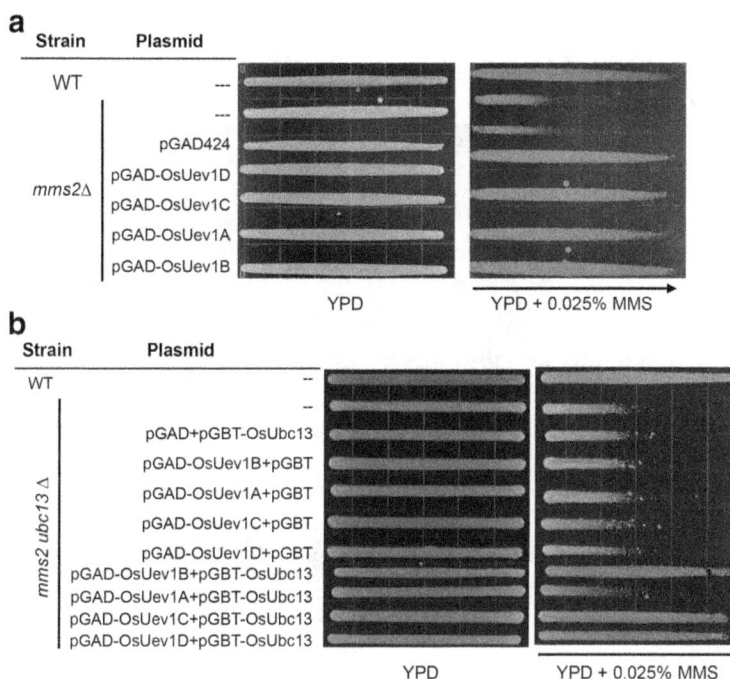

Fig. 4 Functional complementation of yeast single and double mutants. **a** Functional complementation of the *mms2* single mutant by *OsUEV1* genes. WXY902 (*mms2Δ*) transformants were incubated overnight and printed onto YPD and YPD + MMS gradient plates. The plates were incubated at 30 °C for 2 days before being photographed. Only one selected MMS-containing plate is shown. The arrow indicates increased MMS concentration. **b** Functional complementation of the *mms2Δ ubc13Δ* double mutant (WXY955) by *OsUBC13* along with *OsUEV1s*. Experimental conditions were as described in **a**

Since *OsUBC13* is also able to complement the yeast *ubc13* null mutant [28], we next asked whether the OsUbc13-OsUev1 complexes are able to functionally complement the yeast *mms2 ubc13* double mutant. The yeast *mms2 ubc13* double mutant cells were co-transformed with two yeast plasmids expressing *OsUBC13* and *OsUEV1*, or a corresponding empty vector. As expected, neither *OsUBC13* nor *OsUEV1s* with corresponding empty vectors is able to rescue the yeast *mms2 ubc13* double mutant. Surprisingly, while the combination of *OsUBC13* with *OsUEV1B*, *OsUEV1C* or *OsUEV1D* restored the yeast *mms2 ubc13* double mutant sensitivity to MMS to the wild-type level, the combination of *OsUBC13* and *OsUEV1A* did not provide *mms2 ubc13* mutant cells with MMS resistance (Fig. 4b).

Roles of the OsUev1A C-terminal domain and its putative prenylation site

To ask whether the C-terminal extension in general or the prenylation motif in particular is responsible for the above observed distinct phenotypes of OsUev1A over other OsUev1s, we made two OsUev1A-derived constructs, namely OsUev1A-ΔCT that removes the C-terminal 18 amino-acid tail and OsUev1A-C158S in which the conserved C158S amino acid substitution prevents potential prenylation [33, 34]. In a yeast two-hybrid assay (Fig. 5a), both OsUev1A mutant derivatives reduced the interaction

capacity with OsUbc13 in comparison to OsUev1A, since they were unable to grow in the high-stringent -Ade plate [37]. Meanwhile, together with *OsUBC13*, the two *OsUEV1A* mutant derivatives restored the MMS resistance in the yeast *mms2 ubc13* mutant strain (Fig. 5b), reminiscent of *OsUEV1B*, *OSUEV1C* and *OsUEV1D*. From the above observations, we conclude that the C-terminal tail of OsUev1A and its putative prenylation is responsible for its unique phenotypes in yeast cells.

The C-terminal tail and putative prenylation of OsUev1A determines its subcellular localization and membrane association

To further understand cellular functions of OsUev1s, we monitored the subcellular localization of selected OsUev1s in *Nicotiana benthamiana* leaves. In this experiment, GFP-fused OsUev1D is found in both cytoplasm and nucleus (Fig. 6a, top row). In contrast, GFP-OsUev1A is clearly excluded from the nucleus (Fig. 6a, 2nd row). As all four OsUev1s are highly conserved in their core region except that OsUev1A contains an additional C-terminal tail, we asked whether the unique localization pattern of OsUev1A is caused by its C-terminus. As shown in the third row of Fig. 6a, after removal of its C-terminal tail sequence, the subcellular localization pattern of GFP-OsUev1A-ΔCT appears to be different from that of

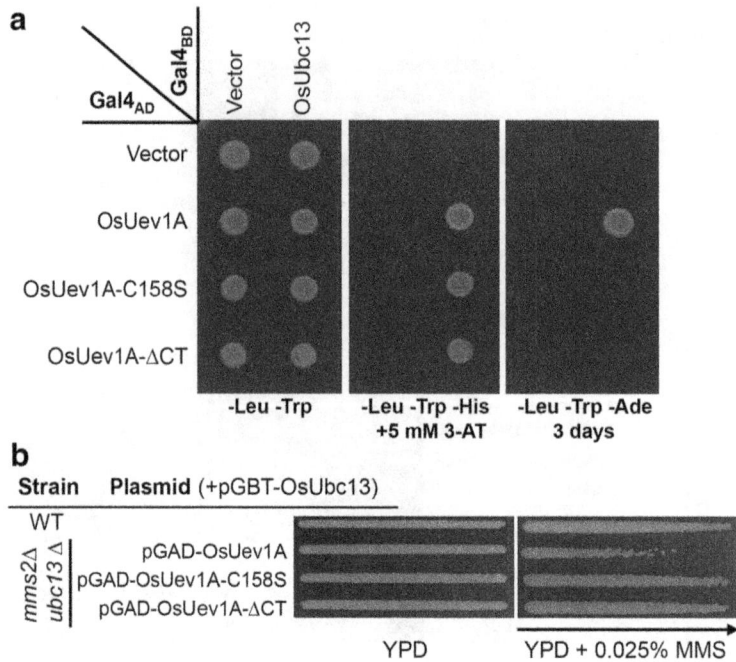

Fig. 5 Effects of OsUev1A C-terminal domain and the putative prenylation site on the Ubc13-Uev1 complex formation and functional complementation in yeast. **a** Yeast two-hybrid analysis of physical interaction between OsUbc13 and OsUev1A or its derivatives. Experimental conditions were as described in Fig. 2a. **b** Functional complementation of the yeast *mms2Δ ubc13Δ* mutant by OsUev1A and its derivatives in the presence of pGBT-OsUbc13. Experimental conditions were as described in Fig. 4

OsUev1A and comparable to that of OsUev1D, particularly for the nuclear localization. Furthermore, GFP-OsUev1A-C158S behaves like GFP-OsUev1A-ΔCT and differs from OsUev1A (Fig. 6a, 4th row). These observations indicate that the C-terminal tail and most likely the prenylation of OsUev1A is responsible for its subcellular distribution.

Since protein prenylation has been reported to facilitate protein-protein interaction and/or protein-membrane interaction [33, 34], we asked whether GFP-tagged OsUev1 variants are indeed associated with the membrane. Tobacco leaves transformed with GFP-tagged OsUev1 variants were subject to a Triton X-114 based protein-partitioning assay. As shown in Fig. 6b, GFP-tagged OsUev1A is almost exclusively found in the detergent (D) phase, whereas C-terminally truncated GFP-OsUev1A is partially diffused to the aqueous (A) phase. The partial dissociation of Os-Uev1A-ΔCT from membrane fraction is because either it still contains another membrane association motif, or the ectopically expressed GFP-tagged OsUev1 level is higher than the native Uev1A. Nevertheless, these results collectively indicate that OsUev1A preferentially associates with membrane and that this association is dependent on its C-terminal sequence and probably on its prenylation, whereas other three OsUev1s are soluble proteins spread in both cytoplasm and the nucleus.

Expression of *OsUEV1*s in different tissues and during different developmental stages

Since we have previously shown that the expression of *UBC13* both in *Arabidopsis* [25] and rice [28] remains constitutive in different tissues and even under stresses, we speculated that *OsUEV1*s, like their *Arabidopsis* counterparts [26], may be regulated at the transcriptional level to modulate the Ubc13-Uev complex activity. We searched microarray databases online by utilizing Genevestigator. The retrieved data as shown in Fig. 7a indicate that the four *OsUEV1*s display various expression patterns in different tissues. Both *OsUEV1B* and *OsUev1C* maintain a constant and relatively high-level expression in different tissues, whereas the expression of *OsUEV1A* and *OsUEV1D* fluctuates rather dramatically. For example, OsUEV1D is expressed at very high level in various parts of the leaf, but its expression is extremely low in pollen and sperm cells. During rice development, *OsUEV1B* and *OsUev1C* still maintain a stable and high-level expression, while the expression of *OsUEV1A* is also stable but the transcript level is relatively low. In contrast, the expression of *OsUEV1D* fluctuates dramatically during development (Fig. 7b).

We also analyzed *OsUEV1* expression patterns in response to various environmental stresses as reported from the database (Fig. 8), in which the expression of *OsUEV1B* and *OsUEV1C* is remarkably constant. In contrast, *OsUEV1D* is highly sensitive to essentially all

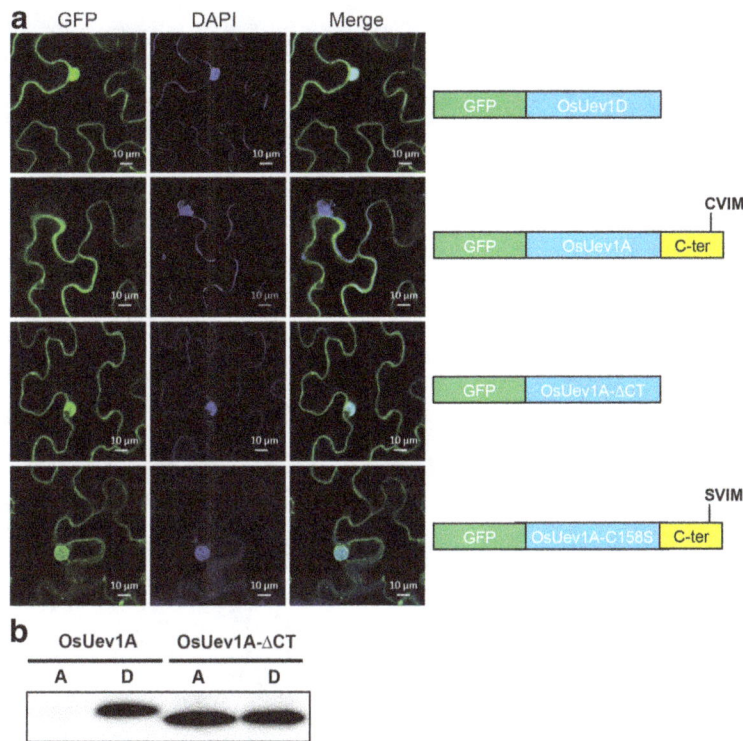

Fig. 6 Subcellular distribution of selected OsUev1s. **a** Subcellular localization of GFP-OsUev1s and their derivatives as illustrated in the right panel. The GFP-tagged OsUev1s were expressed in *N. benthamiana* leaves by an *Agrobacterium*-mediated infiltration method. Photos were taken after 2–3 days of infiltration. A representative image is shown for each transformant. **b** The above transformed *N. benthamiana* leaves were analyzed by a protein-partitioning assay as described in Methods. GFP-OsUev1A and GFP-OsUev1A-ΔCT were detected by an anti-GFP antibody (B-2, sc-9996, Santa Cruz). A: aqueous phase; D: detergent phase

perturbations examined. For example, its expression appears to be repressed under drought conditions and highly induced during anaerobic seed germination. Interestingly, *OsUEV1D* is induced when seeds are shifted from aerobic to anaerobic conditions for germination, while its expression is repressed when seeds are shifted from anaerobic to aerobic germination conditions. The expression of *OsUEV1A* is also perturbed in response to various biotic and abiotic stresses to moderate extents, most notably during anaerobic seed germination.

Discussion

In this study, we identified and cloned four highly conserved *UEV* genes from the rice genome and our in vitro studies confirm that these Uevs are able to interact with OsUbc13 to form a stable heterodimer and mediate Lys63-linked polyubiquitination. Functional studies indicate that these rice *UEV* genes can restore cellular activity of the yeast *mms2* null mutant for resistance to a DNA-damaging agent, reminiscent of the ability of *OsUBC13* to restore the corresponding yeast *ubc13* mutant [28]. Furthermore, several observations are consistent with the notion that the four *OsUEV1*s confer different functions in vivo. Firstly, when both yeast *MMS2* and

UBC13 genes are replaced by different combinations of *OsUBC13* and *OsUEV1*s, three of them can fully restore the DNA-damage tolerance activity, while *OsUBC13-OsUEV1A* cannot. Secondly, in a yeast two-hybrid assay the OsUbc13-OsUev1A interaction appears to be stronger than the other three pairs, ruling out a possibility that the lack of functional complementation by *OsUBC13-OsUEV1A* is due to reduced physical interaction. Finally, the subcellular localization of OsUev1A differs from that of OsUev1D (and presumably OsUev1B and OsUev1C) in plants. While OsUev1D behaves like a small soluble protein and appears to be enriched in the nucleus, OsUev1A is excluded from the nucleus and appears to be membrane-bound. The above observations collectively indicate that OsUev1s confer function(s) in addition to DDT and that different OsUev1s may have distinct physiological functions. This conclusion is not unexpected as in yeast, the regulation of the DDT pathway is the only known function of Ubc13-Mms2; however, the two distinct Ubc13-Uev complexes turn out to be multifunctional in multi-cellular organisms like mammals [38]. Hence, it is reasonable to speculate that Uevs are also multi-functional in plants. Indeed, plant Ubc13 has been implicated to function in apical dominance [39], iron

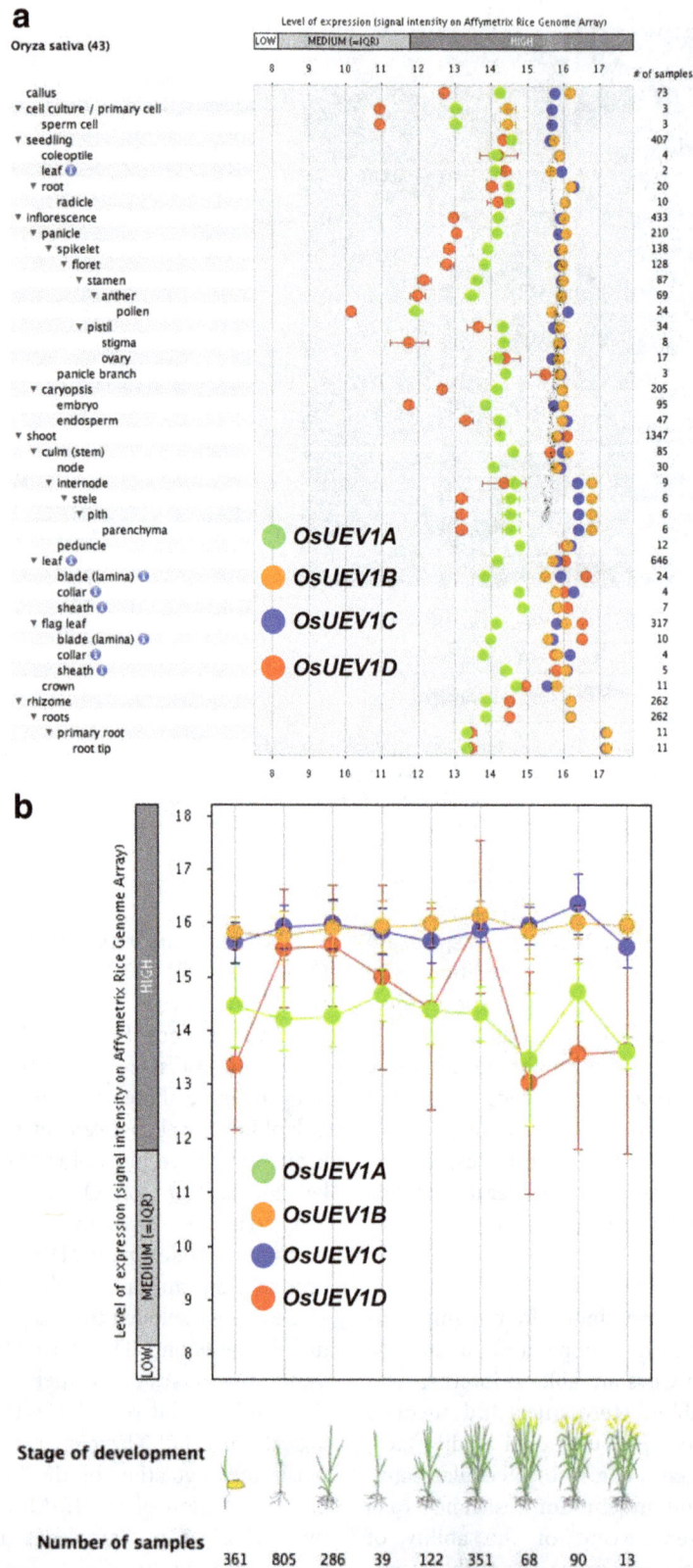

Fig. 7 Quantitative analysis of *OsUEV1* expression. **a** Expression of *OsUEV1A*, *OsUEV1B*, *OsUEV1C* and *OsUEV1D* in different tissues. Samples were taken from different tissues as indicated and relative transcript levels of the entire transcriptome were determined by microarray analysis. **b** Expression of the four *OsUEV1* genes during different life stages. The above data were retrieved from Genevestigator (www.genevestigator.com)

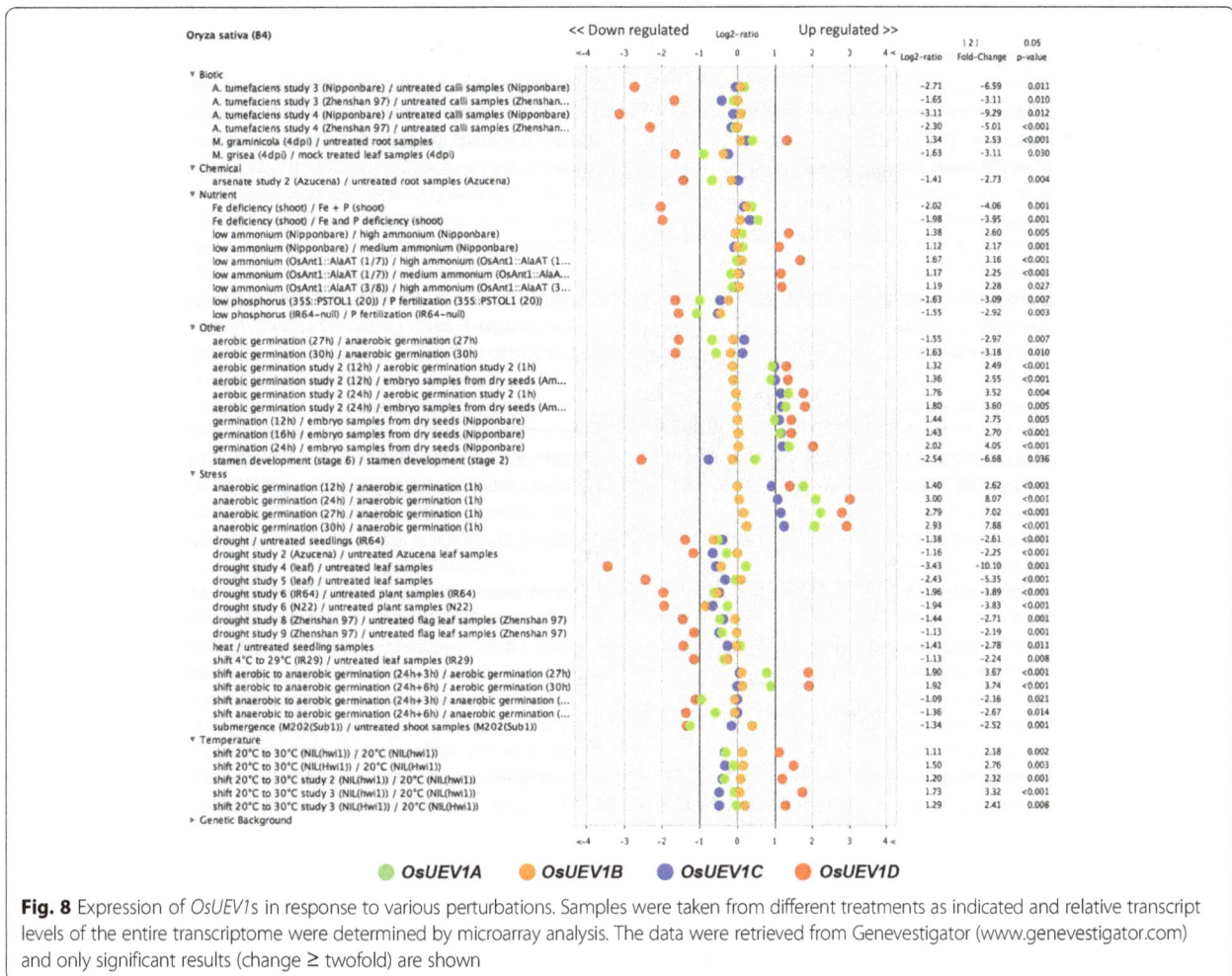

Fig. 8 Expression of *OsUEV1s* in response to various perturbations. Samples were taken from different treatments as indicated and relative transcript levels of the entire transcriptome were determined by microarray analysis. The data were retrieved from Genevestigator (www.genevestigator.com) and only significant results (change ≥ twofold) are shown

metabolism [40], innate immunity [41] and auxin signaling [42], and at least some of the above functions may require the Ubc13-Uev E2 complex and Lys63-linked ubiquitination.

Mammalian genomes contain one *UBC13* gene and at least two *UEV* genes, and the *UEV* genes often confer distinct functions. For example, mammalian Ubc13 regulates the DDT pathway by interacting with Mms2 and mediates NF-κB signaling by associating with Uev1A [38]. In this study, we identified four *OsUEV* genes in rice and at least one of them, *OsUEV1A*, functions differently from other *OsUEV* genes. Similar results are also observed in *Arabidopsis*, in which AtUev1A and AtUev1B contain additional C-terminal sequences [26]. A novel finding in this study is that OsUev1A distributes differently in the cellular compartment than other OsUev1s, and that its unique localization and membrane-bound property can be abolished by removal of the C-terminus or simply mutating the predicted prenylation site. Although exactly which cellular role(s) it plays remains unknown, it can be cautiously predicted based on this study that it is membrane-related and non-nuclear. This function must be critical for the plant development and/ or environmental response, as essentially all known plant genomes contain at least one OsUev1A ortholog with predicted CaaX motif at their C-terminus (data not shown). On the other hand, the remaining three OsUev1s may function in DNA-damage response like their *Arabidopsis* Uev1D counterparts [26], consistent with the observed OsUev1D nuclear localization. Given that AtUbc13 has been implicated in several cellular processes and these functions are likely conserved with OsUbc13, it is of great interest to investigate which Ubc13-mediated cellular process involves which OsUev1 and to discover additional cellular processes in which Ubc13-Uev participates.

As a non-canonical ubiquitination, Lys63-linked ubiquitination is most likely involved in stress response signaling, where Ubc13-Uev plays a critical role in assembling Lys63-linked poly-Ub chains on the target protein. Therefore, it is conceivable that its activity is tightly regulated in response to different environmental signals. To date, no report has found altered activity of Ubc13 in plant species examined [25, 28, 43]. Instead, its activity and specificity are largely determined by the cognate

Uev, and the cellular levels of Uev appear to fluctuate in different tissues and in response to various environmental stresses [44–46]. Furthermore, the pathway involvement of Lys63-linked Ub chain is mainly determined by the Uev that interacts with Ubc13 [38]. In this study, four distinct *UEV* genes in rice also display different expression patterns among different tissues, life stages and environmental stresses. In addition to the constitutively expressed *OsUEV1B* and *OsUEV1C* genes, the *OsUEV1A* and *OsUEV1D* expression fluctuates under all the above conditions, suggesting that these two gene products play regulatory roles under different biological processes. Hence, the regulation of Uev activity appears to be evolutionarily preferred and Uevs serve as regulatory subunits of the Ubc13-Uev E2 complex in response to distinct cellular and environmental signals.

Conclusions

In this article, we report the molecular cloning and functional characterization of four rice *UEV1* genes. Like other plant species, rice also contains two classes of *UEV1* genes with their encoded proteins differ in the C-terminal extension. This study reveals that OsUev1A contains a C-terminal tail not found in other three OsUev1s, that the tail sequences are highly conserved within higher plants, from both monocotyledon and dicotyledon, and that a putative posttranslational modification site is also conserved. Our limited experimental results showed that the two classes of *OsUEV1s* genes function differently in a heterologous yeast host and that their protein subcellular distribution patterns are also different in plants. Furthermore, the above differences are attributed to the OsUev1A C-terminal tail and most likely to its putative prenylation. Unlike the *OsUBC13* gene that is constitutively expressed, database analyses reveal that the expression of four *OsUEV1* genes fluctuates dramatically in different tissues, during different developmental stages as well as in response to various biotic and abiotic stresses, suggesting that these *OsUEV1* gene products regulate the Ubc13-Uev1 activity.

Methods

Plant materials and yeast cell culture

Rice (*Oryza sativa* L. cv. Japonica) seeds were surface sterilized with 2% NaClO for 30 min after a pre-wash by sterile distilled water, followed by washing seven times in sterile water. The sterilized rice seeds were plated in Murashige and Skoog (MS) plates containing 2.2 g/l minimal organics, 10 g/l sucrose and 1% agar. They were cultured in a growth chamber (16 h light/8 h dark and 30 °C).

Yeast strains used in this study include PJ69-4A [37] for the yeast two-hybrid assay, HK578-10D (*MATa ade2–1 can1–100 his3–11,15 leu2–3, 112 trp1–1 ura3–1*)

and its *mms2Δ::HIS3* derivative WXY902 and *mms2Δ::HIS3 ubc13Δ::hisG-URA3-hisG* derivative WXY955 for the functional analysis. Yeast cells were grown at 30 °C in either rich YPD or a synthetic dextrose (SD) medium supplemented with nutrients as instructed [47]. To make plates, 2% agar was added to YPD or SD medium prior to autoclaving. Yeast cells were transformed by a LiAc method [48].

Cloning rice *UEV1* cDNAs and plasmid construction

To clone the full-length *OsUEV1* open reading frames (ORFs), total RNA was extracted from rice seedlings with TRIzol reagents (Invitrogen, Carlsbad), which was used as a template for RT-PCR using the RevertAid First Strand cDNA Synthesis Kit (Fermentas). Gene-specific primers are as follows: *OsUEV1A*: 5′-taaccg*gaattc*ATGGGGTCCGAGGGATC-3′ and 5′-ggcacgc*gtcgac*TTACATGATGACACACCTA-3′; *OsUEV1A-C158S*: 5′-gg cacgc*gtcgac*TTACATGATGACAC**T**CCTA-3′; *OsUEV1A-ΔCT*: 5′-ggcacgc*gtcgac*TTAGCCATCATGGGGTTGATG-3′; *OsUEV1B*: 5′-gaacc*ggaattc*ATGGCGTCGAGTGGAGAT-3′ and 5′-gcacgc*gtcgac*CTAGAAGAATGTCCCCTC-3′; *OsUE V1C*: 5′-tgacc*ggaattc*ATGACGCTGGGGAGCTC-3′ and 5′-gcacgc*gtcgac*CTAGAAGAACGTCCCTTC-3′; *OsUEV1D*: 5′-taact*ggaattc*ATGACGATCGGCGGCG-3′ and 5′-tccc*gcgtcgac*CTAGAAGAAGGTCCCTTC-3′. The forward primers contain the *Eco*RI restriction site and the reverse primers contain the *Sal*I site, as italicized. The PCR product of *OsUEV1A-1D* ORFs were cloned into a yeast two-hybrid vector pGAD424Bg, which was derived from pGAD424 [49].

Yeast two-hybrid analysis

The yeast two-hybrid strain PJ69-4A [37] was used for this assay. The co-transformation, selection and two-hybrid detection steps were as previously described [28].

Recombinant protein purification and ubiquitination assay

OsUEV1 ORFs were isolated from pGAD-OsUev1s and cloned into pGEX6p-1. The resulting pGEX-OsUev1s were transformed into *E. coli* BL21 CodonPlus (DE3)-RIL cells. The pGEX-OsUev1 fusion proteins were purified following a previously published protocol [38]. Meanwhile, GST and His$_6$-OsUbc13 were produced and purified as previously described [28]. For an in vitro ubiquitination assay, a previously described protocol [28] was followed.

GST pull-down assay

The *E. coli* BL21 competent cells were transformed with either pGEX6p-1, pGEX-OsUev1s alone, or co-transformed with pET-OsUbc13. The whole-cell extracts were incubated with Glutathione Sepharose 4B Microspin™ beads (17–0756-01, GE Healthcare) at 4 °C for 2 h, which were then harvested by centrifugation, washed 5 times with a lysis buffer and boiled with 2 × loading buffer. The products were analyzed on a 12% SDS-PAGE gel.

Yeast gradient plate assay

Yeast strain HK578-10D and its isogenic *mms2Δ* single or *ubc13Δ mms2Δ* double mutants were either singly transformed with pGAD-OsUev1A-1D or co-transformed with pGAD-OsUev1s and pGBT-OsUbc13. The transformants were selected on SD-Leu (for *mms2Δ*) or SD-Leu-Trp (for *ubc13Δ mms2Δ*) plates. The gradient plate assay was conducted as described [50].

Subcellular localization

The ORFs of *OsUEV1*s and derivatives were amplified and cloned into the pCAMBIA1302 vector containing an N-terminal GFP tag. These *GFP-OsUev1s* constructs were transformed into the *Agrobacterium tumefaciens* (GV3101/pMP90), and positive colonies were cultured overnight and infiltrated into *Nicotiana benthamiana* leaves as described [51]. After 2–3 day incubation, epidermal cells of the transformed tobacco leaf were viewed by confocal microscopy (Zeiss LSM 780, Germany). Excitation parameters are 488 nm and 405 nm for GFP and DAPI, respectively.

Protein partitioning assay

For protein partitioning assay, total protein was extracted from transformed *N. benthamiana* leaves using a buffer containing 50 mM Tris-HCl pH 8.0, 0.3 M NaCl, 1% TritonX-114, 10 mM PMSF, 3 mM DDT and 1 tablet (for 50 ml buffer) protease inhibitors (Roche). The extract was incubated in Triton X-114 containing buffer for 1 h at 4 °C before centrifugation at 12,000 g for 10 min at 4 °C. Samples were then incubated at 37 °C for 5 min and centrifuged at 12,000 g. The aqueous upper phase and detergent-enriched lower phase were separated and extracted once again with detergent and aqueous solutions, respectively. The resulting four samples were adjusted to equal volume and proteins were precipitated with chloroform/methanol prior to Western blot analysis.

Acknowledgments
The authors wish to thank Guang Yang for technical assistance and Michelle Hanna for proofreading the manuscript.

Funding
This work is supported by the National Natural Science Foundation of China (31270823) and Beijing Municipal Commission of Education to WX.

Authors' contributions
QW, YZ and WX conceived and designed experiments; QW, YZ and XZ performed experiments. QW, YZ and WX analyzed data and wrote the article. All authors read and approved the final manuscript.

Competing interests
The authors declare that they have no competing interests.

References

1. Glickman MH, Ciechanover A. The ubiquitin-proteasome proteolytic pathway: destruction for the sake of construction. Physiol Rev. 2002;82(2):373–428.
2. Hershko A, Ciechanover A. The ubiquitin system. Annu Rev Biochem. 1998; 67:425–79.
3. Komander D, Rape M. The ubiquitin code. Annu Rev Biochem. 2012;81:203–29.
4. Jackson SP, Durocher D. Regulation of DNA damage responses by ubiquitin and SUMO. Mol Cell. 2013;49(5):795–807.
5. Chen ZJ, Sun LJ. Nonproteolytic functions of ubiquitin in cell signaling. Mol Cell. 2009;33(3):275–86.
6. Komander D. The emerging complexity of protein ubiquitination. Biochem Soc Trans. 2009;37(Pt 5):937–53.
7. Wickliffe KE, Williamson A, Meyer HJ, Kelly A, Rape M. K11-linked ubiquitin chains as novel regulators of cell division. Trends Cell Biol. 2011;21(11):656–63.
8. Jentsch S, McGrath JP, Varshavsky A. The yeast DNA repair gene RAD6 encodes a ubiquitin-conjugating enzyme. Nature. 1987;329(6135):131–4.
9. Hoege C, Pfander B, Moldovan GL, Pyrowolakis G, Jentsch S. RAD6-dependent DNA repair is linked to modification of PCNA by ubiquitin and SUMO. Nature. 2002;419(6903):135–41.
10. Stelter P, Ulrich HD. Control of spontaneous and damage-induced mutagenesis by SUMO and ubiquitin conjugation. Nature. 2003;425(6954):188–91.
11. Broomfield S, Chow BL, Xiao W. MMS2, encoding a ubiquitin-conjugating-enzyme-like protein, is a member of the yeast error-free postreplication repair pathway. Proc Natl Acad Sci U S A. 1998;95(10):5678–83.
12. Brusky J, Zhu Y, Xiao W. UBC13, a DNA-damage-inducible gene, is a member of the error-free postreplication repair pathway in *Saccharomyces cerevisiae*. Curr Genet. 2000;37(3):168–74.
13. Xiao W, Chow BL, Broomfield S, Hanna M. The *Saccharomyces cerevisiae* RAD6 group is composed of an error-prone and two error-free postreplication repair pathways. Genetics. 2000;155(4):1633–41.
14. Ulrich HD, Jentsch S. Two RING finger proteins mediate cooperation between ubiquitin-conjugating enzymes in DNA repair. EMBO J. 2000;19(13):3388–97.
15. Xu X, Blackwell S, Lin A, Li F, Qin Z, Xiao W. Error-free DNA-damage tolerance in *Saccharomyces cerevisiae*. Mutat Res Rev Mutat Res. 2015;764:43–50.
16. Ball LG, Zhang K, Cobb JA, Boone C, Xiao W. The yeast Shu complex couples error-free post-replication repair to homologous recombination. Mol Microbiol. 2009;73(1):89–102.
17. Andersen PL, Xu F, Xiao W. Eukaryotic DNA damage tolerance and translesion synthesis through covalent modifications of PCNA. Cell Res. 2008;18(1):162–73.
18. Pastushok L, Xiao W. DNA postreplication repair modulated by ubiquitination and sumoylation. Adv Protein Chem. 2004;69:279–306.
19. Curtis MJ, Hays JB. Tolerance of dividing cells to replication stress in UVB-irradiated Arabidopsis roots: requirements for DNA translesion polymerases eta and zeta. DNA Repair (Amst). 2007;6(9):1341–58.
20. Anderson HJ, Vonarx EJ, Pastushok L, Nakagawa M, Katafuchi A, Gruz P, Di Rubbo A, Grice DM, Osmond MJ, Sakamoto AN, et al. *Arabidopsis thaliana* Y-family DNA polymerase eta catalyses translesion synthesis and interacts functionally with PCNA2. Plant J. 2008;55(6):895–908.
21. Garcia-Ortiz MV, Ariza RR, Hoffman PD, Hays JB, Roldan-Arjona T. *Arabidopsis thaliana* AtPOLK encodes a DinB-like DNA polymerase that extends mispaired primer termini and is highly expressed in a variety of tissues. Plant J. 2004;39(1):84–97.
22. Sakamoto A, Lan VT, Hase Y, Shikazono N, Matsunaga T, Tanaka A. Disruption of the AtREV3 gene causes hypersensitivity to ultraviolet B light and gamma-rays in Arabidopsis: implication of the presence of a translesion synthesis mechanism in plants. Plant Cell. 2003;15(9):2042–57.
23. Takahashi S, Sakamoto A, Sato S, Kato T, Tabata S, Tanaka A. Roles of Arabidopsis AtREV1 and AtREV7 in translesion synthesis. Plant Physiol. 2005;138(2):870–81.
24. Kunz BA, Xiao W. DNA damage tolerance in plants via translesion synthesis. Genes, Genomes & Genomics. 2007;1(1):89–99.

25. Wen R, Newton L, Li G, Wang H, Xiao W. *Arabidopsis thaliana* UBC13: implication of error-free DNA damage tolerance and Lys63-linked polyubiquitylation in plants. Plant Mol Biol. 2006;61(1–2):241–53.

26. Wen R, Torres-Acosta JA, Pastushok L, Lai XQ, Pelzer L, Wang H, Xiao W. Arabidopsis UEV1D promotes lysine-63-linked polyubiquitination and is involved in DNA damage response. Plant Cell. 2008;20(1):213–27.

27. Wang S, Wen R, Shi X, Lambrecht A, Wang H, Xiao W. RAD5a and REV3 function in two alternative pathways of DNA-damage tolerance in Arabidopsis. DNA Repair. 2011;10(6):620–8.

28. Zang YP, Wang Q, Xue CY, Li MN, Wen R, Xiao W. Rice UBC13, a candidate housekeeping gene, is required for K63-linked polyubiquitination and tolerance to DNA damage. Rice. 2012;5:24.

29. Pastushok L, Moraes TF, Ellison MJ, Xiao W. A single Mms2 "key" residue insertion into a Ubc13 pocket determines the interface specificity of a human Lys63 ubiquitin conjugation complex. J Biol Chem. 2005;280(18):17891–900.

30. McKenna S, Spyracopoulos L, Moraes T, Pastushok L, Ptak C, Xiao W, Ellison MJ. Noncovalent interaction between ubiquitin and the human DNA repair protein Mms2 is required for Ubc13-mediated polyubiquitination. J Biol Chem. 2001;276(43):40120–6.

31. Hofmann RM, Pickart CM. Noncanonical MMS2-encoded ubiquitin-conjugating enzyme functions in assembly of novel polyubiquitin chains for DNA repair. Cell. 1999;96(5):645–53.

32. Guo H, Wen R, Wang Q, Datla R, Xiao W. Three *Brachypodium distachyon* Uev1s promote Ubc13-mediated Lys63-linked polyubiquitination and confer different functions. Front Plant Sci. 2016;7:1551.

33. Yalovsky S, Rodr Guez-Concepcion M, Gruissem W. Lipid modifications of proteins - slipping in and out of membranes. Trends Plant Sci. 1999;4(11):439–45.

34. Wang M, Casey PJ. Protein prenylation: unique fats make their mark on biology. Nat Rev Mol Cell Bio. 2016;17(2):110–22.

35. Pastushok L, Spyracopoulos L, Xiao W. Two Mms2 residues cooperatively interact with ubiquitin and are critical for Lys63 polyubiquitination in vitro and in vivo. FEBS Lett. 2007;581(28):5343–8.

36. Lewis MJ, Saltibus LF, Hau DD, Xiao W, Spyracopoulos L. Structural basis for non-covalent interaction between ubiquitin and the ubiquitin conjugating enzyme variant human MMS2. J Biomol NMR. 2006;34(2):89–100.

37. James P, Halladay J, Craig EA. Genomic libraries and a host strain designed for highly efficient two-hybrid selection in yeast. Genetics. 1996;144(4):1425–36.

38. Andersen PL, Zhou HL, Pastushok L, Moraes T, McKenna S, Ziola B, Ellison MJ, Dixit VM, Xiao W. Distinct regulation of Ubc13 functions by the two ubiquitin-conjugating enzyme variants Mms2 and Uev1A. J Cell Biol. 2005;170(5):745–55.

39. Yin XJ, Volk S, Ljung K, Mehlmer N, Dolezal K, Ditengou F, Hanano S, Davis SJ, Schmelzer E, Sandberg G, et al. Ubiquitin lysine 63 chain forming ligases regulate apical dominance in Arabidopsis. Plant Cell. 2007;19(6):1898–911.

40. Li W, Schmidt W. A lysine-63-linked ubiquitin chain-forming conjugase, UBC13, promotes the developmental responses to iron deficiency in Arabidopsis roots. Plant J. 2010;62(2):330–43.

41. Mural RV, Liu Y, Rosebrock TR, Brady JJ, Hamera S, Connor RA, Martin GB, Zeng L. The tomato Fni3 lysine-63-specific ubiquitin-conjugating enzyme and suv ubiquitin E2 variant positively regulate plant immunity. Plant Cell. 2013;25(9):3615–31.

42. Wen R, Wang S, Xiang D, Venglat P, Shi X, Zang Y, Datla R, Xiao W, Wang H. UBC13, an E2 enzyme for Lys63-linked ubiquitination, functions in root development by affecting auxin signaling and aux/IAA protein stability. Plant J. 2014;80(3):424–36.

43. Guo H, Wen R, Liu Z, Datla R, Xiao W. Molecular cloning and functional characterization of two *Brachypodium distachyon* UBC13 genes whose products promote K63-linked polyubiquitination. Front Plant Sci. 2016;6:1222.

44. Fritsche J, Rehli M, Krause SW, Andreesen R, Kreutz M. Molecular cloning of a 1alpha,25-dihydroxyvitamin D3-inducible transcript (DDVit 1) in human blood monocytes. Biochem Biophys Res Commun. 1997;235(2):407–12.

45. Ma L, Broomfield S, Lavery C, Lin SL, Xiao W, Bacchetti S. Up-regulation of CIR1/CROC1 expression upon cell immortalization and in tumor-derived human cell lines. Oncogene. 1998;17(10):1321–6.

46. Sancho E, Vila MR, Sanchez-Pulido L, Lozano JJ, Paciucci R, Nadal M, Fox M, Harvey C, Bercovich B, Loukili N, et al. Role of UEV-1, an inactive variant of the E2 ubiquitin-conjugating enzymes, in in vitro differentiation and cell cycle behavior of HT-29-M6 intestinal mucosecretory cells. Mol Cell Biol. 1998;18(1):576–89.

47. Sherman F, Fink GR, Hicks J. Methods in yeast genetics. Cold Spring Harbor, NY: Cold Spring Harbor Laboratory Press; 1983.

48. Ito H, Fukuda Y, Murata K, Kimura A. Transformation of intact yeast cells treated with alkali cations. J Bacteriol. 1983;153(1):163–8.

49. Bartel PL, Fields S. Analyzing protein-protein interactions using two-hybrid system. Methods Enzymol. 1995;254:241–63.

50. Xu X, Lambrecht AD, Xiao W. Yeast survival and growth assays. Methods Mol Biol. 2014;1163:183–91.

51. Waadt R, Kudla J: In Planta Visualization of Protein Interactions Using Bimolecular Fluorescence Complementation (BiFC). *Cold Spring Harbor Protocols* 2008, 2008(5): pdb.prot4995-pdb.prot4995.

11

Residual transpiration as a component of salinity stress tolerance mechanism: a case study for barley

Md. Hasanuzzaman[1,4], Noel W. Davies[2], Lana Shabala[1], Meixue Zhou[1], Tim J. Brodribb[3] and Sergey Shabala[1*]

Abstract

Background: While most water loss from leaf surfaces occurs via stomata, part of this loss also occurs through the leaf cuticle, even when the stomata are fully closed. This component, termed residual transpiration, dominates during the night and also becomes critical under stress conditions such as drought or salinity. Reducing residual transpiration might therefore be a potentially useful mechanism for improving plant performance when water availability is reduced (e.g. under saline or drought stress conditions). One way of reducing residual transpiration may be via increased accumulation of waxes on the surface of leaf. Residual transpiration and wax constituents may vary with leaf age and position as well as between genotypes. This study used barley genotypes contrasting in salinity stress tolerance to evaluate the contribution of residual transpiration to the overall salt tolerance, and also investigated what role cuticular waxes play in this process. Leaves of three different positions (old, intermediate and young) were used.

Results: Our results show that residual transpiration was higher in old leaves than the young flag leaves, correlated negatively with the osmolality, and was positively associated with the osmotic and leaf water potentials. Salt tolerant varieties transpired more water than the sensitive variety under normal growth conditions. Cuticular waxes on barley leaves were dominated by primary alcohols (84.7–86.9%) and also included aldehydes (8.90–10.1%), n-alkanes (1.31–1.77%), benzoate esters (0.44–0.52%), phytol related compounds (0.22–0.53%), fatty acid methyl esters (0.14–0.33%), β-diketones (0.07–0.23%) and alkylresorcinols (1.65–3.58%). A significant negative correlation was found between residual transpiration and total wax content, and residual transpiration correlated significantly with the amount of primary alcohols.

Conclusions: Both leaf osmolality and the amount of total cuticular wax are involved in controlling cuticular water loss from barley leaves under well irrigated conditions. A significant and negative relationship between the amount of primary alcohols and a residual transpiration implies that some cuticular wax constituents act as a water barrier on plant leaf surface and thus contribute to salinity stress tolerance. It is suggested that residual transpiration could be a fundamental mechanism by which plants optimize water use efficiency under stress conditions.

Keywords: Residual transpiration, Osmolality, Osmotic potential, Leaf water potential, Cuticular waxes

Background

Under optimal conditions plants lose typically 95–98% water from the leaf surface via stomatal pores in a process termed stomatal transpiration. However, under some environmental conditions, a relatively large portion of evaporated water may bypass the stomata and occur through the cuticle. Depending on the species and conditions, water loss through the cuticle can be as high as 28% of the water transpired through stomata [1, 2]. Moreover, some water can escape the leaf via stomata even when they are fully closed [3, 4]. Because of this, using the term "cuticular transpiration" is not always appropriate, and this process is best described as "residual transpiration". It has been estimated that leaf cuticular water permeability varies extensively among species and ranges from 10^{-7} to 10^{-4} m s^{-1} [2, 5]. Residual transpiration is usually localized to the area surrounding stomata, where there are more and larger

* Correspondence: Sergey.Shabala@utas.edu.au
[1]School of Land and Food, University of Tasmania, Private Bag 54, Hobart, Tas 7001, Australia
Full list of author information is available at the end of the article

cuticular pores [6]. While stomatal conductance is a dynamic process that can be rapidly controlled by ion fluxes into/out of guard cells, residual transpiration depends almost entirely on the existing (passive) lipophilic cuticular pathway of the leaf surface, and, hence cannot rapidly be adjusted to changing conditions [7, 8]. However, when stomata are closed under salinity or drought conditions, the balance between stomatal and non-stomatal transpiration is shifted. Under severe stress conditions, when stomata are closed and stomatal transpiration is reduced to nearly zero, the difference in residual transpiration becomes a significant factor determining water use efficiency. Thus, reducing non-stomatal (residual) transpiration is a potentially useful mechanism for improving plant performance under stress conditions. Genotypes having lower residual transpiration can conserve relatively more water under water stress conditions, and it has therefore been suggested as a selection trait in the breeding of cereals genotypes adapted to a dry environment [9, 10].

Cuticular wax is the outermost hydrophobic layer of the aerial plant tissues, and plays an important role in protecting plants against biotic and abiotic environmental stresses, and acts as a barrier to excessive non-stomatal transpiration [11]. The main functions of cuticular waxes include maintaining equilibrium between the transpirational water loss and root water uptake by transpiration control, defending against attack by insects and pathogens, reducing water retention on plant surfaces by controlling surface wettability, controlling loss and uptake of polar solutes, and regulating the exchange of gases and vapour [12]. Extraction of cuticular waxes from plant parts with organic solvent increases the cuticular water permeability indicating that the wax layer is a fundamental water transport-limiting barrier of the cuticle, especially when stomata are closed [13]. Some reports suggested that plants that have a thicker cuticle or a cuticle containing a larger amount of waxes are more efficient in reducing non-stomatal transpiration and thus better adapted to water stress conditions [14], and in some species total wax loads increased by 30 to 70% under water stress conditions [15]. However, the correlation between residual transpiration and the thickness of cuticle and/or amount of total cuticular waxes is still not clear-cut. Some researchers found that the total amount of cuticular waxes and cuticular thickness are negatively correlated with residual transpiration in different plants [16–19]. However, some authors reported no correlation between residual transpiration and waxes [2, 20, 21].

Residual transpiration could be influenced by the characteristics of the leaf surface and morphological structure of the plant. Some studies argued [2] that residual transpiration did not relate to the amount of wax coverage and thickness of the cuticle but could be dependant on physical properties, orientation of wax crystal structure and wax composition. It is not clear however if this conclusion can be extrapolated to all species. The cuticle layer is a cutin-rich domain with embedded polysaccharides and an overlying layer that is less abundant in polysaccharides but enriched in waxes referred to as the cuticle proper [11]. The waxes are either deposited within the cutin matrix known as intracuticular wax or accumulate on its surface known as epicuticular wax crystals, or films. Cuticular waxes is a general term for the complex mixture of homologous series of very-long-chain fatty acids, primary n-alcohols, secondary n-alcohols, n-aldehydes, n-alkanes, n-alkyl esters, and cyclic organic compounds like pentacyclic triterpenoids, flavonoids, tocopherols and hydroxycinnamic acids derivatives [22]. Specific chemical compounds of the cuticle may be related to the water barrier. Higher levels of nonpolar long chain aliphatic wax compounds of cuticular wax such as hydrophobic alcohols, n-alkanes, and aldehydes tend to be associated with a barrier against cuticular water loss while alicyclic wax components including triterpenoids and sterol derivatives are less effective as a water barrier [23–26].

It is also not clear whether residual transpiration is only related to the cuticular wax on the leaf surface or it is also associated with the plant water relations. It was suggested that residual transpiration is correlated with leaf water status such as leaf water content, osmotic potential and leaf water potential [9]. Other evidence however shown that residual transpiration is not related to relative water content or osmotic potential [27].

The objectives of this study were to investigate the effect of residual transpiration on salinity tolerance and the relationship of residual transpiration to plant water relations, and cuticular wax load at three different leaf positions under irrigated conditions of two salt tolerant and two salt sensitive barley genotypes.

Methods
Plant materials and growth conditions
Four barley (*Hordeum vulgare* L.) genotypes contrasting in their salt tolerance were used in this study. Cultivars Franklin and Gairdner were salt sensitive and failed to produce any grain when grown under highly saline (300 mM NaCl) conditions in the glasshouse [28], while cultivars TX9425 and ZUG293 were salt tolerant and managed to produce ~30% increased grain yield (compared with control) under same conditions. Seeds were obtained from the Australian Winter Cereal Collection and multiplied in the field at Tasmanian Institute of Agriculture facilities in Launceston. Seeds were surface sterilized with 10% commercial bleach and thoroughly rinsed with tap water, and sown in 2 L plastic pots using

standard potting mixture containing 70% composted pine bark; 20% coarse sand; 10% sphagnum peat; Limil at 1.8 kg m^{-3}, dolomite at 1.8 kg m^{-3}. The plant nutrient balance was maintained by adding the slow release Osmocote Plus™ fertilizer (at 6 kg m^{-3}), plus ferrous sulphate (at 500 g m^{-3}). Plants were grown under controlled glasshouse condition (day length 14 h; day/night temperatures 25/15 °C; relative humidity 65%) at the University of Tasmania (Hobart, Australia) in January 2015. The plants were irrigated automatically twice per day.

Residual transpiration measurement

Two different methods were used for the determination of residual transpiration from the excised leaf under dark conditions as follows:

Method-1

Residual transpiration was determined following Clarke and McCaig [29] with modification. Three fully expanded leaves from each genotype at three positions (old leaf, intermediate leaf and young flag leaf) were selected for sampling (Fig. 1a). The leaves were excised and sealed with vacuum grease on the cut end immediately. Then collected leaves were immediately transported to the laboratory. Fresh weights (W$_0$) were determined by an

electronic balance. The leaves were then placed in a controlled dark room at 20–21 °C and 50% relative humidity (RH). The leaves were weighed at 2, 4 and 6 h (W$_2$, W$_4$ and W$_6$ respectively) intervals and then placed in dry oven at 60 °C for 24 h and reweighed (W$_d$). Residual transpiration was measured per dry weight basis by using the following formula

$$\text{Residual transpiration} = \frac{(W_0-W_2) + (W_2-W_4) + (W_4-W_6)}{3 \times W_d(T_2-T_1)}$$

where T$_1$-T$_2$ = time interval between two subsequent measurements (2 h).

The measured residual transpiration was then recalculated per projected leaf area basis and expressed in mg H$_2$O cm^{-2} h^{-1}.

Method-2

Residual transpiration was measured according to Clarke and co-authors [9] with modification. Leaf sampling was the same as for Method-1. Initial weights were determined immediately after excision of leaves. The leaves were maintained in darkness for stomatal closure under ambient room conditions at 20–21 °C and 50% RH. The leaves were weighed again after 24 h. The leaves were dried at 60 °C for 24 h and then dry weight

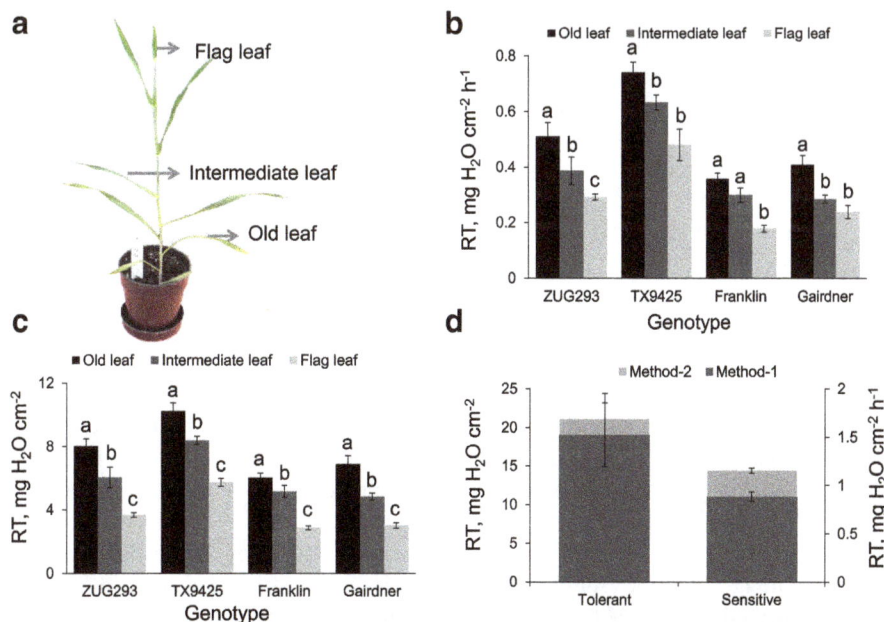

Fig. 1 Quantifying the residual transpiration (RT) from leaves of three different positions in barley. **a** sampled leaves; **b**, **c** RT values measured from leaves of three different positions from 4 barley varieties contrasting in salinity stress tolerance by Method-1 and Method-2, respectively. Data is mean ± SE (*n* = 6). **d** mean RT values for plants in salt-tolerant (ZUG293, TX9425) and salt-sensitive (Gairdner, Franklin) groups estimated by two different methods. Data labelled with different *lower case letters* in panels (**b**) and (**c**) are significantly different at *P* < 0.05

was determined. Residual water loss was determined per dry weight basis by using the following formula

$$\text{Residual transpiration} = \frac{(W_i - W_d) - (W_{24} - W_d)}{W_d}$$

where W_i = Initial fresh weight; W_{24} = Fresh weight after 24 h; W_d = Dry weight

The measured residual water loss was then recalculated per leaf area basis and expressed in mg H_2O cm^{-2}.

Measurement of leaf osmolality and osmotic potential

Three leaves at three leaf position e.g. old, intermediate and young flag leaves were taken from each genotype. Representative leaf samples were taken in centrifuge tubes and frozen at –20 °C overnight and then squeezed to extract sap. An amount of 10 µl sap was taken from each sample for measuring leaf osmolality (c) using a vapour pressure osmometer (Vapro model 5520, Wescor Inc., Logan, Utah). The osmotic potential was calculated by Van't Hoff's equation from the osmolality (mmol kg^{-1}): osmotic potential (MPa) = –c (mmol kg^{-1}) × 2.4789 × 10^{-3} at 25 °C.

Measurement of leaf water potential

Two leaves were excised from each genotype from three positions of the stem for leaf water potential determinations. The leaf blades were cut with a sharp blade and immediately sealed in an elliptical grass compression gland gasket. The leaf blades were sealed in a pressure chamber (Model 615; PMS Instruments, Albany, OR, USA), and the chamber was pressurized using compressed air at a rate of 0.1 MPa s^{-1} until water first appeared at the cut surface of the leaf. The total elapsed time from when the leaf was cut from the plant to the initial pressurisation of the chamber was 5–10 s. The leaf water potential data were reported in MPa.

Scanning electron microscopy (SEM)

After sampling the leaves were stored at –20 °C overnight and then lyophylised in a pre-cooled freeze drier (Mini-ultra cold, Dynavac, Aus, Techno lab). The dried samples (3–5 mm long) were mounted on SEM specimen stubs with double-sided carbon tape (one half with adaxial and the other with abaxial surface uppermost) and then coated with a thin film (2–3 nm) of Pt for 20 min using a sputter coater (BalTec SCD 050) in an atmosphere of argon to improve the electrically conducting properties of leaf and high resolution of images. Three replicates of coated samples were examined with a Hitachi SU-70 UHR field emission scanning electron microscope setting with 1.5 kV, 17.2 mm × 2.00 k SE (M). The imaging was performed in the Central Science Laboratory, University of Tasmania.

Wax extraction and analysis

Three fresh leaves at three positions of the plant from each genotype were excised and ten 0.64 cm^2 disks were sampled from each by leaf punch.. The leaf segments were soaked in 5 mL of solvent (dichloromethane with n-docosane (C_{22} alkane, 20 mg/L) as an internal standard) for 5 min with gentle stirring [30]. The extract contained waxes from both abaxial and adaxial leaf surfaces. The extracts were evaporated to dryness under a nitrogen stream for 30 min at 58 °C. The samples were redissolved in 0.5 ml dichloromethane for analysis by combined gas chromatography-mass spectrometry (GC-MS) on a Varian 3800 gas chromatograph coupled to a Bruker-300 triple quadrupole mass spectrometer. One microlitre injections in splitless mode were made with an injector temperature of 275 °C. The column was a 30 m × 0.25 mm DB5 (0.25 µm film thickness) (Agilent, Australia) and the oven temperature program was 60 °C (held for 1 min) to 220 °C at 30 °C per minute, then to 310 °C at 10 °C per minute with a final hold time of 5 min. The carrier gas was helium at a constant flow of 3.5 ml min^{-1}. Mass spectra were collected over the range m/z 40 to 600 every 0.3 s. Compounds were identified through a combination of MS reference databases (NIST MS database and an in-house database of relevant compounds), and Kovats' retention indices. The individual components and total wax were expressed in terms of µg equivalents of n-docosane cm^{-2}. All subsequent µg cm^{-2} values are in terms of n-docosane equivalents in the text and figures.

Statistical analysis

All data were analyzed by using SPSS 20.0 for Windows (SPSS Inc.). Significant differences between different genotypes were determined by one-way analysis of variance based on Duncan's multiple range tests. Different lower case letters in the figures represent significant differences. The significance of correlations between different parameters was determined by bivariate correlations based on Pearson's correlation (two-tailed).

Results
Residual transpiration

As both stomata density and amount of cuticular waxes depends on the leaf age, we hypothesised that a significant variation in residual transpiration should exist between leaves of different positions. A significant variation was seen in the different leaf positions for all varieties ($P < 0.05$; Fig. 1a and b). Old leaves transpired more water than the intermediate and flag leaves for all varieties using both methods. In Method-1, significant variation was observed between old leaves and intermediate leaves but not in intermediate and flag leaves in most genotypes. Old leaves of TX9425 (0.74 ± 0.04 mg

H_2O cm^{-2} h^{-1}) genotype transpired the highest amount of water and Franklin transpired the lowest amount of water (0.36 ± 0.02 mg H_2O cm^{-2} h^{-1}). In Method-2, significant differences were seen between the three leaf position in all genotypes. Old leaves of TX9425 (10.24 ± 0.53 mg H_2O cm^{-2}) transpired the highest amount of water followed by old leaves of ZUG293 (8.01 ± 0.48 mg H_2O cm^{-2}), Gairdner (6.88 ± 0.52 mg H_2O cm^{-2}) and Franklin (6.02 ± 0.28 mg H_2O cm^{-2}), respectively. Young flag leaves of TX9425 (5.73 ± 0.25 mg H_2O cm^{-2}) transpired the highest amount of water followed by ZUG293 (3.68 ± 0.14 mg H_2O cm^{-2}), Gairdner (3.02 ± 0.17 mg H_2O cm^{-2}) and Franklin (2.86 ± 0.12 mg H_2O cm^{-2}), respectively. Salt tolerant varieties transpired more water through the cuticle than that of sensitive varieties under normal growth conditions (Fig. 1c). The cumulative loss of water of the three leaf positions of two tolerant genotypes (TX9425 and ZUG293) was higher than two sensitive genotypes (Gairdner and Franklin) in both methods. The two tolerant genotypes transpired 43% and 32% more water respectively than the two sensitive genotypes in the two methods under normal growth condition.

Leaf sap osmolality correlates negatively with residual transpiration

A significant difference of leaf sap osmolality was observed among different leaf positions ($P < 0.05$; Fig. 2a). Leaf sap osmolality decreased with increasing leaf age for all genotypes. The osmotic potential was highest in old leaf and lowest in flag leaf in all genotypes ($P < 0.05$; Fig. 3a). The highest decrease (60%) was observed in TX9425 followed by ZUG293 (43%), whereas the lowest decrease (20%) was measured in Franklin followed by Gairdner (28%), in old and young leaves respectively. A strong negative correlation ($R^2 = -0.86$ for Method-1 and -0.92 for Method-2; significant at $P < 0.01$) was found between the overall leaf sap osmolality in plants grown under normal growth conditions and residual transpiration.

Osmotic potential and leaf water potential correlate positively with residual transpiration

The osmotic potential was highest in old leaves and lowest in flag leaves in all genotypes ($P < 0.05$; Fig. 3a). ZUG293 and TX9425 followed the order old > intermediate > young flag leaf, whereas Franklin and Gairdner followed old > intermediate = young flag leaf. A strong positive correlation ($R^2 = 0.86$ for Method-1 and 0.92 for Method-2; significant at $P < 0.01$) was found between the overall leaf osmotic potential in plants grown under normal growth conditions and residual transpiration. A significant variation of leaf water potential was found among the three leaf positions in all four genotypes ($P < 0.05$; Fig. 4a). Leaf water potential increased with increasing the plant leaf age, the highest and lowest leaf water potential was found at old leaf and young flag leaf, respectively. A positive correlation ($R^2 = 0.59$; significant at $P < 0.01$) was found (in Method 2) between the overall leaf water potential in plants grown under normal growth condition and residual transpiration.

Structure and distribution of cuticular waxes on leaf epidermis

SEM analysis showed similar cuticular waxes structure in three different leaf positions of four barley genotypes. The cuticular waxes formed combined coatings of different arrangement of minute crystallised plates about 1–2 μm in size, relatively vertically oriented to the leaf epidermal surface (Fig. 5; Additional file 1: Figure S1). Cuticular wax structures were a less dense covering of adaxial surface of old leaves compared to the intermediate and young flag leaves for all genotypes. The epidermis of three different leaf positions of four genotypes

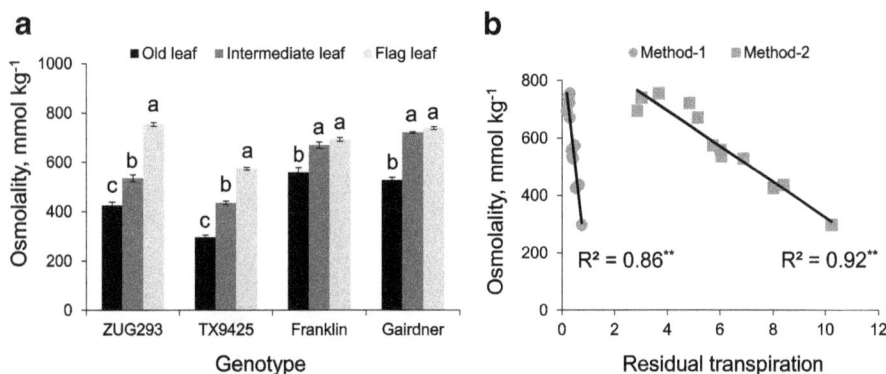

Fig. 2 a genetic variability in osmolality of barley leaves at three positions in plants grown under normal (no salt) growth condition. Mean ± SE (*n* = 6). **b** correlations (Pearson's R^2 values) between leaf sap osmolality and residual transpiration measured by two different methods. Data labelled with different *lower case letters* are significantly different at $P < 0.05$ and *asterisks* are significant at $P < 0.01$

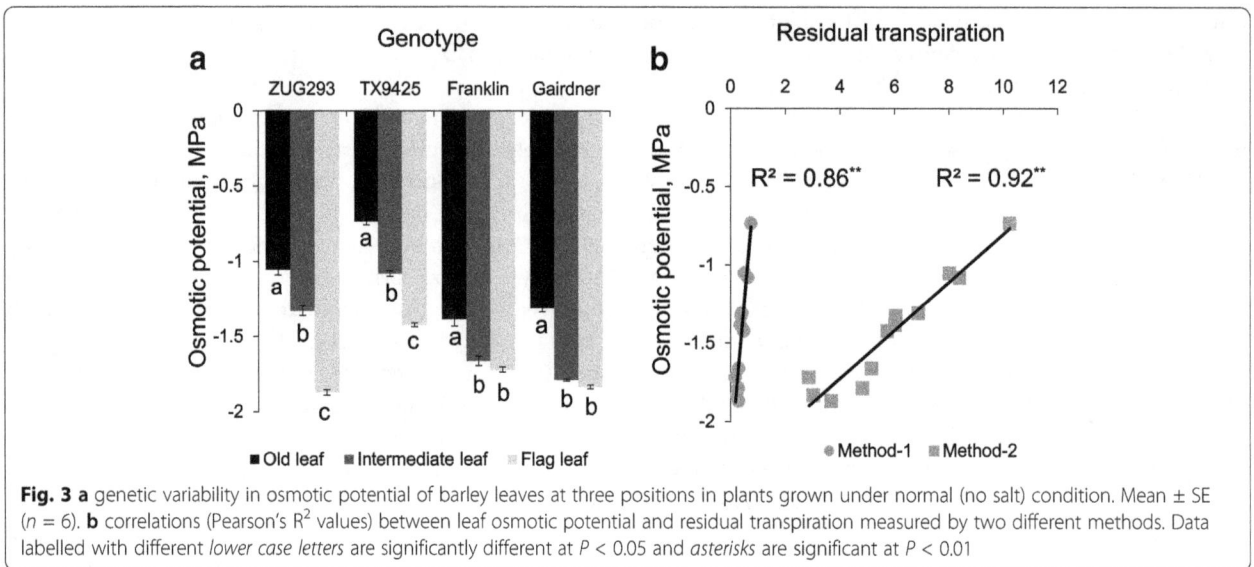

Fig. 3 a genetic variability in osmotic potential of barley leaves at three positions in plants grown under normal (no salt) condition. Mean ± SE ($n = 6$). **b** correlations (Pearson's R^2 values) between leaf osmotic potential and residual transpiration measured by two different methods. Data labelled with different *lower case letters* are significantly different at $P < 0.05$ and *asterisks* are significant at $P < 0.01$

was covered with waxy plates, but not fully over the guard cell of all genotypes (Fig. 6). In the case of TX9425 and ZUG293 genotypes, the guard cells of stomata were not fully covered with waxy plates, whereas the guard cells of Franklin and Gairdner were fully covered with waxy plates. No differences were found for adaxial and abaxial surface of leaves in all genotypes regarding to cuticular wax structure and density (data not shown).

Total wax content of leaves correlates negatively with residual transpiration

A significant negative correlation ($R^2 = -0.41$ for Method-1 and -0.34 for Method-2; significant at $P < 0.05$) was found between the total cuticular wax content of leaves and residual transpiration measured by two different methods in plants grown under normal growth conditions (Fig. 7a).

Cuticular wax constituents, contents and effect on residual transpiration

Across all four barley varieties the average of total leaf cuticular wax was found to be 5.37 μg cm^{-2} under normal growth condition. The averages of total cuticular wax of old leaves, intermediate leaves and flag leaves of all genotypes studied were 5.06 μg cm^{-2}, 5.06 μg cm^{-2} and 5.98 μg cm^{-2} respectively. Cuticular waxes on barley leaves were dominated by primary alcohols (84.7–86.9%), aldehydes (8.90–10.1%), *n*-alkanes (1.31–1.77%), benzoate esters (0.44–0.52%), a phytol related compound (0.22–0.53%), fatty acid methyl esters (0.14–0.33%), β-diketones (0.07–0.23%) and alkylresorcinols constituents (1.65–3.58%). Primary alcohols consisted of odd and even numbers of carbon from C$_{22}$ to C$_{29}$, particularly *n*-docosanol (C$_{22}$), *n*-tetracosanol (C$_{24}$), *n*-hexacosanol (C$_{26}$), and *n*-octasonanol (C$_{28}$), and much smaller amount of odd numbered carbons. The higher *n*-alkane

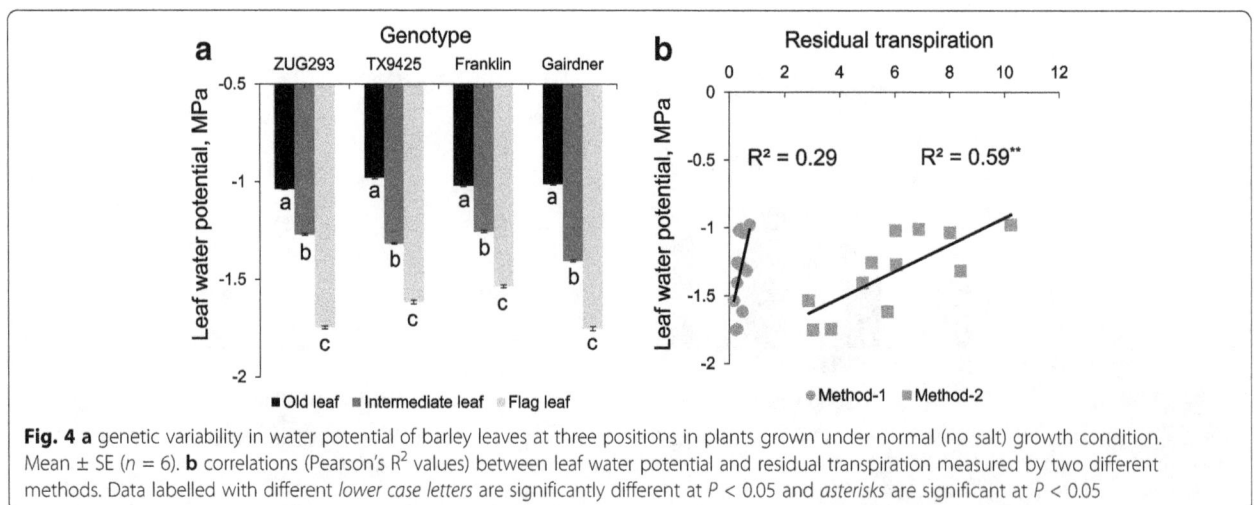

Fig. 4 a genetic variability in water potential of barley leaves at three positions in plants grown under normal (no salt) growth condition. Mean ± SE ($n = 6$). **b** correlations (Pearson's R^2 values) between leaf water potential and residual transpiration measured by two different methods. Data labelled with different *lower case letters* are significantly different at $P < 0.05$ and *asterisks* are significant at $P < 0.05$

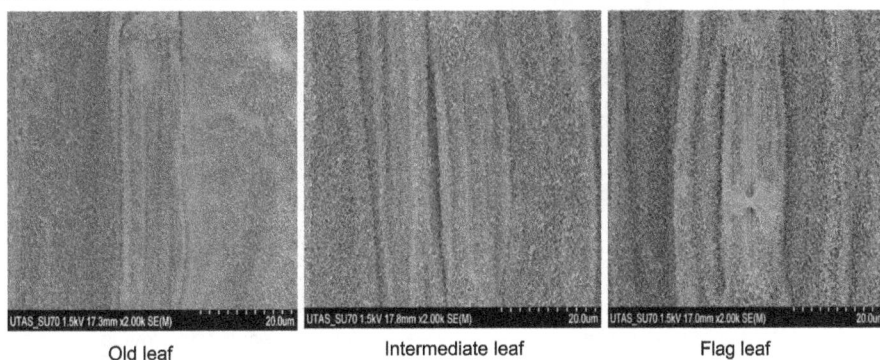

Fig. 5 Representative SEM images showing cuticular wax on the adaxial surface in leaves of three different positions in variety Franklin grown under control condition. One (of six) typical images is shown for each position

component on barley leaf consisted mainly of n-hentria-contane (C_{31}) and n-tritriacontane (C_{33}). The main aldehydes were n-hexacosanal (C_{26}), n-octacosanal (C_{28}) and n-triacontanal (C_{30}). Benzoate esters included n-docosyl benzoate (C_{22}), n-tetracosyl benzoate (C_{24}) and n-hexacosyl benzoate (C_{26}). Major fatty acid methyl esters were methyl n-octacosanoate (C_{28}), methyl n-triacontanoate (C_{30}) and methyl n-dotriacontanoate (C_{32}).

Old leaves for all genotypes studied showed the average highest absolute amount of alcohols (4.39 μg cm^{-2}) followed by aldehydes (0.45 μg cm^{-2}) and the lowest β-diketones (Table 1). Similar results were found at intermediate and flag leaves for all genotypes (Tables 2 and 3; Additional file 2: Table S1). Among the genotypes,

ZUG293 old leaves contained the highest amount of alcohols followed by Franklin. The same results were found for intermediate leaf for all genotypes (Table 2; Additional file 2: Table S1). For flag leaves of all genotypes the average highest alcohols were measured from Franklin followed by ZUG293 (Table 3).

A negative significant correlation (R^2 = −0.44 for Method-1; $P < 0.05$ and R^2 = −0.36 for Method-2; significant at $P < 0.05$) was found between residual transpiration and primary alcohols of cuticular wax component of barley genotypes (Fig. 7b). No significant correlations were found between residual transpiration measured by two different methods and other cuticular wax components (Table 4).

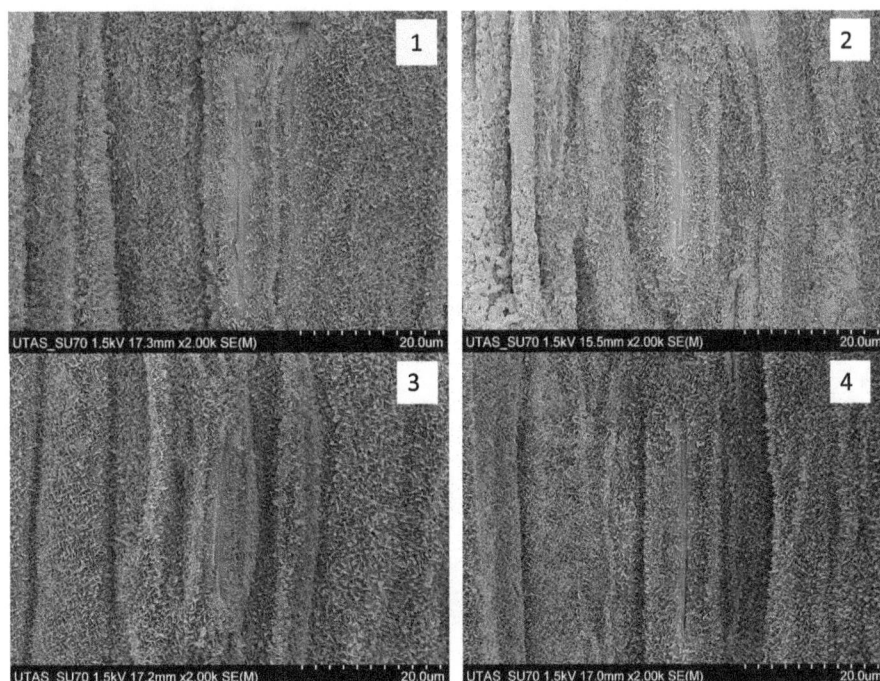

Fig. 6 Representative SEM images showing cuticular wax on the adaxial surface of the flag leaf in barley varieties ZUG293 (**1**), TX9425 (**2**), Franklin (**3**) and Gairdner (**4**) grown under control conditions. One (of six) typical images is shown for each genotype

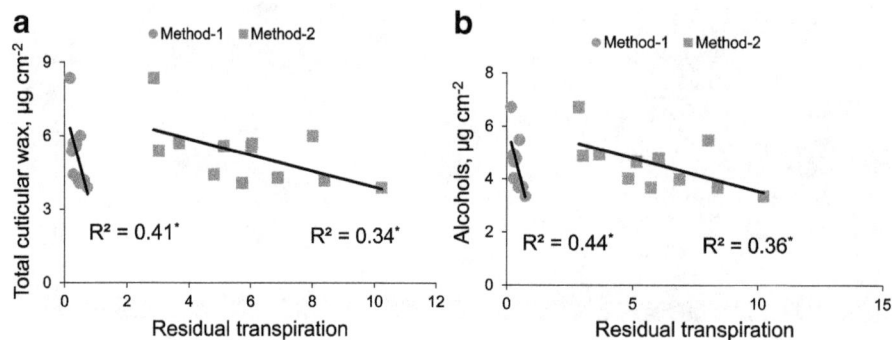

Fig. 7 a correlations (Pearson's R^2 values) between total cuticular wax and residual transpiration measured by Method-1 (mg H_2O cm^{-2} h^{-1}) and Method-2 (mg H_2O cm^{-2}). **b** correlations (Pearson's R^2 values) between alcohols and residual transpiration measured by Method-1 (mg H_2O cm^{-2} h^{-1}) and Method-2 (mg H_2O cm^{-2}). Data labelled with *asterisks* are significant at $P < 0.05$

Discussion

Residual transpiration and plant water relations

To maintain proper growth and leaf expansion, the growing shoot needs to maintain positive turgor which can be achieved by maintaining osmotic cellular adjustment by either increasing the production of compatible solutes or inorganic ions. As plants accumulate more organic osmolytes in young leaves than old leaves to maintain turgor pressure [31], it was hypothesised that residual transpiration should be less in young leaves due to the fact that they have higher osmolality and hence better water retention, and this was found to be the case. As shown in Fig. 2a and b, young flag leaves had a higher osmolality than the older leaves, and increased osmolality had a strong negative correlation with the residual transpiration under normal growth conditions indicating that the increase of leaf sap osmolality might decrease the water transpiration through plant cuticle. An effective osmotic adjustment mechanism may maintain water status in the leaf tissue by decreasing in the cell sap osmotic potential resulting from a net increase of intracellular solutes [32].

A leaf can increase its resistance to dehydration through a reduction in cellular osmotic potential by a net accumulation of cellular solutes. In this study, young flag leaves possessed significantly lower osmotic potential than the intermediate and older leaves; a trend that was correlated positively with residual transpiration (Fig. 3a and b). This indicated that a leaf with lower osmotic potential had more turgor pressure to spend and could resist greater loss of water through the cuticle. Lower negative leaf water potential was measured with increasing leaf age for all varieties, which was negatively correlated with residual transpiration (Fig. 4a and b). Young leaves maintained less turgor at more negative leaf water potentials and tended to have less residual transpiration. Increased turgor in the epidermis stretches cuticles and causes a change in gas exchange of the cuticle. A leaf with less turgor would have a tighter cuticle, thus inhibiting gas exchange [33]. Burghardt and Riederer [14] observed that cuticle gas exchange was affected when leaf water potentials decreased. Thus, leaf water potential affects the diffusion of water vapour through the cuticular barrier, and residual transpiration is negatively correlated with lower leaf water potential [33].

Table 1 Absolute amount (μg cm^{-2}) of different compounds of cuticular wax on old leaf position of four barley genotypes grown under normal growth conditions ($n = 4$)

Compound	Genotype				
	ZUG293	TX9425	Franklin	Gairdner	Average
Alcohols	5.48 ± 0.16	3.35 ± 0.65	4.76 ± 0.55	3.99 ± 0.27	4.40
Aldehydes	0.38 ± 0.06	0.46 ± 0.09	0.64 ± 0.06	0.32 ± 0.01	0.45
Alkanes	0.07 ± 0.01	0.08 ± 0.01	0.07 ± 0.01	0.05 ± 0.00	0.07
Benzoate esters	0.02 ± 0.00	0.03 ± 0.01	0.03 ± 0.00	0.01 ± 0.00	0.02
Phytol related	0.01 ± 0.00	0.03 ± 0.01	0.04 ± 0.00	0.02 ± 0.00	0.03
Methyl esters	0.03 ± 0.01	0.00 ± 0.00	0.00 ± 0.00	0.00 ± 0.00	0.01
Diketones	0.01 ± 0.00	0.02 ± 0.01	0.00 ± 0.00	0.00 ± 00	0.01
Alkylresorcinols	0.15 ± 0.05	0.00 ± .00	0.15 ± 0.06	0.04 ± 0.01	0.09

Table 2 Absolute amount (µg cm^{-2}) of different compounds of cuticular wax on intermediate leaf position of four barley genotypes grown under normal growth conditions ($n = 4$)

Compound	Genotype				
	ZUG293	TX9425	Franklin	Gairdner	Average
Alcohols	4.78 ± 0.08	3.69 ± 0.44	4.65 ± 0.29	4.02 ± 0.32	4.29
Aldehydes	0.41 ± 0.02	0.45 ± 0.06	0.65 ± 0.07	0.38 ± 0.03	0.47
Alkanes	0.06 ± 0.01	0.06 ± 0.00	0.07 ± 0.01	0.05 ± 0.00	0.06
Benzoate esters	0.03 ± 0.00	0.03 ± 0.01	0.04 ± 0.01	0.02 ± 0.00	0.03
Phytol related	0.04 ± 0.01	0.01 ± 0.00	0.03 ± 0.00	0.01 ± 0.00	0.02
Methyl esters	0.00 ± 0.00	0.01 ± 0.00	0.00 ± 0.00	0.02 ± 0.00	0.01
Diketones	0.01 ± 0.00	0.01 ± 0.00	0.00 ± 0.00	0.00 ± 0.00	0.01
Alkylresorcinols	0.45 ± 0.01	0.03 ± 0.01	0.21 ± 0.06	0.04 ± 0.01	0.18

Change in residual transpiration to improve water use efficiency

Salinity stress is often referred to as a "physiological drought", so some correlation between salinity and drought stress tolerance is expected. The most salinity tolerant varieties showed the highest residual transpiration under unstressed conditions (Fig. 1d). Being somewhat counterintuitive, this is in a good agreement with Bengston et al. [34] who showed that drought stress resistant oat genotypes generally transpired the highest amount of water through the cuticle under unstressed conditions, whereas it was strongly reduced under stress conditions. In addition, higher (33 to 38%) residual transpiration in wheat and cotton leaves was reported from irrigated than rainfed field-grown wheat plants [9]. On the other hand, deposition of cuticular waxes increased in tolerant genotypes during prolonged drought stress, leading to a reduced rate of residual transpiration [16, 35].

Water use efficiency can be expressed as the ratio of leaf net carbon assimilation to total transportation water loss. Plants exhibit higher water use efficiency with higher CO_2 assimilation than the stomatal conductance, when non-stomatal water loss is negligible [36]. Salt tolerant genotypes transpired more water through cuticle

under well irrigated condition that reveals their water use efficiency is lower than sensitive genotypes. Generally stress tolerant barley genotypes have a lower biomass and yield performance under control conditions [37]. This could be due to their higher non-stomatal transpiration under irrigated conditions resulting in lower water use efficiency. Conversely, tolerant genotypes could reduce residual water loss under water deficit conditions when stomata are closed and/or partially closed, and this increased water use efficiency could be a significant factor determining their survival capacity to hostile environmental conditions compared to the standard cultivated genotypes. It has been documented that wheat genotypes having lower residual transpiration adapted and performed better under water stress conditions [38]. Genotypes with normally low residual transpiration are at a functional advantage in water-limited environments since they make more efficient use of the water available. Thus, under conditions of water deficit, residual conductance to water vapour may be an important determinant of plant water balance and stress reactivity.

On the other hand, transpiration is the most effective way of leaf cooling of well-irrigated plants. In plants with adequate water supply stomata may regulate leaf temperature

Table 3 Absolute amount (µg cm^{-2}) of different compounds of cuticular wax on flag leaf position of four barley genotypes grown under normal growth condition ($n = 4$)

Compound	Genotype				
	ZUG293	TX9425	Franklin	Gairdner	Average
Alcohols	4.93 ± 0.21	3.68 ± 0.41	6.71 ± 0.41	4.88 ± 0.17	5.05
Aldehydes	0.40 ± 0.02	0.32 ± 0.05	1.26 ± 0.12	0.45 ± 0.03	0.61
Alkanes	0.06 ± 0.01	0.06 ± 0.00	0.11 ± 0.01	0.06 ± 0.00	0.07
Benzoate esters	0.03 ± 0.00	0.02 ± 0.01	0.05 ± 0.00	0.02 ± 0.00	0.03
Phytol related	0.02 ± 0.00	0.01 ± 0.00	0.02 ± 0.00	0.01 ± 0.00	0.02
Methyl esters	0.01 ± 0.00	0.02 ± 0.00	0.00 ± 0.00	0.05 ± 0.00	0.02
Diketones	0.01 ± 0.00	0.02 ± 0.00	0.01 ± 0.00	0.02 ± 0.00	0.02
Alkylresorcinols	0.36 ± 0.01	0.03 ± 0.00	0.28 ± .00	0.04 ± 0.00	0.18

Table 4 Correlations (Pearson's R^2 values) between residual transpiration measured by two different methods and different cuticular wax compounds of three different leaf positions of four barley genotypes grown under normal growth condition. Values labelled with asterisk are significant at $P < 0.05$

Compound	R^2 values with residual transpiration				Correlation
	Method-1		Method-2		
	R^2 Value	P value	R^2 value	P value	
Aldehydes	0.21	0.15	0.17	0.18	Negative
Alkanes	0.00	0.93	0.00	0.88	Negative
Benzoates	0.16	0.21	0.15	0.21	Negative
Phytols	0.00	0.86	0.05	0.50	Positive
Methyl esters	0.02	0.63	0.06	0.43	Negative
Diketones	0.04	0.52	0.00	0.89	Positive
Alkylresorcinols	0.21	0.15	0.16	0.19	Negative

close to the optimum for metabolic processes, including photosynthesis or to prevent tissue heat damage under excessive radiation or temperature [39]. Moreover, under water limited conditions, stomatal closure and decreased transpiration, associated with high water use efficiency, may lead to a dramatic increase in leaf temperature (up to 7 °C above air temperature) [40]. At this condition, high temperatures may disrupt the photosynthetic-related enzymes and produce reactive oxygen species which would challenge the plant cell [41].

Relationship between residual transpiration and amount of cuticular waxes

Our working hypothesis in this study was that reduced residual transpiration should be positively correlated with hydrophobicity of the leaf surface (hence, amount of cuticular waxes deposited). A significant negative correlation (Fig. 7a) between the total amount of cuticular wax and residual transpiration was found in the present investigation, which indicated that amount of cuticular wax may create a protecting barrier to reduce the loss of water through the cuticle. Previous studies have reported a weak but significant negative correlation between the cuticular wax and residual transpiration in sorghum [18], wheat [17], and barley [42]. This weak correlation may be due to the protecting barrier to the diffusion of water through the cuticle depends on the structure, orientation of wax plates on epidermis, variation of epicuticular and intracuticular wax compositions and distribution of wax plates. Both intracuticular [43] and epicuticular [44] wax layer may contribute to the formation of residual transpiration barrier depending on the plant species and specific cuticle constituents. Plants generally exhibited a significant increase in the amount of cuticular wax amount per unit area of leaves under different stress condition such as water deficit and

salinity [20]. The quantity of cuticular wax, however, is not the sole contributor to residual transpiration due to the complexity of water flow through the cuticle [45].

Cuticular waxes have different types of structural morphology including granules, filaments, plates and tubes [12]. According to the SEM images analysis, plate type cuticular wax observed on the leaf surface consisted of aliphatic compounds in which the primary alcohols n-hexacosanol and n-octacosanol were predominant in different leaf positions for all the barley genotypes.

Cuticular waxes on barley leaves consisted of alcohols, aldehydes, alkanes, benzoate esters, phytol related compounds, fatty acid methyl esters, β-diketones and alkylresorcinols (Tables 1, 2 and 3). Generally, plate type primary alcohol based cuticular waxes always dominate on the leaf surface in the Fabaceae and Poaceae (wheat, barley) [42] and constitute the major barrier to water loss. This was also the case in our study reported here (Fig. 7b) [45]. However, such findings could be not generalized to all species. The hydrophobic long chain alcohol, hydrocarbon and aldehyde fractions are the active components of cuticle in controlling residual transpiration in different plant species [44]. The main portion of the transpiration barrier in tomato fruits and *Rhazya stricta* leaves is located in the intracuticular wax layer containing large amount of pentacyclic triterpenoids whereas cuticular very long chain aliphatics play a minor role [46, 47]. Plant species containing fatty acid with very long aliphatic chain (alcohols, aldehydes and alkanes) in the epicuticular wax, together with high amount of alicyclic compounds such as triterpenoids, steroids, or tocopherols in the intracuticular wax contribute equally to the formation of residual transpiration barrier (44). In general, it is accepted that higher levels of long chain aliphatic components in the wax can lead to a higher hydrophobicity of the residual transpiration barrier and thus decrease cuticular water loss [26]. This should be kept in mind while targeting this trait in the breeding programs.

Conclusions

Both leaf osmotic potential and the amount of cuticular waxes are involved in controlling water loss from barley leaves under well irrigated conditions. A significant and negative relationship between the amount of primary alcohols and cuticular transpiration implies that primary alcohols may influence the water barrier more than other constituents on plant leaf surface and thus contribute to salinity stress tolerance, at least in barley.

Additional files

Additional file 1: Figure S1. SEM images showing cuticular wax on the adaxial surface in three different positions of leaf in varieties ZUG293 (A), TX9425 (B) and Gairdner (C) grown under control conditions.

Additional file 2: **Table S1.** Amount (µg cm^{-2}) of different components of cuticular wax in three different positions of leaf of four barley genotypes ($n = 4$).

Abbreviations
MS: Mass spectrometry; RH: Relative humidity; RT: Residual transpiration; SEM: Scanning electron microscopy

Acknowledgements
We cordially thank Dr. Sandrin T. Feig and Dr. Karsten Goemann from the Central Science Laboratory, University of Tasmania, for the assistance with SEM. We also thank Norhawa Puniran-Hartley, Joseph Hartley, Koushik Chakraborty and Akhikun Nahar for their help with freeze drying samples and wax extraction.

Funding
This work was funded by the Grain Research and Development Corporation (GRDC) grants to Sergey Shabala and Meixue Zhou. The funders had no role in study design, data collection, analysis and interpretation of data and in writing the manuscript.

Authors' contributions
MH conducted the bulk of experiments and wrote the paper draft. SS was responsible for experimental design and data interpretation, and took the leading role in writing. NWD was leading the cuticular wax analysis, critically assessed all the data and commented on the manuscript. LS, MZ and TJB provided a logistical support for this work and contributed to data interpretation and writing. All authors read and approved the final manuscript.

Competing interests
The authors declare that they have no competing interests.

Author details
[1]School of Land and Food, University of Tasmania, Private Bag 54, Hobart, Tas 7001, Australia. [2]Central Science Laboratory, University of Tasmania, Hobart, Tas 7001, Australia. [3]School of Biological Science, University of Tasmania, Private Bag 55, Hobart, Tas 7001, Australia. [4]Department of Agronomy, Faculty of Agriculture, Sher-e-Bangla Agricultural University, Sher-e-Bangla Nagar, Dhaka -1207, Bangladesh.

References
1. Boyer JS, Wong SC, Farquhar GD. CO_2 and water vapor exchange across leaf cuticle (epidermis) at various water potentials. Plant Physiol. 1997;114(1):185–91.
2. Riederer M, Schreiber L. Protecting against water loss: analysis of the barrier properties of plant cuticles. J Exp Bot. 2001;52(363):2023–32.
3. McAdam SA, Brodribb TJ. Separating active and passive influences on stomatal control of transpiration. Plant Physiol. 2014;164(4):1578–86.
4. Caird MA, Richards JH, Donovan LA. Nighttime stomatal conductance and transpiration in C_3 and C_4 plants. Plant Physiol. 2007;143(1):4–10.
5. Burghardt M, Riederer M. Cuticular transpiration. In: Riederer M, Müller C, editors. Annual Plant Reviews, Biology of the Plant Cuticle. Blackwell publishing. 2008;23:292.
6. Marschner H. Mineral nutrition of higher plant. New York: Academic Press; 1995.
7. Blatt MR. Cellular signaling and volume control in stomatal movements in plants. Annu Rev Cell Dev Biol. 2000;16(1):221–41.
8. Popp C, Burghardt M, Friedmann A, Riederer M. Characterization of hydrophilic and lipophilic pathways of Hedera helix L. cuticular membranes: permeation of water and uncharged organic compounds. J Exp Bot. 2005;56(421):2797–806.
9. Clarke J, Richards R, Condon A. Effect of drought stress on residual transpiration and its relationship with water use of wheat. Can J Plant Sci. 1991;71(3):695–702.
10. Petcu E. The effect of water stress on cuticular transpiration and relationships with winter wheat yield. Rom Agric Res. 2005;22:15–9.
11. Yeats TH, Rose JK. The formation and function of plant cuticles. Plant Physiol. 2013;163(1):5–20.
12. Riederer M, Muller C. Annual Plant Reviews, Biology of the Plant Cuticle. Blackwell publishing. 2008;23:292–309.
13. Šantrůček J, Šimáňová E, Karbulková J, Šimková M, Schreiber L. A new technique for measurement of water permeability of stomatous cuticular membranes isolated from Hedera helix leaves. J Exp Bot. 2004;55(401):1411–22.
14. Burghardt M, Riederer M. Ecophysiological relevance of cuticular transpiration of deciduous and evergreen plants in relation to stomatal closure and leaf water potential. J Exp Bot. 2003;54(389):1941–9.
15. Kim KS, Park SH, Jenks MA. Changes in leaf cuticular waxes of sesame (Sesamum indicum L.) plants exposed to water deficit. J Plant Physiol. 2007;164(9):1134–43.
16. González A, Ayerbe L. Effect of terminal water stress on leaf epicuticular wax load, residual transpiration and grain yield in barley. Euphytica. 2010;172(3):341–9.
17. Premachandra GS, Saneoka H, Fujita K, Ogata S. Leaf water relations, osmotic adjustment, cell-membrane stability, Epicuticular wax load and growth as affected by increasing water deficits in sorghum. J Exp Bot. 1992;43(257):1569–76.
18. Jordan W, Shouse P, Blum A, Miller F, Monk R. Environmental physiology of sorghum. II. Epicuticular wax load and cuticular transpiration. Crop Sci. 1984;24(6):1168–73.
19. Ni Y, Sun Z, Huang X, Huang C, Guo Y. Variations of cuticular wax in mulberry trees and their effects on gas exchange and post-harvest water loss. Acta Physiol Plant. 2015;37(6):1–9.
20. Sánchez FJ. Manzanares ma, de Andrés EF, Tenorio JL, Ayerbe L. residual transpiration rate, epicuticular wax load and leaf colour of pea plants in drought conditions. Influence on harvest index and canopy temperature. Eur J Agron. 2001;15(1):57–70.
21. Ni Y, Guo Y, Han L, Tang H, Conyers M. Leaf cuticular waxes and physiological parameters in alfalfa leaves as influenced by drought. Photosynthetica. 2012;50(3):458–66.
22. Jetter R, Kunst L, Samuels AL. Composition of plant cuticular waxes. In: Riederer M, Müller C, editors. Annual Plant Reviews, Biology of the Plant Cuticle. Blackwell publishing. 2008;23:145.
23. Buschhaus C, Jetter R. Composition and physiological function of the wax layers coating Arabidopsis leaves: β-amyrin negatively affects the intracuticular water barrier. Plant Physiol. 2012;160(2):1120–9.
24. Leide J, Hildebrandt U, Vogg G, Riederer M. The positional sterile (ps) mutation affects cuticular transpiration and wax biosynthesis of tomato fruits. J Plant Physiol. 2011;168(9):871–7.
25. Leide J, Hildebrandt U, Reussing K, Riederer M, Vogg G. The developmental pattern of tomato fruit wax accumulation and its impact on cuticular transpiration barrier properties: effects of a deficiency in a β-ketoacyl-coenzyme a synthase (LeCER6). Plant Physiol. 2007;144(3):1667–79.
26. Macková J, Vašková M, Macek P, Hronková M, Schreiber L, Šantrůček J. Plant response to drought stress simulated by ABA application: changes in chemical composition of cuticular waxes. Environ Exp Bot. 2013;86:70–5.
27. Rawson H, Clarke J. Nocturnal transpiration in wheat. Funct Plant Biol. 1988;15(3):397–406.
28. Chen Z, Zhou M, Newman IA, Mendham NJ, Zhang G, Shabala S. Potassium and sodium relations in salinised barley tissues as a basis of differential salt tolerance. Funct Plant Biol. 2007;34(2):150–62.
29. Clarke JM, McCaig TN. Excised-leaf water-retention capability as an indicator of drought resistance of Triticum genotypes. Can J Plant Sci. 1982;62(3):571–8.

30. Wu H, Shabala L, Barry K, Zhou M, Shabala S. Ability of leaf mesophyll to retain potassium correlates with salinity tolerance in wheat and barley. Physiol Plantarum. 2013;149(4):515–27.

31. Puniran-Hartley N, Hartley J, Shabala L, Shabala S. Salinity-induced accumulation of organic osmolytes in barley and wheat leaves correlates with increased oxidative stress tolerance: in planta evidence for cross-tolerance. Plant Physiol Biochem. 2014;83:32–9.

32. Silva EN, Vieira SA, Ribeiro RV, Ponte LF, Ferreira-Silva SL, Silveira JA. Contrasting physiological responses of *Jatropha curcas* plants to single and combined stresses of salinity and heat. J Plant Growth Regul. 2013;32(1): 159–69.

33. Boyer JS. Turgor and the transport of CO_2 and water across the cuticle (epidermis) of leaves. J Exp Bot. 2015;66(9):2625–33.

34. Bengtson C, Larsson S, Liljenberg C. Effects of water stress on cuticular transpiration rate and amount and composition of epicuticular wax in seedlings of six oat varieties. Physiol Plant. 1978;44(4):319–24.

35. Shepherd T, Wynne GD. The effects of stress on plant cuticular waxes. New Phytol. 2006;171(3):469–99.

36. Yoo CY, Pence HE, Hasegawa PM, Mickelbart MV. Regulation of transpiration to improve crop water use. CRC Crit Rev Plant Sci. 2009;28(6):410–31.

37. Munns R, James RA, Läuchli A. Approaches to increasing the salt tolerance of wheat and other cereals. J Exp Bot. 2006;57(5):1025–43.

38. David M. Water loss from excised leaves in a collection of *Triticum aestivum* and *Triticum durum* cultivars. Rom Agric Res. 2010;27:27–34.

39. Chaves M, Costa J, Zarrouk O, Pinheiro C, Lopes C, Pereira J. Controlling stomatal aperture in semi-arid regions—the dilemma of saving water or being cool? Plant Sci. 2016.

40. Blum A. Towards a conceptual ABA ideotype in plant breeding for water limited environments. Funct Plant Biol. 2015;42(6):502–13.

41. Shabala S, Munns R. Salinity stress: physiological constraints and adaptive mechanisms. Plant Stress Physiol. 2012:59–93.

42. Larsson S, Svenningsson M. Cuticular transpiration and epicuticular lipids of primary leaves of barley (*Hordeum vulgare*). Physiol Plant. 1986;68(1):13–9.

43. Zeisler V, Schreiber L. Epicuticular wax on cherry laurel (*Prunus laurocerasus*) leaves does not constitute the cuticular transpiration barrier. Planta. 2016; 243(1):65–81.

44. Jetter R, Riederer M. Localization of the transpiration barrier in the epi-and intracuticular waxes of eight plant species: water transport resistances are associated with fatty acyl rather than alicyclic components. Plant Physiol. 2016;170(2):921–34.

45. Ristic Z, Jenks MA. Leaf cuticle and water loss in maize lines differing in dehydration avoidance. J Plant Physiol. 2002;159(6):645–51.

46. Vogg G, Fischer S, Leide J, Emmanuel E, Jetter R, Levy AA, et al. Tomato fruit cuticular waxes and their effects on transpiration barrier properties: functional characterization of a mutant deficient in a very-long-chain fatty acid β-ketoacyl-CoA synthase. J Exp Bot. 2004;55(401):1401–10.

47. Schuster AC, Burghardt M, Alfarhan A, Bueno A, Hedrich R, Leide J, et al. Effectiveness of cuticular transpiration barriers in a desert plant at controlling water loss at high temperatures. AoB Plants. 2016;8:27.

12

Mapping of a major QTL for salt tolerance of mature field-grown maize plants based on SNP markers

Meijie Luo[1†], Yanxin Zhao[1†], Ruyang Zhang[1], Jinfeng Xing[1], Minxiao Duan[1], Jingna Li[1], Naishun Wang[1], Wenguang Wang[1], Shasha Zhang[1], Zhihui Chen[2], Huasheng Zhang[1], Zi Shi[1*], Wei Song[1*] and Jiuran Zhao[1*]

Abstract

Background: Salt stress significantly restricts plant growth and production. Maize is an important food and economic crop but is also a salt sensitive crop. Identification of the genetic architecture controlling salt tolerance facilitates breeders to select salt tolerant lines. However, the critical quantitative trait loci (QTLs) responsible for the salt tolerance of field-grown maize plants are still unknown.

Results: To map the main genetic factors contributing to salt tolerance in mature maize, a double haploid population (240 individuals) and 1317 single nucleotide polymorphism (SNP) markers were employed to produce a genetic linkage map covering 1462.05 cM. Plant height of mature maize cultivated in the saline field (SPH) and plant height-based salt tolerance index (ratio of plant height between saline and control fields, PHI) were used to evaluate salt tolerance of mature maize plants. A major QTL for SPH was detected on Chromosome 1 with the LOD score of 22.4, which explained 31.2% of the phenotypic variation. In addition, the major QTL conditioning PHI was also mapped at the same position on Chromosome 1, and two candidate genes involving in ion homeostasis were identified within the confidence interval of this QTL.

Conclusions: The detection of the major QTL in adult maize plant establishes the basis for the map-based cloning of genes associated with salt tolerance and provides a potential target for marker assisted selection in developing maize varieties with salt tolerance.

Keywords: Salt tolerance, QTL mapping, SNP, Plant height, Maize

Background

Elevated salt content in the soil leads to the suppression of plant growth and the reduction of productivity. About 6% of the land on earth and 20% of the total cultivated land worldwide are affected by high salt [1, 2]. In many areas, salinity problem is further aggravated by unsustainable agricultural practices. Salt stress causes ion and hyperosmotic imbalance in plants, which causes secondary oxidative damage. These changes occur at the molecular, cellular, and whole-plant levels, resulting in the plant growth arrest and death [3, 4]. Therefore, understanding the genetic architecture of salt tolerance in plant is of great significance for the selection, utilization, and breeding of salt tolerant varieties.

A major salt tolerance strategy in plant is to re-establish cellular ion homeostasis [4]. A high concentration of sodium ions (Na^+) inhibits many key enzymes, so Na^+ influx to the cell cytoplasm and organelles is sophisticatedly regulated. Therefore, tremendous efforts have been made to identify the transporters modulating the influx and efflux of Na^+ in plant cells. Vacuolar Na^+/H^+ antiporters manage the compartmental Na^+ in the vacuole to prevent Na^+ toxicity in cytosol, which has been shown as a strategy in many naturally salt tolerant plants (halophytes) [5]. In addition to the control of Na^+ influx, Na^+/H^+ antiporters on the plasma membrane are also important in exporting

* Correspondence: shizi_baafs@126.com; songwei1007@126.com; maizezhao@126.com
†Equal contributors
[1]Beijing Key Laboratory of Maize DNA Fingerprinting and Molecular Breeding, Maize Research Center, Beijing Academy of Agriculture and Forestry Sciences (BAAFS), Beijing, China
Full list of author information is available at the end of the article

Na^+ to maintain low Na^+ concentration in the cytoplasm. In *Arabidopsis*, a salt overly sensitive (SOS) signal transduction pathway has been identified to mediate ion homeostasis and Na^+ tolerance [6]. *SOS1* encodes a plasma membrane Na^+/H^+ antiporter, and *sos1* mutation renders *Arabidopsis* the extreme sensitivity to Na^+ stress [7]. Besides the SOS-dependent pathway, other mechanisms have also been reported to play roles in salt tolerance of plants, including the accumulation of osmolytes to establish osmotic homeostasis and the increase of antioxidants to mediate oxidative protection [3, 4].

Quantitative trait locus (QTL) mapping has been applied to detect the genetic basis of salt tolerance in many plants [8], which provides valuable information for further map-based cloning of salt tolerance genes and marker-assisted selection (MAS) in crop breeding. Using restriction fragment length polymorphism (RFLP) markers and an $F_{2:3}$ population derived from the cross between salt tolerant and salt sensitive rice varieties, two major QTLs explaining 48.5% and 40.1% of the total phenotypic variance (PVE) were detected in rice [9]. The QTL detection led to the cloning of the gene responsible for salt tolerance, *SKC1*, which encoded a high affinity K^+/Na^+ transporter [10]. Taking advantage of the amplified fragment length polymorphism (AFLP), RFLP and simple sequence repeat (SSR) markers, a QTL, *Nax1*, was identified on the Chromosome 2AL of durum wheat using an F_2 and an $F_{2:3}$ population, which accounted for 38% of the phenotypic variation, and the SSR marker closely linked to the QTL was proven to be useful for the MAS in the breeding program [11]. Later, the *Nax2* locus was discovered and a Na^+-selective transporter gene *TmHKT1;5-A* was subsequently characterized in durum wheat [12]. In soybean, utilizing random amplified polymorphic DNA (RAPD), insertion-deletion (InDel), and SSR markers with an $F_{2:3}$ population and recombinant inbred lines (RILs), a salt tolerance QTL was mapped within a 209-kb region on Chromosome 3 and its flanking markers were used for MAS in soybean breeding [13]. In maize, nine conditional QTLs for salt tolerance at the seedling stage were identified on Chromosomes 1, 3, and 5 using single nucleotide polymorphism (SNP) markers and $F_{2:5}$ RILs, three of which explained more than 20% of phenotypic variation [14]. However, studies of QTLs for salt tolerance in maize are still very limited, and QTLs for salt tolerance has not been reported in mature field-grown maize.

In this study, we identified a major QTL for salt tolerance in mature maize grown in a saline field using a permanent double haploid (DH) population and high-density SNP markers, and two candidate genes harbored in this QTL might be involved in the SOS pathway. Our results not only shed light on the mechanism of salt tolerance in field-grown maize, but will also facilitate the breeding of maize varieties with salt tolerance.

Methods

Plant materials and treatment

The parental maize inbred lines, PH6WC and PH4CV, were obtained from DuPont Pioneer (Johnston IA, USA). A DH population consisting 240 lines derived from the F_1 hybrid of PH6WC × PH4CV was developed by pollinating with the parthenogenetic-inducing line Jingkeyou006 to obtain the haploid plants, and then followed by artificial doubling with colchicine [15]. Jingkeyou006 was obtained from the Maize Research Center of Beijing Academy of Agriculture and Forestry Sciences.

The 240 DH lines and their parents PH6WC and PH4CV were used in field experiments. For the salt stress treatment, plant materials were planted in the saline field at Tongzhou (TZ, N39°41′49.70″, E116°40′50.75′ in Google Earth™), Beijing, China, in the spring of 2014, 2015, and 2016. Plants cultivated in the normal field at Changping (CP, N40°10′50.38″, E116°27′15.40″), Beijing, in the spring of 2014, 2015 and 2016 served as controls. All field experiments were performed in accordance with local legislation. At each location, the experiments were conducted in a randomized complete block design [16–18] with two replicates in each year. For each block, 20 plants of each DH line were planted for a whole row, with the row length of 5 m and the spacing between rows of 60 cm.

Soil sampling and analyses

According to the five-point sampling method [19], soil samples were collected from five representative locations (upper left, upper right, lower left, lower right and the middle) of the fields in both TZ and CP for composition analysis. Total salt content was determined using the residue-drying method [20]. Soil pH was measured using a pH meter (Hach, Loveland, CO, USA), and Na^+ content was determined by flame emission spectroscopy (PerkinElmer, Norwalk, CT, USA) [21].

Phenotype analysis

At harvest stage, in the summer of 2014, 2015, and 2016, plant height of the DH lines and their parents in the saline field (SPH) and control field (NPH) were recorded. Plant height was measured from the top of the main inflorescence down to the ground. For 240 DH lines, five randomly selected plants in each row were measured as one replicate, and two replicates were performed each year. After collecting plant height data in TZ and CP, salt tolerance index (PHI) based on plant height for each line was calculated using the following formula, PHI = H_{SPH}/H_{NPH}, where H_{SPH} represents the average height of five mature maize plants of each DH line grown in TZ, H_{NPH} represents the average height of

10 mature maize plants (5 of each replicate) of the same line grown in CP.

Linkage map construction and QTL identification

Total genomic DNA was extracted from leaves using the CTAB method and then the 240 DH lines were genotyped using the MaizeSNP3072 chip [22]. A comparative linkage map was constructed using the Kosambi function in JoinMap4 software with a minimum LOD score of 2.0. Composite interval mapping was carried out using Windows QTL Cartographer software V2.5 which was developed by the Department of Statistics, NCSU with a walk speed of 1.0 cM and the LOD threshold of 3.0.

Candidate gene analysis

Salt tolerance-related genes previously characterized in other plant species, such as *AtSOS1* [7], *AtSOS2* [23], *AtSOS3* [24], *OsSKC1* [10], and *TmHKT1;5* [12], were employed to query the maize genome database (MaizeGDB, http://www.maizegdb.org/, B73 RefGen_v2) using the tool of local BLASTP with the e-value cutoff of 1e-4 [25]. The maize homologs fell in the confidence interval of the identified major QTL were considered as candidate genes associated with salt tolerance.

qRT-PCR analysis

Seeds of PH6WC and PH4CV were surface sterilized with 1% NaClO for 10 min, following by rinsing with sterile water for three times. The resulting sterile seeds were sown in maize seedling identifying instrument (Chinese patent, patent number: ZL200920177285.0) according to its manufacturer's instructions, and then were placed in a greenhouse under 12 h light/ 12 h dark at 25 °C with the light density of 150–180 μmol m^{-2} s^{-1} and the relative humidity of 70% [26]. Seeds were grown in sterile water for three days and then in the Hoagland's nutrient solution (Phyto Technology Laboratories Co., Ltd., USA) which was replaced with fresh Hoagland's solution for every 2 days. After 7 days, non-germinated seeds and seedlings exhibited abnormal growth were discarded and the remaining seedlings were treated with 100 mM NaCl at day 12. Shoot and root samples were collected in three biological replicates each with five seedlings, and were immediately frozen in liquid nitrogen for RNA extraction.

Total RNA of shoot and root was extracted using the TRIZOL reagent (Invitrogen, Carlsbad, CA, USA). After RNA was treated with RNase-free DNase1 (Fermentas, Thermo scientific, USA), reverse transcription was conducted using the PrimeScript™ II 1st strand cDNA Synthesis Kit (Takara Bio Inc., Shiga, Japan) according to the user's manual. qRT-PCR was carried out using the QuantStudio™ 6 Flex (ABI Life Technologies, USA).

Gene-specific forward and reverse primers were designed (Additional file 1) using the PrimerQuest tool on IDTDNA (http://sg.idtdna.com/Primerquest/Home/Index). The reaction was performed in the 20 μl PCR system using the SYBR Premix Ex TaqII (Takara Bio Inc., Shiga, Japan) according to the user's manual. *ZmActin1* served as the internal reference and mRNA relative expression levels were calculated using the $2^{-\triangle\triangle Ct}$ method.

Statistical analysis

Using the Graphpad Prism software (http://www.graphpad.com/), the average of total salt content, Na$^+$ content, and pH in fields of TZ and CP was compared by t-test. Two-way ANOVA with Bonferroni test was carried out for plant height of two parental lines grown in two soil conditions as well as the gene expression analysis. The frequency distribution, linear regression, and correlation analysis of the average of SPH, NPH and PHI across three environments were determined with the non-linear Gaussian regression, linear regression and correlation function in the column analysis, respectively. The combined ANOVA of SPH, NPH and PHI for the DH population and the heritability of three traits were obtained using the ANOVA tool in IciMapping 4.1 program [27].

Results

Soil composition

Soil in agricultural land is usually subject to the compound effect of salt and alkali. To understand the main stress factor in soil in this study, we analyzed the soil composition in TZ and CP fields at the depths of 0–20 and 20–40 cm (Fig. 1). The total salt content and Na$^+$ content of soil from both the 0–20 cm and 20–40 cm layers in TZ were significantly higher than those in CP, but no substantial difference in the pH was observed between two locations, indicating that salt, rather than pH, was the major stress factor in TZ soil.

Phenotypic variation and correlations

To evaluate salt tolerance of mature maize in the field, plant height in the saline field (SPH) and salt tolerance index (PHI) were used as salt tolerance indicators. At the mature stage, although the plant height of both PH6WC and PH4CV were significantly reduced by salt stress (Fig. 2a and b), the PHI of PH6WC was significantly higher than that of PH4CV, indicating that PH6WC is less sensitive to salt stress compared to PH4CV (Fig. 2c). The variance effects of genotype (G), environment (E), and their interaction were extremely significant for all three traits (Table 1), and variance of the replicates within seasons for SPH and PHI was also significant. Broad sense heritability (H^2) of SPH, NPH and PHI was 74.7, 86.4 and 74.9% respectively,

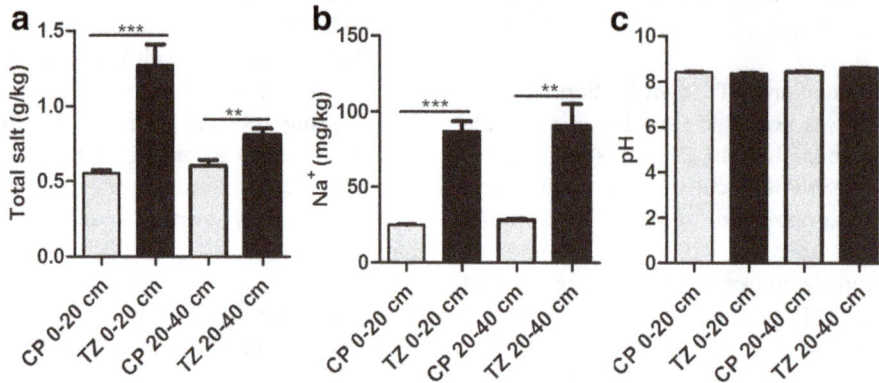

Fig. 1 Soil composition in CP and TZ fields at depths of 0–20 and 20–40 cm. The bar charts represent the mean ± SE of five soil samples from each field. **a** Total salt content. **b** Na$^+$ content. **c** pH value. *, **, and *** indicate significant difference at $P < 0.05$, $P < 0.01$, and $P < 0.001$, respectively. CP indicates Changping. TZ indicates Tongzhou

indicating that genotypes play an important role in the determination of these phenotypes. Phenotypic frequencies of all three traits showed a normal or near-normal distribution (Fig. 3). The correlation coefficient of the three-year average of SPH and NPH was low ($r = 0.397$), but it was as high as 0.686 between SPH and PHI, suggesting that the phenotypic results of SPH and PHI were highly positively correlated. As expected, NPH was negatively correlated with PHI with a low correlation coefficient of −0.229 (Fig. 3).

Genetic linkage map construction based on SNP markers
A genetic linkage map was developed using the 240 DH lines and MaizeSNP3072 chip [22]. Of the 3072 SNPs, 1337 were polymorphic between the parents. For the 1337 SNPs, twenty SNPs were removed because of inconsistencies between their physical and genetic positions. With the 1317 SNPs, the linkage map constructed covered 10 chromosomes and spanned 1462.05 cM of the maize genome with an average distance of 1.11 cM between marker loci.

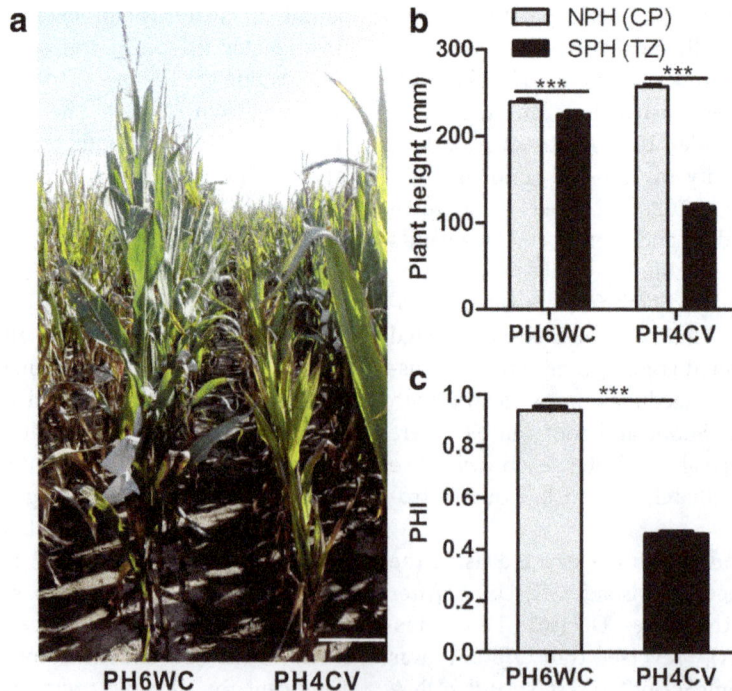

Fig. 2 Comparison of mature maize plant height of PH6WC and PH4CV in saline field at TZ. **a** The representative image of field grown PH6WC and PH4CV in TZ. **b** Plant height of field grown PH6WC and PH4CV at Changping (CP) (control) and TZ (salt stress) in 2016. Bar charts represent the mean ± SE of 15 maize plants. **c** Effect of salt stress on plant height of PH6WC and PH4CV. Scale bar = 25 cm

Table 1 Analysis of variance (ANOVA) for plant height of DH population in three environments

Trait	Source of variation	F	H^2
SPH	Environment (E)	31.8862[***]	0.7471
	Genotype (G)	14.1894[***]	
	Replication	3.0553[*]	
	G × E	4.4658[***]	
NPH	Environment (E)	4.7790[**]	0.8639
	Genotype (G)	39.2024[***]	
	Replication	0.7548	
	G × E	6.0168[***]	
PHI	Environment (E)	12.3846[***]	0.7492
	Genotype (G)	16.2502[***]	
	Replication	3.6290[*]	
	G × E	5.1059[***]	

H^2 indicates broad-sense heritability. [*], [**] and [***] represent significant levels at $P < 0.05$, $P < 0.01$ and $P < 0.001$, respectively

Identification of salt tolerance related QTLs in maize

Based on the average plant height across three seasons at TZ, we identified one major QTL on Chromosome 1 with a LOD score of 22.40 and the PVE of 31.24%, which was designated as $qSPH1$ (Table 2; Fig. 4). The LOD peak of $qSPH1$ was mapped at 88.51 cM on Chromosome 1 and its confidence interval covered the region of 77.61–98.18 cM between the SNP markers $PZE101094436$ and $PZE101150513$. In fact, $qSPH1$ was detected across all seasons (Additional file 2), indicating that it is highly stable and not sensitive to environmental factors. Five QTLs controlling NPH were detected on chromosomes 4, 5, 8 and 9, respectively, with the PVE ranged from 5.0% to 11.9%. Except for $qNPH8$ (LOD, 3.88; PVE, 6.36) which was identified across all three environments, $qNPH4$, $qNPH5$, $qNPH9–1$ and $qNPH9–2$ were detected only in single or two environments (Table 2; Fig. 4). There was no QTL controlling NPH mapped on Chromosome 1, revealing that $qSPH1$ is the locus associated with salt tolerance, but not the plant height of maize.

In addition, to exclude the impact of inherited factor on plant height under salt stress, we compared the plant height between the saline and control fields for each maize line to obtain PHI. Using the average of PHI across three seasons, a major QTL, named $qPHI1$, was identified on Chromosome 1 with the LOD score of 16.2, which accounted for 25.94% of the total phenotypic variation (Table 2; Fig. 4). The LOD peak of $qPHI1$ was mapped at 90.21 cM on Chromosome 1 and its confidence interval spanned from 80.11 to 100.71 cM of the genetic map between the SNP markers $PZE101109084$ and $SYN25920$. Similar to $qSPH1$, $qPHI1$ was also identified across all three seasons (Additional file 2). The similar confidence intervals, positions, and LOD scores

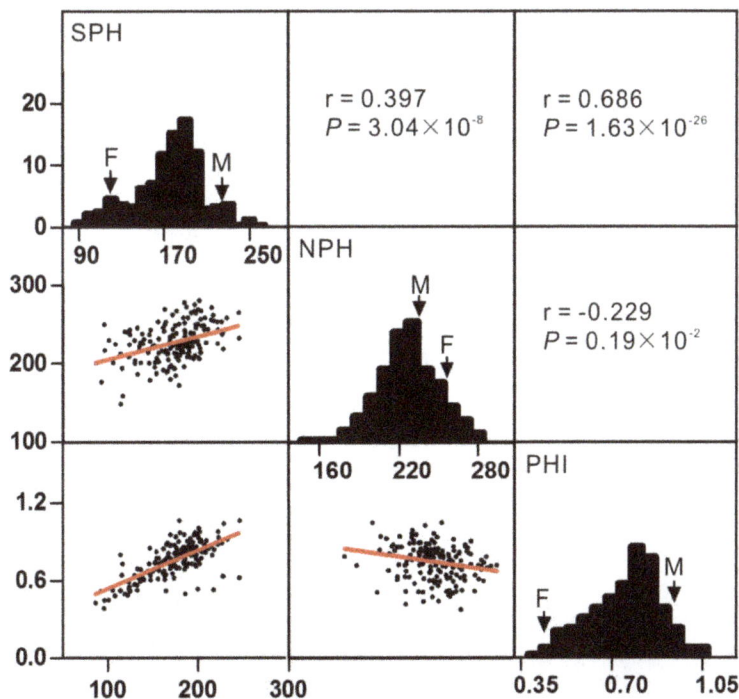

Fig. 3 Frequency distribution and correlation analysis of NPH, SPH and PHI traits across three seasons. Frequency distribution and correlation analysis were conducted using Graphpad Prism software (http://www.graphpad.com/). r indicates correlation coefficient. F represents PH4CV and M represents PH6WC

Table 2 QTLs controlling plant height of mature maize grown in saline soil in three planting seasons

Traits[a]	Years	QTL[b]	Chr.	Position (cM)	Marker interval (Coordinates)	LOD	PVE[c] (%)	Add[d]
NPH	2014	qNPH4	4	49.71	PZE104023902 - SYN4889 (27048623–39,360,273)	4.39	8.33	8.40
		qNPH8	8	56.01	PZE108041337- PZE108090114 (66870236–146,632,672)	3.67	6.90	8.12
	2015	qNPH4	4	46.01	PZE104023902 - PZE104081530 (27048623–155,939,590)	6.28	10.11	7.78
		qNPH5	5	73.51	PZE105045981 - PZE105128581 (34001093–185,707,663)	3.19	4.98	5.55
		qNPH8	8	61.61	PZE108028588 - PZE108103365 (26352213–158,556,937)	5.03	7.07	6.94
		qNPH9–2	9	71.31	PZE109064469 - SYN27201 (107573682–139,631,117)	4.56	7.24	6.61
	2016	qNPH8	8	57.31	PZE108041337 - PZE108097446 (66870236–152,623,958)	4.29	7.75	8.11
		qNPH9–1	9	45.81	PZE109011840 - PZE109040519 (12582417–62,366,526)	5.79	10.74	8.93
	mean	qNPH4	4	49.71	PZE104023902 - PZE104081530 (27048623–155,939,590)	6.97	11.90	8.13
		qNPH8	8	57.31	PZE108028588 - PZE108090114 (26352213–146,632,672)	3.88	6.36	6.40
		qNPH9–2	9	71.31	PZE109064469 - SYN27201 (107573682–139,631,117)	4.80	7.97	6.66
SPH	2014	qSPH1	1	95.21	SYN309 - SYN25920 (91198154–194,798,124)	4.47	13.28	13.79
	2015	qSPH1	1	88.51	PZE101094436 - PZE101150513 (92353978–194,525,458)	11.46	16.99	16.86
		qSPH 5–1	5	77.31	SYN16675 - PZE105128581 (165361318–185,707,663)	3.50	5.01	8.82
	2016	qSPH1	1	90.21	PZE101109084 - SYN25920 (116462939–194,798,124)	19.47	35.03	24.69
		qSPH5–2	5	71.01	PZE105049283 - PZE105117757 (40588124–174,670,738)	4.26	6.17	19.86
	mean	qSPH1	1	88.51	PZE101094436 - PZE101150513 (92353978–194,525,458)	22.40	31.24	19.14
		qSPH5–1	5	77.31	SYN1390 - PZE105128581 (163779167–185,707,663)	3.32	3.96	6.55
PHI	2014	qPHI1	1	94.81	PZE101109084 - SYN25920 (116462939–194,798,124)	8.94	28.10	0.10
		qPHI3	3	23.51	SYN28626 - PZE103019163 (2813337–11,407,445)	4.63	13.64	0.67
	2015	qPHI1	1	90.21	SYN5444 - SYN25920 (96632104–194,798,124)	10.59	19.51	0.08
	2016	qPHI1	1	90.21	PZE101109084 - SYN25920 (116462939–194,798,124)	14.27	27.51	0.09
	mean	qPHI1	1	90.21	PZE101109084 - SYN25920 (116462939–194,798,124)	16.20	25.94	0.07
		qPHI4	4	56.91	SYN4889 - SYN4250 (39360273–170,439,029)	4.14	5.63	0.03
		qPHI9	9	130.34	SYN24345 - SYN5732 (153690286–155,054,745)	3.39	5.59	0.03
		qPHI10	10	108.16	PZE110100655- SYN19213 (144384594–147,029,157)	3.29	4.47	0.03

[a]NPH, plant height under normal condition (Changping, Beijing). SPH, plant height under salt stress (Tongzhou, Beijing). PHI, plant height-based salt tolerance index. [b]QTLs are designated as "q" followed by trait name: "NPH" (plant height under normal field), "SPH" (plant height under salt stress) and "PHI" (plant height-based salt tolerance index), and then the chromosome number. [c]Phenotypic variation explained by QTL. [d]Additive effect of tolerance allele

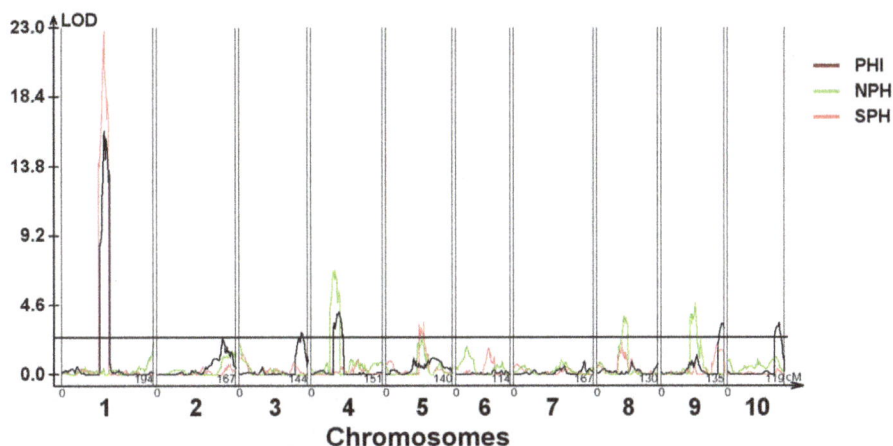

Fig. 4 QTL locations of salt tolerance in field grown maize. *qSPH* represents QTL based on average plant height under salt stress across three seasons; *qNPH* represents QTL based on average plant height under normal condition across three seasons; *qPHI* represents QTL based on the average plant height salt tolerance index across three seasons

of *qSPH1* and *qPHI1*, along with the fact that no QTL detected for NPH on Chromosome 1 indicate that *qSPH1* is a major QTL responsible for the salt tolerance of mature maize but not the loci controlling the plant height.

Several minor QTLs were also detected under saline field condition, such as *qSPH5−1* and *qSPH5−2* on Chromosome 5. *qSPH5−2* (LOD, 4.26; PVE, 6.17%) was only identified in 2016, and *qSPH5−1* was detected both in 2015 and the mean of three-year SPH data (Table 2). Four additional minor QTLs responsible for PHI, named *qPHI3*, *qPHI4*, *qPHI9*, and *qPHI10*, were identified on Chromosomes 3, 4, 9 and 10, respectively, with the LOD score ranged from 3.29 to 4.63, which contributed 4.47% to 13.64% of the phenotypic variation. *qPHI3* was only detected in 2014, while other three minor QTLs were identified only based on the three-year average of PHI (Table 2). Their inconsistent detection in different planting seasons indicated that these QTLs were susceptible to environmental conditions.

Candidate genes in the major QTL

AtSOS1, *AtSOS2*, *AtSOS3*, *OsSKC1*, and *TmHKT1;5* have been shown to play major roles in the salt tolerance of plants [4, 10, 12]. Using the protein sequences of these genes as queries, we screened the annotated maize genome database, and discovered two candidate genes within the confidence interval of the detected major QTL: *GRMZM2G007555* and *GRMZM2G098494* (Additional files 3 and 4). No homologs of *AtSOS2*, *OsSKC1*, and *TmHKT1;5* were identified within the QTL region. The product of maize gene *GRMZM2G007555* showed 61.38% identity to AtSOS3 which was predicted to encode a calcineurin B-like (CBL) protein, and thus it might be involved in regulating calcium signaling under salt stress.

GRMZM2G098494 was a homolog of AtSOS1 and was annotated as a Na^+/H^+ antiporter.

To further validate the potential candidate genes, their relative expression levels in the shoot and root of salt treated seedlings were determined by qRT-PCR (Fig. 5). In shoot, the expression of both candidate genes in the salt tolerant parent PH6WC was extremely significantly higher than that of the salt susceptible parent PH4CV at all three tested time points, 0 h, 1 h and 2 h after salt treatment, though the expression showed fluctuation in both lines (Fig. 5a and b). In root, although the basal expression of *GRMZM2G098494* in PH4CV was comparable to that in PH6WC, its induced expression was exceptionally higher in PH6WC compared to that of PH4CV (Fig. 5c). While the expression of *GRMZM2G07555* in PH6WC held at a steadily high level, it was dramatically decreased in PH4CV as the duration increase of salt treatment (Fig. 5d). Therefore, the detection of these two candidate genes in the major QTL along with their expression pattern upon salt treatment in both parents suggests that ion homeostasis regulation may play an important role in the salt tolerance of field-grown maize.

Discussion

Utilizing salt tolerant maize germplasms is an important practice to battle the challenge of field saline stress, so understanding the genetic basis controlling salt tolerance in the mature field-grown maize can guide the breeding of maize varieties with salt tolerance. Therefore, QTL mapping for salt tolerance of mature maize in field conditions is of great significance, as it is closely related to crop production. In the present study, we identified a major QTL associated with salt tolerance in field-grown maize using a DH population (Table 2; Fig. 4).

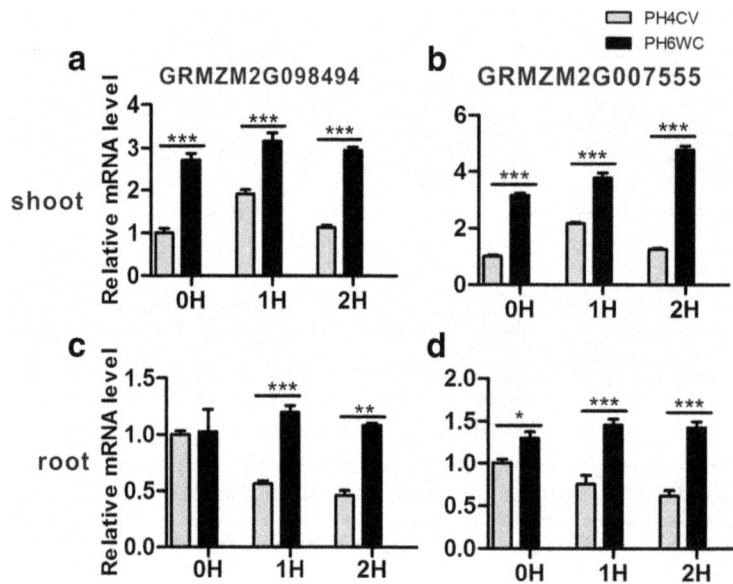

Fig. 5 The relative expression level of potential candidate genes in maize seedlings of two *inbred lines*. Total RNA samples were collected at 0 h, 1 h and 2 h after salt treatment from shoot and root tissues, respectively. qRT-PCR was performed using maize *Actin* as the internal reference. Bar charts represent the mean ± SE of three biological replicates, each containing tissues from five individual seedlings. *Asterisks* show statistically significant difference in two lines by ANOVA. *, ** and *** indicate $P < 0.05$, $P < 0.01$ and $P < 0.001$, respectively. Expression of *GRMZM2G098494* in shoot (**a**), *GRMZM2G007555* in shoot (**b**), *GRMZM2G098494* in root (**c**) and *GRMZM2G007555* in root (**d**)

Salt tolerance in plant is a complicated phenomenon and many parameters have been used as indicators to evaluate its effect, including agronomic traits [28], physiological indicators [26, 29] and phenotypic classification [14]. As an important index for plant salt tolerance, yield served as one of the phenotypic factors in the QTL analysis of salt tolerance in tomato [30, 31] and bread wheat [16]. However, grain yield in maize typically exhibits low heritability [32] and is affected by many external conditions. On the contrary, plant height is an important agronomic trait with relatively high heritability, and it has been shown to positively correlate to grain yield, such as in the V12 growth stage of maize [33], coffee [17] and wheat [34]. Therefore, along with its easy determination at the early growing stage, plant height may serve as a desired proxy of the grain yield. Actually, it has been employed in the identification of QTLs conditioning salt tolerance in other crop species, including a barley DH population [35] and rice RILs [36]. Thus, we believe that plant height of maize growing in the saline soil is the appropriate trait to investigate QTLs associated with salt tolerance.

In this study, the broad sense heritability of plant height in saline fields was 74.7%, indicating that phenotypic variation of this trait was largely determined by genotype. It is comparable with the high heritability of plant height in maize under heat stress [37] and the tall fescue under drought stress [38]. Based on this trait, a major QTL of *qSPH1* was identified on Chromosome 1,

which explained 31.2% of phenotypic variation (Table 2). Only five minor QTLs controlling plant height with PVE less than 12% were identified under normal environment and none of them was co-localized with *qSPH1*, suggesting that the loci of *qSPH1* was linked to salt tolerance of mature field-grown maize.

To eliminate the inherent effect of agronomic traits, the ratio between stress and normal conditions has been used to evaluate the degree of stress tolerance in maize [39], rice [40], and cowpea [41]. Accordingly, the PHI for each DH line was calculated in this study and was used to map QTLs for salt tolerance. A major QTL, *qPHI1*, explaining 25.94% of PVE was identified at the same location as *qSPH1*, further verifying that this main QTL is responsible for salt tolerance in maize.

Although seedlings and mature plants are different at developmental stages, and their physiological characters and stress resistance mechanisms usually vary [42], but the genetic factors controlling a certain trait can still be the same in different stages [43]. Using an $F_{2:5}$ RILs population, Cui et al., identified a major QTL, *QFgr1* or *QFstr1*, flanked by the marker *PZE101140869* and *PZE101138116*, which was responsible for the maize germination rate in saline fields or seedling salt tolerance ranking [14]. Comparing the QTL locations, we found that the position of *qSPH1* and *qPHI1* was comparable to *QFgr1* or *QFstr1*. Thus, it seems that this major QTL confers salt tolerance in two different populations, and it controls salt tolerance of both maize seedling stage and mature stage. In this regard, this QTL location may serve as a good target site for MAS in maize breeding to increase the salt tolerance of salt susceptible and elite inbred lines during both seedling and adult stages.

Tremendous efforts have been made to identify genes controlling salt tolerance in plants. In *Arabidopsis*, the SOS pathway has been demonstrated to mediate salt tolerance which comprises three members: SOS1, SOS2, and SOS3. After binding Ca^{2+}, SOS3 interacts with SOS2 to form a kinase complex, which further activates SOS1, a plasma membrane Na^+/H^+ antiporter that exports Na^+ out of the cell [3, 4]. In addition, *OsSKC1* [10] and *TmHKT1;5* [12] were responsible for salt tolerance in rice and wheat, respectively, and they were both cloned as Na^+ transporters. To investigate whether genes in the confidence interval of the major QTL in the present study share any similarity to these genes cloned in other species, blastp was performed and two proteins GRMZM2G007555 and GRMZM2G098494 were shown to be homologous to AtSOS3 and AtSOS1 with the e-value of 1.740e-62 and 0, respectively. Such low e-values showed substantial protein identity, suggesting that two candidate genes may share similar function to *AtSOS3* and *AtSOS1* in the resistance to salt stress. The induced level of both genes was not substantially elevated at 1 h and 2 h after salt treatment (Fig. 5), which probably due to the sample collecting time points and the possible discrepancy in the salt response time between *Arabidopsis* [44] and maize seedlings, but the expression of both *GRMZM2G098494* and *GRMZM2G007555* in the salt tolerant PH6WC was significantly higher than that of PH4CV in both shoot and root tissues of maize seedlings in basal and induced conditions, except for *GRMZM2G098494* at 0 h in root. In addition, the expression pattern of both genes in PH4WC was declined as the increase in the treatment duration, whereas PH6WC exhibited a stable or slightly increased level, suggesting that these two genes may be associated with the difference in salt tolerance of both parents. Further investigation with more time points is required to fully understand their expression in the course of salt treatment. Therefore, we speculate that the regulation of ion homeostasis by the SOS pathway may be the genetic control for salt tolerance of mature field-grown maize. However, further research is needed to verify the function of these genes in salt tolerance, which will potentially facilitate the breeding of salt tolerant maize varieties.

Conclusions

In the present study, we have constructed a SNP-based maize genetic linkage map and mapped a major QTL at the 88–95 cM region on Chromosome 1 based on plant height of mature maize grown in a saline field, which accounted for 31.24% of PVE. This main effect QTL for salt tolerance was further validated by QTL mapping based on plant height salt tolerance index, and two candidate genes with predicted functions in the SOS pathway have been identified in the locus, indicating that ion homeostasis regulation may play an important role in the salt tolerance of mature maize. Therefore, this locus will be valuable for marker assisted selection in maize breeding.

Additional files

Additional file 1: Table S1. The primers used in the qRT-PCR.

Additional file 2: Figure S1. Chromosomal locations and logarithm of odds (LOD) scores of the major QTL conditioning SPH and PHI on Chromosome 1 using data from individual years and overall means. Mean represents the average of three growing seasons.

Additional file 3: Figure S2. Protein sequences alignments by BLAST against B73 filtered gene set translations 5b.60 for RefGen_v2 (maizesequence.org) database. **(a)** Alignment of AtSOS1 and GRMZM2G098494. **(b)** Alignment of AtSOS3 and GRMZM2G007555.

Additional file 4: Table S2. The identity of two candidate genes identified in the major QTL region to *Arabidopsis SOS* genes and their positions and annotations.

Abbreviations

AFLP: Amplified fragment length polymorphism; CBL: Calcineurin B-like; CP: Changping; DH: Double haploid; E: Environment; G: Genotype; H^2: Broad sense heritability; InDel: Insertion-deletion; LOD: Logarithm of odds; MAS: Marker-assisted selection; Na^+: Sodium ions; NPH: Plant height in control field; PHI: Salt tolerance index; PVE: Phenotypic variance; QTL: Quantitative trait loci; RAPD: Random amplified polymorphic DNA; RFLP: Restriction fragment length polymorphism; RILs: Recombinant inbred lines; SNP: Single nucleotide polymorphism; SOS: Salt overly sensitive; SPH: Plant height in the saline field; SSR: Simple sequence repeat; TZ: Tongzhou

Acknowledgements
Not applicable

Funding
This research was financially supported by the BAAFS Innovation Team of Corn Germplasm Innovation and Breeding of New Varieties (JNKYT201603 to Jiuran Zhao) and the Postdoctoral Fellow Fund of BAAFS (2016 to Meijie Luo). The funding body does not play roles in the design of the study and collection, analysis, and interpretation of data and in writing the manuscript.

Authors' contributions
MJL, YXZ, ZS and JRZ conceived the experiment and made the revision of the manuscript. MJL, RYZ, YXZ, WS, MXD, JFX, JNL, NSW, WGW, SSZ and HSZ involved in the data acquisition. MJL, ZHC, YXZ, ZS and JRZ analyzed the data and wrote the manuscript. All authors revised and approved the final manuscript, and agreed to be accountable for this work.

Competing interests
The authors declare that they have no competing interests.

Author details
[1]Beijing Key Laboratory of Maize DNA Fingerprinting and Molecular Breeding, Maize Research Center, Beijing Academy of Agriculture and Forestry Sciences (BAAFS), Beijing, China. [2]Institute of Crops Research, Hunan Academy of Agricultural Sciences, Changsha, China.

References

1. Munns R, Tester M. Mechanisms of salinity tolerance. Annu Rev Plant Biol. 2008;9(1):651–81.
2. Cheng TL, Chen JH, Zhang JB, Shi SQ, Zhou YW, Lu L, et al. Physiological and proteomic analyses of leaves from the halophyte *Tangut Nitraria* reveals diverse response pathways critical for high salinity tolerance. Front Plant Sci. 2015;6:30.
3. Zhu JK. Plant salt tolerance. Trends Plant Sci. 2001;6(2):66–71.
4. Zhu JK. Abiotic stress signaling and responses in plants. Cell. 2016;167(2):313–24.
5. Wang J, Meng Y, Li B, Ma X, Lai Y, Si E, et al. Physiological and proteomic analyses of salt stress response in the halophyte *Halogeton glomeratus*. Plant Cell Environ. 2015;38(4):655–69.
6. Zhu JK. Genetic analysis of plant salt tolerance using *Arabidopsis*. Plant Physiol. 2000;124(3):941–8.
7. Shi H, Quintero FJ, Pardo JM, Zhu JK. The putative plasma membrane Na$^+$/H$^+$ antiporter SOS1 controls long-distance Na$^+$ transport in plants. Plant Cell. 2002;14(2):465–77.
8. Tiwari S, SI K, Kumar V, Singh B, Rao AR, Mithra SA, et al. Mapping QTLs for salt tolerance in rice (*Oryza sativa L.*) by bulked segregant analysis of recombinant inbred lines using 50K SNP chip. PLoS One. 2016;11(4): e0153610.
9. Lin HX, Zhu MZ, Yano M, Gao JP, Liang ZW, Su WA, et al. QTLs for Na$^+$ and K$^+$ uptake of the shoots and roots controlling rice salt tolerance. Theor Appl Genet. 2004;108(2):253–60.
10. Ren ZH, Gao JP, Li LG, Cai XL, Huang W, Chao DY, et al. A rice quantitative trait locus for salt tolerance encodes a sodium transprter. Nat Genet. 2005; 37(10):1141–6.
11. Lindsay MP, Lagudah ES, Hare RA, Munns R. A locus for sodium exclusion (*Nax1*), a trait for salt tolerance, mapped in durum wheat. Funct Plant Biol. 2004;31(11):1105–14.
12. Munns R, James RA, Xu B, Athman A, Conn SJ, Jordans C, et al. Wheat grain yield on saline soils is improved by an ancestral Na$^+$ transporter gene. Nat Biotechnol. 2012;30(4):360–4.
13. Guan R, Chen J, Jiang J, Liu G, Liu Y, Tian L, et al. Mapping and validation of a dominant salt tolerance gene in the cultivated soybean (*Glycine max*) variety Tiefeng 8. The Crop Journal. 2014;2(6):358–65.
14. Cui D, Wu D, Somarathna Y, Xu C, Li S, Li P, et al. QTL mapping for salt tolerance based on snp markers at the seedling stage in maize (*Zea mays L.*). Euphytica. 2015;203(2):273–83.
15. Zhang RY, Duan MX, Zhao JR. Difference in the haploid induction rates of six maize inducers. Journal of Maize Sciences. 2013;21(2):6–10.
16. Azadi A, Mardi M, Hervan EM, Mohammadi SA, Moradi F, Tabatabaee MT, et al. QTL mapping of yield and yield components under normal and salt-stress conditions in bread wheat (*Triticum aestivum L.*). Plant Mol Biol Rep. 2014;33(1):102–20.
17. Moncada MDP, Tovar E, Montoya JC, González A, Spindel J, Mccouch S. A genetic linkage map of coffee (*Coffea arabica L.*) and QTL for yield, plant height, and bean size. Tree Genet Genomes. 2016;12(1):1–17.
18. Yang X, Wang Y, Zhang G, Wang X, Wu L, Ke H, et al. Detection and validation of one stable fiber strength QTL on c9 in tetraploid cotton. Mol Gen Genomics. 2016;291(4):1625–38.
19. Zhai Y, Hou M, Jin Q. Influence of land treatment for rural sewage on soil environment and flue-cured tobacco quality. Advance Journal of Food Science & Technology. 2016;11(1):88–94.
20. Wüst RAJ, Schlüchter C. The origin of soluble salts in rocks of the thebes mountains, Egypt: the damage potential to ancient egyptian wall art. J Archaeol Sci. 2000;27(12):1161–72.
21. Miller JJ, Beasley BW, Larney FJ, Olson BM. Soil salinity and sodicity after application of fresh and composted manure with straw or wood-chips. Can J Soil Sci. 2005;85(3):427–38.
22. Tian HL, Wang FG, Zhao JR, Yi HM, Wang L, Wang R, et al. Development of maizeSNP3072, a high-throughput compatible SNP array, for DNA fingerprinting identification of Chinese maize varieties. Mol Breeding. 2015;35(6):136.
23. Liu J, Ishitani M, Halfter U, Kim CS, Zhu JK. The *Arabidopsis thaliana* SOS2 gene encodes a protein kinase that is required for salt tolerance. P Natl Acad Sci USA. 2000;97(7):3730.
24. Halfter U, Ishitani M, Zhu JK. The *Arabidopsis* SOS2 protein kinase physically interacts with and is activated by the calcium-binding protein SOS3. P Natl Acad Sci USA. 2000;97(7):3735–40.
25. Zhao Y, Cai M, Zhang X, Li Y, Zhang J, Zhao H, et al. Genome-wide identification, evolution and expression analysis of *mTERF* gene family in maize. PLoS One. 2014;9(4):e94126.
26. Luo MJ, Zhao YX, Song W, Zhang RY, Su AG, Li CH, et al. Effect of saline stress on the physiology and growth of maize hybrids and their related inbred lines. Maydica. 2017;62(2):11.
27. Meng L, Li H, Zhang L, Wang J. QTL IciMapping: integrated software for genetic linkage map construction and quantitative trait locus mapping in biparental populations. The Crop Journal. 2015;3(3):269–83.
28. Guo H, Ding W, Chen J, Chen X, Zheng Y, Wang Z, et al. Genetic linkage map construction and QTL mapping of salt tolerance traits in Zoysiagrass (*Zoysia japonica*). PLoS One. 2014;9(9):e107249.
29. Frary A, Göl D, Keleş D, Okmen B, Pinar H, Siğva HO, et al. Salt tolerance in *Solanum pennellii*: antioxidant response and related QTL. BMC Plant Biol. 2010;10(1):1–16.
30. Estañ MT, Villalta I, Bolarín MC, Carbonell EA, Asins MJ. Identification of fruit yield loci controlling the salt tolerance conferred by solanum rootstocks. Theor Appl Genet. 2009;118(2):305–12.
31. Villalta I, Bernet GP, Carbonell EA, Asins MJ. Comparative QTL analysis of salinity tolerance in terms of fruit yield using two solanum populations of F 7 lines. Theor Appl Genet. 2007;114(6):1001–17.
32. Bočanski J, Srečkov Z, Nastić A. Genetic and phenotypic relationship between grain yield and components of grain yield of maize (*Zea mays L.*). Genetika. 2009;41(2):145–54.
33. Kelly J, Crain JL, Raun WR. By-plant prediction of corn (*Zea mays L.*) grain yield using height and stalk diameter. Communications in Soil Science & Plant Analysis. 2011;46(5):564–75.
34. Costa JM, Bollero GA. Stability analysis of grain yield in barley (*Hordeum vulgare*) in the US mid-Atlantic region. Ann Appl Biol. 2001;139:137–43.
35. Xue D, Huang Y, Zhang X, Kang W, Westcott S, Li C, et al. Identification of QTLs associated with salinity tolerance at late growth stage in barley. Euphytica. 2009;169(2):187–96.
36. Wang Z, Cheng J, Chen Z, Huang J, Bao Y, Wang J, et al. Identification of QTLs with main, epistatic and QTL × environment interaction effects for salt tolerance in rice seedlings under different salinity conditions. Theor Appl Genet. 2012;125(4):807–15.
37. Johnson EC, Fischer KS, Edmeades GO, Palmer AFE. Recurrent selection for reduced plant height in lowland tropical maize. Crop Sci. 1986;26(2):253–60.
38. Ebrahimiyan M, Majidi MM, Mirlohi A. Genotypic variation and selection of traits related to forage yield in tall fescue under irrigated and drought stress environments. Grass & Forage Science. 2013;68(1):59–71.
39. Cakir R. Effect of water stress at different development stages on vegetative and reproductive growth of corn. Field Crop Res. 2004;89(1):1–16.
40. Ali S, Gautam RK, Mahajan R, Krishnamurthy SL, Sharma SK, Singh RK, et al. Stress indices and selectable traits in SALTOL QTL introgressed rice genotypes for reproductive stage tolerance to sodicity and salinity stresses. Field Crop Res. 2013;154(3):65–73.
41. Win KT, Oo AZ. Genotypic difference in salinity tolerance during early vegetative growth of cowpea (*Vigna unguiculata L.* Walp.) from Myanmar. Biocatalysis & Agricultural Biotechnology. 2015;4(4):449–55.
42. Farooq M, Hussain M, Wakeel A, Siddique KHM. Salt stress in maize: effects, resistance mechanisms, and management. A review Agro Sustain Dev. 2015; 35(2):461–81.
43. Yang C, Guo W, Li G, Gao F, Lin S, Zhang T. QTLs mapping for *Verticillium* wilt resistance at seedling and maturity stages in *Gossypium barbadense L.* Plant Sci. 2008;174(3):290–8.
44. Shi H, Ishitani M, Kim C, Zhu JK. The *Arabidopsis thaliana* salt tolerance gene SOS1 encodes a putative Na$^+$/H$^+$ antiporter. P Natl Acad Sci USA. 2000; 97(12):6896–901.

13

Maize network analysis revealed gene modules involved in development, nutrients utilization, metabolism, and stress response

Shisong Ma[1]*[iD], Zehong Ding[2] and Pinghua Li[3]*

Abstract

Background: The advent of big data in biology offers opportunities while poses challenges to derive biological insights. For maize, a large amount of publicly available transcriptome datasets have been generated but a comprehensive analysis is lacking.

Results: We constructed a maize gene co-expression network based on the graphical Gaussian model, using massive RNA-seq data. The network, containing 20,269 genes, assembles into 964 gene modules that function in a variety of plant processes, such as cell organization, the development of inflorescences, ligules and kernels, the uptake and utilization of nutrients (e.g. nitrogen and phosphate), the metabolism of benzoxazionids, oxylipins, flavonoids, and wax, and the response to stresses. Among them, the inflorescences development module is enriched with domestication genes (like *ra1*, *ba1*, *gt1*, *tb1*, *tga1*) that control plant architecture and kernel structure, while multiple other modules relate to diverse agronomic traits. Contained within these modules are transcription factors acting as known or potential expression regulators for the genes within the same modules, suggesting them as candidate regulators for related biological processes. A comparison with an established Arabidopsis network revealed conserved gene association patterns for specific modules involved in cell organization, nutrients uptake & utilization, and metabolism. The analysis also identified significant divergences between the two species for modules that orchestrate developmental pathways.

Conclusions: This network sheds light on how gene modules are organized between different species in the context of evolutionary divergence and highlights modules whose structure and gene content can provide important resources for maize gene functional studies with application potential.

Keywords: Comparative genomics, Gene network analysis, Maize development, Maize metabolism pathways, Plant nutrient uptake and utilization

Background

Advances over the past two decades have generated numerous transcriptome datasets. Increasingly, ever more complete transcriptome data can be merged and integrated with gene and genome structures. They provide unbiased snapshots of gene expression dynamics within organisms under various conditions. While cell-, organ-, or condition-specific expression profiles abound, it remains a key challenge to deduce the underlying gene regulatory circuits that control and give rise to the observed gene expression dynamics. To address this, gene network analysis has emerged as a tool that can filter and refine the analysis of large gene expression datasets. Such gene networks consist of genes (nodes) and connections (edges) between genes that represent co-expression dynamics or association patterns underlying the expression data. According to a 'guilt-by-association'

* Correspondence: sma@ustc.edu.cn; pinghuali@sdau.edu.cn
[1]School of Life Sciences, University of Science and Technology of China, Hefei, Anhui, China
[3]State Key Laboratory of Crop Biology, College of Agronomy, Shandong Agricultural University, Tai'an, Shandong, China
Full list of author information is available at the end of the article

paradigm, connected genes may have similar functions, be part of the same complex or pathway, or participate in the same signaling circuits [1]. Gene networks can assign putative functions to unknown genes based on functions revealed by their associates, or to identify novel genes for existing pathways [2, 3].

Different from random networks, gene networks generally identify gene modules categorizing and tracing groups of highly inter-connected genes that share similar expression patterns and, often, are recognizable by their relationship to a function in a particular biological process. Each module can be viewed as a unit in control of one or several biological functions. As such, modules identify and bring together segments of a biological system. Gene network analysis has been used to detect gene modules associated with, for example, human diseases or plant seed germination, and to study the transcriptome landscapes and gene module organization in yeasts, plants, and animals [4–10].

Gene networks derived from microarray data have been described for plant species such as Arabidopsis, maize, and rice [6, 8, 11–13]. These networks have been constructed via a variety of approaches that employed different ways to measure interactions between genes. Most common are co-expression networks that utilize the Pearson correlation coefficient (PCC) to measure expression similarity between genes, where gene pairs with PPC larger than a chosen threshold value are considered to interact with each other. Examples included two Arabidopsis networks: one identified clusters of genes involved in processes such as photosynthesis, vitamin metabolism, or cell cycle-regulation, while a second network revealed groupings with cellular organelles and tissue-specific functions [8, 11]. Rice and maize networks were also assembled via the Weighted Gene Coexpression Network Analysis (WGCNA) method, which utilizes a power function of PCC to assess expression similarity [6, 12–14]. Specifically, Downs et al. used WGCNA to construct a maize developmental gene co-expression network that captured modules with tissue and developmental stages specificity, while Ficklin and Feltus compared WGCNA networks from maize and rice to identify conserved modules [6, 13].

Yet another way to approach co-expression network analysis uses the graphical Gaussian model (GGM), which utilizes partial correlation (Pcor) to identify association relationships between genes [15–17]. Significantly, Pcor determines the correlation that remains between two genes after removing the effects of all other genes. Pcor measures direct association between genes, while PCC often fails to differentiate between direct and indirect associations [15, 16]. Thus, Pcor is deemed as a better metric than PCC for gene network analysis [16, 18]. However, the utilization of GGM has been impeded by a

requirement that the number of samples must be far larger than the number of genes. The original design was able to calculate Pcor data among a few thousand genes only [16]. Previously, we developed a random sampling-based method to overcome this obstacle and constructed the first genome-wide GGM network for Arabidopsis, followed by an updated network model termed AtGGM2014 [7, 17]. Compared to other networks, AtGGM2014 contained more genes and identified additional modules participating in a large variety of plant processes, like development, metabolism, response to stresses, and response to hormones [7]. For example, among many informative gene modules, it included hormonal signaling modules for phytohormones like auxin, abscisic acid, jasmonic acid, gibberellins, cytokinins, ethylene, and salicylic acid, demonstrating the network's potential to facilitate systems biology studies on Arabidopsis gene functions.

More recently, a large collection of RNA-Seq based maize gene expression datasets have been generated by different groups and deposited in the public domain. Among these datasets are a gene expression atlas for 79 maize tissues [19, 20], as well as expression datasets for specific organs like inflorescences [21], leaves [22], ligules [23], embryos and endosperms [24], and transcriptome datasets for different compartments in the endosperm [25, 26]. Others resulted from monitoring maize responses to abiotic stresses [27, 28], fungal infections [29], and different nutrition regimes [30]. These datasets provide extremely valuable information for maize functional genomics research. A maize RNA-Seq-based gene network focused exclusively on development has been constructed from expression data of 23 different tissues, which identified 19 gene modules via the WGCNA method [31]. However, to our knowledge, missing so far is a comprehensive gene network analysis that combines the wealth of many different maize datasets and merges them into an inclusive network to outline future research opportunities.

Here, we constructed a maize GGM gene co-expression network using expression data from 787 RNA-Seq runs deposited in the NCBI SRA database. 964 gene modules were identified from this network, highlighting functions in various cell organization, development, nutrients, metabolism, and stress responses pathways. As examples, we describe in detail modules involved in the development of inflorescences, ligules, and kernels, the uptake and utilization of nitrogen and phosphate, the metabolism of benzoxazionids, oxylipins, flavonoids and wax, and the response to heat stress, endoplasmic reticulum (ER) stress, and fungal infections. These modules provide a general picture for the relevant biological processes and identified both known genes as well as potential, so far unconnected, candidate genes for future functional studies. Importantly, many of these modules contain transcription factor genes that act as potential gene expression regulator for the genes within the same modules. In addition, the maize network has been compared to a previous published Arabidopsis network [7]. This juxtaposition revealed conserved as well as diverging modules in the two species. The identified gene modules were further used to analyze a dataset on a maize leaf developmental series [22], demonstrating the usefulness of this network for systems biology analysis.

Results

Overview of the maize GGM gene network

A maize gene co-expression network based on GGM was constructed using maize RNA-Seq transcriptome datasets deposited in the NCBI Sequence Read Archive (SRA) database. The publicly available raw data files (.sra) were downloaded, mapped against the maize reference genome (AGP_v3.22), and processed into gene expression values. After removing files with mapping rates <70%, 787 RNA-Seq runs from 36 different studies (Additional file 1: Table S1) were retained and their Fragments Per Kilobase of transcript per Million mapped reads (FPKM)-normalized gene expression values were combined into a large gene expression matrix. These datasets monitor maize transcriptomes from various tissues and developmental stages, after a variety of biotic and abiotic stress treatments, or from well-characterized mutants. The genes with maximum FPKM values <20 were filtered out, resulting in a gene expression matrix that included 29,316 maize genes (rows) and 787 RNA-Seq runs (columns). Similar to other RNA-Seq data, these maize RNA-Seq data showed a mean-variance dependency [32], i.e. genes with higher

mean expression values were more likely to have larger variances. The gene expression matrix was transformed via log-transformation to reduce such mean-variance dependency (Additional file 2: Fig. S1), making the dataset more suitable for correlation analysis [32]. Log-transformation has been used before for gene co-expression and gene clustering analysis with RNA-Seq data [33–40]. The log-transformed gene expression matrix was then used for the calculation of partial correlation coefficients (Pcor), following a procedure described before [17]. Figure 1a shows the distribution of the Pcors between all genes pairs, 98.4% of which are in the range between −0.01 and 0.01. The gene pairs with |Pcor| > = 0.035 (pValue = 2.22E-16) were kept for network construction. Additionally, Pearson Correlation Coefficients (PCC) was also calculated for all gene pairs and those with |PCC| < 0.35 were removed, reasoning that a low PCC very likely indicates independence between genes. As a result, 123,093 genes pairs with Pcor > = 0.035 and PCC > = 0.35 and 573 gene pairs with Pcor <= −0.035 and PCC < = −0.35 were selected for gene network construction (Additional file 3: Table S2).

This reasoning resulted in a maize GGM network including 20,269 genes and 123,666 co-expressed gene pairs (0.06% of all possible gene pairs). The clustering coefficient (C) of the network is 0.209, while the expected C of a random network of the same size is only 0.0006 [41]. Since C measures a network's potential modularity [42], the large C of the maize GGM network indicates high characteristics of modularity. Indeed, a network clustering procedure using the MCL program [43] identified 964 gene modules that contain 5 or more genes each, and these 964 modules encompass 16,668 genes in total. These modules can be viewed as co-expression modules, whose genes share similar

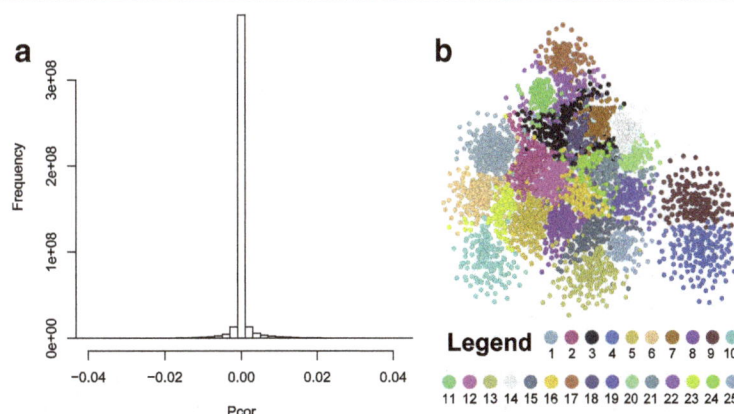

Fig. 1 An overview of the maize GGM network. **a** A histogram showing the distribution of the partial correlation coefficients of all the gene pairs. Most genes pairs have their Pcors in the range between −0.01 and 0.01, indicating no interactions. **b** A sub-network for the largest 25 modules identified from the network. *Each dot* represents a gene, and a connection between two genes indicates interaction between the two. The *color of a node* indicates its module identity, as shown in the legend

co-expression patterns and, according to the 'guilt-by-association' paradigm, might have similar functions or participate in the same pathways. The rest 3601 genes were assigned to small modules with 4 or less genes, which were not considered in further analysis. Within our network, the largest 222 modules containing between 15 and 306 genes (Fig. 1b, Additional file 4: Table S3). Based on Gene Ontology (GO) enrichment analysis, many of these modules fall into 5 distinct categories, with functions in: **i.** Cell organization; **ii.** Development; **iii,** Nutrient uptake & utilization; **iv.** Metabolism, including primary & secondary metabolism; and **v.** Stress responses (Table 1 & Additional file 1: Table S4). Modules for cell organization and primary metabolism include those for cell cycle regulation (Module #60), DNA replication (#64), cytoskeleton organization (#69), nucleosome assembly (#49), photosynthesis (#20 and 129), and mitochondria-related functions (#14 and 86) (Table 1). Several of these modules are highly conserved when compared to similar modules reported in a published Arabidopsis network, AtGGM2014 [7].

Shown in Fig. 2a is **Module #129** for photosynthesis, and Fig. 2b is a subnetwork for Arabidopsis homologues of the genes in this module extracted from the Arabidopsis network AtGGM2014. The similarity of the structure and gene content in both species is significant. To enable direct comparison between the two networks, we consider a maize gene to have conserved interactions between the two networks if the maize gene and its Arabidopsis homologue share at least one homologous neighboring gene across the networks (Fig. 2c). According to this criteria, 9.7% of the genes in the whole maize network have conserved interactions. In contrast, 65% of the genes in Module #129 show conserved interactions in Arabidopsis, much higher than the network-wide level, indicating high conservation for the genes within this module. Within the modules for cell organization and mitochondria-related functions, 19% to 80% of the genes also revealed conserved interactions. We will describe modules in other GO categories with varied degrees of conservation between the two species.

Gene modules revealing developmental features

Maize plant architecture is a major yield-determining trait. Our gene network identified three particularly relevant modules. **Module #16** is enriched with genes involved in the development of ears and tassels, and contributes to shaping the maize inflorescences architecture (Fig. 3a, Table 1, and see Additional file 4: Table S3 for the complete list of genes within this module and the modules discussed below). According to published gene expression datasets on maize development [19–21, 23–26], the 139 genes within this modules are mainly expressed in ears, tassels, and immature cobs (Additional file 5: Fig. S2), and 56 of them encode transcription factors (TF). Included

within the module are *ra1* (*ramosa1*) and *ba1* (*barren stalk1*), two TF genes with opposite roles in tassels development [44, 45]. Maize recessive mutants of *ra1* and *ba1* have increased and zero number of tassel branches respectively. A number of genes for the development of ears and tassels were included as well: *ra3* (*ramosa3*), *sk1* (*silkless ears1*), *spi1* (*sparse inflorescence1*), *tb1* (*teosinte branched1*), *gt1* (*grassy tillers1*), and *bd1* (*branched silkless1*), [46–51]. Also contained is *tga1* (*teosinte glume architecture1*), a gene conferring the naked kernel phenotype in maize [52]. Uncharacterized genes within the module include *GRMZM2G022606* and *GRMZM2G026556*, two genes homologous to *BOP2* in Arabidopsis. In Arabidopsis, *BOP1*, *BOP2*, and *PUCHI* (a *bd1* homologue) redundantly promote floral meristem fate [53], raising the possibility that the two maize genes might have a similar function. Also uncharacterized are *ereb161* and *nactf114*, homologous to Arabidopsis genes *ANT* and *CUC3* that regulate ovule and embryonic apical meristem formation, respectively [54, 55]. In addition, the module contains other potential development related genes as well, including 8 YABBY, 4 AP2/EREBP, 3 SHI RELATED SEQUENCE (SRS), and 3 SQUAMOSA PROMOTER BINDING PROTEIN (SBP) type TF genes. The genes in Module#16 could possibly form an elaborate yet balanced network orchestrating maize inflorescences development. And at least five of them, *ra1*, *ba1*, *gt1*, *tb1*, and *tga1*, have been subjected to selection during maize domestication and were designated as domestication genes [44, 45, 47, 51, 52, 56]. Interestingly, when compared to the Arabidopsis network, only 18 of the 139 genes within the maize module show conserved interaction(s), among them 7 YABBY TFs. Thus, only 8.4% of the genes revealed conserved interaction(s) when the YABBY genes are not considered. This indicates extensive pathways re-shuffling between the two species for inflorescences development.

Ligules are fringe-like tissues located at the junction of the leaf blade and leaf sheath. In maize ligules control leaf angles and affect vegetative architecture [57, 58]. Two modules involved in ligules development were identified. **Module #48** (Fig. 3b, Table 1 & Additional file 4: Table S3) is specifically expressed in the pre-ligule region of the leaf primordia (Additional file 5: Fig. S2). A key gene within the module is *lg1* (*liguleless1*), encoding a SBP TF that acts as a mater regulator of ligule development [58]. The recessive *lg1* mutations in maize erases ligules and renders leaves more upright compared to wild type leaves [57]. Interestingly, the genes within this module have their promoters enriched with a SBP TF binding motif "CGTAC" (pValue = 1.04E-6) [59], indicating they might be targets of Lg1. Two other SPB genes, *sbp3* and *sbp28*, are also included within the module, although their functions remain uncharacterized.

Table 1 Selected gene modules identified from the network

Category	Module no.	Number of genes within the module	Percentage of genes sharing conserved interactions with Arabidopsis	Module annotation	Selected genes from the module[a]	Enriched GO term	pValue
Cell Organization	49	71	80.28%	chromatin organization	his2a1/his2b1/his2b2/pcna2	nucleosome organization	2.15E-132
	60	59	44.07%	cell division	cyc1/hmg9/GRMZM2G061287(CYCB2;4)	cell division	1.32E-07
	64	58	63.79%	DNA replication	rfa1/rrb3/GRMZM2G139894(MCM2)	DNA replication	3.16E-39
	69	53	30.19%	cytoskeleton	krp2/krp4/krp8/cyc3/cyc8	microtubule motor activity	1.64E-46
Development	6	182	35.71%	seed maturation	ole1/ole3/vp1	monolayer-surrounded lipid storage body	1.75E-10
	16	139	12.95%	inflorescences development	ra1/ba1/ra3/spi1/tb1/bd1/yab10	abaxial cell fate specification	5.25E-10
	46	73	2.74%	kernels development	o2/pbf1/az19D2/az19D1/de30/zp1/zp22.1	nutrient reservoir activity	6.32E-64
	48	72	1.39%	ligules development	lg1/sbp28/ns1/myb43/sbp3	DNA binding	9.53E-06
	210	15	60.00%	ligules development	knox3/kn1/rs1/hb123/lg3/lg4/hb76/hb8/hb48/gn1/knox5	sequence-specific DNA binding	8.77E-17
Nutrients	36	88	23.86%	response to phosphate starvation	ppck2/ppck3/ppck4/pht7	cellular response to phosphate starvation	9.62E-17
	72	51	43.14%	nitrate assimilation	nii2/nnr1/gln3/gln6	nitrate assimilation	1.18E-09
Primary Metabolism	14	151	18.54%	respiratory electron transport chain	GRMZM2G450825(COX1)/GRMZM2G173833(COX2)	mitochondrion	9.10E-92
	20	130	69.23%	photosynthesis	cyb6/GRMZM2G448174(PETA)/GRMZM2G433927(PSBE)	photosynthesis	5.46E-30
	86	40	42.50%	mitochondrial electron transport / ATP synthesis	nad1/nad4/GRMZM2G156068(ATP5)	mitochondrial membrane	9.35E-28
	129	26	65.38%	photosynthesis	gpa1/lhcb2/lhcb3/lhcb7	photosynthesis	1.69E-38
Secondary Metabolism	40	84	27.38%	wax biosynthesis	gl1/gl2/gl3/fdl1	wax biosynthetic process	2.25E-13
	65	57	29.82%	flavonoids biosynthesis	pr1/c2/whp1/chi1	flavonoid metabolic process	1.89E-12
	73	49	42.86%	fatty acid biosynthesis	wri1/wri2/acc1/acc2	monocarboxylic acid metabolic process	2.64E-27
	80	43	6.98%	benzoxazinoids and oxylipins biosynthesis	bx1/bx2/bx3/bx4/bx5/lox2/lox3/lox5/lox10	oxylipin metabolic process	6.12E-12
	154	20	30.00%	flavonoids biosynthesis	bz2/a1/bz1/a2/r1	flavonoid biosynthetic process	2.26E-08
Stress	5	188	15.96%	response to fungus	nactf7/chn1/chn2/sip1	defense response to fungus, incompatible interaction	3.27E-09
	10	157	47.13%	response to heat stress	hsp18c/hsp18f/hsp1/hsftf8/hsftf12	response to heat	8.75E-49
	95	35	65.71%	response to endoplasmic reticulum stress	bip1/bip2/der1/bzip60	response to endoplasmic reticulum stress	6.89E-15

[a]Shown in parenthesis are the maize genes' homologues in Arabidopsis

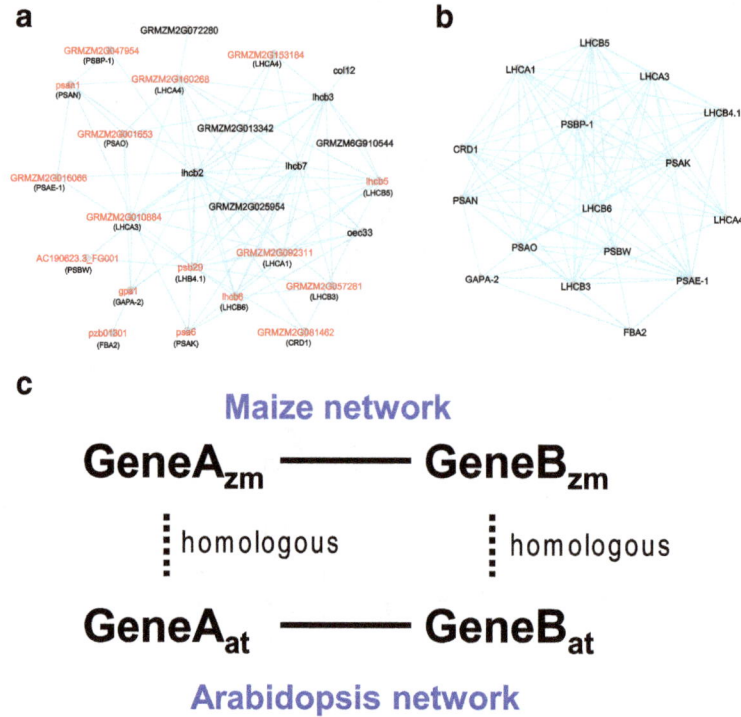

Fig. 2 Network comparison between maize and Arabidopsis. **a** A maize module functioning in photosynthesis. *Red color* indicates the gene has conserved interaction(s) in the Arabidopsis network. The maize genes' Arabidopsis homologues are shown in parenthesis. **b** An Arabidopsis sub-network for the homologues of the maize genes in (**a**). **c** A maize gene is considered to have conserved interaction(s) in Arabidopsis if its homologues and at least one of its interactors' homologue also interact within the Arabidopsis network

Other potential development regulators included *myb44*, a homologue of Arabidopsis *LOF1* that functions in organ boundary specification [60], *GRMZM2G480687*, encoding a MEMBRANE-ASSOCIATED KINASE REGULATOR (MARK) proteins, and *GRMZM2G145909*, homologous to an Arabidopsis atypical bHLH gene *IBL1*. MARKs and atypical bHLH TFs in Arabidopsis participate in developmental processes mediated through brassinosteroids (BR) [61, 62], while BR-signaling also regulates ligules development in maize [63]. **Module #210** contains 15 genes, eight of which encode TFs in the KNOX family (Table 1 & Additional file 4: Table S3). Among these TFs are *gn1* (*gnarley1*), *kn1* (*knotted1*), *rs1* (*rough sheath1*), *lg3* (*liguleless3*), and *lg4* (*liguleless4*), none of which show any expression in the pre-ligule regions of the wild type plants (Additional file 5: Fig. S2). However, ectopic expression of any of these genes in their corresponding dominant mutant background affects and distorts leaf and ligule development [64–66]. It remains to be tested if the other 3 KNOX TFs within the module have similar functions and if and how these TFs function together to regulate ligules development.

Kernels development represents yet another critical process determining maize grain yield and quality. Five relevant gene modules are identified. Among them,

Module #46 (Fig. 3c, Table 1 & Additional file 4: Table S3), including 73 genes, is specifically expressed in the endosperm (Additional file 5: Fig. S2) and enriched with genes indicating nutrient reservoir activity (GO pValue = 6.32E-64). The module contains 16 genes encoding α-, δ-, or γ-zein proteins, the major seed storage proteins in maize, and 4 genes for starch biosynthesis (*bt1*, *bt2*, *sh2*, and *wx1*) [67, 68]. Notably, also included are a bZIP TF gene *o2* (*opaque endosperm2*) and a Dof TF gene *pbf1* (*prolamin-box binding factor1*), two master regulators of zein gene expression [69, 70]. It is recognized that within the maize network transcription factors and their target genes are often contained within the same modules, providing an edifying way to identify those modules' expression regulator(s). Module #42 includes other TF genes as well, such as *ereb167*, *platz12*, *nrp1*, *nactf130*, as potential transcription regulators. **Module #6** is expressed in both embryo and endosperm tissues at late developmental stages (Additional file 5: Fig. S2) and enriched with genes functioning in seed maturation (pValue = 7.39E-6). The module includes many lipid storage genes (e.g. *oleosin1*), desiccation tolerance genes (e.g. genes encoding late embryogenesis abundant proteins), and the TF gene *vp1* (*viviparous1*), a master regulator of seed maturation and

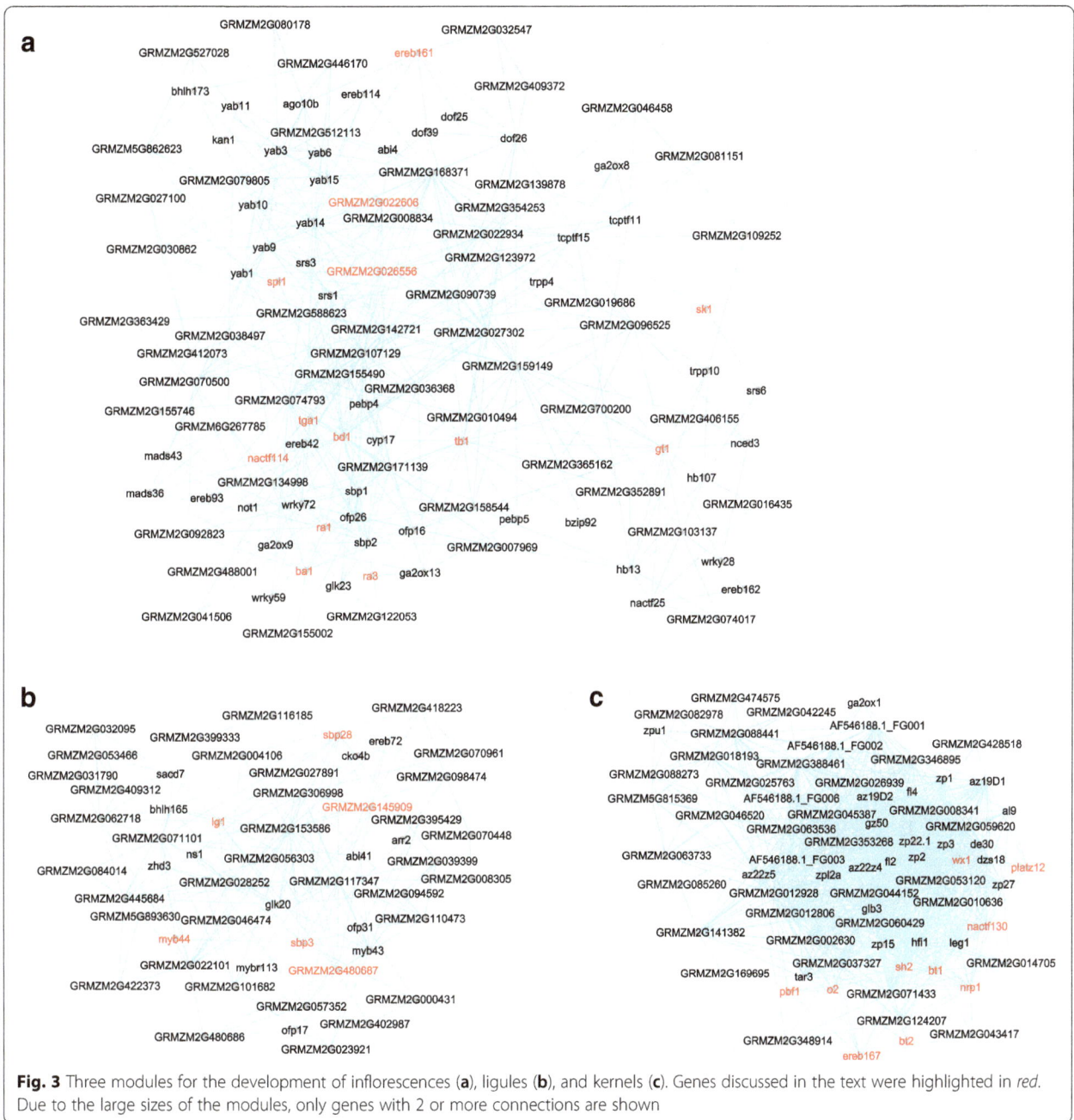

Fig. 3 Three modules for the development of inflorescences (**a**), ligules (**b**), and kernels (**c**). Genes discussed in the text were highlighted in *red*. Due to the large sizes of the modules, only genes with 2 or more connections are shown

dormancy (Additional file 4: Table S3) [71]. 36% of the genes within this modules possess conserved interactions when compared with the Arabidopsis network, including the key seed development genes like *vp1*, *ole1*, *ole3*, and *mlg3*. Additional modules were recovered with specific expression in different compartments of the maize endosperm, such as the basal endosperm transfer layer (BETL) (#11), the embryo-surrounding region (ESR) (#32), and the placento-chalazal region (PC) (#53) (Additional file 6: Fig. S3), similar to a previous report [26]. These examples indicate that our network delineates different modules

corresponding to different functional domains of kernels development and provides a general picture of the process.

Other modules were identified that draw attention to the development of other tissues and organs (Additional file 1: Table S4, Additional file 7: Fig. S4), such as anthers (#3, 17 and 22), meiotic tassels (#18), roots (#30), carpels (#92), Casparian strip (#76), and epidermal cells (#155). Furthermore, identified were modules for primary (#543) and secondary cell wall biosynthesis (#13), and for the signaling pathways of development related hormones, such as auxin (#158) and cytokinins (#183) (Additional file 1: Table S4). These modules are valuable resources for future functional studies on distinct and related developmental processes.

Modules for nutrients uptake and utilization

Nutrient use efficiency is a key objective for crop improvement. Modules can be pinpointed from the maize network as functioning in nitrogen, phosphate, iron, and sulfate uptake and utilization. In general, these modules are conserved between maize and Arabidopsis. For example, **Module #72** (Fig. 4a, Table 1 & Additional file 4: Table S3) shows enrichment for nitrate-responsive genes that encode the key enzymes for reducing and incorporating nitrate into glutamine, including nitrate reductases (*GRMZM5G878558* and *nnr1*), nitrite reductase (*nii2*),

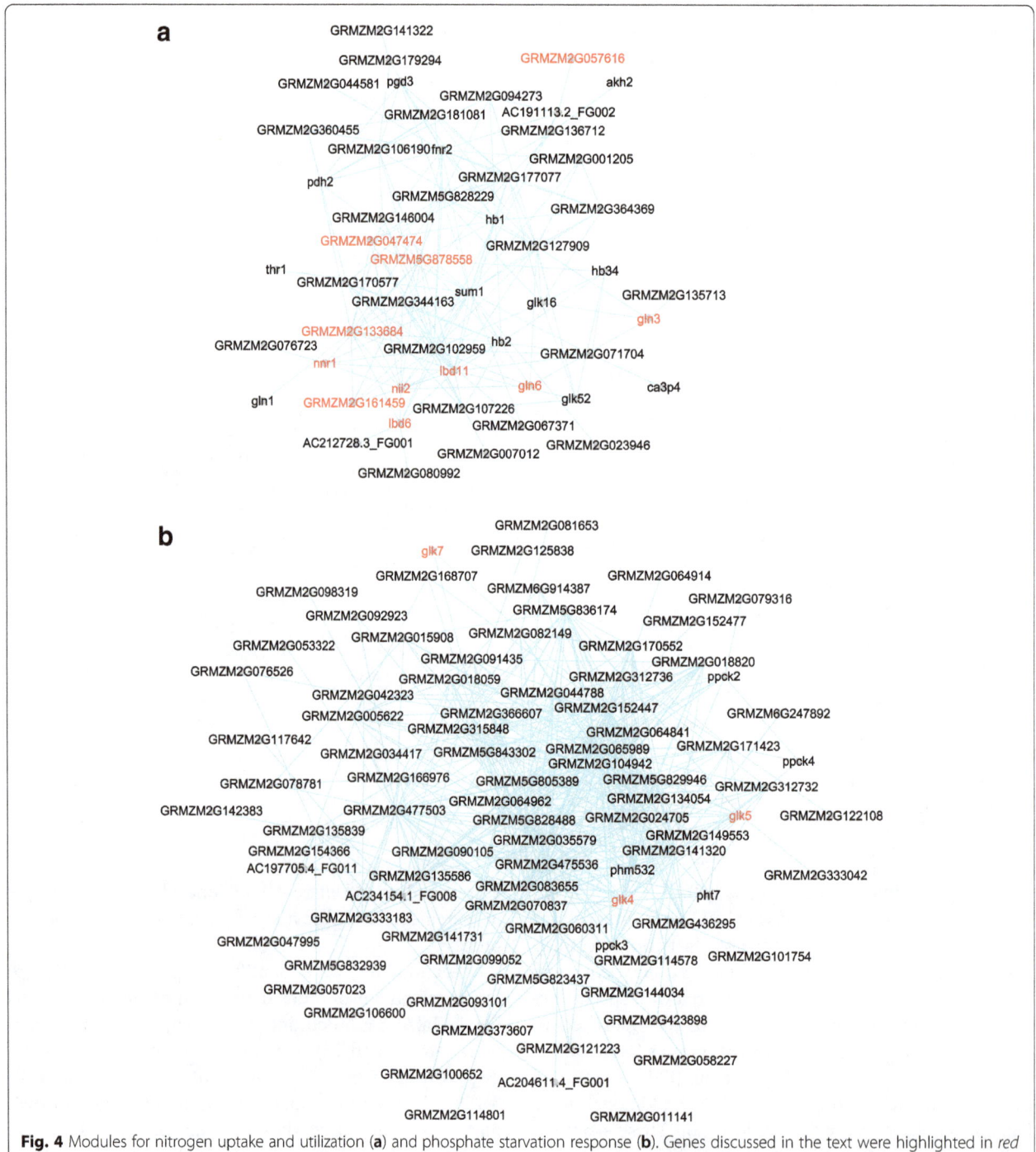

Fig. 4 Modules for nitrogen uptake and utilization (**a**) and phosphate starvation response (**b**). Genes discussed in the text were highlighted in *red*

and glutamine synthetase (*gln1*, *gln3*, *gln6*) [72–74]. Also contained are *GRMZM2G161459* and *GRMZM2G057616*, homologues of the Arabidopsis genes *NRT1.1* (encoding a nitrate transporter and sensor) and *CLC-A* (encoding a vacuole nitrate transporter required for high nitrate uptake capacity) [75, 76]. Compared with the Arabidopsis network, 22 out of the 51 genes in this module showed conserved interaction(s). Among them are two unknown genes, *GRMZM2G047474* and *GRMZM2G133684*, whose homologues in Arabidopsis *AT5G39590* (encoding a TLD domain containing protein) and *AT5G62720* (encoding an integral membrane HPP family protein) are also included in an Arabidopsis nitrate-responsive module but they remain uncharacterized. Their appearance in nitrate responsive modules in both species strongly suggests functions in nitrate sensing, uptake and/or utilization. Also conserved are two uncharacterized LOB-type TF genes *lbd6* and *lbd11*, homologous to one Arabidopsis nitrate response regulator gene *LBD37* [77], indicating *lbd6* and *lbd11* arguably could have similar functions in maize. Additionally, the maize module includes genes encoding G2-like, HB, and C2H2 type TFs, whose functions in nitrate response have not yet been tested.

Module #36 includes genes involved in the phosphate starvation response (pValue = 9.62E-17) (Fig. 4b, Table 1 & Additional file 4: Table S3). The maize module includes homologues of Arabidopsis phosphate starvation response genes, such as *SPX2*, *SPX3*, *SQD1*, *SQD2*, *MGD2*, and *PS2* [78], although their functions have not been characterized in maize. A P1BS motif "GNATATNC", the binding site of the Arabidopsis G2-like transcription factor PHR1, a master regulator of phosphate starvation response gene expression [79], is enriched in the promoters of the genes in this module (pValue = 9.40E-24). Interestingly, the maize module

contains three G2-like TFs, *glk4*, *glk5*, *glk7*, which could also act as master regulators.

Maize modules involved in the uptake and utilization of iron (#43) and sulfate (#79) were also identified (Additional file 1: Table S4). It should be noted that, compared to Arabidopsis, fewer transcriptome datasets are available for maize. As more data become available, more maize modules for nutrient uptake and usage should be revealed in future analysis.

Modules for metabolic processes

Gene modules were identified for various processes in maize metabolism. In addition to those involved in primary metabolisms, numerous modules functioning in secondary metabolism emerge as well. Some of these modules are unique to maize, while others share considerable similarities between Arabidopsis and maize. Among them, genes in **Module #80** (Fig. 5a, Table 1 & Additional file 4: Table S3) provide functions in the production of benzoxazinoids and oxylipins, which are secondary metabolites effective in anti-herbivore defense [80]. Benzoxazinoids are mainly found in Poaceae species, including maize, wheat, and rye, but appear only infrequently in dicots and are absent in Arabidopsis [81]. Module #80 includes most known genes of the maize benzoxazinoids biosynthesis pathway – *bx1* (*benzoxazinless1*), *bx2*, *bx3*, *bx4*, *bx5*, *bx6*, and *bx8* that form a gene cluster in chromosome 4, and *bx9* in chromosome 1 [81]. Interestingly, none of these *bx* genes share any conserved interactions with the Arabidopsis network, consistent with the absence of benzoxazinoids biosynthesis pathway in Arabidopsis. Additionally, the module is enriched with oxylipins biosynthesis genes (pValue = 2.22E-10), such as *lox2*, *lox3*, *lox5*, *lox6*, and *lox10*. The inclusion of both benzoxazinoids and oxylipins biosynthesis genes implies this module's function in defense response.

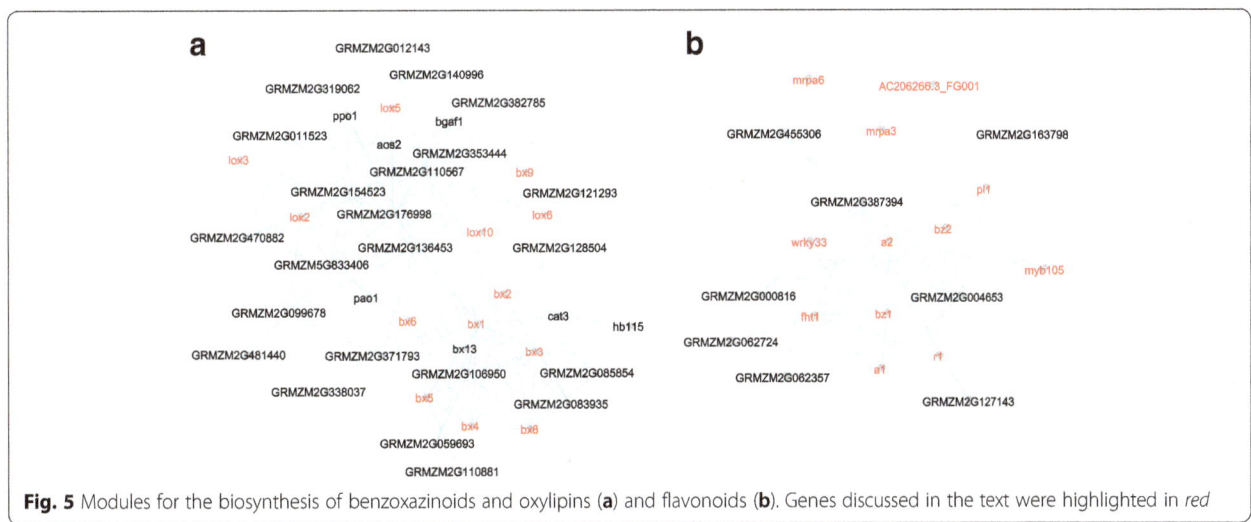

Fig. 5 Modules for the biosynthesis of benzoxazinoids and oxylipins (**a**) and flavonoids (**b**). Genes discussed in the text were highlighted in *red*

Module #154 (Fig. 5b, Table 1 & Additional file 4: Table S3) is enriched with flavonoids biosynthesis pathway genes (pValue = 2.26E-08), including *fht1, a1, a2, bz1, bz2* [82]. The module also contains *pl1* (*purple plant1*) and *r1* (*colored1*), two TF genes controlling anthocyanin biosynthesis [82], and two potential regulatory TFs *myb105* and *wrky33*. In Arabidopsis, *TTG2*, a homolog of *wrky33*, regulates tannin level in the seed coat [83], while the function of maize *wrky33* remains to be tested. The maize module also includes the gene *mrpa3*, encoding a tonoplast-localized anthocyanin transporter [84], and two uncharacterized transporter genes *mrpa6* and *AC206266.3_FG001*. In addition to Module #154, involvement in flavonoids biosynthesis is indicated for Module #65 as well.

Waxes, deposited on the aerial surface of plants as a water-proof layer, are essential for plants in that they not only significantly limit water loss, but also counteract environmental stresses [85]. **Module #40** (Additional file 8: Fig. S5, Table 1 & Additional file 4: Table S3) of the maize network is enriched with wax biosynthesis genes (pValue = 2.25E-13), i.e. *gl1, gl2,* and *gl3* [86]. Among the 84 genes within this module, 23 have conserved interactions when compared with the corresponding Arabidopsis network. Conservation extends to uncharacterized genes that are homologous to Arabidopsis wax biosynthesis genes like *KCS1, KCS6, KCS12, CER3,* and *LACS1* [86]. Another conserved gene pair are *fdl1* (*fused leaves1*) in maize and *MYB94* in Arabidopsis, homologous to each other, both identified recently as TF regulators of wax biosynthesis [87, 88]. The maize module also contains *gl3* (*glossy3*), another MYB TF as a master regulator of wax production [89]. Other potential TF regulators within the module include 3 AP-EREBP type TFs, *ereb12, ereb60,* and *ereb143*. Among them, *ereb60* shows homology to Arabidopsis *WRI1* that regulates the accumulation of fatty acids – precursors for wax biosynthesis [90]. Additionally, another related module for fatty acid biosynthesis (#73) was also identified from the network (Table 1).

In addition to benzoxazinoids, oxylipins, flavonoids, wax, and fatty acids, modules were also identified for the metabolism of suberin (#42), trehalose (#61), glucose (#234), glucan (#278), and lignins (#712) (Additional file 1: Table S4). Based on these modules, promising candidate genes can be selected for future functional studies.

Modules for stress responses

Modules involved in abiotic stress responses were also identified. For example, the genes in **Module #10** (Additional file 8: Fig. S6, Table 1 & Additional file 4: Table S3) show relationships to heat stress responses. Enriched are heat shock stress response genes

(pValue = 8.75E-49). 74 of the 157 genes within this module share conserved gene interaction(s) with an Arabidopsis heat shock related gene module [7]. The conserved genes include 6 heat shock transcription factor genes (*hsft7/8/12/20/24*) and a co-activator gene *MBF1C (GRMZM2G051135)*, highlighting their overarching importance in regulating heat activated gene expression regulation. The splicing regulator genes *SR45a (GRMZM2G073567)* and *SR30 (GRMZM2G331811)* are also included in the list of conserved genes, suggesting alternative splicing could play important roles in the heat shock response in both species. Similarly, **Module #95** (Additional file 8: Fig. S7, Table 1 & Additional file 4: Table S3) is enriched with ER stress response genes (pValue = 6.89E-15). 23 out of the 35 genes within the module shared conserved interactions with Arabidopsis, including *bip1, bip2, pdi1,* and *der1* [91]. Interestingly, contained within the maize module is a putative master regulatory TF gene *bzip60* [92], whose homologue in Arabidopsis, *bZIP60*, is a major regulator of the ER stress response [93], indicating *bzip60* might have similar function in maize.

Also identified from the network were modules related to biotic stress responses. For example, **Module #5** (Additional file 8: Fig. S8, Table 1 & Additional file 4: Table S3) is enriched with genes functioning in defense responses to fungal infections (pValue = 3.27E-09), such as *cta1, wip1, prp1,* and *tps6*. It contains a NAC type TF gene *nactf7*, a homologue of the Arabidopsis *NAC042* gene. In Arabidopsis, *NAC042* is a master TF that regulates the biosynthesis of the anti-fungal compound camalexin [94]. Although maize does not produce camalexin, the inclusion of *nactf7* in this anti-fungal module indicates that it may modulate other secondary metabolism processes to produce anti-fungal compounds. Indeed, within the module are many metabolism-related genes, including 9 genes encoding cytochrome P450 enzymes, whose roles in maize anti-fungal defenses remain to be studied. Interestingly, the module also encompasses genes for gibberellin biosynthesis (*ks1, ks4, ko2,* and *cpps2*), consistent with previous reports that infection by certain fungal pathogens upregulates gibberellin related genes in maize [95].

Additional defense related modules were identified from the network, for example Module #2 and #47 (Additional file 1: Table S4). These modules provide useful targets for future functional studies.

Gene network comparison between maize and Arabidopsis

The examples discussed above demonstrate that some maize gene modules are conserved between maize and Arabidopsis, while others display greater divergence. The percentage of genes with conserved interaction(s) ranged from 0% to 83% for different modules (Fig. 6, Additional file 1: Table S5). Among them, the cell organization related modules retain the highest degree of conservation, in particular those for DNA replication and nucleosome assembly. Many modules in the nutrient uptake, stress-response, and metabolism categories display conservation above the network-wide average level, except for Module #80 with focus on benzoxazinoids and oxylipins biosynthesis. For the development modules, conservation varies significantly. Those involved in carpels, Casparian strip, epidermis, and cell wall development are conserved, all of which constitute the basic building blocks of the plant body. However, much more divergence is seen for modules for kernel/seed metabolism, anthers, inflorescences structure, or root development. Such difference can be attributed to the obvious biological differences between the two species, which define development, growth habitus and structure defining the two species. Nevertheless, both differences and similarities of genes identified in our high-stringency module structure can provide clues about unidentified functions in either one of the two species. The monocot/dicot comparison may further be exploited for additional gains in knowledge.

Gene expression dynamics of gene modules in maize leaves

The maize gene modules identified from our network were used to re-analyze a previously published dataset that had measured transcriptomes from 15 segments of maize leaves, from the base to the tip, representing pronounced developmental gradients [22]. Owing to the pronounced differences in development and physiological functions in maize leaf segments, distinct expression dynamics are revealed in these modules. The average gene expression levels in each module for every segment was computed (Fig. 7). As expected, Module #60, involved in cell cycle regulation, is mainly expressed in Segment 1 at the base of the leaves, where cell

division is most active. Other modules with peak expression at the base include those for DNA replication (#64), cytoskeleton organization (#69), nucleosome assembly (#49), ribosomal functions (#125, 140), plastid development (#308), primary cell wall biosynthesis (#543), and ER stress response (#95). These modules appear to be involved in the early development and building of leaf cells, and high expression of ER related genes might be indicators of intense protein synthesis. Peak gene expression for four modules is observed in the mid part of the leaves (Segments 2 to 5), including those for the biosynthesis of secondary cell walls (#13), waxes (#40), suberin (#42), and lignin (#331), likely representing maturation of the leaf tissue. Module #129, containing mainly light reaction-related photosynthesis genes, was highly expressed and peaked around Segments 9 and 10, which can be an indicator of overall light reaction intensity. Several modules functioning in carbohydrate metabolism or transport show peak expression at late stages (Segments 12 to 14), including #61 for trehalose biosynthesis, #278 for glucan metabolism, and #105 for carbohydrate transport (containing genes such as *sweet13a/b/c*). Thus, combining the modular arrangements revealed by the gene network with detailed leaf segment RNA-Seq datasets provided a coherent and detailed picture of maize leaf development, maturation, and biochemical activities. The results were consistent with the previous report [22, 96], and similar analysis can be applied to other maize datasets as well.

Discussion

We report on the construction of a maize GGM gene co-expression network that includes 20,269 genes based on large-scale RNA-Seq transcriptome data. The resulting gene network was then analyzed and clustered via the MCL clustering algorithm [43]. Although the algorithm partitioned the network purely based on its topology, the analysis resulted in 964 distinct and informative gene modules that included functions in a

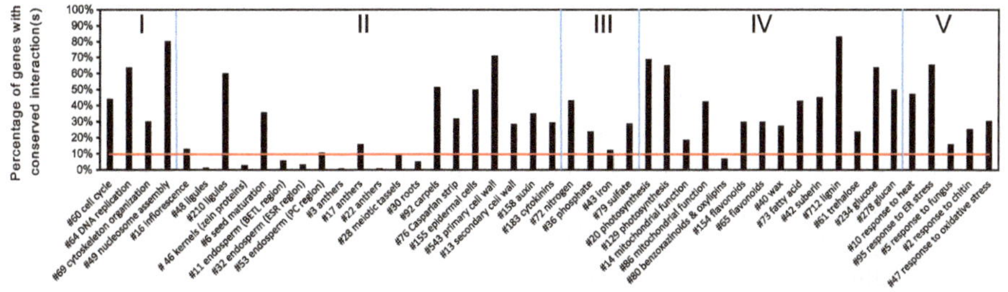

Fig. 6 Network comparison between maize and Arabidopsis. For selected maize gene modules, the percentage of genes with conserved interaction in Arabidopsis were shown. Modules are organization by 5 categories: I, cell organization; II, development; III, nutrient uptake & utilization; IV, metabolism; V, stress response. The *red line* indicates the percentage of genes with conserved interaction(s) in the whole maize network

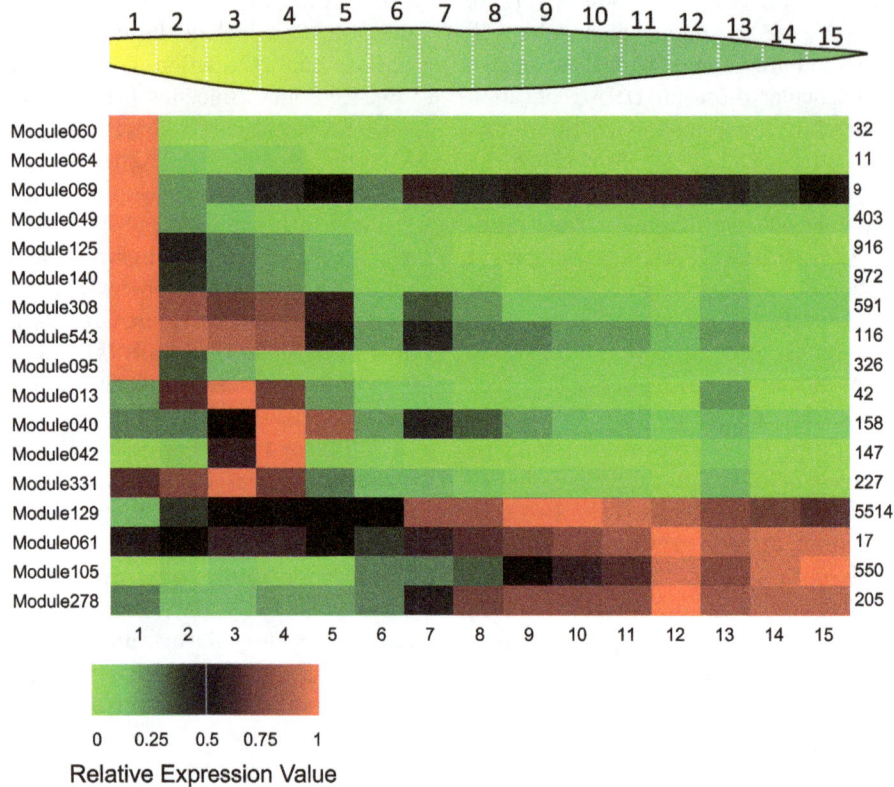

Fig. 7 The average expression level for the genes within selected modules in the maize leaves transcription datasets. On *top* is a diagram of the 15 leaf segments. In the *middle* is a heatmap showing the relative expression level of each module in each segments compared to the highest expression level for that module over all segments, as indicated by the number on the right. A legend indicating the relative expression level is shown on the *bottom*

wide range of maize physiological processes. These modules are particularly useful in that they can assign putative functions to unknown/uncharacterized genes and identify participating genes (including novel genes) for specific developmental or physiological processes, as demonstrated by the selected examples. Module structure and the nature of the genes assembled in a module may then be used to analyze individual gene expression datasets, for example the expression data on maize leaf segments.

The modules identified in our network analysis covered many aspects of maize biology. Compared to a previously published maize transcriptional network that contained 49 modules [6], our model defined 964 modules in total. Sizes of previously identified modules were large. Considering that 87% of all genes in previous network [6] were included in modules with more than 1000 genes it was difficult to pinpoint potential key regulatory genes as candidates for future studies. In contrast, our GGM network identified modules containing between 5 and 306 genes, which facilitated ranking of potentially interesting genes, as demonstrated by the presentation above for various modules. Also, we were particularly

encouraged by the numerous examples in which individual GGM modules identified control genes that had previously been revealed and verified experimentally by mutants analysis, such as *ra1/3* in Module #16, *o2*, *pbf1* in Module #46, *bx1/2/3/4/5/6/8* in Module #80, and *pl1*, *r1* in Module #154 [44, 50, 69, 70, 81, 82]. In an additional contrast to Downs et al. (2013), where the focus on developmental microarray data sets recovered modules with tissue-specific expression, our maize GGM gene network is condition-independent, constructed from transcriptome datasets related to development, stresses, nutrition, as well as other treatments. This facilitated the distinction between modules involved in cell organization, development, nutrients utilization, metabolism, and stress responses. These modules could potentially play critical roles in determining important maize agronomic traits. For example, the inflorescences development related Module #16 includes 5 genes *ra1*, *ba1*, *gt1*, *tga1*, and *tb1* that have been shown by mutants analysis to regulate maize inflorescences architecture, and these 5 genes are designated as domestication genes since they were subjected to selections in the history of maize domestication as they control desirable traits

[44, 45, 47, 51, 52, 56], while other genes within this module might also control maize architecture and performance. As well, modules for ligules and kernels development, for nitrogen and phosphate uptake and utilization, for primary and secondary metabolisms, and for responses to fungal infections include a wealth of information potentially useful for crop improvement.

An important feature of our network is that transcription factors and their target genes are often contained within the same module, suggesting shared expression characteristics. One example is the endosperm development Module #46, which is enriched with zein storage protein genes. The regulatory genes for zein biosynthesis, TF *o2* and *pbf1*, are contained within this modules as well. In yet another example, the master regulator TF of seed maturation, *vp1*, is included within the seed maturation Module #6. The ligule-related Module #48 contains the master regulator gene *lg1*, encoding a SPL TF. The genes within this module are enriched a SPL binding motif, indicating they are targets of Lg1. Similarly, the heat-induced Module #10 contains 6 heat shock TF genes *hsft7/8/12/20/24*, the ER-stress response Module #95 includes a putative maser regulator *bzip60*, and the phosphate starvation response Module #38 possesses three putative master regulator *glk4, glk5, glk7*. Significantly, many identified modules also included genes encoding unknown and uncharacterized TFs that may represent novel gene expression regulators. The modular structures, as revealed by our network, provide an expedient and edifying way to identify putative TF regulators for various maize pathways.

The maize gene network also enables cross-species comparison between maize and Arabidopsis. The comparison revealed an unexpected degree of insight into different degrees of conservation in different pathways. Not surprisingly, cell organization modules showed the highest percentage of conserved genes, indicating the evolutionary stability of such basic cellular pathways. Among development related modules, those involved in generating basic building blocks of the plant body are shown to be conserved as well, i.e. carpels, Casparian strip, epidermis, and cell walls. However, divergence was found for functions in modules related to the overall architecture of the plants, for example the development of inflorescence structures. Such comparison sheds light on pathways that might have been "hot targets" for evolutionary changes. In the near future, with RNA-Seq transcriptome datasets rapidly accumulating, such network comparison analysis can be extended into more plant species to identify steps that highlight and determine plant evolutionary trajectories. As shown here for maize and Arabidopsis, conversation and similarity in modular comparisons will assist in pinpointing key

regulators in various modules that can then be analyzed in detailed studies.

Conclusions

In conclusion, the maize GGM network presented here - in juxtaposition with a corresponding Arabidopsis network [7] - sheds light on similarities and differences in the organization of gene modules between different species in the context of evolutionary separation and different life histories. Additionally, our analysis highlights modules whose structure and gene content can provide important new resources for maize gene functional studies with application potential.

Methods

Maize RNA-Seq data collection

The publicly available maize RNA-Seq transcriptome datasets deposited in the NCBI SRA database were used in the analysis. These datasets were organized by studies. The studies were manually inspected to filter out those focusing on non-coding RNAs or those measuring transcriptome of the same tissues from a large number of maize varieties. Also removed were the studies with less than five RNA-Seq runs or without published articles. As a result, 36 studies were kept and their raw data files (sra) were downloaded. For RNA-Seq data processing, adapter sequences, if present, were removed from raw sequence reads using FASTX-toolkit pipeline version 0.0.13 (http://hannonlab.cshl.edu/fastx_toolkit/). Sequence quality was examined using FastQC (http://www.bioinformatics.babraham.ac.uk/projects/fastqc/), and low quality read was filtered by FASTX-toolkit. The remaining reads were then mapped to the maize genome AGP v3.22 (Ensembl Plants, http://plants.ensembl.org) using Tophat v2.0.10 [97] with default settings. After removing files with mapping rate smaller than 70%, the bam files from 787 RNA-Seq runs were analyzed to obtain gene expression values (FPKM) via Cufflinks v2.1.1 [98].

Maize GGM network construction

The gene expression data were then merged into a single gene expression matrix with 787 columns, and the low expressed genes (maximum FPKM values among all samples being less than 20) were filtered out, resulting in a matrix with 29,316 genes and 787 columns. The matrix was log-transformed [32–40] via the $\log_2$ (FPKM + 1) function, a procedure that significantly reduced the dataset's mean-variance dependency (Additional file 2: Fig. S1). The log-transformed gene expression matrix was then used for partial correlation calculation, following a method described before [17]. Briefly, the calculation involved a procedure with 25,000 iterations. In each iteration, 2000 genes were randomly selected and the partial correlation coefficients between gene pairs were estimated via the "ggm.estimate.pcor" function in the GeneNet v1.2.13 package in R [16]. The Pcors were recorded in every iteration. After 25,000 iterations, for every gene pair, the Pcor with the lowest absolute value was chosen as its final Pcor. The PCC between all gene pairs were also calculated. The gene pairs with Pcor > = 0.035 and PCC > = 0.35 and those with Pcor <= −0.035 and PCC < = −0.35 were selected for gene network construction (Additional file 2: Table S2), resulting in a maize GGM gene network based on the log-transformed gene expression data.

To evaluate the effect of log-transformation on the network quality, another gene network was also constructed directly from the gene expression matrix with the original FPKM values without data transformation, designated as non-transformed FPKM network, keeping all other parameters the same as in the log-transformation-based network outlined above. These two networks were then evaluated and compared via the EGAD package in R regarding their capacities to connect maize genes with shared GO terms [99]. The results indicated the log-transformation-based network outperformed the non-transformed FPKM network (Additional file 8: Fig. S9). Additionally, the log-transformation-based network identified gene modules that were not recovered by the non-transformed FPKM network, such as those related to ER stress response (#95) and nitrate response & assimilation (#72) (data not shown). These modules identified only by the log-transformation-based network contained genes that have been identified in different analyses to be related to the modules in question [73, 74, 91, 92].

Thus, we considered the higher power of the log-transformation-based network, and only results using the log-transformation-based network, designated as the Maize GGM Network, were further analyzed and discussed.

Gene network properties and gene module identification

The R package of RBGL v 1.44.0 (http://bioconductor.org/packages/RBGL/) is used to calculate the clustering coefficient of the maize GGM network. The network was clustered via the MCL clustering algorithm, using these parameters "-I 1.5 -Scheme 7" [43]. The genes within each module were then analyzed for Gene Ontology enrichment via GOStats [100], with GO annotation file downloaded from the Gramene database (ftp://ftp.gramene.org/). The maize genes and their Arabidopsis homologues were further annotated with annotation files from MaizeGDB, TAIR, and PlnTFDB [101–103]. Selected modules were also tested for promoter motifs enrichment via the binomial distribution. An R script, included in the accompanied program MaizeGGM2016, was developed to extract

sub-networks for gene modules and to draw development heatmaps for the genes within selected modules, with expression data from published datasets [19–21, 23–26]. The whole GGM network and the extracted sub-networks were layout and visualized with BioLayout Express 3D and Cytoscape 3.3, respectively [104, 105].

Gene network comparison between the maize network and the Arabidopsis network

To enable comparison between the maize GGM network and the Arabidopsis network AtGGM2014, the InParanoid program (v 4.1) [106] was used to identify the maize genes', if present, most similar homologues in Arabidopsis. For any gene within the maize network, if there exists a homologous gene within the Arabidopsis AtGGM2014 network, the maize gene's immediate neighboring genes within the maize network were extracted as group A. Also extracted, as group B, were its homologous gene's neighbors within the Arabidopsis network. If any of the gene within group A has a homologous gene in group B, the original maize gene was considered to have conserved interaction within the Arabidopsis network. For any given module, the percentage of genes with conserved interaction was calculated as an indicator of evolution conserveness or divergence.

Additional files

Additional file 1: Table S1. The RNA-Seq studies used in the analysis. Table S4. Selected gene modules identified from the network. Table S5. The percentage of genes with conserved interactions within each module.

Additional file 2: Figure S1. Log-transformation reduced the mean-variance dependency of the maize RNA-Seq data. **a** The standard deviation of each gene's non-transformed FPKM expression values across all 787 RNA-seq runs are shown against the rank of genes. Genes are ranked by their mean expression values, from low (left) to high (right). The red line depicts a trend line of standard deviation, which indicates a clear mean-variance dependency. **b** The standard deviation of each gene's expression values after log transformation across all RNA-Seq runs against the rank of genes. The mean-variance dependency is greatly reduced.

Additional file 3: Table S2. The 123,666 gene pairs used for the maize GGM gene network construction.

Additional file 4: Table S3. The 964 gene modules identified from the network. Listed are the genes' names and their module identity, and the results of gene ontology enrichment analysis.

Additional file 5: Figure S2. A heatmap showing the tissue-specific gene expression patterns for the inflorescences, ligules, and kernels development related modules. The data source of the SRA studies were labeled in the sample names. Listed are the maize genes names and its symbol in lowercase letters, or, if it has no symbols, the names of its Arabidopsis homologues.

Additional file 6: Figure S3. A heatmap for modules related to endosperm development.

Additional file 7: Figure S4. A heatmap for additional modules related to development.

Additional file 8: Figure S5. A module for wax biosynthesis. Figure S6. A module for heat-shock stress response. Figure S7. A module for ER-stress response. Figure S8. A module for stress response to fungus. Figure S9. Comparison between the non-transformed FPKM network and the log-transformation-based network. Both networks were evaluated via the EGAD package [99] in R regarding their capacities to connect genes with shared GO terms. For each GO term, the maize genes with that GO were considered as a gene set, and an AUROC value was calculated for each network using the EGAD package. A higher AUROC value indicates genes within that gene set are more likely to have each other as neighbors, and thus a better performance of the network. The histogram shows the overall distribution of the AUROC values for 1728 GO terms calculated for the non-transformed FPKM network (green bar) and for the log-transformation-based network (transparent bar with black border). The log-transformation-based network has more GO terms with higher AUROC values, thus it performs better than the other network.

Acknowledgements
We thank Dr. Hans Bohnert for his critical reading of the manuscript. We are indebted to his invaluable suggestions.

Funding
S.M. is supported by grants from Thousand Youth Talents Program and University of Science and Technology of China (Start-up fund). P.L. is supported by grants from NSFC (91435108), National Key Research and Development Program of China (2016YFD0101003) and Taishan Scholarship.

Authors' contributions
SM and PL designed experiments. SM, ZD, and PL performed experiments. SM and PL wrote and edited the manuscript. All authors read and approved the final manuscript.

Competing interests
The authors declare that they have no competing interests.

Author details
[1]School of Life Sciences, University of Science and Technology of China, Hefei, Anhui, China. [2]The Institute of Tropical Bioscience and Biotechnology, Chinese Academy of Tropical Agricultural Sciences, Haikou, Hainan, China. [3]State Key Laboratory of Crop Biology, College of Agronomy, Shandong Agricultural University, Tai'an, Shandong, China.

References
1. Usadel B, Obayashi T, Mutwil M, Giorgi FM, Bassel GW, Tanimoto M, Chow A, Steinhauser D, Persson S, Provart NJ. Co-expression tools for plant biology: opportunities for hypothesis generation and caveats. Plant Cell Environ. 2009;32:1633–51.
2. Obayashi T, Kinoshita K, Nakai K, Shibaoka M, Hayashi S, Saeki M, Shibata D, Saito K, Ohta H. ATTED-II: a database of co-expressed genes and cis elements for identifying co-regulated gene groups in Arabidopsis. Nucleic Acids Res. 2007;35:D863–9.
3. Persson S, Wei H, Milne J, Page GP, Somerville CR. Identification of genes required for cellulose synthesis by regression analysis of public microarray data sets. Proc Natl Acad Sci U S A. 2005;102:8633–8.
4. FANTOM Consortium and the RIKEN PMI and CLST (DGT), Forrest AR, Kawaji H, Rehli M, Baillie JK, de Hoon MJ, Haberle V, Lassmann T, et al. A promoter-level mammalian expression atlas. Nature. 2014;507:462–70.
5. Dekkers BJ, Pearce S, van Bolderen-Veldkamp RP, Marshall A, Widera P, Gilbert J, Drost HG, Bassel GW, Muller K, King JR, et al. Transcriptional dynamics of two seed compartments with opposing roles in Arabidopsis seed germination. Plant Physiol. 2013;163:205–15.
6. Downs GS, Bi YM, Colasanti J, Wu W, Chen X, Zhu T, Rothstein SJ, Lukens LN. A developmental transcriptional network for maize defines coexpression modules. Plant Physiol. 2013;161:1830–43.
7. Ma S, Bohnert HJ, Dinesh-Kumar SP. AtGGM2014, an Arabidopsis gene co-expression network for functional studies. Sci China Life Sci. 2015;58:276–86.
8. Mao L, Van Hemert JL, Dash S, Dickerson JA. Arabidopsis gene co-expression network and its functional modules. BMC Bioinformatics. 2009;10:346.
9. Mistry M, Gillis J, Pavlidis P. Meta-analysis of gene coexpression networks in the post-mortem prefrontal cortex of patients with schizophrenia and unaffected controls. BMC Neurosci. 2013;14:105.
10. Stuart JM, Segal E, Koller D, Kim SK. A gene-coexpression network for global discovery of conserved genetic modules. Science. 2003;302:249–55.
11. Mentzen WI, Wurtele ES. Regulon organization of Arabidopsis. BMC Plant Biol. 2008;8:99.
12. Childs KL, Davidson RM, Buell CR. Gene coexpression network analysis as a source of functional annotation for rice genes. PLoS One. 2011;6:e22196.
13. Ficklin SP, Feltus FA. Gene coexpression network alignment and conservation of gene modules between two grass species: maize and rice. Plant Physiol. 2011;156:1244–56.
14. Zhang B, Horvath S. A general framework for weighted gene co-expression network analysis. Stat Appl Genet Mol Biol 2005;4:Article17.
15. Wille A, Zimmermann P, Vranova E, Furholz A, Laule O, Bleuler S, Hennig L, Prelic A, von Rohr P, Thiele L, et al. Sparse graphical Gaussian modeling of the isoprenoid gene network in Arabidopsis Thaliana. Genome Biol. 2004;5:R92.
16. Schäfer J, Strimmer K. A shrinkage approach to large-scale covariance matrix estimation and implications for functional genomics. Stat Appl Genet Mol Biol 2005;4:Article32.
17. Ma S, Gong Q, Bohnert HJ. An Arabidopsis gene network based on the graphical Gaussian model. Genome Res. 2007;17:1614–25.
18. Zuo Y, Yu G, Tadesse MG, Ressom HW. Biological network inference using low order partial correlation. Methods. 2014;69:266–73.
19. Sekhon RS, Briskine R, Hirsch CN, Myers CL, Springer NM, Buell CR, de Leon N, Kaeppler SM. Maize gene atlas developed by RNA sequencing and comparative evaluation of transcriptomes based on RNA sequencing and microarrays. PLoS One. 2013;8:e61005.
20. Stelpflug SC, Sekhon RS, Vaillancourt B, Hirsch CN, Buell CR, de Leon N, Kaeppler SM: An expanded maize gene expression atlas based on RNA sequencing and its use to explore root development. The Plant Genome 2016;9:0.
21. Eveland AL, Goldshmidt A, Pautler M, Morohashi K, Liseron-Monfils C, Lewis MW, Kumari S, Hiraga S, Yang F, Unger-Wallace E, et al. Regulatory modules controlling maize inflorescence architecture. Genome Res. 2014;24:431–43.
22. Wang L, Czedik-Eysenberg A, Mertz RA, Si Y, Tohge T, Nunes-Nesi A, Arrivault S, Dedow LK, Bryant DW, Zhou W, et al. Comparative analyses of C(4) and C(3) photosynthesis in developing leaves of maize and rice. Nat Biotechnol. 2014;32:1158–65.
23. Johnston R, Wang M, Sun Q, Sylvester AW, Hake S, Scanlon MJ. Transcriptomic analyses indicate that maize ligule development recapitulates gene expression patterns that occur during lateral organ initiation. Plant Cell. 2014;26:4718–32.
24. Chen J, Zeng B, Zhang M, Xie S, Wang G, Hauck A, Lai J. Dynamic transcriptome landscape of maize embryo and endosperm development. Plant Physiol. 2014;166:252–64.
25. Yi G, Neelakandan AK, Gontarek BC, Vollbrecht E, Becraft PW. The naked endosperm genes encode duplicate INDETERMINATE domain transcription factors required for maize endosperm cell patterning and differentiation. Plant Physiol. 2015;167:443–56.
26. Zhan J, Thakare D, Ma C, Lloyd A, Nixon NM, Arakaki AM, Burnett WJ, Logan KO, Wang D, Wang X, et al. RNA sequencing of laser-capture

microdissected compartments of the maize kernel identifies regulatory modules associated with endosperm cell differentiation. Plant Cell. 2015;27:513–31.

27. Makarevitch I, Waters AJ, West PT, Stitzer M, Hirsch CN, Ross-Ibarra J, Springer NM. Transposable elements contribute to activation of maize genes in response to abiotic stress. PLoS Genet. 2015;11:e1004915.

28. Ding Y, Virlouvet L, Liu N, Riethoven JJ, Fromm M, Avramova Z. Dehydration stress memory genes of Zea Mays; comparison with Arabidopsis Thaliana. BMC Plant Biol. 2014;14:141.

29. Lanubile A, Ferrarini A, Maschietto V, Delledonne M, Marocco A, Bellin D. Functional genomic analysis of constitutive and inducible defense responses to Fusarium verticillioides infection in maize genotypes with contrasting ear rot resistance. BMC Genomics. 2014;15:710.

30. Bi YM, Meyer A, Downs GS, Shi X, El-Kereamy A, Lukens L, Rothstein SJ. High throughput RNA sequencing of a hybrid maize and its parents shows different mechanisms responsive to nitrogen limitation. BMC Genomics. 2014;15:77.

31. Walley JW, Sartor RC, Shen Z, Schmitz RJ, Wu KJ, Urich MA, Nery JR, Smith LG, Schnable JC, Ecker JR, Briggs SP. Integration of omic networks in a developmental atlas of maize. Science. 2016;353:814–8.

32. Zwiener I, Frisch B, Binder H. Transforming RNA-Seq data to improve the performance of prognostic gene signatures. PLoS One. 2014;9:e85150.

33. Specht AT, Li J. LEAP: constructing gene co-expression networks for single-cell RNA-sequencing data using pseudotime ordering. Bioinformatics. 2017;33:764–6.

34. Israel JW, Martik ML, Byrne M, Raff EC, Raff RA, McClay DR, Wray GA. Comparative developmental Transcriptomics reveals rewiring of a highly conserved gene regulatory network during a major life history switch in the sea urchin genus Heliocidaris. PLoS Biol. 2016;14:e1002391.

35. Kim B, Suo B, Emmons SW. Gene function prediction based on developmental Transcriptomes of the two sexes in C. Elegans. Cell Rep. 2016;17:917–28.

36. Pfeifer M, Kugler KG, Sandve SR, Zhan B, Rudi H, Hvidsten TR, International Wheat Genome Sequencing C, Mayer KF, Olsen OA. Genome interplay in the grain transcriptome of hexaploid bread wheat. Science. 2014;345:1250091.

37. Sekhon RS, Hirsch CN, Childs KL, Breitzman MW, Kell P, Duvick S, Spalding EP, Buell CR, de Leon N, Kaeppler SM. Phenotypic and transcriptional analysis of divergently selected maize populations reveals the role of developmental timing in seed size determination. Plant Physiol. 2014;165:658–69.

38. Fertuzinhos S, Li M, Kawasawa YI, Ivic V, Franjic D, Singh D, Crair M, Sestan N. Laminar and temporal expression dynamics of coding and noncoding RNAs in the mouse neocortex. Cell Rep. 2014;6:938–50.

39. Xiao X, Moreno-Moral A, Rotival M, Bottolo L, Petretto E. Multi-tissue analysis of co-expression networks by higher-order generalized singular value decomposition identifies functionally coherent transcriptional modules. PLoS Genet. 2014;10:e1004006.

40. Langfelder P, Horvath S: WGCNA package FAQ. https://labs.genetics.ucla.edu/horvath/CoexpressionNetwork/Rpackages/WGCNA/faq.html. Accessed 11 May 2017.

41. Watts DJ, Strogatz SH. Collective dynamics of 'small-world' networks. Nature. 1998;393:440–2.

42. Ravasz E, Somera AL, Mongru DA, Oltvai ZN, Barabasi AL. Hierarchical organization of modularity in metabolic networks. Science. 2002;297:1551–5.

43. van Dongen S. Graph clustering by flow simulation. Dissertation: University of Utrecht; 2000.

44. Vollbrecht E, Springer PS, Goh L, Buckler ES, Martienssen R. Architecture of floral branch systems in maize and related grasses. Nature. 2005;436:1119–26.

45. Gallavotti A, Zhao Q, Kyozuka J, Meeley RB, Ritter MK, Doebley JF, Pe ME, Schmidt RJ. The role of barren stalk1 in the architecture of maize. Nature. 2004;432:630–5.

46. Chuck G, Muszynski M, Kellogg E, Hake S, Schmidt RJ. The control of spikelet meristem identity by the branched silkless1 gene in maize. Science. 2002;298:1238–41.

47. Doebley J, Stec A, Hubbard L. The evolution of apical dominance in maize. Nature. 1997;386:485–8.

48. Gallavotti A, Barazesh S, Malcomber S, Hall D, Jackson D, Schmidt RJ, McSteen P. Sparse inflorescence1 encodes a monocot-specific YUCCA-like gene required for vegetative and reproductive development in maize. Proc Natl Acad Sci U S A. 2008;105:15196–201.

49. Hayward AP, Moreno MA, Howard TP 3rd, Hague J, Nelson K, Heffelfinger C, Romero S, Kausch AP, Glauser G, Acosta IF, et al. Control of sexuality by the sk1-encoded UDP-glycosyltransferase of maize. Sci Adv. 2016;2:e1600991.

50. Satoh-Nagasawa N, Nagasawa N, Malcomber S, Sakai H, Jackson D. A trehalose metabolic enzyme controls inflorescence architecture in maize. Nature. 2006;441:227–30.

51. Whipple CJ, Kebrom TH, Weber AL, Yang F, Hall D, Meeley R, Schmidt R, Doebley J, Brutnell TP, Jackson DP. Grassy tillers1 promotes apical dominance in maize and responds to shade signals in the grasses. Proc Natl Acad Sci U S A. 2011;108:E506–12.

52. Wang H, Nussbaum-Wagler T, Li B, Zhao Q, Vigouroux Y, Faller M, Bomblies K, Lukens L, Doebley JF. The origin of the naked grains of maize. Nature. 2005;436:714–9.

53. Karim MR, Hirota A, Kwiatkowska D, Tasaka M, Aida M. A role for Arabidopsis PUCHI in floral meristem identity and bract suppression. Plant Cell. 2009;21:1360–72.

54. Klucher KM, Chow H, Reiser L, Fischer RL. The AINTEGUMENTA gene of Arabidopsis required for ovule and female gametophyte development is related to the floral homeotic gene APETALA2. Plant Cell. 1996;8:137–53.

55. Vroemen CW, Mordhorst AP, Albrecht C, Kwaaitaal MA, de Vries SC. The CUP-SHAPED COTYLEDON3 gene is required for boundary and shoot meristem formation in Arabidopsis. Plant Cell. 2003;15:1563–77.

56. Swanson-Wagner R, Briskine R, Schaefer R, Hufford MB, Ross-Ibarra J, Myers CL, Tiffin P, Springer NM. Reshaping of the maize transcriptome by domestication. Proc Natl Acad Sci U S A. 2012;109:11878–83.

57. Becraft PW, Bongard-Pierce DK, Sylvester AW, Poethig RS, Freeling M. The liguleless-1 gene acts tissue specifically in maize leaf development. Dev Biol. 1990;141:220–32.

58. Moreno MA, Harper LC, Krueger RW, Dellaporta SL, Freeling M. Liguleless1 encodes a nuclear-localized protein required for induction of ligules and auricles during maize leaf organogenesis. Genes Dev. 1997;11:616–28.

59. Birkenbihl RP, Jach G, Saedler H, Huijser P. Functional dissection of the plant-specific SBP-domain: overlap of the DNA-binding and nuclear localization domains. J Mol Biol. 2005;352:585–96.

60. Lee DK, Geisler M, Springer PS. LATERAL ORGAN FUSION1 and LATERAL ORGAN FUSION2 function in lateral organ separation and axillary meristem formation in Arabidopsis. Development. 2009;136:2423–32.

61. Jaillais Y, Hothorn M, Belkhadir Y, Dabi T, Nimchuk ZL, Meyerowitz EM, Chory J. Tyrosine phosphorylation controls brassinosteroid receptor activation by triggering membrane release of its kinase inhibitor. Genes Dev. 2011;25:232–7.

62. Wang H, Zhu Y, Fujioka S, Asami T, Li J. Regulation of Arabidopsis brassinosteroid signaling by atypical basic helix-loop-helix proteins. Plant Cell. 2009;21:3781–91.

63. Kir G, Ye H, Nelissen H, Neelakandan AK, Kusnandar AS, Luo A, Inze D, Sylvester AW, Yin Y, Becraft PW. RNA interference knockdown of BRASSINOSTEROID INSENSITIVE1 in maize reveals novel functions for Brassinosteroid signaling in controlling plant architecture. Plant Physiol. 2015;169:826–39.

64. Becraft PW, Freeling M. Genetic analysis of rough sheath1 developmental mutants of maize. Genetics. 1994;136:295–311.

65. Fowler JE, Freeling M. Genetic analysis of mutations that alter cell fates in maize leaves: dominant Liguleless mutations. Dev Genet. 1996;18:198–222.

66. Foster T, Yamaguchi J, Wong BC, Veit B, Hake S. Gnarley1 is a dominant mutation in the knox4 Homeobox gene affecting cell shape and identity. Plant Cell. 1999;11:1239–52.

67. Giroux MJ, Boyer C, Feix G, Hannah LC. Coordinated transcriptional regulation of storage product genes in the maize endosperm. Plant Physiol. 1994;106:713–22.

68. Beatty MK, Rahman A, Cao H, Woodman W, Lee M, Myers AM, James MG. Purification and molecular genetic characterization of ZPU1, a pullulanase-type starch-debranching enzyme from maize. Plant Physiol. 1999;119:255–66.

69. Hartings H, Maddaloni M, Lazzaroni N, Di Fonzo N, Motto M, Salamini F, Thompson R. The O2 gene which regulates zein deposition in maize endosperm encodes a protein with structural homologies to transcriptional activators. EMBO J. 1989;8:2795–801.

70. Vicente-Carbajosa J, Moose SP, Parsons RL, Schmidt RJ. A maize zinc-finger protein binds the prolamin box in zein gene promoters and interacts with the basic leucine zipper transcriptional activator Opaque2. Proc Natl Acad Sci U S A. 1997;94:7685–90.

71. McCarty DR, Carson CB, Stinard PS, Robertson DS. Molecular analysis of viviparous-1: an Abscisic acid-insensitive mutant of maize. Plant Cell. 1989;1:523–32.

72. Xu G, Fan X, Miller AJ. Plant nitrogen assimilation and use efficiency. Annu Rev Plant Biol. 2012;63:153–82.

73. Lahners K, Kramer V, Back E, Privalle L, Rothstein S. Molecular cloning of complementary DNA encoding maize nitrite reductase: molecular analysis and nitrate induction. Plant Physiol. 1988;88:741–6.

74. sakakibara H, Kawabata S, Takahashi H, Hase T, Sugiyama T. Molecular Cloning of the Family of Glutamine Synthetase Genes from Maize: Expression of Genes for Glutamine Synthetase and Ferredoxin-Dependent Glutamate Synthase in Photosynthetic and Non-Photosynthetic Tissues. Plant Cell Physiol. 1992;33:49–58.

75. Ho CH, Lin SH, Hu HC, Tsay YF. CHL1 functions as a nitrate sensor in plants. Cell. 2009;138:1184–94.

76. De Angeli A, Monachello D, Ephritikhine G, Frachisse JM, Thomine S, Gambale F, Barbier-Brygoo H. The nitrate/proton antiporter AtCLCa mediates nitrate accumulation in plant vacuoles. Nature. 2006;442:939–42.

77. Rubin G, Tohge T, Matsuda F, Saito K, Scheible WR. Members of the LBD family of transcription factors repress anthocyanin synthesis and affect additional nitrogen responses in Arabidopsis. Plant Cell. 2009;21:3567–84.

78. Hammond JP, Bennett MJ, Bowen HC, Broadley MR, Eastwood DC, May ST, Rahn C, Swarup R, Woolaway KE, White PJ. Changes in gene expression in Arabidopsis shoots during phosphate starvation and the potential for developing smart plants. Plant Physiol. 2003;132:578–96.

79. Rubio V, Linhares F, Solano R, Martin AC, Iglesias J, Leyva A, Paz-Ares J. A conserved MYB transcription factor involved in phosphate starvation signaling both in vascular plants and in unicellular algae. Genes Dev. 2001;15:2122–33.

80. Tzin V, Fernandez-Pozo N, Richter A, Schmelz EA, Schoettner M, Schafer M, Ahern KR, Meihls LN, Kaur H, Huffaker A, et al. Dynamic maize responses to aphid feeding are revealed by a time series of Transcriptomic and Metabolomic assays. Plant Physiol. 2015;169:1727–43.

81. Frey M, Schullehner K, Dick R, Fiesselmann A, Gierl A. Benzoxazinoid biosynthesis, a model for evolution of secondary metabolic pathways in plants. Phytochemistry. 2009;70:1645–51.

82. Petroni K, Pilu R, Tonelli C. Anthocyanins in corn: a wealth of genes for human health. Planta. 2014;240:901–11.

83. Johnson CS, Kolevski B, Smyth DR. TRANSPARENT TESTA GLABRA2, a trichome and seed coat development gene of Arabidopsis, encodes a WRKY transcription factor. Plant Cell. 2002;14:1359–75.

84. Goodman CD, Casati P, Walbot V. A multidrug resistance-associated protein involved in anthocyanin transport in Zea Mays. Plant Cell. 2004;16:1812–26.

85. Post-Beittenmiller D. Biochemistry and molecular biology of wax production in plants. Annu Rev Plant Physiol Plant Mol Biol. 1996;47:405–30.

86. Lee SB, Suh MC. Advances in the understanding of cuticular waxes in Arabidopsis Thaliana and crop species. Plant Cell Rep. 2015;34:557–72.

87. La Rocca N, Manzotti PS, Cavaiuolo M, Barbante A, Dalla Vecchia F, Gabotti D, Gendrot G, Horner DS, Krstajic J, Persico M, et al. The maize fused leaves1 (fdl1) gene controls organ separation in the embryo and seedling shoot and promotes coleoptile opening. J Exp Bot. 2015;66:5753–67.

88. Lee SB, Suh MC. Cuticular wax biosynthesis is up-regulated by the MYB94 transcription factor in Arabidopsis. Plant Cell Physiol. 2015;56:48–60.

89. Liu S, Yeh CT, Tang HM, Nettleton D, Schnable PS. Gene mapping via bulked segregant RNA-Seq (BSR-Seq). PLoS One. 2012;7:e36406.

90. Baud S, Wuilleme S, To A, Rochat C, Lepiniec L. Role of WRINKLED1 in the transcriptional regulation of glycolytic and fatty acid biosynthetic genes in Arabidopsis. Plant J. 2009;60:933–47.

91. Kirst ME, Meyer DJ, Gibbon BC, Jung R, Boston RS. Identification and characterization of endoplasmic reticulum-associated degradation proteins differentially affected by endoplasmic reticulum stress. Plant Physiol. 2005;138:218–31.

92. Li Y, Humbert S, Howell SH. ZmbZIP60 mRNA is spliced in maize in response to ER stress. BMC Research Notes. 2012;5:144.

93. Iwata Y, Fedoroff NV, Koizumi N. Arabidopsis bZIP60 is a proteolysis-activated transcription factor involved in the endoplasmic reticulum stress response. Plant Cell. 2008;20:3107–21.

94. Saga H, Ogawa T, Kai K, Suzuki H, Ogata Y, Sakurai N, Shibata D, Ohta D. Identification and characterization of ANAC042, a transcription factor family gene involved in the regulation of camalexin biosynthesis in Arabidopsis. Mol Plant-Microbe Interact. 2012;25:684–96.

95. Skibbe DS, Doehlemann G, Fernandes J, Walbot V. Maize tumors caused by Ustilago Maydis require organ-specific genes in host and pathogen. Science. 2010;328:89–92.

96. Li P, Ponnala L, Gandotra N, Wang L, Si Y, Tausta SL, Kebrom TH, Provart N, Patel R, Myers CR, et al. The developmental dynamics of the maize leaf transcriptome. Nat Genet. 2010;42:1060–7.

97. Trapnell C, Pachter L, Salzberg SL. TopHat: discovering splice junctions with RNA-Seq. Bioinformatics. 2009;25:1105–11.

98. Trapnell C, Roberts A, Goff L, Pertea G, Kim D, Kelley DR, Pimentel H, Salzberg SL, Rinn JL, Pachter L. Differential gene and transcript expression analysis of RNA-seq experiments with TopHat and cufflinks. Nat Protoc. 2012;7:562–78.

99. Ballouz S, Weber M, Pavlidis P, Gillis J. EGAD: ultra-fast functional analysis of gene networks. Bioinformatics. 2017;33:612–14.

100. Falcon S, Gentleman R. Using GOstats to test gene lists for GO term association. Bioinformatics. 2007;23:257–8.

101. Andorf CM, Cannon EK, Portwood JL 2nd, Gardiner JM, Harper LC, Schaeffer ML, Braun BL, Campbell DA, Vinnakota AG, Sribalusu VV, et al. MaizeGDB update: new tools, data and interface for the maize model organism database. Nucleic Acids Res. 2016;44:D1195–201.

102. Lamesch P, Berardini TZ, Li D, Swarbreck D, Wilks C, Sasidharan R, Muller R, Dreher K, Alexander DL, Garcia-Hernandez M, et al. The Arabidopsis information resource (TAIR): improved gene annotation and new tools. Nucleic Acids Res. 2012;40:D1202–10.

103. Perez-Rodriguez P, Riano-Pachon DM, Correa LG, Rensing SA, Kersten B, Mueller-Roeber B. PlnTFDB: updated content and new features of the plant transcription factor database. Nucleic Acids Res. 2010;38:D822–7.

104. Shannon P, Markiel A, Ozier O, Baliga NS, Wang JT, Ramage D, Amin N, Schwikowski B, Ideker T. Cytoscape: a software environment for integrated models of biomolecular interaction networks. Genome Res. 2003;13:2498–504.

105. Theocharidis A, van Dongen S, Enright AJ, Freeman TC. Network visualization and analysis of gene expression data using BioLayout express(3D). Nat Protoc. 2009;4:1535–50.

106. Sonnhammer EL, Ostlund G. InParanoid 8: orthology analysis between 273 proteomes, mostly eukaryotic. Nucleic Acids Res. 2015;43:D234–9.

14

Genetic diversity, population structure and marker-trait associations for agronomic and grain traits in wild diploid wheat *Triticum urartu*

Xin Wang[1,2], Guangbin Luo[1,2], Wenlong Yang[1], Yiwen Li[1], Jiazhu Sun[1], Kehui Zhan[3], Dongcheng Liu[1]*⊙ and Aimin Zhang[1,3]*

Abstract

Background: Wild diploid wheat, *Triticum urartu* (*T. urartu*) is the progenitor of bread wheat, and understanding its genetic diversity and genome function will provide considerable reference for dissecting genomic information of common wheat.

Results: In this study, we investigated the morphological and genetic diversity and population structure of 238 *T. urartu* accessions collected from different geographic regions. This collection had 19.37 alleles per SSR locus and its polymorphic information content (PIC) value was 0.76, and the PIC and *Nei's* gene diversity (GD) of high-molecular-weight glutenin subunits (HMW-GSs) were 0.86 and 0.88, respectively. UPGMA clustering analysis indicated that the 238 *T. urartu* accessions could be classified into two subpopulations, of which Cluster I contained accessions from Eastern Mediterranean coast and those from Mesopotamia and Transcaucasia belonged to Cluster II. The wide range of genetic diversity along with the manageable number of accessions makes it one of the best collections for mining valuable genes based on marker-trait association. Significant associations were observed between simple sequence repeats (SSR) or HMW-GSs and six morphological traits: heading date (HD), plant height (PH), spike length (SPL), spikelet number per spike (SPLN), tiller angle (TA) and grain length (GL).

Conclusions: Our data demonstrated that SSRs and HMW-GSs were useful markers for identification of beneficial genes controlling important traits in *T. urartu*, and subsequently for their conservation and future utilization, which may be useful for genetic improvement of the cultivated hexaploid wheat.

Keywords: *Triticum urartu*, Genetic diversity, SSR markers, HMW-GS, Marker-trait association

Background

Bread wheat (*Triticum aestivum* L.) is one of the most important crops in the world, providing about 20% of the calories consumed globally (FAOSTAT 2011; http://faostat.fao.org/). To meet the world's growing demand for food, it is urgent to develop high yielding varieties with good end-product making quality [1]. A better understanding of the genetic basis of yield and its components is a pre-requisite, though genomic research in bread wheat remains a major challenge because of the complexity associated with its hexaploid structure and huge genome size [2, 3]. Wild diploid wheat, *T. urartu* (2n =2× =14; AA), the A-genome donor of cultivated tetraploid (2n =4× =28; genome AABB) and hexaploid wheat (2n =6× =42; AABBDD) [4], played an important role in the development and evolution of cultivated bread wheat. With the available reference genome [5], it is more feasible to exploit *T. urartu* as the reference sub-genome of common wheat, which will obviously provide considerable valuable information for the improvement of the latter.

* Correspondence: dcliu@genetics.ac.cn; amzhang@genetics.ac.cn
[1]State Key Laboratory of Plant Cell and Chromosome Engineering, National Center for Plant Gene Research, Institute of Genetics and Developmental Biology, Chinese Academy of Sciences, 1 West Beichen Road, Chaoyang District, Beijing 100101, China
Full list of author information is available at the end of the article

Although *T. urartu* possesses the A genome in common with bread wheat, its genetic diversity has not been well investigated as Einkorn wheat (*T. monococcum*), another diploid progenitor [6]. Nowadays, the genetic diversity within wheat cultivars has been drastically reduced in the process of domestication and breeding [7, 8], and it is essential to apply new contributing genes for wheat improvement. As a wild diploid progenitor of hexaploid wheat, *T. urartu* harbors rich allelic diversity for numerous important traits, including agronomic characteristics, grain quality and biotic stress tolerance [9–11]. Genetic variation of *T. urartu* has been investigated using various markers such as isozyme, restriction fragment length polymorphism (RFLP), amplified fragment length polymorphism (AFLP), and random amplified polymorphic DNA (RAPD) markers [12–16]. Such genetic diversities can be exploited to elucidate the genetic basis of natural variation of important quantitative traits. Hence, more accessions widespread should be employed to provide a more comprehensive on the characteristic of the whole population.

Most agronomic traits in common wheat, such as yield and its components, dough quality and resistant characters, are controlled by multiple genes and influenced substantially by the environment, which hinders the dissection of their genetic basis [17]. Classical linkage mapping based on bi-parental populations was a conventional approach to dissect the genetic bases of complex traits [18]. Various studies have identified a set of major effect quantitative trait loci (QTL) for important agronomic traits in wheat, such as kernel weight and dough quality [17, 19–24]. As an alternative way to QTL mapping, association mapping uses diverse material to associate genetic markers with a phenotype of interest, which presents higher mapping resolution of the phenotypes at a population level [25]. It has been exploited successfully to identify genomic regions contributing to numerous traits in diverse crops, such as maize [26, 27], rice [28], sorghum [29] and soybean [30]. There is increasing interests in identifying novel marker-trait associations using association mapping in wheat [31, 32]. For example, Breseghello and Sorrells [17] found significant associations between some simple sequence repeats (SSR) markers and wheat kernel traits, including weight, length and width of kernels.

The objectives of this study were to investigate the genetic diversity, population structure and relationships among a collection of 238 *T. urartu* accessions collected from the Fertile Crescent region and to identify marker loci associated with important agronomic and grain traits. Our results would provide further insights into the utility of association mapping for marker-assisted selection and its potential application in bread wheat breeding.

Methods

Plant material

A total of 238 *T. urartu* accessions, which covered most of the original areas, were subjected to SSR and high-molecular-weight glutenin subunits (HMW-GS) analysis with SDS-PAGE. This panel was obtained from the Institute of Genetics and Developmental Biology, Chinese Academy of Sciences (IGDB, CAS), the National Small Grain collection (USDA-ARS) and the International Center for Agricultural Research in the Dry Areas (ICARDA). Among these accessions, 84 were originated from Lebanon, 80 from Turkey, 37 from Syria, 12 from Armenia, 11 from Jordan, 11 from Iran, and three from Iraq (Fig. 1; Additional file 1: Table S1).

Field experiment and phenotyping

The collected *T. urartu* accessions were planted at the experimental station of the Institute of Genetics and Developmental Biology, Chinese Academy of Sciences, Beijing (40°5′56″N and 116°25′8″E), that of Henan Agricultural University, Zhengzhou (34°51′52″N, 113°35′45″E) and that of Dezhou Academy of Agricultural Sciences, Dezhou (37°45′69″N, 116°30′23″E) in two consecutive years (2013-2014 and 2014-2015 cropping seasons). These environments were designated as E1 (Beijing, 2013), E2 (Zhengzhou, 2013), E3 (Dezhou, 2013), E4 (Beijing, 2014), E5 (Zhengzhou, 2014), and E6 (Dezhou, 2014), respectively.

Each field trial was managed in a completely randomized block design with two replications. All the plants were grown in a single 2-m row with 40 cm between rows and 20 cm between individuals. Nine traits were evaluated and analyzed, including heading date (HD), plant height (PH), spike length (SPL), spikelet number per spike (SPLN), tiller angle (TA), grain length (GL), grain width (GW), grain length/width ratio (GLW) and thousand-grain weight (TGW). The HD was counted as days from sowing to heading, and the date of heading was subsequently recorded when half of the spikes emerged from the flag leaf in each accession. The TA was measured between the last developed tillers and the ground level with a protractor at the maximum tillering stage, while the other agronomic traits were determined on the primary tiller. After the harvest, a minimum of 200 grains from each sample was photographed on a flat-bed scanner and the images were obtained as the aerial view of the ventral side of the grains. The GL, GW and GLW were calculated using the grain analyzer software (SC-G Scanner, Wanshen Detection Technology Inc., Hangzhou, China), and the TGW was measured by the average of two 1000 kernel-weights.

DNA isolation and nested-PCR amplification

Genomic DNA was isolated from young leaf tissue of two-week-old seedlings (one individual per accession)

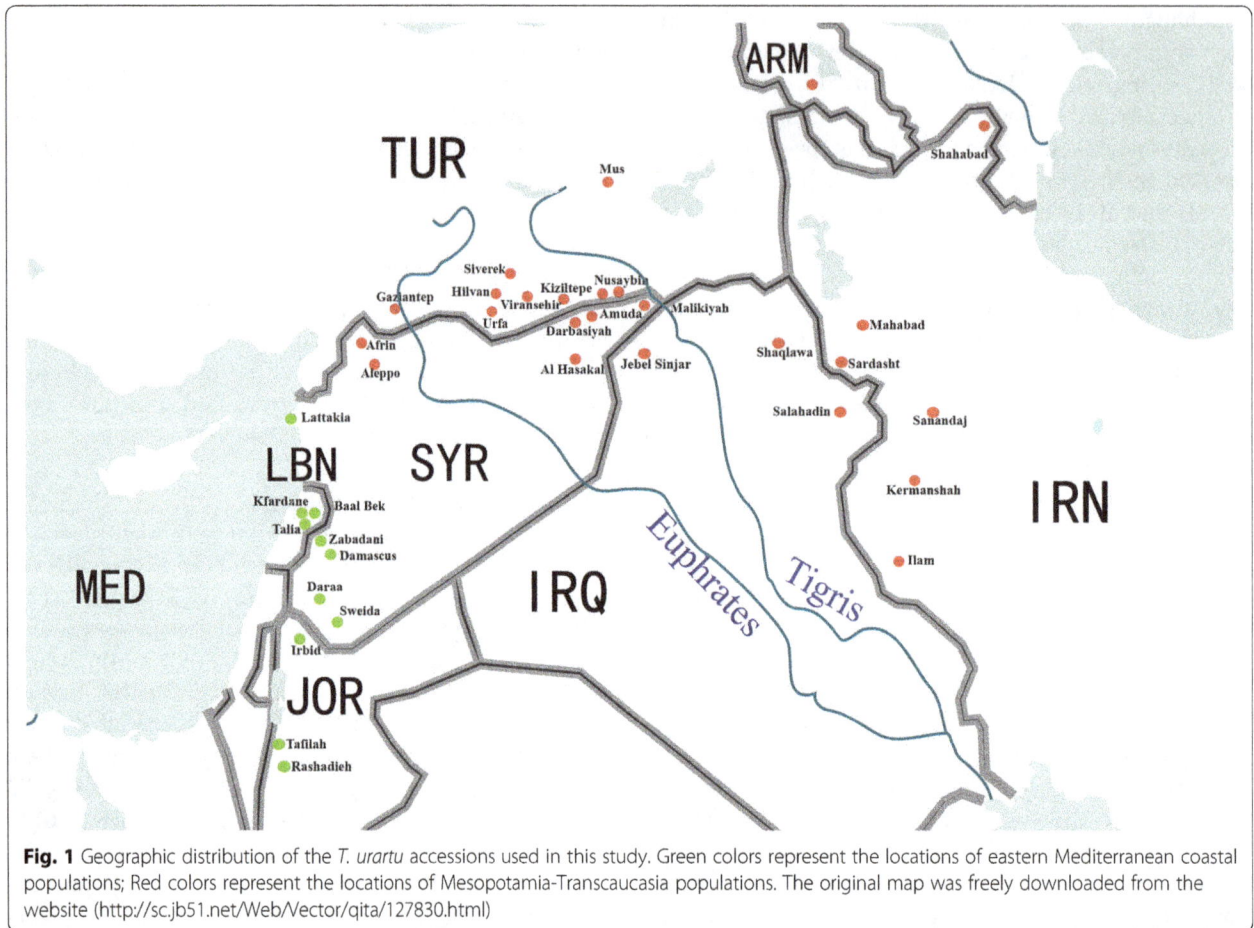

Fig. 1 Geographic distribution of the *T. urartu* accessions used in this study. Green colors represent the locations of eastern Mediterranean coastal populations; Red colors represent the locations of Mesopotamia-Transcaucasia populations. The original map was freely downloaded from the website (http://sc.jb51.net/Web/Vector/qita/127830.html)

using the cetyl trimethyl ammonium bromide (CTAB) method [33]. DNA concentration was determined and diluted to a working solution of 50 ng/μL. Primer sequences of 62 SSR markers (Additional file 1: Table S2) used in this experiment were obtained from the GrainGenes database (https://wheat.pw.usda.gov/cgi-bin/graingenes/browse.cgi?class=marker). The nested-PCR amplifications were performed using fluorescent dye labeling system according to Schuelke [34]. In brief, amplification reactions were performed in a final volume of 15 μL, containing 3 μL template DNA, 0.2 μL forward primer with the M13 tail at its 5′-end (2.0 μM), 1.0 μL M13 primer (labeled with 6-FAM, NED, VIC or PET), 1.2 μL reverse primer (2.0 μM), 7.5 μL Mix-Taq (CWBIO, China) and 2.1 μL H_2O (CWBIO, China). PCR was performed as follows: initial denaturing at 95 °C for 4 min, 15 cycles of 94 °C for 30 s, 60 °C for 45 s, and 72 °C for 45 s, followed by 25 cycles of 95 °C for 30s, 52 °C for 45 s, 72 °C for 45 s, and a final extension at 72 °C for 10 min. All reactions were conducted using a thermal cycler, Veriti 96 (Applied Biosystems, Foster City, USA).

Amplified PCR products were pooled and then purified with 3.0 M sodium acetate and 70% ethanol before adding HiDi-formamide. Fluorescently labeled DNA fragments were separated by capillary electrophoresis in an ABI 3730*xl*DNA Analyzer with GeneScan-500 LIZ size standard (Applied Biosystems, Foster City, CA, USA). SSR polymorphism was analyzed by GeneMapper software version 4.0 (Applied Biosystems, Foster City, CA, USA) according to the manufacturer's instructions.

Protein extraction and SDS-PAGE analysis
In each accession, the HMW-GS proteins were extracted from three seeds according to previous procedure [35], and SDS-PAGE was performed to fractionate the HMW-GSs using 10% (*w/v*) separating and 3% (*w/v*) stacking gels. Electrophoresis was conducted at a constant current of 18 mA at room temperature for about 8 h, and then the SDS-PAGE gels were stained overnight with Coomassie Brilliant Blue G-250 in a solution containing 20% (*v/v*) methanol and 10% (*v/v*) acetic acid. De-staining was carried out with tap water and then the gels were subjected to image capturing on a high-resolution scanner (GE ImageScanner III). The identification of HMW-GSs and alleles were based on the methodology described by [36].

Genetic diversity and phylogenetic analysis

For 62 SSR polymorphic loci, the genetic parameters of variability were estimated via the POPGENE1.32 software [37], including number of observed (Ao) and expected (Ae) alleles per locus, observed (Ho) and expected (He) heterozygosity, *Shannon's* information index (I), *Nei's* Gene diversity (GD) and Polymorphism information content (PIC). The accuracy of these genotyping data was also manually checked for the scoring errors and null alleles using the Micro-Checker software [38]. All the parameters were computed for both the whole collection and the subsets considering their geographic origin and population structure category. Genetic distance and phylogenetic analyses of *T. urartu* accessions were conducted based on the Jaccard's coefficient similarity matrix obtained from the proportion of shared SSR fragments and HMW-GS bands, then a dendrogram was drawn with NTSYSpc 2.11 program [39] using UPGMA algorithm. The fitness of the dendrogram was assessed by bootstrap analysis running 1000 replications.

Population structure

The model-based (Bayesian) STRUCTURE version 2.3.1 was applied to identify clusters of genetically similar individuals on the basis of their genotypes [40]. The program was run five times independently for *K* value (number of subpopulations) ranging from 1 to 15, adopting the admixture model and correlated allele frequencies, with a burn-in period of 50,000 and the number of replications at 100,000. The normal logarithm of the probability was calculated against each *K* value, and the optimal number of subpopulations was determined using the *ΔK* approach described by Evanno et al. [41]. Then each *T. urartu* individuals could be assigned to the putative subpopulations according to their average membership coefficient, which was calibrated using CLUMPP [42, 43]. A nested analysis of population structure with the same software and parameters was carried out to distinguish the next level of subpopulations.

Analysis of molecular variance (AMOVA)

Analysis of molecular variance (AMOVA) was performed to estimate the genetic variance within and among inferred populations in *T. urartu* accessions. Population differentiation was assessed by calculating pairwise *F*st values and *Nei's* genetic distances for different regional population pairs. The threshold for statistical significance was determined by running 10,000 permutations. Principal coordinate analysis (PCoA) was also carried out based on binary genetic distance. The generated eigenvalues and accumulated percentage of the variation were applied to plot the scatter diagram of these representative accessions, with the first two principal coordinates which accounted for the highest variation. All the calculations mentioned above were implemented in the GenAlex 6.5 software [44].

Association analysis

The markers with minor allele frequency less than 5% were removed in order to reduce false positive associations. The pairwise kinship (K matrix) among samples was generated from the program SPAGeDi [45]. Linkage disequilibrium estimates for each pair of loci and marker-trait association analysis were conducted using TASSEL 2.1 software [46], via a general linear model (GLM) and a mixed linear model (MLM). In GLM model, population structure of the *T. urartu* mapping panel was included as fixed effects, while association was estimated by simultaneous accounting of multiple levels of population structure (Q matrix) and relative kinship among the individuals (K matrix) in MLM. The significant threshold for associations between loci and traits was set at $P < 0.01$. The Bonferroni correction of multiple testing was performed based on the q value using false discovery rate (FDR, $\alpha_c = 0.05$). For all associated markers, the average phenotypic effects of different alleles were estimated using the method proposed by Breseghello and Sorrells [17].

Results

Phenotypic variation and correlations among traits

The phenotypic data of investigated traits across the six environments, including mean values, minimum and maximum values, standard deviations and the heritability estimates (h^2), were calculated (Table 1). The data revealed a broad variation for all traits in these *T. urartu* accessions, e.g. PH had an average of 112.43 cm (minimum 86.10 cm and maximum 149.10 cm) with 12.23 cm standard deviation in E1, and TGW ranged from 4.29 to 18.87 g with 2.96 g standard deviation in E5. The phenotypic data for HD, PH, SPL, SPLN, GL, GW, GLW and TGW followed the normal distribution, suggesting that these traits were controlled by multiple loci. The broad-sense heritability (h^2) of all the traits was relatively high, ranging from 68.91% for TA (E2) to 94.05% for SPLN (E4), which confirmed that most of the phenotypic variance was genetically determined. Moreover, significant correlations coefficients among different environments were also detected, implying the less genotype x environment interactions of these traits (Additional file 1: Table S3). Depending on the collection site, all of the *T. urartu* accessions could be roughly split as Eastern Mediterranean coastal and Mesopotamia-Transcaucasia group, and the details of phenotypic performances illustrated that most of the traits investigated in this study differed greatly between the two major subsets (Additional file 1: Table S4).

Table 1 Phenotypic performances and distribution parameters for the investigated traits of 238 *T. urartu* accessions in six environments

Trait	Env. [a]	Mean [b]	Min. [c]	Max. [d]	SD [e]	h^2 (%)[f]
HD	E1	207.43	196.00	229.00	8.57	78.18
Heading date (days)	E2	205.45	192.00	225.00	8.85	82.36
	E3	200.42	182.00	218.00	9.22	80.54
	E4	216.57	204.00	239.00	8.89	83.77
	E5	208.16	195.00	228.00	10.15	78.95
	E6	209.05	199.00	227.00	8.06	81.14
PH	E1	112.43	86.10	149.10	12.23	76.43
Plant height (cm)	E2	117.02	82.70	151.40	13.68	73.82
	E3	116.34	89.00	154.60	15.25	75.47
	E4	122.58	92.33	157.00	14.07	79.15
	E5	116.70	92.66	148.00	10.87	78.56
	E6	106.64	67.67	136.00	12.24	72.84
SPL	E1	11.10	7.38	14.92	1.62	87.22
Spike length (cm)	E2	13.02	7.02	18.43	2.31	86.61
	E3	11.19	5.00	16.03	2.05	84.89
	E4	13.09	9.00	18.67	2.07	89.27
	E5	12.52	8.38	16.90	2.00	87.33
	E6	10.63	7.30	14.80	1.65	89.06
SPLN	E1	27.29	19.20	36.00	3.76	92.73
Spikelet number/spike	E2	29.96	20.00	41.70	4.32	91.46
	E3	27.95	20.20	39.30	4.09	90.94
	E4	32.44	23.00	49.50	5.11	94.05
	E5	26.96	17.00	38.70	4.32	92.55
	E6	26.34	16.80	36.00	3.70	89.49
TA	E2	60.43	25.00	72.00	15.29	68.91
Tiller angle (°)	E4	65.14	33.00	78.00	17.60	70.87
	E5	59.32	28.00	75.00	12.59	71.53
GL	E1	6.87	5.11	8.39	0.61	90.65
Grain length (mm)	E2	7.52	6.16	8.95	0.55	87.46
	E3	6.96	5.17	8.48	0.62	89.08
	E4	7.39	5.46	8.52	0.51	90.41
	E5	7.41	6.04	8.58	0.55	88.70
	E6	7.56	6.07	9.24	0.58	87.16
GW	E1	1.58	1.05	2.07	0.21	88.78
Grain width (cm)	E2	1.80	1.32	2.28	0.22	90.52
	E3	1.58	1.12	2.04	0.21	89.67
	E4	1.74	1.22	2.23	0.19	91.33
	E5	1.78	1.41	2.11	0.20	89.29
	E6	1.77	1.33	2.24	0.17	87.06
GLW	E1	4.49	3.51	5.71	0.40	88.49
Grain length/width ratio	E2	4.26	3.50	5.19	0.41	86.95
	E3	4.49	3.76	5.55	0.41	89.42
	E4	4.37	3.55	5.47	0.42	90.13
	E5	4.26	3.69	5.05	0.35	88.35

Table 1 Phenotypic performances and distribution parameters for the investigated traits of 238 *T. urartu* accessions in six environments *(Continued)*

	E6	4.37	3.69	5.04	0.38	87.29
TGW	E1	7.86	2.28	15.61	2.86	81.03
Thousand-grain weight (g)	E2	11.43	5.15	21.84	3.10	83.80
	E3	8.12	1.73	15.32	2.81	78.38
	E4	10.26	2.50	15.40	2.41	85.11
	E5	10.06	4.29	18.87	2.96	82.04
	E6	9.77	3.91	18.23	2.64	81.90

[a] Environment: E1, E2, E3, E4, E5 and E6 represent Beijing 2013, Zhengzhou 2013, Dezhou 2013, Beijing 2014, Zhengzhou 2014 and Dezhou 2014, respectively
[b] Mean value for *T. urartu* accessions
[c] Minimum value among *T. urartu* accessions
[d] Maximum value among *T. urartu* accessions
[e] Standard deviation of each set of phenotypic data
[f] Broad sense heritability

The correlation coefficient (*r*) analysis revealed that several traits were correlated (Additional file 1: Table S5). The highest positive correlation was detected between TGW and GW (*r* ranging from 0.68 to 0.81 in different environments, significant at $P < 0.01$), followed by TGW versus GL (0.60–0.69, $P < 0.01$), indicating that grain weight was largely influenced by grain size in *T. urartu*. In addition, significant positive correlations of HD with PH, SPL and SPLN were observed in all environments, suggesting that late-heading varieties were prone to have a high statue and large spike with many spikelets. On the other hand, HD was also negatively correlated with GL and GLW, demonstrating that a *T. urartu* accession with late heading date almost followed with small grain length and low grain length/width ratio. Notably, TA had a significant positive correlation with PH and SPLN, which reflected the tendency for prostrate type to have a high plant and a large number of spikelets.

Genetic diversity revealed by SSR markers

To evaluate the genetic diversity of the *T. urartu* population, 62 SSR primers (loci) distributed on seven chromosomes (Table 2) were selected to perform the nested-PCR amplifications and detected with fluorescent dye labeling system. In 238 *T. urartu* accessions, a total of 1201 alleles ranging from 4 (*Xbarc138*) to 42 (*Xgwm136*) were amplified, with an average of 19.37 alleles per locus (Ao). 881 rare alleles (73.36%) were detected with the frequency lower than 5%, but none was fixed with the frequency more than 90%, resulting in an average expected allele (Ae) of 7.29. This suggested that the higher genetic variations of alleles were present in *T. urartu* accessions. Consequently, the major allele frequencies varied from 0.13 (*Xcfd15*) to 0.86 (*Xbarc206*), with the overall mean of 0.32.

We observed 18 heterozygous loci for the SSR markers assayed particularly, with the observed heterozygosity (Ho) and expected heterozygosity (He) ranged between 0.04-0.72 and 0.24-0.85, respectively. As a result, a total of 1357 genotypes were deduced with an average of 21.89 per SSR marker. According to the polymorphic information content (PIC), 53 SSR loci (85.48%) were highly informative (PIC >0.5), eight (12.90%) were reasonably informative (0.5 > PIC >0.25) and only one (1.61%) was slightly informative (PIC <0.25). Calculation of the *Nei's* gene diversity (GD) for 62 loci demonstrated that *Xcfa2134* preserved the highest GD (GD = 0.94) and *Xbarc206* did the lowest (GD = 0.25), with the mean value of 0.80. Hardy–Weinberg equilibrium testing of these markers indicated that *T. urartu* population is not mating randomly, probably owing to the self-pollination in diploid wheat (Additional file 1: Table S6).

The panel of analyzed *T. urartu* accessions possessed high polymorphic information, covering most original areas. In order to explore and compare the variability inherent in genetic diversity, variability parameters were calculated in eight sample subsets (Table 3). The number of alleles amplified in the Mesopotamia-Transcaucasia group was higher than that in the Eastern Mediterranean coastal group (16.38 versus 11.02, $P < 0.01$), resulting in a general decrease in GD from 0.76 to 0.64. Likewise, the *Shannon's* information indices (I) were counted as 1.97 versus 1.48 (Table 3). Beyond that, the Mesopotamia-Transcaucasia group preserved more rare alleles (frequency < 5%) (498) than the Eastern Mediterranean coastal group (377). Our data demonstrated that the Mesopotamia-Transcaucasia group had much higher genetic diversity, and these regions might be the diversity center of *T. urartu*.

With respect to the geographic regions analyzed separately, the subpopulation of accessions collected from Turkey exhibited the highest diversity, followed by that from Northern Syria, Southwestern Syria, Lebanon, Iran, Jordan, Armenia and Iraq, due to their differences in number of alleles and genetic diversity (Table 3). GD in each population showed a similar trend ranging from

Table 2 Diversity parameters revealed by SSR markers in 238 *T. urartu* accessions

Loci	Chromosome	Position (cM) [a]	Repeat pattern	Size range (bp)	Ao [b]	Ae [c]	Gn [d]	MAF [e]	PIC [f]	Ho [g]	He [h]	GD [i]
Xgwm136	1A	14	(CT)n	224–308	42	11.43	51	0.51	0.91	0.21	0.79	0.93
Xcfd15	1A	23	(CT)n(TGTA)n	166–226	26	14.85	28	0.13	0.93	0.07	0.76	0.91
Xbarc148	1A	43	(CT)n	197–209	8	2.08	8	0.65	0.45	0	0.32	0.51
Xgwm357	1A	51	(GA)n	122–158	17	9.73	19	0.15	0.89	0	0.62	0.80
Xgwm164	1A	57	(CT)n	118–148	15	5.87	16	0.28	0.81	0	0.62	0.83
Xcfa2129	1A	83	(GA)n	154–206	25	9.05	25	0.23	0.91	0	0.77	0.89
Xcfa2219	1A	126	(GT)n	214–250	18	7.77	22	0.19	0.88	0.23	0.68	0.87
Xbarc17	1A	136	(TAA)n	268–325	19	5.72	19	0.32	0.86	0	0.50	0.83
Xgwm210.1	2A	4	(GA)n	183–187	5	1.70	5	0.73	0.35	0	0.39	0.41
Xgwm614	2A	10	(GA)n	120–192	36	13.25	36	0.30	0.87	0	0.70	0.88
Xgwm328	2A	43	(GT)n	199–212	15	8.05	26	0.28	0.87	0.36	0.72	0.88
Xgwm249.1	2A	59	(GA)n(GGA)n	199–238	24	6.56	24	0.24	0.83	0	0.56	0.85
Xcfa2043	2A	71	(GA)n	197–233	21	6.86	21	0.24	0.83	0	0.73	0.85
Xcfa2058	2A	76	(TC)n	252–290	24	10.34	24	0.21	0.84	0	0.73	0.90
Xcfa2121	2A	82	(CA)n	146–184	22	14.98	42	0.17	0.90	0.49	0.79	0.93
Xgwm265	2A	100	(GT)n	185–195	10	3.21	10	0.42	0.94	0	0.52	0.68
Xgwm382.1	2A	117	(GA)n	110–170	30	14.41	30	0.16	0.63	0	0.76	0.93
Xcfa2086	2A	133	(CA)n	217–284	32	11.37	32	0.22	0.93	0	0.75	0.91
Xbarc57	3A	0	(TTA)n	205–265	22	8.25	22	0.19	0.91	0	0.73	0.88
Xbarc12	3A	25	(TAA)n	167–227	20	9.08	20	0.16	0.87	0	0.72	0.89
Xgwm369	3A	36	(CT)nTT(CT)n	170–204	20	10.05	20	0.23	0.88	0	0.67	0.90
Xcfa2076	3A	61	(TG)n	184–218	14	4.02	14	0.42	0.89	0	0.55	0.75
Xgwm674	3A	80	(CT)nCCC(GT)n	168–178	7	1.92	7	0.66	0.42	0	0.24	0.47
Xcfa2134	3A	101	(TC)n	207–283	38	14.71	38	0.19	0.38	0	0.78	0.94
Xgwm480	3A	105	(CT)n(CA)n	176–194	12	3.71	12	0.42	0.93	0	0.46	0.73
Xgwm247	3A	117	(GA)n	202–214	27	6.32	27	0.29	0.69	0	0.74	0.84
Xcfa2193	3A	171	(GT)n	198–246	24	6.31	24	0.30	0.83	0	0.56	0.84
Xbarc206	4A	0	(CT)n	234–252	9	1.37	12	0.86	0.29	0.04	0.24	0.25
Xgwm192.1	4A	29	(CT)n	151–169	9	2.89	9	0.49	0.24	0	0.34	0.65
Xbarc138	4A	39	(CT)n	192–196	4	2.03	4	0.56	0.48	0	0.27	0.36
Xgwm397	4A	69	(CT)n	195–235	30	14.63	38	0.17	0.43	0.25	0.76	0.93
Xgwm269.2	4A	76	(CA)n	113–197	24	10.68	24	0.19	0.93	0	0.63	0.91
Xcfd88	4A	108	(CCG)n	179–182	7	1.94	7	0.61	0.90	0	0.36	0.49
Xbarc70	4A	187	(TATCTA)n(TCTA)n	202–274	21	8.85	21	0.43	0.38	0	0.41	0.74
Xbarc180	5A	12	(ATT)n	184–226	17	8.93	17	0.19	0.71	0	0.74	0.89
Xbarc117	5A	18	(CA)n	235–248	8	2.65	11	0.50	0.88	0.04	0.44	0.62
Xgwm293	5A	27	(CA)n	178–205	17	3.73	17	0.45	0.68	0	0.57	0.84
Xbarc1	5A	30	(TAA)n	255–297	14	2.19	14	0.38	0.61	0	0.46	0.76
Xbarc165	5A	35	(ATT)n	207–236	19	7.94	28	0.24	0.54	0.69	0.85	0.87
Xbarc141	5A	39	(GA)n	264–290	13	4.84	13	0.30	0.86	0	0.60	0.79
Xbarc330	5A	48	(CT)n	108–160	28	11.45	28	0.17	0.77	0	0.75	0.91
Xbarc151	5A	54	(CT)n	202–224	12	4.65	12	0.38	0.76	0	0.61	0.78
Xgwm639	5A	64	(GA)n	146–169	26	8.34	26	0.19	0.87	0	0.70	0.88
Xgwm179	5A	96	(GT)n	194–208	8	3.55	8	0.33	0.67	0	0.23	0.72

Table 2 Diversity parameters revealed by SSR markers in 238 *T. urartu* accessions *(Continued)*

Xgwm410.1	5A	109	(CA)n	288–345	37	11.58	54	0.25	0.91	0.72	0.78	0.91
Xgwm334	6A	28	(GA)n	127–153	26	8.85	42	0.23	0.88	0.28	0.74	0.89
Xbarc3	6A	66	(CCT)n	206–256	30	9.67	30	0.33	0.84	0	0.63	0.85
Xcfd80	6A	84	(GA)n	165–183	11	4.04	11	0.40	0.72	0	0.53	0.75
Xgwm132	6A	111	(GA)n(GAA)n	112–146	20	7.01	33	0.33	0.85	0.69	0.81	0.86
Xgwm570	6A	118	(CT)n(GT)n	130–152	11	3.86	11	0.38	0.70	0	0.49	0.74
Xbarc104	6A	125	(TAA)n	174–213	13	3.11	13	0.40	0.63	0	0.54	0.78
Xgwm427	6A	137	(CA)n	154–228	31	12.14	31	0.17	0.91	0	0.72	0.92
Xgwm617	6A	140	(GA)n	106–166	11	5.28	11	0.28	0.78	0	0.42	0.75
Xgwm471	7A	20	(CA)n	120–186	32	9.95	32	0.27	0.89	0	0.76	0.90
Xcfd242	7A	39	(GTT)n(AGC)n	219–237	8	3.47	12	0.39	0.66	0.30	0.58	0.75
Xbarc127	7A	47	(CT)n	196–264	22	9.18	37	0.19	0.88	0.21	0.73	0.89
Xbarc154	7A	49	(CT)n	237–285	22	5.60	26	0.29	0.80	0.07	0.61	0.82
Xbarc174	7A	71	(ATT)n	170–227	21	8.85	21	0.21	0.88	0	0.55	0.89
Xgwm276	7A	84	(CT)n	97–129	16	4.93	20	0.37	0.77	0.06	0.61	0.80
Xcfd20	7A	103	(GGAA)n(CTAC)n	288–459	11	4.17	14	0.35	0.73	0.13	0.57	0.76
Xgwm63	7A	106	(CA)n(TA)n	157–183	16	6.61	24	0.30	0.83	0.18	0.69	0.85
Xcfa2040	7A	124	(CA)n	293–341	24	11.45	24	0.17	0.91	0	0.73	0.91
Mean					19.37	7.29	21.89	0.32	0.76	0.08	0.61	0.80

[a] Indicated map positions in bread wheat according to Song et al. [48]
[b] Number of observed alleles
[c] Expected number of alleles
[d] Number of genotype at each locus
[e] Frequency of major allele
[f] Polymorphism information content
[g] Observed heterozygosity
[h] Expected heterozygosity
[i] *Nei's* Gene diversity

Table 3 Genetic diversity for different *T. urartu* subsets based on 62 SSR markers

Subset of accessions	Sample Size	Ao [a]	Ae [b]	PIC [c]	GD [d]	I [e]
All samples	238	19.37	7.29	0.76	0.80	2.04
Eastern Mediterranean coast	112	11.02	3.91	0.61	0.64	1.48
Lebanon	84	6.41	2.83	0.49	0.52	1.18
Southwestern Syria	17	7.50	4.89	0.66	0.69	1.59
Jordan	11	3.27	2.25	0.40	0.45	0.83
Mesopotamia-Transcaucasia	126	16.38	6.71	0.74	0.76	1.97
Turkey	81	11.71	5.18	0.72	0.75	1.73
Northern Syria	19	8.73	5.07	0.69	0.71	1.71
Iraq	3	2.69	2.55	0.45	0.52	1.08
Iran	11	5.53	4.28	0.66	0.69	1.41
Armenia	12	3.08	1.77	0.33	0.37	0.68

[a] Number of observed alleles
[b] Expected number of alleles
[c] Polymorphism information content
[d] *Nei's* Gene diversity
[e] *Shannon's* information indices

0.75 (Turkey) to 0.37 (Armenia). The maximum and the minimum *Shannon's* information indices (I) were observed as well in Turkish (1.73) and Armenian (0.68) accessions, respectively.

Genetic diversity revealed by HMW-GSs

The pattern of HMW-GSs from most of the *T. urartu* accessions is formed by two distinct electrophoretic moving zones, including one major 1Ax subunit zone with slower mobility and one major 1Ay subunit zone with faster mobility (Fig. 2). Among the 238 *T. urartu* accessions, the 1Ay subunit was not expressed in 69 accessions (28.99%), and only one (0.42%) was found silenced for the 1Ax subunit. All the 1Ax subunits in *T. urartu* showed faster electrophoretic mobility than the subunit 1Ax1 detected in bread wheat Xiaoyan 54 (XY), and four 1Ax subunits in 46 accessions displayed slower electrophoretic mobility than the 1Ax2* present in bread wheat Cheyenne (CNN) (Fig. 2). In most *T. urartu* accessions, the electrophoretic mobility of the 1Ay subunits was faster than the 1Dy10 subunit of Cheyenne, except for one Armenian and two Turkish accessions (U17).

A total of eleven 1Ax and eight 1Ay subunits were detected, resulting in 18 HMW-GS genotypes (U1-U18) (Table 4). U1, U6 and U10 appeared exclusively in Turkish accessions, U14, U15 and U16 only in Lebanese accessions, U5 merely in southwestern Syria, and U12 and U18 were solely present in northern Syria. Concerning the frequencies of these HMW-GS genotypes, U7 was the most abundant (65 accessions, 27.31%), followed by U8 (42 accessions, 17.65%), U2 (35 accessions, 14.71%) and U14 (21 accessions, 8.82%). The remaining 14 patterns totally accounted for 31.51% of accessions, of which U1, U6 and U18 were each present in only one accession. When considering the HMW-GS locus as a co-dominant marker, its PIC and GD were 0.86 and 0.88, respectively, which were comparable with these of the SSR markers.

Genetic relationships among the accessions

The information about genetic variation determined from SSR data combining with the SDS-PAGE analysis

of HMW-GS was employed to estimate similarity matrix value. Based on Jaccard's coefficient, an UPGMA dendrogram was constructed to reveal the genetic relationships (Fig. 3). The phylogenetic tree clearly assigned the 238 *T. urartu* accessions into two clusters: Cluster I mainly distributed in the east of Mediterranean coastal regions, including Lebanon, Jordan and southwestern Syria, and Cluster II tended to occur in the Mesopotamia and Transcaucasia regions, including Turkey, Iraq, Iran, Armenia and northern Syria. The genetic distance in the dendrogram revealed that Cluster II exhibited more diversity than Cluster I, which was consistent with its extensive geographic distribution. In Cluster I, Lebanese accessions (84) gathered tightly and were further distinguished from the other accessions, thus this cluster split into two major subclasses. Cluster II was also divided into two subclasses, one of which contained all the Iraqi and Iranian accessions (14), most of Turkish and northern Syrian accessions (84), and one Armenian accession, whereas the other one contained eight Turkish accessions, six northern Syrian accessions and eleven Armenian accessions. Typically, the sequenced accession, PI428198 (G1812) [5] grouped with most Turkish materials at the first subclass of Cluster II. The Mantel test revealed a high and significant cophenetic correlation ($r = 0.92$, $P < 0.001$), indicating a good fit between the dendrogram and its original similarity matrix. The range of similarity coefficients (0.03-0.97) showed abundant genetic variation in this collection, which is supported by the high means observed for the number of alleles per locus and the PIC values.

Principal coordinates analysis (PCoA) was also performed in order to assess the individual differences (Fig. 4). The first two axes accounted for 42.25 and 17.28% of the genetic variation, respectively, and occupied 59.53% in total. The first coordinate clearly discriminated Eastern Mediterranean coastal accessions from Mesopotamia-Transcaucasia accessions, while the second coordinate separated the two large clusters gradually into small groups due to the latitude variation. The PCoA data confirmed the UPGMA analysis.

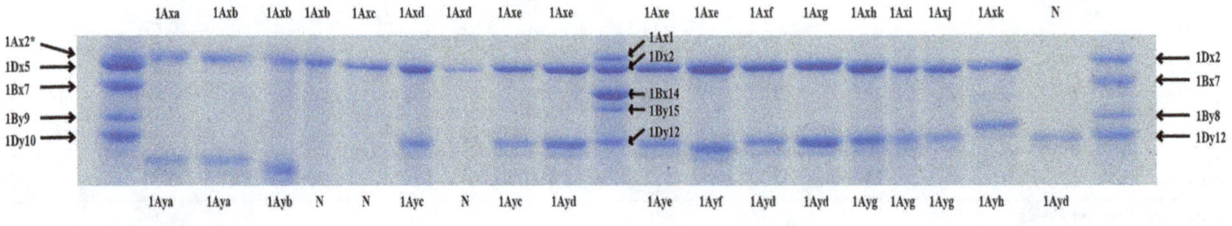

Fig. 2 SDS-PAGE electrophoretic separations of HMW-glutenin subunits. Lanes from U1 to U18 indicate the *Glu-A1* allele band patterns in *T. urartu* accessions. Lanes CNN, XY and CS represent HMW-GS patterns in bread wheat Cheyenne, Xiaoyan 54 and Chinese Spring, respectively. N represents no allelic variants detected

Table 4 Summary of the HMW-GS patterns analyzed by SDS-PAGE

HMW pattern	1Ax	1Ay	Original region								Accession Number	Frequency (%)
			Lebanon	Jordan	Southwestern Syria	Turkey	Northern Syria	Iraq	Iran	Armenia		
U1	a	a				1					1	0.42
U2	b	a			3	27	1	1	3		35	14.71
U3	b	b		7	1						8	3.36
U4	b	N[a]				1			1		2	0.84
U5	c	N[a]			2						2	0.84
U6	d	c	1								1	0.42
U7	d	N[a]	5	4	12	31	2		1	10	65	27.31
U8	e	c	40		1				1		42	17.65
U9	e	d				5	8	1	2		16	6.72
U10	e	e				5					5	2.10
U11	e	f				7	2	1	4		14	5.88
U12	f	d					3				3	1.26
U13	g	d				1	1				2	0.84
U14	h	g	21								21	8.82
U15	i	g	2								2	0.84
U16	j	g	15								15	6.30
U17	k	h				2				1	3	1.26
U18	N[a]	d					1				1	0.42

[a] Not detected allelic variants

Population structure

In order to understand the population stratification of *T. urartu*, the model-based Bayesian cluster analysis was tested using the software program STRUCTURE (Fig. 5). The data was analyzed by successively increasing the number of subpopulations (K) from two to fifteen, with five independent runs for each K value. The CLUMPP alignment of independent solutions showed the high posterior probability in assignment of accessions among runs. At $K = 2$, we detected a division between 112 eastern Mediterranean coastal accessions (Cluster I) and 126 Mesopotamia-Transcaucasia accessions (Cluster II). At $K = 3$, Cluster II split into 61 Turkish accessions and the remainders, while 58 Lebanese accessions were segregated from the rest of Cluster I when $K = 4$. With increasing K, minor subpopulations from different geographic regions such as Jordan, Iran, Iraq and Armenia could be separated gradually. The optimum number of subpopulations (K) was identified based on lnP (D) value and the delta K (ΔK) method suggested $K = 2$ as the best fitting one in our study (Additional file 2: Fig. S1). This structure-based data was mainly consistent with the genetic relationships of the traditional clusters.

Based on AMOVA analysis, most genetic variation was detected among individuals (60.04%), followed by variation among geographic regions (18.93%), variation within individuals (12.87%) and variation among the large two Clusters (8.16%). The overall *Fst* among the Clusters and geographic regions were 0.1790 and 0.2506, respectively ($P < 0.05$). The pairwise *Fst* value, interpreted as standardized population distances between two populations, ranged from 0.0658 (between Turkish and northern Syrian populations) to 0.4419 (between Jordanian and Armenian populations). The *Nei's* genetic distance data consisted with the *Fst* estimates, in which Turkish population showed the smallest genetic distance with northern Syrian population (0.0289), whereas the greatest genetic distance was observed between Jordanian and Armenian populations (0.7947) (Table 5).

Linkage disequilibrium and marker-trait association analysis

The SSR and HWM-GS data were subjected to evaluate the linkage disequilibrium (LD) on a whole genome level. Across all the 63 loci, 1953 locus pairs were detected in the entire *T. urartu* collection, with 257 possible linked (in the same linkage groups) and 1696 unlinked locus pairs (from different linkage groups), respectively. Among these linked locus pairs, the pairwise r^2 values varied from 0.00 to 0.25 with a median of 0.04, and only nine (3.50%) marker pairs were at significant LD level ($r^2 > 0.1$ and $P < 0.001$), which suggested seldom LD among the analyzed loci.

The association analysis between markers and phenotypic data was carried out using GLM and MLM models

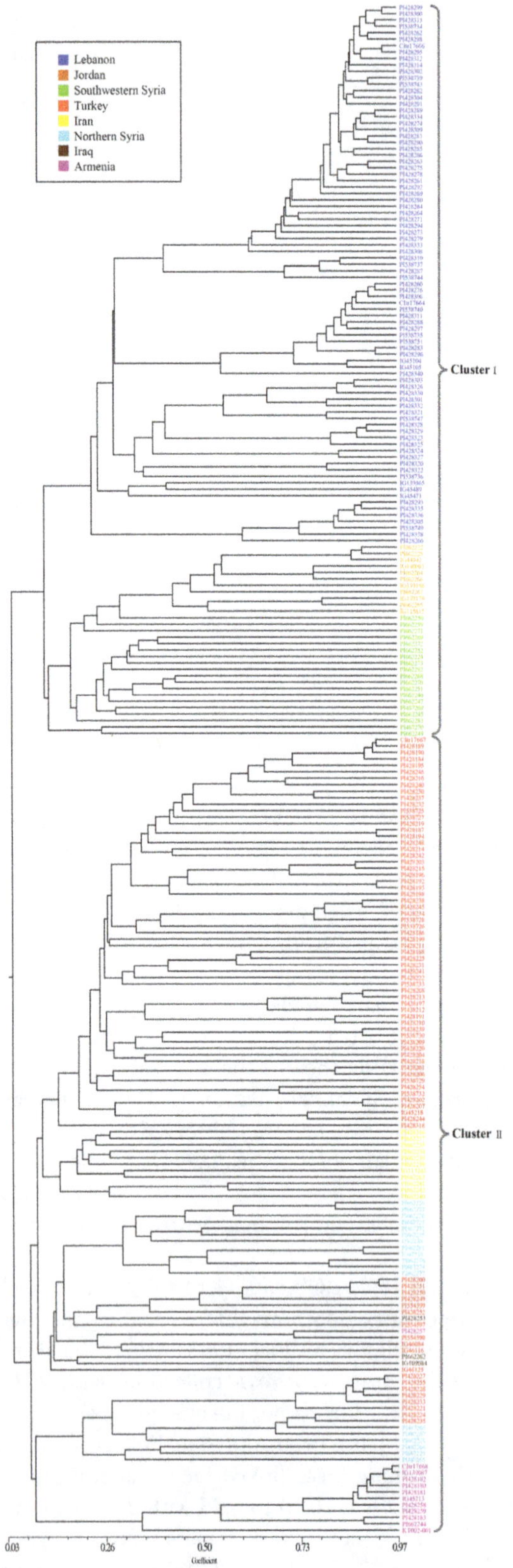

Fig. 3 UPGMA dendrogram showing the genetic relationships among the 238 *T. urartu* accessions collected from various countries. The different subpopulations are shown in different colors

in the software TASSEL. Among the 63 × 9 marker-trait comparisons, 67 significant associations (11.82%) were identified using GLM approach with Q-matrix. However, the MLM analysis reduced to 10 significant associations (1.76%) once considering K-matrix as co-variate. The Q-matrix ($K = 2$) inferred from STRUCTURE defined the ancestry coefficient of individuals in the population and the K-matrix was subjected to correct for their genetic relatedness. Besides, the averages of the phenotypic variations calculated by the two models for association mapping were 15.89% (Q) and 22.74% (Q + K), indicating that the MLM could explain more genetic variation than GLM. Henceforth, our attention will focus on associations incorporating both Q-matrix and K-matrix since they were more conservative and accurate.

Considering the MLM statistics, significant marker-trait associations for six traits were identified with six SSRs and protein marker HMW-GSs (HD: *Xcfa2193-3A*; PH: *Xgwm328-2A*; SPL: *Xgwm328-2A* and *Xgwm63-7A*; SPLN: *Xgwm328-2A*, *Xgwm63-7A* and *Xbarc138-4A*; TA: HMW-GS; GL: *Xcfa2219-1A* and *Xgwm293-5A*) in *T. urartu* accessions (Table 6). The *Xcfa2193-3A* on chromosome 3A was highly associated with HD in four environments, explaining 9.39 to 15.74% of the phenotypic variation, while *Xgwm328-2A* on chromosome 2A was closely associated with PH in all the six environments, accounting for 23.47 to 32.15% of the phenotypic variation. *Xgwm328-2A* was also simultaneously associated with SPL and SPLN in at least five environments, having the phenotypic variations of 32.43 to 37.90% and 28.42 to 38.77%, respectively. This multiple marker-trait association might be attributed to the pleiotropic effects of the genetic locus or the consequential relationship among these associated traits. *Xgwm63-7A* was significantly associated with SPL and SPLN in five environments, with the phenotypic variations of 20.67 to 27.80% and 19.24 to 23.93%, respectively, and *Xbarc138-4A* showed a stable association with SPLN in four environments, with the phenotypic variations of 14.43 to 18.18%. In particular, the HMW-GSs, encoded by *Glu-1* locus on the long arm of chromosome 1A, were associated with TA in three environments, and it explained phenotypic variations ranging from 28.18 to 34.46%. As for grain traits, *Xcfa2219-1A* and *Xgwm293-5A* were associated with grain length (GL) in five environments, and could explain 15.99 to 20.33% and 23.55 to 28.23% of the phenotypic variation, respectively.

We further investigated the relationships between the genotype and haplotype with the phenotypic traits analyzed (Fig. 6). For HD, the 214-bp genotype of *Xcfa2193-3A*, which was observed in 10 Armenian accessions, exerted a positive effect on delaying the heading date, whereas the 202-bp genotype in 72 accessions was linked to medium heading date. Furthermore, the 200-bp allele present at *Xgwm328-2A* locus in 11 accessions

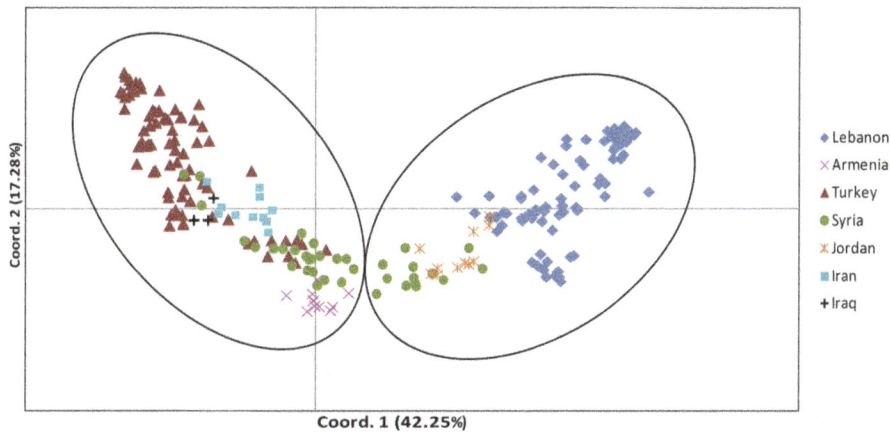

Fig. 4 Principal coordinates analysis of 238 *T. urartu* accessions. The first and second principal coordinates account for 42.25% and 17.28% of the total variation respectively. The different colors represent the accessions of different geographical origins

was strongly associated with high values of PH, which also correlated with the increase of SPL and SPLN. In contrast, the 208-bp allele was preferentially shared by genotypes in 29 accessions with low PH, SPL and SPLN. Similarly, 55 accessions carrying the 170-bp allele at *Xgwm63-7A* produced remarkably longer SPL and more SPLN than accessions with other alleles, and the 196-bp allele at *Xbarc138-4A* could also bring more SPLN in 17 accessions than others. At *Glu-1* locus, the HMW-GS encoded U7 pattern was associated with the erect plant architecture (65 accessions), whereas the U14 pattern

with the prostrate type (21 accessions). Regarding GL, the significant associations were attributed to the 224-bp allele at *Xcfa2219-1A* and the 205-bp allele at *Xgwm293-5A* being specific to genotypes with large kernel length.

Discussion

Nowadays, the genetic variability of wheat cultivars is decreasing as a consequence of the genetic erosion of cultivated hexaploid wheat. As a wild diploid progenitor of hexaploid wheat, *T. urartu* harbors rich allelic diversity for numerous important traits, including agronomic

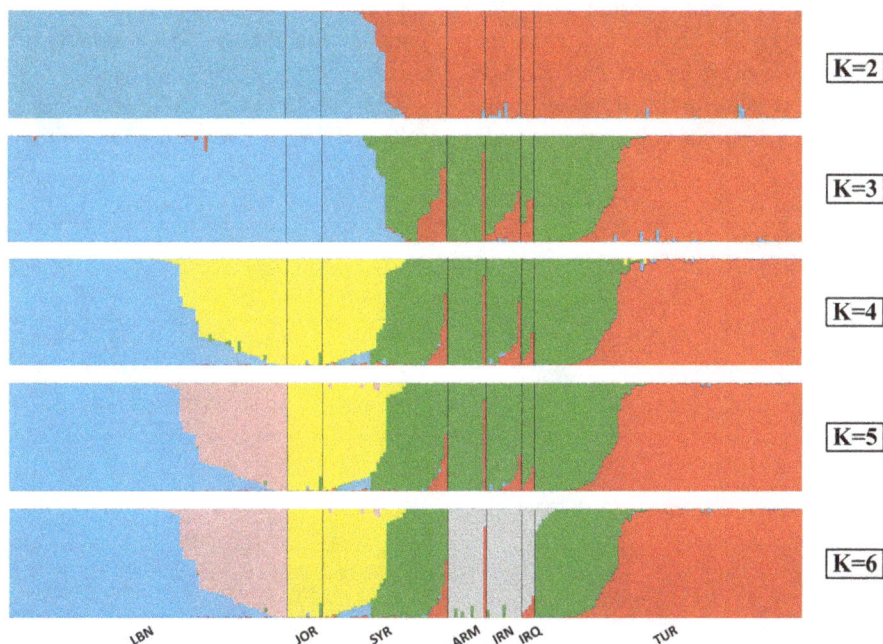

Fig. 5 Estimated population structure from 62 nuclear SSR loci based on Bayesian clustering approaches for *K* = 2 to *K* = 6 using STRUCTURE. Each accession is represented by a vertical line. The different subpopulations are separated by a black line and shown in different colors. The bottom row indicates the geographic region. LBN, Lebanon; JOR, Jordan; SYR, Syria; ARM, Armenia; IRN, Iran; IRQ, Iraq; TUR, Turkey

Table 5 Pairwise estimates of *Fst* and *Nei's* genetic distance between populations from different regions

Cluster	Lebanon	Jordan	Southwestern Syria	Turkey	Armenia	Iran	Iraq	Northern Syria
Lebanon		0.3060	0.3283	0.4628	0.7244	0.4340	0.6343	0.5932
Jordan	0.2044		0.0334	0.4419	0.7947	0.4399	0.6596	0.5129
Southwestern Syria	0.1590	0.1284		0.2507	0.5222	0.1944	0.4769	0.2061
Turkey	0.1813	0.2022	0.1058		0.4803	0.0312	0.1696	0.0289
Armenia	0.3290	0.4419	0.2361	0.2251		0.3852	0.4790	0.4587
Iran	0.1876	0.2131	0.1043	0.0790	0.2180		0.1731	0.0434
Iraq	0.2151	0.2343	0.1333	0.0985	0.2567	0.1081		0.0295
Northern Syria	0.1826	0.1968	0.0953	0.0658	0.2164	0.0788	0.0690	

Nei's genetic distance estimates appear above the diagonal and pairwise *Fst* appears below the diagonal

characteristics, dough quality and biotic stress tolerance [9–11]. Thus, this wild species could be exploited in yield potential and quality improvement of bread wheat.

Morphological and genetic diversity

In recent years, SSRs markers have been proven to be an efficient tool for molecular and genetic studies in wheat [47–49]. Many SSRs covering the A genome of common wheat were employed in diploid wheat, due to their transferability among closely related species [50]. In this work, we observed a higher level of polymorphisms in *T. urartu* using SSR markers developed in common wheat. A total of 1201 alleles were identified from the 62 SSR loci in the 238 accessions, with an average of 19.37 alleles per locus, which is much higher than that observed in an earlier study (19.37 vs. 8.00) of 23 *T. urartu* accessions using 25 SSRs [51]. In addition, low genetic variation of *T. urartu* has also been reported with RFLP and RAPD markers [12–14]. This is the first time that such a high level of genetic diversity was characterized in *T. urartu* using SSR markers, probably because of a relatively large number of accessions collected in a wider geographic region. Moreover, compared to other wheat species, *T. urartu* possesses higher level of genetic diversity than the A genome of tetraploid and hexaploid wheat [52, 53].

HMW-GSs are the major seed storage proteins that determine dough viscoelastic properties and bread-making quality [54, 55], which could also represent useful markers for assessment of genetic variability in wild diploid wheat collections [56, 57]. In present study, a total of eleven 1Ax and eight 1Ay subunits were detected, resulting in 18 HMW-GS combinations in 238 accessions. Even though 1Ay subunit is totally inactive in common wheat [58], 71.01% *T. urartu* accessions were found to express the 1Ay subunit, which was consistent with previous reports [56, 59, 60]. The polymorphic information content (PIC) of our collections was 0.86, and the GD value was 0.88, which suggested that genetic variation revealed by HMW-GSs was comparable with that by SSR markers, probably because of the post-translational modification of protein markers.

The genetic diversity observed in this study is well reflected by the variation in multiple biological traits.

Table 6 Common loci and significant markers associated with agronomic and grain shape related traits

Trait	Loci	Chromosome	P value[b]						R^2 (%)[c]					
			2014[a]			2015			2014			2015		
			E1	E2	E3	E4	E5	E6	E1	E2	E3	E4	E5	E6
HD	Xcfa2193	3A	4.30×10^{-4}	0.0032	–	9.62×10^{-4}	–	1.71×10^{-5}	10.62	15.74	–	9.39	–	9.75
PH	Xgwm328	2A	1.45×10^{-5}	1.04×10^{-6}	0.0044	9.84×10^{-5}	2.07×10^{-4}	2.34×10^{-6}	30.54	28.21	25.81	23.47	27.23	32.15
SPL	Xgwm328	2A	8.67×10^{-10}	5.06×10^{-8}	5.35×10^{-4}	1.58×10^{-10}	6.85×10^{-5}	4.74×10^{-6}	35.10	37.90	32.43	36.60	35.32	35.44
	Xgwm63	7A	5.19×10^{-5}	–	6.57×10^{-5}	3.48×10^{-4}	0.0015	1.97×10^{-4}	26.92	–	20.67	27.80	25.11	21.59
SPLN	Xgwm328	2A	0.0024	9.95×10^{-4}	3.99×10^{-4}	0.0072	7.07×10^{-4}	–	34.19	38.77	33.97	28.42	35.45	–
	Xbarc138	4A	–	6.84×10^{-4}	0.0027	8.74×10^{-4}	–	0.0076	–	15.72	18.18	14.43	–	15.18
	Xgwm63	7A	1.24×10^{-4}	–	9.42×10^{-5}	2.82×10^{-5}	0.0039	5.53×10^{-4}	23.93	–	21.85	19.24	22.25	21.85
TA	HWM-GS	1A	–	2.39×10^{-4}	–	0.0059	6.25×10^{-4}	–	–	34.46	–	28.18	31.09	–
GL	Xcfa2219	1A	0.0050	4.45×10^{-6}	0.0098	6.87×10^{-5}	–	9.89×10^{-4}	19.89	16.92	20.33	15.99	–	17.93
	Xgwm293	5A	2.29×10^{-6}	0.0081	3.22×10^{-4}	9.11×10^{-6}	0.0035	–	28.23	26.95	27.34	24.83	23.55	–

Marker–trait association was performed with linear mixed-effects model (MLM) incorporating structure Q-matrix and kinship K-matrix in TASSEL 2.1
[a] Missing result is represented by '–' due to unavailable data
[b] Marker–trait association is significant at $P < 0.01$ with FDR correction at $\alpha c = 0.05$
[c] R^2 is the percentage of phenotypic variation explained by the marker

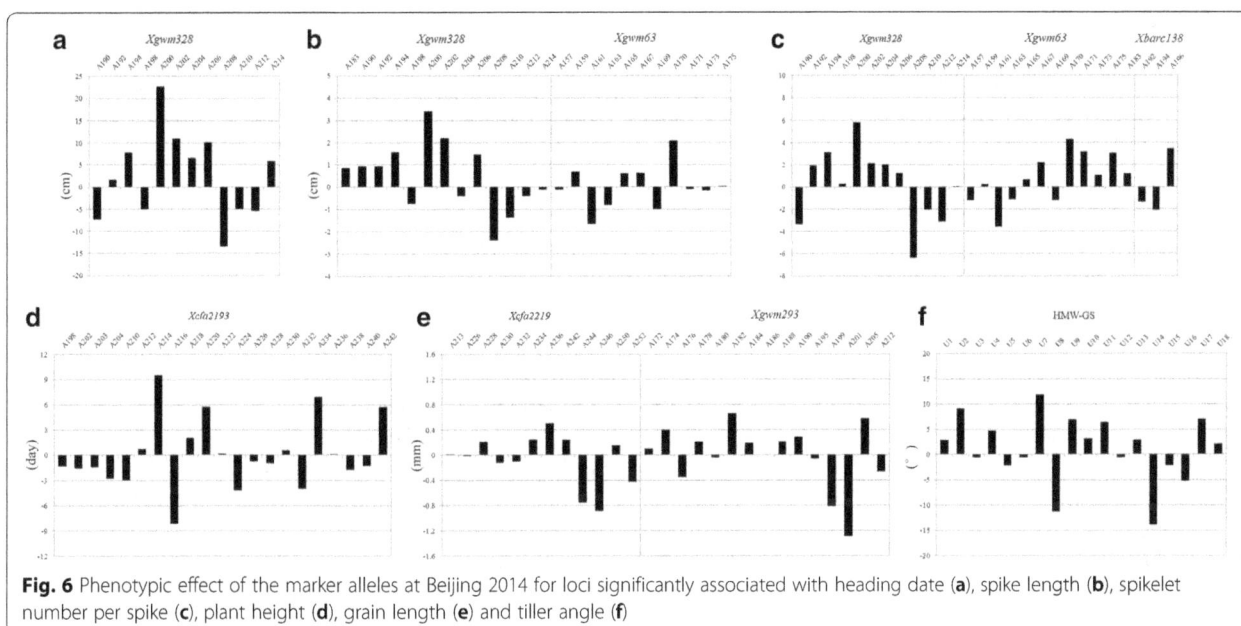

Fig. 6 Phenotypic effect of the marker alleles at Beijing 2014 for loci significantly associated with heading date (**a**), spike length (**b**), spikelet number per spike (**c**), plant height (**d**), grain length (**e**) and tiller angle (**f**)

For example, PH showed a broad variation with an average of 112.43 cm ranging from 86.10 cm to 149.10 cm. All these data demonstrated that the collection of *T. urartu* possessed high genetic variation, which make them suitable for mining valuable genes based on association mapping.

Genetic relationships and population structure

T. urartu species are endemic to the major geographic regions of the Fertile Crescent [61, 62]. In our study, the UPGMA dendrogram divided the diverse panel into two major subpopulations consistent with their geographic origin and ecological distribution (Fig. 1). The accessions from Eastern Mediterranean coast belonged to Clusters I, spreading from Lebanon to Jordan, of where the climate was described as poor rainfall and low temperature [63]. The levels of population differentiation in the Bekaa valley of Lebanon, southwestern Syria and the plains alongside Jordan River were low owing to high degrees of gene flow, occasional migration and cross-pollination among these accessions [15]. The other possibility is that Lebanese, southwestern Syrian and Jordan populations most probably originated from the same ancestral population, which could be inferred from the rare alleles shared by multiple loci in this study. The rest of accessions from Mesopotamia and Transcaucasia belonged to Clusters II, including Turkey, Iraq, Iran, Armenia and northern Syria. Of them, most Turkish accessions collected along the east-west road from Nusaybin to Viransehir, were in the area of South Eastern Anatolian basin characterized as mild winter and warm-dry summer [64]. The Iranian samples, located in the Zagros mountain range with a similar climatic environment, were also included in the same branch of phylogenetic tree. Other places, such as Urfa, Gaziantep and Mus of Turkey, northern Syria and part of Iraq, had a continental cold and long winter with mild summer-dry climate [64], thus, these *T. urartu* accessions were grouped together. Specially, populations originated from Armenian, Nusaybin east of Turkey and Malikiyah of northern Syria exhibited high frequencies for the rare alleles such as 127-bp at *Xgwm614-2A* and 214-bp at *Xcfa2193-3A*, which separated the populations from other regions. The restricted distribution of these alleles indicated that variation at the associated loci possibly has some adaptive significance, which was supported by previous report [14]. On the whole, genetic differentiation among the accessions was also clearly demonstrated by the first and the second principal coordinates (Fig. 4).

In order to avoid distortion of the relationships among members and the spurious association mapping between phenotype and genotype, we examined population structure of the representative *T. urartu* collection used in this study. The result separated these accessions into Eastern Mediterranean coastal and Mesopotamia-Transcaucasia populations, generally in agreement with the phylogenetic tree and PCoA analysis. However, the loss of genetic diversity and decrease of population size were detected from Cluster II to Cluster I, since bottleneck and genetic drift may generate in the duration of natural selections. Given the subdivision, accessions from Turkey and northern Syria exhibited higher diversity than others, which had also been reported previously [15, 51], and huge genetic variations were subsequently emerged among different originated populations. Our results provided inspirations for preservation and sampling of natural *T. urartu* populations in these regions.

Marker-trait associations

Agronomic traits such as PH, HD, SPL and TA play important roles in wheat life cycle and environment adaptability, which are closely associated with the yield potential [19, 65]. Grain shape, as specified by GL, GW and GLW, is a crucial determinant of grain appearance quality and grain weight in wheat [66]. Therefore, these quantitative traits have drawn major attention in the process of wheat breading over the world. Compared to traditional QTL mapping, association mapping is less time-consuming as no segregating population needs to be developed and no segregating offspring needs to be grown [25]. Under such circumstances, a number of marker-trait associations have already been identified for significant meta-QTLs in bread wheat [67–70]. Nevertheless, a genome-wide association mapping study of agronomic traits based on elite diploid wheat germplasm is still lacking. In this study, six SSR and HMW-GS markers were detected to be highly associated with six agronomic and grain shape related traits in *T. urartu* (Table 5). The *Xcfa2193-3A* on chromosome 3A associated with HD explained >35.41% of phenotypic variation in four environments, which had also been reported in common wheat [71]. *Xgwm328-2A* and *Xgwm63-7A* associated with SPL/SPLN in our study appeared to increase both SPL and SPLN in common wheat [72, 73]. As for GL, we also detected a closely associated SSR marker *Xcfa2219-1A*, which had been reported to locate in an additive-effect QTL controlling GL in common wheat [74]. *Xgwm293* on chromosome 5A was recently characterized associated with GL and TKW in a doubled haploid (DH) mapping population [75]. Moreover, *Xbarc138-4A* for SPLN was associated with yellow rust in ITMI-mapping population [76]. The data demonstrated that these genetic regions may be conservative between *T. urartu* and hexaploid wheat, and *T. urartu* could be explored in underlying gene characterization through mapping based cloning for its relative simple genome [5]. Interestingly, the marker of HMW-GSs showed a significant association with TA after the correction of FDR. In this case, the U7 pattern containing only one 1Ax subunit was associated with the erect type, whereas the U14 pattern containing both 1Ax and 1Ay subunits was associated with the prostrate type. However, the common wheat varieties rarely express 1Ay subunit and tend to develop an erect plant, this differentiation may be due to the evolution and domestication from *T. urartu* to hexaploid wheat.

Except the phenotypic traits mentioned above, the A genome of wheat is also known to contain QTLs or genes affecting other agronomic traits, heat and drought tolerance, pathogen resistance and so on [77–79]. The allelic variation between *T. urartu* and *T. aestivum* indicates the great potential for discovery and utilization of wild diploid relatives in wheat breeding. The associations determined in this study would be very useful for marker-assisted selection (MAS) in wheat breeding programs, although more effort is needed to validate these associations in other populations. Along with the progress and wide applications of comparative genomics approaches, further work to elucidate the genetic basis of complex traits will be accelerated.

Conclusion

Genetic diversity and population structure of 238 *T. urartu* accessions were analyzed through SSR and HMW-GS markers and their associations with phenotypes were detected. Six markers, associated with HD, PH, SPL, SPLN, TA and GL were determined, which should be beneficial to effectively exploit new genetic variations of the wild diploid *T. urartu* in yield and quality improvement of common wheat using MAS programs. Our results also provide further insights into conservation and future utilization of wild wheat resources.

Additional files

> **Additional file 1: Table S1.** *T. urartu* accessions used in the present. **Table S2.** List of SSR primers used for genetic diversity and association analysis. **Table S3.** Correlations coefficients for the investigated traits of 238 *T. urartu* accessions in six environments. **Table S4.** Details of phenotypic performances and ANOVA analysis of differences between the Eastern Mediterranean coastal and Mesopotamia-Transcaucasia groups. **Table S5.** Correlation coefficients between the investigated traits in 238 *T. urartu* accessions in six environments. **Table S6.** Summary of Hardy-Weinberg equilibrium testing for SSR markers used in this study.
>
> **Additional file 2: Figure S1.** Optimization of the number of subpopulations (*K* value) for 238 *T. urartu* accessions by the method of Delta *K* (Evanno et al. 2005). The peak represents the appropriate number of subpopulations.

Abbreviations

Ae: Number of expected alleles per locus; AFLP: Amplified fragment length polymorphism; AMOVA: Analysis of Molecular Variance; Ao: Number of observed alleles per locus; CTAB: Cetyl trimethyl ammonium bromide; FDR: False discovery rate; GD: *Nei's* Gene diversity; GL: Grain length; GLM: General linear model; GLW: Grain length/width ratio; GW: Grain width; HD: Heading date; He: Expected heterozygosity; HMW-GS: High-molecular-weight glutenin subunit; Ho: Observed heterozygosity; I: *Shannon's* information index; MAS: Marker-assisted selection; MLM: Mixed linear model; PCoA: Principal coordinate analysis; PH: Plant height; PIC: Polymorphic information content; QTL: Quantitative trait locus or loci; RAPD: Random amplified polymorphic DNA; RFLP: Restriction fragment length polymorphism; SDS-PAGE: Sodium dodecyl sulfate polyacrylamide gel electrophoresis; SPL: Spike length; SPLN: Spikelet number per spike; SSR: Simple sequence repeat; TA: Tiller angle; TGW: Thousand-grain weight; UPGMA: Unweighted pair-group method with arithmetic means

Acknowledgments

We thank Dr. Shancen Zhao (BGI-Shenzhen) for his critical review.

Funding

This research was financially supported by the Ministry of Science and Technology of China (2014CB138101) and the Chinese Academy of Sciences (XDA08010104).

Authors' contributions

XW carried out most experiments and analyzed the data. XW, DL and WY wrote the manuscript. GL and JS run the SDS-PAGE and analyze the data. JS, KZ and YL carried out the field experiments, collected the phenotypic data and analyze the data. AZ and DL co-designed the experiments and revised the manuscript. All authors read and approved the final manuscript.

Competing interests

The authors declare that they have no competing interests.

Author details

[1]State Key Laboratory of Plant Cell and Chromosome Engineering, National Center for Plant Gene Research, Institute of Genetics and Developmental Biology, Chinese Academy of Sciences, 1 West Beichen Road, Chaoyang District, Beijing 100101, China. [2]University of Chinese Academy of Sciences, Beijing 100049, China. [3]College of Agronomy/The Collaborative Innovation Center of Grain Crops in Henan, Henan Agricultural University, No. 95 Wenhua Road, Zhengzhou 450002, China.

References

1. Curtis T, Halford NG. Food security: the challenge of increasing wheat yield and the importance of not compromising food safety. Ann Appl Biol. 2013; 164(3):354–72.
2. Gupta PK, Mir RR, Mohan A, Kumar J. Wheat genomics: present status and future prospects. Int J Plant Genome. 2008;2008:896451.
3. Brenchley R, Spannagl M, Pfeifer M, Barker GL, D'Amore R, Allen AM, et al. Analysis of the bread wheat genome using whole-genome shotgun sequencing. Nature. 2012;491(7426):705–10.
4. Dvořák J, Terlizzi PD, Zhang HB, Resta P. The evolution of polyploid wheats: identification of the a genome donor species. Genome. 1993;36(1):21–31.
5. Ling HQ, Zhao SC, Liu DC, Wang JY, Sun H, Zhang C, et al. Draft genome of the wheat A-genome progenitor Triticum urartu. Nature. 2013;496:87–90.
6. Kimber G, Sears ER. Evolution in the genus Triticum and the origin of cultivated wheat. In: Heyne EG, editor. Wheat and Wheal Improvement. Madison: American Society of Agronomy; 1987. p. 154–64.
7. Peng JH, Sun DF, Nevo E. Domestication evolution, genetics and genomics in wheat. Mol Breeding. 2011;28(3):281–301.
8. Reif JC, Gowda M, Maurer HP, Longin CFH, Korzun V, Ebmeyer E, et al. Association mapping for quality traits in soft winter wheat. Theor Appl Genet. 2011;122(5):961–70.
9. Qiu YC, Zhou RH, Kong XY, Zhang SS, Jia JZ. Microsatellite mapping of a Triticum urartu Tum. Derived powdery mildew resistance gene transferred to common wheat (Triticum aestivum L.). Theor Appl Genet. 2005;111(8):1524–31.
10. Rouse MN, Jin Y. Stem rust resistance in A-genome diploid relatives of wheat. Plant Dis. 2011;95(8):941–4.
11. Guzmán C, Alvarez JB. Molecular characterization of a novel waxy allele (Wx-Au1a) from Triticum urartu Thum. Ex Gandil. Genet Resour Crop Ev. 2012;59(6):971–9.
12. Vierling RA, Nguyen HT. Use of RAPD markers to determine the genetic diversity of diploid wheat genotypes. Theor Appl Genet. 1992;84(7):835–8.
13. Dhaliwal HS, Sidhu JS, Minocha JL. Genetic diversity in diploid and hexaploid wheats as revealed by RAPD markers. Crop Improv. 1993; 20:17–20.
14. Castagna R, Gnocchi S, Perenzin M, Heun M. Genetic variability of the wild diploid wheat Triticum urartu revealed by RFLP and RAPD markers. Theor Appl Genet. 1997;94(3):424–30.
15. Moghaddam M, Ehdaie B, Waines JG. Genetic diversity in populations of wild diploid wheat Triticum urartu tum. Ex. gandil. Revealed by isozyme markers. Genetic Res Crop Ev. 2000;47(3):323–34.
16. Baum BR, Bailey LG. Genetic diversity in the red wild einkorn: T. urartu gandilyan (Poaceae: Triticeae). Genetic Resour Crop Ev. 2013;60(1):77–87.
17. Breseghello F. Sorrells ME association mapping of kernel size and milling quality in wheat (Triticum aestivum L.) cultivars. Genetics. 2006;172:1165–77.
18. Risch N, Merikangas K. The future of genetic studies of complex human diseases. Science. 1996;273(5281):516–1517.
19. Kato K, Miura H, Sawada S. Mapping QTLs controlling grain yield and its components on chromosome 5A of wheat. Theor Appl Genet. 2000;101(7): 1114–21.
20. Gegas VC, Nazari A, Griffiths S, Simmonds J, Fish L, Orford S, et al. A genetic frame work for grain size and shape variation in wheat. Plant Cell. 2010; 22(4):1046–56.
21. Sun X, Marza F, Ma H, Carver BF, Bai G. Mapping quantitative trait loci for quality factors in an inter-class cross of US and Chinese wheat. Theor Appl Genet. 2010;120(5):1041–51.
22. Okamoto Y, Nguyen AT, Yoshioka M, Iehisa JC, Takumi S. Identification of quantitative trait loci controlling grain size and shape in the D genome of synthetic hexaploid wheat lines. Breeding Sci. 2013;63(4):423–9.
23. Williams K, Munkvold J, Sorrells M. Comparison of digital image analysis using elliptic Fourier descriptors and major dimensions to phenotype seed shape in hexaploid wheat (Triticum aestivum L.). Euphytica. 2013;190(1):99–116.
24. Yu K, Liu DC, Wu WY, Yang WL, Sun JZ, Li X, et al. Development of an integrated linkage map of einkorn wheat and its application for QTL mapping and genome sequence anchoring. Theor Appl Genet. 2017; 130(1):53–70.
25. Addington J, Cornblatt BA, Cadenhead KS, Cannon TD, McGlashan TH, Perkins DO, et al. At clinical high risk for psychosis: outcome for nonconverters. Am J Psy. 2011;168(8):800–5.
26. Li H, Peng ZY, Yang XH, Wang WD, Fu JJ, Wang JH, et al. Genome-wide association study dissects the genetic architecture of oil biosynthesis in maize kernels. Nat Genet. 2012;45(1):43–50.
27. Ding JQ, Ali F, Chen GS, Li HH, Mahuku G, Yang N, et al. Genome-wide association mapping reveals novel sources of resistance to northern corn leaf blight in maize. BMC Plant Biol. 2015;15(1):206.
28. Huang XH, Zhao Y, Wei XH, Li CY, Wang AH, Zhao Q, et al. Genome-wide association study of flowering time and grain yield traits in a worldwide collection of rice germplasm. Nat Genet. 2012;44(1):32–9.
29. Morris GP, Ramu P, Deshpande SP, Hash CT, Shah T, Upadhyaya HD, et al. Population genomic and genome-wide association studies of agroclimatic traits in sorghum. Proc Natl Acad Sci. 2013;110(2):453–8.
30. Zhou ZK, Yu J. Z heng W, Gou ZH, Lyu J, Li WY, et al. Resequencing 302 wild and cultivated accessions identifies genes related to domestication and improvement in soybean. Nat Biotech. 2015;33(4):408–14.
31. Mizumoto K, Hirosawa S, Nakamura C, Takumi S. Nuclear and chloroplast genome genetic diversity in the wild einkorn wheat, Triticum urartu, revealed by AFLP and SSLP analyses. Hereditas. 2002;137(3):208–14.
32. Kulwal PL, Kumar N, Gaur A, Khurana P, Khurana JP, Tyagi AK, et al. Mapping of a major QTL for pre-harvest sprouting tolerance on chromosome 3A in bread wheat. Theor Appl Genet. 2005;111(6):1052–9.
33. Saghai-Maroof MA, Soliman KM, Jorgensen RA, Allard RW. Ribosomal DNA spacer-length polymorphisms in barley: Mendelian inheritance, chromosomal location, and population dynamics. Proc Natl Acad Sci. 1984; 81(24):8014–8.
34. Schuelke M. An economic method for the fluorescent labeling of PCR fragments. Nature Biotech. 2000;18(2):233–4.
35. Singh SP, Gutierrez JA, Molina A, Urrea C, Gepts P. Genetic diversity in cultivated common bean: II. Marker-based analysis of morphological and agronomic traits. Crop Sci. 1991;31(1):23–9.
36. Payne PI, Lawrence GJ. Catalogue of alleles for the complex gene loci, Glu-A1, Glu-B1, and Glu-D1 which code for high-molecular-weight subunits of glutenin in hexaploid wheat. Cereal Res Commun. 1983; 11(1):29–35.
37. Yeh FC, Yang RC, Boyle T. POPGENE Version 1.32, Microsoft Window-based freeware for population genetic analysis. Molecular Biology and Biotechnology Centre, University of Alberta, Canada, 1999.

38. Van OC, Hutchinson WF, Wills DPM, Shipley P. MICRO-CHECKER: software for identifying and correcting genotyping errors in microsatellite data. Mol Ecol Notes. 2004;4(3):535–8.

39. Rohlf FJ. NTSYS-PC, numerical taxonomy system for the PC ExeterSoftware, version 2.1. Setauket USA: Applied Biostatistics Inc; 2000.

40. Pritchard JK, Stephens M, Donnelly P. Inference of population structure using multilocus genotype data. Genetics. 2000;155(2):945–59.

41. Evanno G, Regnaut S, Goudet J. Detecting the number of clusters of individuals using the software STRUCTURE: a simulation study. Mol Ecol. 2005;14(8):2611–20.

42. Jakobsson M, Rosenberg NA. CLUMPP: a cluster matching and permutation program for dealing with label switching and multimodality in analysis of population structure. Bioinformatics. 2007;23(14):1801–6.

43. Zanetti E, Marchi MD, Dalvit C, Cassandro M. Genetic characterization of local italian breeds of chickens undergoing in situ conservation. Poultry Sci. 2010;89(3):420–7.

44. Peakall ROD, Smouse PE. GENALEX 6: genetic analysis in excel. Population genetic software for teaching and research. Mol Ecol Notes. 2006;6(1):88–295.

45. Hardy OJ, Vekemans X. SPAGeDi: a versatile computer program to analyse spatial genetic structure at the individual or population levels. Molecul Ecol Notes. 2002;2(4):618–20.

46. Bradbury PJ, Zhang Z, Kroon DE, Casstevens TM, Ramdoss Y, Buckler ES. TASSEL: software for association mapping of complex traits in diverse samples. Bioinformatics. 2007;23(19):2633–5.

47. Gupta P, Balyan H, Edwards K, Isaac P, Korzun V, Röder MS, et al. Genetic mapping of 66 new microsatellite (SSR) loci in bread wheat. Theor Appl Genet. 2002;105(2):413–22.

48. Song QJ, Shi JR, Singh S, Fickus EW, Costa JM, Lewis J, et al. Development and mapping of microsatellite (SSR) markers in wheat. Theor Appl Genet. 2005;110(3):550–60.

49. Periyannan S, Bansal U, Bariana H, Deal K, Luo MC, Dvorak J, et al. Identification of a robust molecular marker for the detection of the stem rust resistance gene Sr45 in common wheat. Theor Appl Genet. 2013;127(4):947–55.

50. Bai JR, Liu KF, Jia X, Wang DW. An analysis of homoeologous microsatellites from Triticum urartu, and Triticum monococcum. Plant Sci. 2004;166(2):341–7.

51. Hammer K, Filatenko AA, Korzun V. Microsatellite markers – a new tool for distinguishing diploid wheat species. Genet Resour Crop Ev. 2000;47(5):497–505.

52. Ren J, Chen L, Sun DK, You FM, Wang JR, Peng YL, et al. SNP-revealed genetic diversity in wild emmer wheat correlates with ecological factors. BMC Evol Biol. 2013;13(1):1–15.

53. Huang XQ, Börner A, Röder MS, Ganal MW. Assessing genetic diversity of wheat (Triticum aestivum, L.) germplasm using microsatellite markers. Theor Appl Genet. 2000;105(5):699–707.

54. Shewry PR, Halford NG, Tatham AS. High molecular weight subunits of wheat glutenin. J Cereal Sci. 1992;15(2):105–20.

55. Shewry PR, Tatham AS, Barro F, Barcelo P, Lazzeri P. Biotechnology of bread making: unraveling and manipulating the multi-protein gluten complex. Bio/Technology. 1995;13(11):1185–90.

56. Ciaffi M, Dominici L, Lafiandra D. High molecular weight glutenin subunit variation in wild and cultivated einkorn wheats (Triticum, spp. poaceae). Plant Syst Evol. 1998;209(1):123–37.

57. Luo GB, Zhang XF, Zhang YL, Yang WL, Li YW, Sun JZ, et al. Composition, variation, expression and evolution of low-molecular-weight glutenin subunit genes in Triticum urartu. BMC Plant Biol. 2015;15(1):68.

58. Wan Y, Wang D, Shewry PR, Halford NG. Isolation and characterization of five novel high molecular weight subunit of glutenin genes from Triticum timopheevi and Aegilops cylindrical. Theor Appl Genet. 2002; 104(5):828–39.

59. Waines JG, Payne PI. Electrophoretic analysis of the high molecular weight subunits of Triticum monococcum, T. urartu, and the a genome of bread wheat. Theor Appl Genet. 1987;74(1):71–6.

60. Xu LL, Li W, Wei YM, Zheng YL. Genetic diversity of HMW glutenin subunits in diploid, tetraploid and hexaploid triticum species. Genetic Resour Crop Ev. 2009;56(3):377–91.

61. Johnson BL. Identification of the apparent B-genome donor of wheat. Can J Genet Cyto. 1975;17(1):21–39.

62. Heun M, Haldorsen S, Vollan K. Reassessing domestication events in the near east: einkorn and Triticum urartu. Genome. 2008;51(6):444–51.

63. Wright HE Jr. The environmental setting for plant domestication in the near east. Science. 1976;194(4263):385.

64. Firat AE, Tan A. Ecogeography and distribution of wild cereals in Turkey. In: Zencirci N, Kaya Z, Anikster Y, Adams WT, editors. The Proceedings of International Symposium on In situ Conservation of Plant Genetic Diversity. Ankara: Central Research Institute for Field Crops Publication; 1998. p. 81–6.

65. Maccaferri M. Quantitative trait loci for grain yield and adaptation of durum wheat (Triticum durum Desf.) across a wide range of water availability. Genetics. 2008;178(1):489–511.

66. Ramya P, Chaubal A, Kulkarni K, Gupta L, Kadoo N, Dhaliwal HS, et al. QTL mapping of 1000-kernel weight, kernel length, and kernel width in bread wheat (Triticum aestivum L.). J Appl Genet. 2010;51(4):421–9.

67. Breseghello F, Finney PL, Gaines C, Andrews L, Tanaka J, Penner G, et al. Genetic loci related to kernel quality differences between a soft and a hard wheat cultivar. Crop Sci. 2005;45(5):1685–95.

68. Neumann K, Kobiljski B, Denčić S, Varshney RK, Börner A. Genome-wide association mapping: a case study in bread wheat (Triticum aestivum L.). Mol Breeding. 2011;27(1):37–58.

69. Reif JC, Zhang P, Dreisigacker S, Warburton ML, van Ginkel M, Hoisington D, et al. Wheat genetic diversity trends during domestication and breeding. Theor Appl Genet. 2005;110(5):859–64.

70. Wang GM, Leonard JM, von Zitzewitz J, Peterson CJ, Ross AS, Riera-Lizarazu O. Marker-trait association analysis of kernel hardness and related agronomic traits in a core collection of wheat lines. Mol Breeding. 2014; 34(1):177–84.

71. Francki MG, Shankar M, Walker E, Loughman R, Golzar H, Ohm H. New quantitative trait loci in wheat for flag leaf resistance to Stagonospora nodorum blotch. Phytopathology. 2011;101(11):1278–84.

72. Heidari B, Sayedtabatabaei BE, Saeidi G, Kearsey M, Suenaga K. Mapping QTL for grain yield, yield components, and spike features in a doubled haploid population of bread wheat. Genome. 2011;54(6):517–27.

73. Li WL, Nelson JC, Chu CY, Shi LH, Huang SH, Liu DJ. Chromosomal locations and genetic relationships of tiller and spike characters in wheat. Euphytica. 2002;125(3):357–66.

74. Li MX, Wang ZL, Liang ZY, Shen WN, Sun FL, Xi YJ, et al. Quantitative trait loci analysis for kernel-related characteristics in common wheat (L.). Crop Sci. 2015;55(4):1485–93.

75. Brinton J, Simmonds J, Minter F, Leverington-Waite M, Snape J, Uauy C. Increased pericarp cell length underlies a major quantitative trait locus for grain weight in hexaploid wheat. New Phytol. 2017. doi:10.1111/nph.14624.

76. Kumar A, Chhuneja P, Jain S, Kaur S, Balyan HS, Gupta PK, et al. Mapping main effect QTL and epistatic interactions for leaf rust and yellow rust using high density itmi linkage map. Australian J Crop Sci. 2013;7(4):492–9.

77. Yahiaoui N, Srichumpa P, Dudler R, Keller B. Genome analysis at different ploidy levels allows cloning of the powdery mildew resistance gene Pm3b from hexaploid wheat. Plant J. 2004;37(4):528–38.

78. Pinto RS, Reynolds MP, Mathews KL, McIntyre CL, Olivares-Villegas JJ, Chapman SC. Heat and drought adaptive QTL in a wheat population designed to minimize confounding agronomic effects. Theor Appl Genet. 2010;121(6):1001–21.

79. Zhang LY, Liu DC, Guo XL, Yang WL, Sun JZ, Wang DW, et al. Genomic distribution of quantitative trait loci for yield and yield-related traits in common wheat. J Integr Plant Biol. 2010;52(11):996–1007.

Permissions

The contributors of this book come from diverse backgrounds, making this book a truly international effort. This book will bring forth new frontiers with its revolutionizing research information and detailed analysis of the nascent developments around the world.

We would like to thank all the contributing authors for lending their expertise to make the book truly unique. They have played a crucial role in the development of this book. Without their invaluable contributions this book wouldn't have been possible. They have made vital efforts to compile up to date information on the varied aspects of this subject to make this book a valuable addition to the collection of many professionals and students.

This book was conceptualized with the vision of imparting up-to-date information and advanced data in this field. To ensure the same, a matchless editorial board was set up. Every individual on the board went through rigorous rounds of assessment to prove their worth. After which they invested a large part of their time researching and compiling the most relevant data for our readers.

The editorial board has been involved in producing this book since its inception. They have spent rigorous hours researching and exploring the diverse topics which have resulted in the successful publishing of this book. They have passed on their knowledge of decades through this book. To expedite this challenging task, the publisher supported the team at every step. A small team of assistant editors was also appointed to further simplify the editing procedure and attain best results for the readers.

Apart from the editorial board, the designing team has also invested a significant amount of their time in understanding the subject and creating the most relevant covers. They scrutinized every image to scout for the most suitable representation of the subject and create an appropriate cover for the book.

The publishing team has been an ardent support to the editorial, designing and production team. Their endless efforts to recruit the best for this project, has resulted in the accomplishment of this book. They are a veteran in the field of academics and their pool of knowledge is as vast as their experience in printing. Their expertise and guidance has proved useful at every step. Their uncompromising quality standards have made this book an exceptional effort. Their encouragement from time to time has been an inspiration for everyone.

The publisher and the editorial board hope that this book will prove to be a valuable piece of knowledge for researchers, students, practitioners and scholars across the globe.

List of Contributors

Shihui Ding
Center for Genomics and Biotechnology, Fujian Provincial Key Laboratory of Haixia Applied Plant Systems Biology, Haixia Institute of Science and Technology (HIST), Fujian Agriculture and Forestry University, Fuzhou 350002 China

Liming Wang and Qing Zhang
Center for Genomics and Biotechnology, Fujian Provincial Key Laboratory of Haixia Applied Plant Systems Biology, Haixia Institute of Science and Technology (HIST), Fujian Agriculture and Forestry University, Fuzhou 350002 China
College of Life Science, Fujian Agriculture and Forestry University, Fuzhou 35002, China

Jisen Zhang
Center for Genomics and Biotechnology, Fujian Provincial Key Laboratory of Haixia Applied Plant Systems Biology, Haixia Institute of Science and Technology (HIST), Fujian Agriculture and Forestry University, Fuzhou 350002 China
College of Life Sciences, Fujian Normal University, Fuzhou 350117, China
Key Laboratory of Sugarcane Biology and Genetic Breeding Ministry of Agriculture, Fujian Agriculture and Forestry University, Fuzhou 350002, China

Yuexia Zheng and Youqiang Chen
College of Life Science, Fujian Agriculture and Forestry University, Fuzhou 35002, China

Bin Bai, Bo Shi, Ning Hou, Yanli Cao, Hongwu Bian, Muyuan Zhu and Ning Han
Key Laboratory for Cell and Gene Engineering of Zhejiang Province, Institute of Genetics and Regenerative Biology, College of Life Sciences, Zhejiang University, Zhejiang, Hangzhou 310058, China

Yijun Meng
College of Life and Environmental Sciences, Hangzhou Normal University, Zhejiang, Hangzhou 310036, China

Julien Le Roy, Anne-Sophie Blervacq, Anne Créach, Brigitte Huss, Simon Hawkins and Godfrey Neutelings
University of Lille, CNRS, UMR 8576 - UGSF - Unité de Glycobiologie Structurale et Fonctionnelle, F 59000 Lille, France

Kebede T. Muleta, Sheri Rynearson and Michael O. Pumphrey
Department of Crop and Soil Sciences, Washington State University, Pullman, WA 99164-6420, USA

Matthew N. Rouse
USDA-ARS Cereal Disease Laboratory, Department of Plant Pathology, University of Minnesota, St. Paul, MN 55108, USA

Xianming Chen
USDA-ARS, Wheat Health, Genetics and Quality Research Unit and Department of Plant Pathology, Washington State University, Pullman, WA 99164-6430, Pullman, WA 99164-6430, USA

Bedada G. Buta
Ethiopian Institute of Agricultural Research, Kulumsa Agricultural Research Center, P. O. Box 489, Assela, Ethiopia

An Yang
State Key Laboratory of Vegetation and Environmental Change, Institute of Botany, the Chinese Academy of Sciences, Beijing 100093, People's Republic of China

Qian Li
State Key Laboratory of Vegetation and Environmental Change, Institute of Botany, the Chinese Academy of Sciences, Beijing 100093, People's Republic of China
University of Chinese Academy of Sciences, Beijing 100049, People's Republic of China

Wen-Hao Zhang
State Key Laboratory of Vegetation and Environmental Change, Institute of Botany, the Chinese Academy of Sciences, Beijing 100093, People's Republic of China

University of Chinese Academy of Sciences, Beijing 100049, People's Republic of China
Research Network of Global Change Biology, Beijing Institutes of Life Science, Chinese Academy of Sciences, Beijing 100093, China

Fred Y Peng
Department of Agricultural, Food and Nutritional Science, University of Alberta, 410 Agriculture/ Forestry Centre, Edmonton, AB T6G 2P5, Canada

Rong-Cai Yang
Department of Agricultural, Food and Nutritional Science, University of Alberta, 410 Agriculture/ Forestry Centre, Edmonton, AB T6G 2P5, Canada
Feed Crops Section, Alberta Agriculture and Forestry, 7000 - 113 Street, Edmonton, AB T6H 5T6, Canada

Wenlong Yang Aimin Zhang, Dongcheng Liu, Jiazhu Sun and Xin Li
State Key Laboratory of Plant Cell and Chromosome Engineering, Institute of Genetics and Developmental Biology, Chinese Academy of Sciences, 1 West Beichen Road, Chaoyang District, Beijing 100101, China

Xiaoling Ma
State Key Laboratory of Plant Cell and Chromosome Engineering, Institute of Genetics and Developmental Biology, Chinese Academy of Sciences, 1 West Beichen Road, Chaoyang District, Beijing 100101, China
University of Chinese Academy of Sciences, Beijing 100049, China

Muhammad Sajjad
State Key Laboratory of Plant Cell and Chromosome Engineering, Institute of Genetics and Developmental Biology, Chinese Academy of Sciences, 1 West Beichen Road, Chaoyang District, Beijing 100101, China
Department of Environmental Sciences, COMSATS Institute of Information Technology, Vehari 61100, Pakistan

Jing Wang
State Key Laboratory of Plant Cell and Chromosome Engineering, Institute of Genetics and Developmental Biology, Chinese Academy of Sciences, 1 West Beichen Road, Chaoyang District, Beijing 100101, China

The Institute of Forestry and Pomology, Beijing Academy of Agriculture and Forestry Sciences, Beijing 100093, China

Zhisheng Zhang, Shuan Meng and Nenghui Ye
Southern Regional Collaborative Innovation Center for Grain and Oil Crops in China, Hunan Agricultural University, Changsha, Hunan 410128, China

Xiangyang Li, Lili Cui and Xinxiang Peng
State Key Laboratory for Conservation and Utilization of Subtropical Agro-bioresources, College of Life Sciences, South China Agricultural University, Guangzhou, Guangdong 510642, China

Yuji Yamasaki, Feng Gao and Belay T. Ayele
Department of Plant Science, University of Manitoba, 222 Agriculture Building, Winnipeg, MB R3T 2N2, Canada

Mark C. Jordan
Morden Research and Development Centre, Agriculture and Agri-Food Canada, Morden, MB R6M 1Y5, Canada

Qian Wang, Yuepeng Zang and Xuan Zhou
College of Life Sciences, Capital Normal University, Beijing 100048, China

Wei Xiao
College of Life Sciences, Capital Normal University, Beijing 100048, China
Department of Microbiology and Immunology, University of Saskatchewan, Saskatoon, SK S7N 5E5, Canada

Lana Shabala, Meixue Zhou and Sergey Shabala
School of Land and Food, University of Tasmania, Private Bag 54, Hobart, Tas 7001, Australia

Md. Hasanuzzaman
School of Land and Food, University of Tasmania, Private Bag 54, Hobart, Tas 7001, Australia
Department of Agronomy, Faculty of Agriculture, Sher-e-Bangla Agricultural University, Sher-e Bangla Nagar, Dhaka -1207, Bangladesh

Noel W. Davies
Central Science Laboratory, University of Tasmania, Hobart, Tas 7001, Australia

Tim J. Brodribb
School of Biological Science, University of Tasmania, Private Bag 55, Hobart, Tas 7001, Australia

Meijie Luo, Yanxin Zhao, Ruyang Zhang, Jinfeng Xing, Minxiao Duan, Jingna Li, Naishun Wang, Wenguang Wang, Shasha Zhang, Huasheng Zhang, Zi Shi, Wei Song and Jiuran Zhao
Beijing Key Laboratory of Maize DNA Fingerprinting and Molecular Breeding, Maize Research Center, Beijing Academy of Agriculture and Forestry Sciences (BAAFS), Beijing, China

Zhihui Chen
Institute of Crops Research, Hunan Academy of Agricultural Sciences, Changsha, China

Shisong Ma
School of Life Sciences, University of Science and Technology of China, Hefei, Anhui, China

Zehong Ding
The Institute of Tropical Bioscience and Bio technology, Chinese Academy of Tropical Agricultural Sciences, Haikou, Hainan, China

Pinghua Li
State Key Laboratory of Crop Biology, College of Agronomy, Shandong Agricultural University, Tai'an, Shandong, China

Wenlong Yang, Yiwen Li, Jiazhu Sun and Dongcheng Liu
State Key Laboratory of Plant Cell and Chromosome Engineering, National Center for Plant Gene Research, Institute of Genetics and Developmental Biology, Chinese Academy of Sciences, 1 West Beichen Road, Chaoyang District, Beijing 100101, China

Xin Wang and Guangbin Luo
State Key Laboratory of Plant Cell and Chromosome Engineering, National Center for Plant Gene Research, Institute of Genetics and Developmental Biology, Chinese Academy of Sciences, 1 West Beichen Road, Chaoyang District, Beijing 100101, China
University of Chinese Academy of Sciences, Beijing 100049, China

Aimin Zhang
State Key Laboratory of Plant Cell and Chromosome Engineering, National Center for Plant Gene Research, Institute of Genetics and Developmental Biology, Chinese Academy of Sciences, 1 West Beichen Road, Chaoyang District, Beijing 100101, China
College of Agronomy/The Collaborative Innovation Center of Grain Crops in Henan, Henan Agricultural University, No. 95 Wenhua Road, Zhengzhou 450002, China

Kehui Zhan
College of Agronomy/The Collaborative Innovation Center of Grain Crops in Henan, Henan Agricultural University, No. 95 Wenhua Road, Zhengzhou 450002, China

Index

www.ingramcontent.com/pod-product-compliance
Lightning Source LLC
Chambersburg PA
CBHW082042190326
41458CB00010B/3433